Monographs in Theoretical Computer Science
An EATCS Series

Editors: W. Brauer J. Hromkovič G. Rozenberg A. Salomaa

On behalf of the European Association
for Theoretical Computer Science (EATCS)

Advisory Board:
G. Ausiello M. Broy C.S. Calude A. Condon
D. Harel J. Hartmanis T. Henzinger T. Leighton
M. Nivat C. Papadimitriou D. Scott

Stanisław Gawiejnowicz

Time-Dependent Scheduling

with 26 Figures and 48 Tables

 Springer

Author

Dr. Stanisław Gawiejnowicz
Adam Mickiewicz University
Faculty of Mathematics
and Computer Science
ul. Umultowska 87
61-614 Poznań
Poland
stgawiej@amu.edu.pl

Series Editors

Prof. Dr. Wilfried Brauer
Institut für Informatik der TUM
Boltzmannstr. 3
85748 Garching, Germany
brauer@informatik.tu-muenchen.de

Prof. Dr. Grzegorz Rozenberg
Leiden Institute of Advanced
Computer Science
University of Leiden
Niels Bohrweg 1
2333 CA Leiden, The Netherlands
rozenber@liacs.nl

Prof. Dr. Juraj Hromkovič
ETH Zentrum
Department of Computer Science
Swiss Federal Institute of Technology
8092 Zürich, Switzerland
juraj.hromkovic@inf.ethz.ch

Prof. Dr. Arto Salomaa
Turku Centre of
Computer Science
Lemminkäisenkatu 14 A
20520 Turku, Finland
asalomaa@utu.fi

ISBN 978-3-540-69445-8   e-ISBN 978-3-540-69446-5

Monographs in Theoretical Computer Science. An EATCS Series. ISSN 1431-2654

Library of Congress Control Number: 2008932588

ACM Computing Classification (1998): F.2, I.2

Cover Design: KünkelLopka GmbH, Heidelberg

Printed on acid-free paper

9 8 7 6 5 4 3 2 1

springer.com

To my parents

Preface

The book presented to the reader is devoted to time-dependent scheduling. Scheduling problems, in general, consist in the allocation of resources over time in order to perform a set of jobs. Any allocation that meets all requirements concerning the jobs and resources is called a feasible schedule. The quality of a schedule is measured by a criterion function. The aim of scheduling is to find, among all feasible schedules, a schedule that optimizes the criterion function. A solution to an arbitrary scheduling problem consists in giving a polynomial-time algorithm generating either an optimal schedule or a schedule that is close to the optimal one, if the given scheduling problem has been proved to be computationally intractable. The scheduling problems are subject of interest of the scheduling theory, originated in mid-fifties of the twentieth century. The theory has been developing dynamically and new research areas constantly come into existence. The subject of this book, time-dependent scheduling, is one of such areas.

In time-dependent scheduling, the processing time of a job is variable and depends on the starting time of the job. This crucial assumption allows us to apply the scheduling theory to a broader spectrum of problems. For example, in the framework of the time-dependent scheduling theory we may consider the problems of repayment of multiple loans, fire fighting and maintenance assignments. In this book, we will discuss algorithms and complexity issues concerning various time-dependent scheduling problems.

Time-dependent scheduling is a relatively new subject. Although the first paper from the area appeared in late 1970s most results have been published in the last 10 years. So far, time-dependent scheduling has not gained much attention in books devoted to the scheduling theory. This book, summarizing the results of almost 15 years of the author's research into time-dependent scheduling, hopefully fills this gap.

The book is composed of 14 chapters, organized into four parts.

The first part of the book consists of five chapters and includes the mathematical background used in subsequent chapters. The aim of this part is to give the reader an introductory view of presented topics. Therefore, only fundamental notions and concepts are discussed. In Chap. 1, the mathematical notation, the basic definitions and results used in this book are given. In Chap. 2, the essential concepts related to decision problems and algorithms are recalled. Chapter 3 includes the definitions and the most important results of the theory of $\mathcal{NP}$-completeness. This part is completed by Chaps. 4 and 5, where the basics of the scheduling theory and time-dependent scheduling theory are given, respectively. Each chapter of this part is compleleted with bibliographic notes including a list of selected references in which the reader may find a more comprehensive presentation of the particular topics.

The second part of the book includes a detailed survey of the time complexity of time-dependent scheduling problems. This part is composed of three chapters. In Chap. 6, single-machine time-dependent scheduling problems are discussed. Chapters 7 and 8 cover results concerning time-dependent scheduling on parallel and dedicated machines, respectively. Each chapter of this part is completed with the summary and tables.

The third part of the book is devoted to suboptimal algorithms for $\mathcal{NP}$-hard time-dependent scheduling problems. This part starts with Chap. 9, which presents approximation and heuristic algorithms. Chapter 10 introduces two greedy algorithms, which exploit the properties of the so-called signatures of sequences of job deterioration rates. Finally, local search heuristics for time-dependent scheduling problems are discussed in Chap. 11.

The fourth part of the book includes selected advanced topics in time-dependent scheduling. This part begins with Chap. 12, in which applications of matrix methods to time-dependent scheduling problems are discussed. Chapter 13 is devoted to scheduling proportionally and linearly deteriorating jobs under precedence constraints. In Chap. 14, closing the book, time-dependent scheduling problems with two criteria are studied.

Each chapter of these two parts ends with concluding remarks. Chapters of the fourth part include also comments on selected open problems.

The book is intended for researchers into the scheduling theory, Ph.D. students and everybody interested in recent advances in computer science.

The prerequisites for reading the book are the standard courses in discrete mathematics and calculus, fundamentals of the theory of algorithms and basic knowledge of any high-level programming language. Hence, this book can also be used by students of graduate studies.

The second part of the book can serve as a basis for an introductory course in time-dependent scheduling. The material from the next two parts can be used as a starting point for a research seminar in time-dependent scheduling.

The research presented in the book has been partially supported by grant N519 18889 33 of the Ministry of Science and Higher Education of Poland. While working on the book, I was also supported in different ways by different people. It is my pleasure to list here the names of the people to whom I am mostly indebted for help.

I heartily thank Dr. Alexander Kononov (Sobolev Institute of Mathematics, Novosibirsk, Russia), Dr. Wiesław Kurc and Dr. Lidia Pankowska (both from Adam Mickiewicz University, Poznań, Poland) for many stipulating discussions on different aspects of time-dependent scheduling. Many of the results presented in the book are the effects of my joint work with these researchers.

I sincerely thank Professor Jacek Błażewicz (Poznań University of Technology, Poznań, Poland) for his continuous encouragement and support during the many years of my research into time-dependent scheduling.

I also thank the following people for help in obtaining references concerning time-dependent scheduling: Gerd Finke (Leibniz-IMAG, Grenoble, France), Yi-Chih Hsieh (National Chengchi University, Hsinchu, Taiwan), Shi-Er Ju (Zhongshan University, Guangzhou, P.R. China), Li-Ying Kang (Shanghai University, Shanghai, P.R. China), Mikhail Y. Kovalyov (National Academy of Sciences of Belarus, Minsk, Belarus), Wiesław Kubiak (Memorial University of Newfounland, St. John's, Canada), Bertrand Miao-Tsong Lin (National Chiao Tung University, Hsinchu, Taiwan), Yakov M. Shafransky (National Academy of Sciences of Belarus, Minsk, Belarus), Prabha Sharma (Indian Institute of Technology Kanpur, Kanpur, India), Vitaly A. Strusevich (University of Greenwich, London, United Kingdom), Yoichi Uetake (Adam Mickiewicz University, Poznań, Poland), Ji-Bo Wang (Shenyang Institute of Aeronautical Engineering, Shenyang, P.R. China), Gerhard J. Woeginger (Eindhoven University of Technology, Eindhoven, The Netherlands), Dar-Li Yang (National Formosa University, Yun-Lin, Taiwan).

I direct special thanks to Mrs. Krystyna Ciesielska, M.A., M.Sc., who helped me to improve the English of this book.

Last but not the least, I thank my wife Mirosława and my daughter Agnieszka for their love, patience and support during the many months of my work on this book, when I was not able to be with them.

Poznań, April 2008   *Stanisław Gawiejnowicz*

Contents

Part II COMPLEXITY

Part III ALGORITHMS

Part I

FUNDAMENTALS

1

Preliminaries

The scheduling theory uses notions and methods from different disciplines of mathematics. Therefore, any systematic presentation of an arbitrary branch of the theory needs some mathematical background. The first part of the book introduces this background.

The part is composed of five chapters. In Chap. 1, we present the mathematical notation, the basic definitions and the results. The essential concepts related to decision problems and algorithms are recalled in Chap. 2. The definitions and the most important results of the theory of $\mathcal{NP}$-completeness are presented in Chap. 3. The basics of the scheduling theory and time-dependent scheduling are given in Chap. 4 and Chap. 5, respectively.

Chapter 1 is composed of three sections. In Sect. 1.1, we introduce the notation and terminology used in this book. In Sect. 1.2, we give some mathematical preliminaries used in subsequent chapters. The chapter is completed with bibliographic notes in Sect. 1.3.

1.1 Mathematical notation

We assume that the reader is familiar with basic mathematical notions. Therefore, we explain here only the notation that will be used throughout this book.

1.1.1 Sets and vectors

We will write $a \in A$ ($a \notin A$) if a is (is not) an element of a set A. If $a_1 \in A$, $a_2 \in A, \ldots, a_n \in A$, we will simply write $a_1, a_2, \ldots, a_n \in A$.

If an element a is (is not) equal to an element b, we will write $a = b$ ($a \neq b$). If $a = b$ by definition, we will write $a := b$. In a similar way, we will denote the equality (inequality) of numbers, sets, sequences, etc.

The set composed only of elements $a_1, a_2, \ldots, a_n$ will be denoted by $\{a_1, a_2, \ldots, a_n\}$. The maximal (minimal) element in set $\{a_1, a_2, \ldots, a_n\}$ will be denoted by $\max\{a_1, a_2, \ldots, a_n\}$ ($\min\{a_1, a_2, \ldots, a_n\}$).

If set A is a subset of set B, i.e., every element of set A is an element of set B, we will write $A \subseteq B$. If A is a strict subset of B, i.e., $A \subseteq B$ and $A \neq B$, we will write $A \subset B$. The empty set will be denoted by $\emptyset$.

The number of elements of set A will be denoted by $|A|$. The power set of set A, i.e., the set of all subsets of A, will be denoted by 2^A.

For any sets A and B, the union, intersection and difference of A and B will be denoted by $A \cup B$, $A \cap B$ and $A \setminus B$, respectively.

The Cartesian product of sets A and B will be denoted by $A \times B$. The Cartesian product of $n \geq 2$ copies of a set A will be denoted by A^n.

A partial order, i.e., a reflexive, antisymmetric and transitive binary relation, will be denoted by $\prec$. If $x \prec y$ or $x = y$, we will write $x \preceq y$.

The set-theoretic sum (product) of all elements of a set A will be denoted by $\bigcup_{a_i \in A} a_i$ ($\bigcap_{a_i \in A} a_i$). The union (intersection) of a family of sets A_k, $k \in K$, will be denoted by $\bigcup_{k \in K} A_k$ ($\bigcap_{k \in K} A_k$).

The sets of all natural, integer, rational and real numbers will be denoted by $\mathbb{N}$, $\mathbb{Z}$, $\mathbb{Q}$ and $\mathbb{R}$, respectively. The subsets of positive elements of sets $\mathbb{Z}$, $\mathbb{Q}$ and $\mathbb{R}$ will be denoted by $\mathbb{Z}_+$, $\mathbb{Q}_+$ and $\mathbb{R}_+$, respectively. The subset of $\mathbb{N}$ composed of the numbers that are not greater than a fixed $n \in \mathbb{N}$ will be denoted by $\{1, 2, \ldots, n\}$ or I_n.

Given a set A and a property $\mathfrak{P}$, we will write $B = \{a \in A : \mathfrak{P}$ holds for $a\}$ to denote that B is the set of all elements of set A for which property $\mathfrak{P}$ holds. For example, a closed interval $\langle a, b \rangle$ for $a, b \in \mathbb{R}$, $a \leq b$, can be defined as the set $\langle a, b \rangle = \{x \in \mathbb{R} : a \leq x \leq b\}$.

A ($n \geq 1$)-dimensional vector space over $\mathbb{R}$, its positive orthant and the interior of the orthant will be denoted by $\mathbb{R}^n$, $\mathbb{R}^n_+$ and $\mathrm{int}\mathbb{R}^n_+$, respectively.

A row (column) vector $x \in \mathbb{R}^n$ composed of numbers $x_1, x_2, \ldots, x_n$ will be denoted by $x = [x_1, x_2, \ldots, x_n]$ ($x = [x_1, x_2, \ldots, x_n]^\top$). A norm (the l_p norm) of vector x will be denoted by $\|x\|$ ($\|x\|_p$). The scalar product of vectors x and y will be denoted by $x \circ y$.

The set of all Pareto (weakly Pareto) optimal solutions from a set of all feasible solutions X will be denoted by X_{Par} ($X_{\mathrm{w-Par}}$).

1.1.2 Sequences

A sequence composed of numbers $x_1, x_2, \ldots, x_n$ will be denoted by $(x_j)_{j=1}^n$ or $(x_1, x_2, \ldots, x_n)$. If the range of indices of elements of sequence $(x_j)_{j=1}^n$ is fixed, the sequence will be denoted by (x_j). In a similar way, we will denote sequences of sequences, e.g., the sequence of pairs $(x_1, y_1), (x_2, y_2), \ldots, (x_n, y_n)$ will be denoted by $((x_j, y_j))_{j=1}^n$, $((x_1, y_1), (x_2, y_2), \ldots, (x_n, y_n))$ or $((x_j, y_j))$.

A sequence (z_k) that is a concatenation of sequences (x_i) and (y_j) will be denoted by $(x_i | y_j)$. If A and B are sets of numbers, the sequence composed of elements of A followed by elements of B will be denoted by $(A|B)$.

A sequence (x_j) in which elements are arranged in the non-decreasing (non-increasing) order will be denoted by $(x_j \nearrow)$ ($(x_j \searrow)$). An empty sequence will be denoted by (ϕ).

The algebraic sum (product) of numbers $x_k, x_{k+1}, \ldots, x_m$ for $k, m \in \mathbb{N}$, will be denoted by $\sum_{i=k}^{m} x_i$ ($\prod_{i=k}^{m} x_i$). If the indices of components of the sum (product) belong to a set J, then the sum (product) will be denoted by $\sum_{j \in J} x_j$ ($\prod_{j \in J} x_j$). If $k > m$ or $J = \emptyset$, then $\sum_{i=k}^{m} x_i = \sum_{j \in J} x_j := 0$ and $\prod_{i=k}^{m} x_i = \prod_{j \in J} x_j := 1$.

1.1.3 Functions

A function f from a set X to a set Y will be denoted by $f : X \to Y$. The value of function $f : X \to Y$ for some $x \in X$ will be denoted by $f(x)$.

If a function f is a monotonically increasing (decreasing) function, we will write $f \nearrow$ ($f \searrow$).

For a given set X, the function f such that $f(x) = 1$ if $x \in X$ and $f(x) = 0$ if $x \notin X$ will be denoted $\mathbf{1}_X$.

The absolute value, the binary logarithm and the natural logarithm of $x \in \mathbb{R}$ will be denoted by $|x|$, $\log x$ and $\ln x$, respectively. The largest (smallest) integer number not greater (less) than $x \in \mathbb{R}$ will be denoted by $\lfloor x \rfloor$ ($\lceil x \rceil$).

Given two functions, $f : \mathbb{N} \to \mathbb{R}_+$ and $g : \mathbb{N} \to \mathbb{R}_+$, we will say that function $f(n)$ *is of order* $O(g(n))$, in short $f(n) = O(g(n))$, if there exist constants $c > 0$ and $n_0 \geq 0$ such that for all $n \geq n_0$, there holds the inequality $f(n) \leq cg(n)$.

Permutations of elements of set I_n, i.e., bijective functions from set I_n onto itself, will be denoted by small Greek characters. For example, permutation σ with components $\sigma_1, \sigma_2, \ldots, \sigma_n$, where $\sigma_i \in I_n$ for $1 \leq i \leq n$ and $\sigma_i \neq \sigma_j$ for $i \neq j$, will be denoted by $\sigma = (\sigma_1, \sigma_2, \ldots, \sigma_n)$. In some cases, permutations will also be denoted by small Greek characters with a superscript. For example, σ' and σ'' will refer to two distinct permutations of elements of set I_n. Partial permutations defined on I_n, i.e., bijective functions between two subsets of set I_n, will be denoted by small Greek characters with a superscript in brackets. For example, $\sigma^{(a)} = (\sigma_1^{(a)}, \sigma_2^{(a)}, \ldots, \sigma_k^{(a)})$ is a partial permutation of elements of set I_n. The set of all permutations (partial permutations) of set I_n will be denoted by $\mathfrak{S}_n$ ($\hat{\mathfrak{S}}_n$).

The sequence $(x_j)_{j=1}^{n}$ composed of numbers $x_1, x_2, \ldots, x_n$, ordered according to permutation $\sigma \in \mathfrak{S}_n$, will be denoted by $x_\sigma = (x_{\sigma_1}, x_{\sigma_2}, \ldots, x_{\sigma_n})$.

Because of the nature of the problems considered in this book, we will assume, unless stated otherwise, that all objects (e.g., sets, sequences, etc.) are finite.

1.1.4 Logical notation

In this book, we will use the following logical notation. A negation, conjunction and disjunction will be denoted by $\neg$, $\wedge$ and $\vee$, respectively. The implication of formulae p and q will be denoted by $p \Rightarrow q$. The equivalence of formulae p and q will be denoted by $p \Leftrightarrow q$ or $p \equiv q$. The existential and general quantifiers will be denoted by $\exists$ and $\forall$, respectively.

In the proofs presented in this book, we will use a few proof techniques. The most often applied proof technique is the *pairwise job (element) interchange argument*: we consider two schedules (sequences) that differ only in the order of two jobs (elements) and we show which schedule (sequence) is the better one. A number of proofs are made *by contradiction*: we assume that a schedule (a sequence) is optimal and we show that this assumption leads to a contradiction. Finally, some proofs are made by the *mathematical induction*.

The use of the rules of inference applied in the proofs will be limited mainly to *De Morgan's rules* ($\neg(p \wedge q) \equiv (\neg p \vee \neg q)$, $\neg(p \vee q) \equiv (\neg p \wedge \neg q)$), *material equivalence* ($(p \Leftrightarrow q) \equiv ((p \Rightarrow q) \wedge (q \Rightarrow p))$) and *transposition* ($(p \Rightarrow q) \equiv (\neg q \Rightarrow \neg p)$) rules.

1.1.5 Other notation

Lemmas, theorems and properties will be numbered consecutively in each chapter. In a similar way, we will number definitions, examples, figures and tables. Examples will be ended by the symbol '♦'.

Most results will be followed either by a full proof or by the sketch of a proof. In a few cases, no proof (sketch) will be given and the reader will be referred to the literature. The proofs, sketches and references to the sources of proofs will be ended by symbols '■', '□' and '◇', respectively.

1.2 Basic definitions and results

In this section, we include the definitions and results that are used in proofs presented in this book.

Lemma 1.1. (Elementary inequalities)
(a) *If $y_1, y_2, \ldots, y_n \in \mathbb{R}$, then $\max\{y_1, y_2, \ldots, y_n\} \geq \frac{1}{n}\sum_{j=1}^{n} y_j$.*
(b) *If $y_1, y_2, \ldots, y_n \in \mathbb{R}$, then $\frac{1}{n}\sum_{j=1}^{n} y_j \geq \sqrt[n]{\prod_{j=1}^{n} y_j}$.*
(c) *If $a, x \in \mathbb{R}$, $x \geq -1$, $x \neq 0$ and $0 < a < 1$, then $(1+x)^a < 1 + ax$.*

Proof. (a) This is the *arithmetic-mean inequality*; see Bullen et al. [37, Chap. 2, Sect. 1, Theorem 2].

(b) This is a special case of the *geometric-arithmetic mean inequality*; see Bullen et al. [37, Chap. 2, Sect. 2, Theorem 1].

(c) This is *Bernoulli's inequality*; see Bullen et al. [37, Chap. 1, Sect. 3, Theorem 1].   ◇

Lemma 1.2. (Minimizing or maximizing a sum of products)
(a) *If $x_1, x_2, \ldots, x_n, y_1, y_2, \ldots, y_n \in \mathbb{R}$, then the sum $\sum_{i=1}^{n} x_{\sigma_i} \prod_{j=i+1}^{n} y_{\sigma_j}$ is minimized (maximized) when it is calculated over the permutation $\sigma \in \mathfrak{S}_n$ in which indices are ordered by non-decreasing (non-increasing) values of the $\frac{x_i}{y_i - 1}$ ratios.*

(b) *If $(x_1, x_2, \ldots, x_n)$ and $(y_1, y_2, \ldots, y_n)$ are two sequences of real numbers, then the sum $\sum_{j=1}^{n} x_j y_j$ is minimized if the sequence $(x_1, x_2, \ldots, x_n)$ is ordered non-decreasingly and the sequence $(y_1, y_2, \ldots, y_n)$ is ordered non-increasingly or vice versa, and it is maximized, if the sequences are ordered in the same way.*

Proof. (a) By pairwise element interchange argument; see Kelly [164, Theorems 1-2], Rau [242, Theorem 1].

(b) By pairwise element interchange argument; see Hardy et al. [131, p. 269]. ◇

Definition 1.3. (V-shaped and Λ-shaped sequences)
(a) *A sequence $(x_1, x_2, \ldots, x_n)$ is said to be V-shaped (has a V-shape) if there exists an index k, $1 \leq k \leq n$, such that for $1 \leq j \leq k$ the sequence is non-increasing and for $k \leq j \leq n$ the sequence is non-decreasing.*
(b) *A sequence $(x_1, x_2, \ldots, x_n)$ is said to be Λ-shaped (has a Λ-shape), if the sequence $(-x_1, -x_2, \ldots, -x_n)$ is V-shaped.*

In other words, sequence $(x_j)_{j=1}^{n}$ is V-shaped (Λ-shaped) if the elements which are placed before the smallest (largest) x_j, $1 \leq j \leq n$, are arranged in the non-increasing (non-decreasing) order, and those which are placed after the smallest (largest) x_j are in the non-decreasing (non-increasing) order.

The V-shaped and Λ-shaped sequences will also be called V-*sequences* and Λ-*sequences*, respectively. Moreover, if index k of the minimal (maximal) element in a V-sequence (Λ-sequence) satisfies the inequality $1 < k < n$, we will say that this sequence is *strongly V-shaped (Λ-shaped)*.

Definition 1.4. (The partial order relation $\prec$)
Let $(u, v), (r, s) \in \mathbb{R}^2$. The partial order relation $\prec$ is defined as follows:

$$(u, v) \prec (r, s), \text{ if } (u, v) \leq (r, s) \text{ coordinatewise and } (u, v) \neq (r, s). \quad (1.1)$$

Lemma 1.5. *The relation $(u, v) \prec (0, 0)$ does not hold when either ($u > 0$ or $v > 0$) or ($u = 0$ and $v = 0$).*

Proof. By Definition 1.4, $(u, v) \prec (0, 0)$ if $(u, v) \leq (0, 0)$ coordinatewise and $(u, v) \neq (0, 0)$. By negation of the conjuction, the result follows. ∎

Definition 1.6. (A graph and a digraph)
(a) *A graph (undirected graph) is an ordered pair $G = (N, E)$, where $N \neq \emptyset$ is a finite set of nodes and $E \subseteq \{\{n_1, n_2\} \in 2^N : n_1 \neq n_2\}$ is a set of edges.*
(b) *A digraph (directed graph) is an ordered pair $G = (V, A)$, where $V \neq \emptyset$ is a finite set of vertices and $A \subseteq \{(v_1, v_2) \in V^2 : v_1 \neq v_2\}$ is a set of arcs.*

Example 1.7. Consider graph G_1 and digraph G_2 given in Fig. 1.1.

In the graph $G_1 = (N, E)$, presented in Fig. 1.1a, the set of nodes $N = \{1, 2, 3, 4\}$ and the set of edges $E = \{\{1, 2\}, \{1, 3\}, \{2, 4\}, \{3, 4\}\}$.

In the digraph $G_2 = (V, A)$, presented in Fig. 1.1b, the set of vertices $V = \{1, 2, 3, 4\}$ and the set of arcs $A = \{(1, 2), (1, 3), (2, 4), (3, 4)\}$. ♦

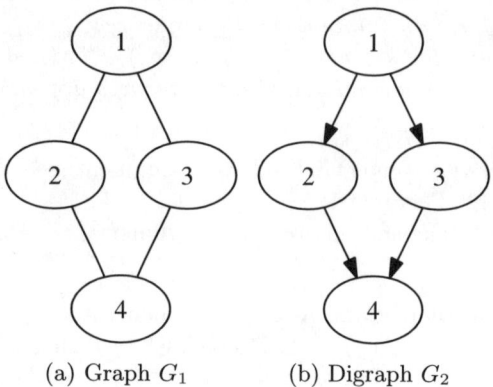

(a) Graph G_1 (b) Digraph G_2

Fig. 1.1: Graph vs. digraph

In this book, we will consider mainly directed graphs. Therefore, all further definitions and remarks will refer to digraphs, unless stated otherwise.

Definition 1.8. (Basic definitions concerning digraphs)
(a) *A digraph $G' = (V', A')$ is called a* subdigraph *of a digraph $G = (V, A)$ if $V' \subseteq V$ and $(x, y) \in A'$ implies $(x, y) \in A$.*
(b) *A* directed path *in a digraph $G = (V, A)$ is a sequence $(v_1, v_2, \ldots, v_m)$ of distinct vertices from V such that $(v_k, v_{k+1}) \in A$ for each $k = 1, 2, \ldots, m - 1$. The number m is called the* length *of the path.*
(c) *A vertex $x \in V$ is called a* predecessor (successor) *of a vertex $y \in V$ if in a digraph $G = (V, A)$ there is a directed path from x to y (from y to x). If the path has unit length, then x is called a* direct predecessor (successor) *of y.*
(d) *A vertex $x \in V$ that has no direct predecessor (successor) is called an* initial (a terminal) *vertex in a digraph $G = (V, A)$. A vertex $x \in V$ that is neither initial nor terminal is called an* internal vertex *in the digraph.*
(e) *A digraph $G = (V, A)$ is* connected *if for every $x, y \in V$ there exists in G a directed path starting with x and ending with y; otherwise, it is* disconnected.

For a given graph (digraph) G and $v \in N$ ($v \in V$), the set of all predecessors and successors of v will be denoted by $Pred(v)$ and $Succ(v)$, respectively.

Definition 1.9. (Parallel and series composition of digraphs)
Let $G_1 = (V_1, A_1)$ and $G_2 = (V_2, A_2)$ be two digraps such that $V_1 \cap V_2 = \emptyset$ and let $Term(G_1) \subseteq V_1$ and $Init(G_2) \subseteq V_2$ denote the set of terminal vertices of G_1 and the set of initial vertices of G_2, respectively. Then
(a) *digraph G_P is said to be a* parallel composition *of digraphs G_1 and G_2, if $G_P = (V_1 \cup V_2, A_1 \cup A_2)$;*
(b) *digraph G_S is said to be a* series composition *of digraphs G_1 and G_2, if $G_S = (V_1 \cup V_2, A_1 \cup A_2 \cup (Term(G_1) \times Init(G_2)))$.*

In other words, digraph G_P is a disjoint union of digraphs G_1 and G_2, while digraph G_S is a composition of digraphs G_1 and G_2 in which the arcs from all terminal vertices in G_1 are connected to all initial vertices in G_2.

Definition 1.10. (Special classes of digraphs)
(a) *A* chain $(v_1, v_2, \ldots, v_k)$ *is a digraph* $G = (V, A)$ *with* $V = \{v_i : 1 \leq i \leq k\}$ *and* $A = \{(v_i, v_{i+1}) : 1 \leq i \leq k - 1\}$.
(b) *An* in-tree (out-tree) *is a digraph which is connected, has a single terminal (initial) vertex called the* root *of this in-tree (out-tree) and in which any other vertex has exactly one direct successor (predecessor). The initial (terminal) vertices of an in-tree (out-tree) are called* leaves.
(c) *A digraph* $G = (V, A)$ *is a* series-parallel *digraph (sp-digraph, in short) if either* $|V| = 1$ *or* G *is obtained by application of parallel or series composition to two series-parallel digraphs* $G_1 = (V_1, A_1)$ *and* $G_2 = (V_2, A_2)$, $V_1 \cap V_2 = \emptyset$.

Remark 1.11. A special type of a tree is a *2-3 tree*, i.e., a balanced tree in which each internal node (vertex) has 2 or 3 successors. In 2-3 trees the operations of insertion (deletion) of a node (vertex) and the operation of searching through the tree can be implemented in $O(\log k)$ time, where k is the number of nodes (vertices) in the tree (see, e.g., Aho et al. [2, Chap. 2]).

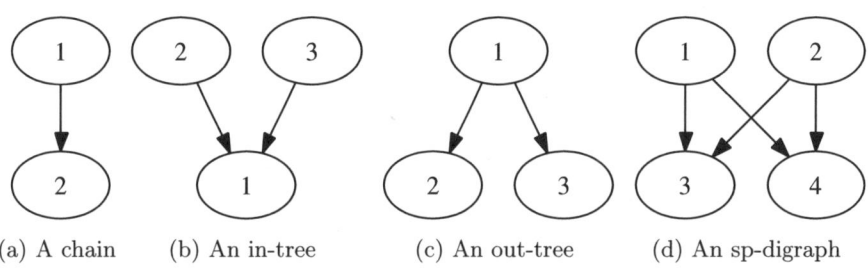

(a) A chain   (b) An in-tree   (c) An out-tree   (d) An sp-digraph

Fig. 1.2: Examples of digraphs from Definition 1.10

Example 1.12. Consider the four digraphs depicted in Fig. 1.2. The chain given in Fig. 1.2a is a digraph $G_1 = (V_1, A_1)$ in which $V_1 = \{1, 2\}$ and $A_1 = \{(1, 2)\}$. The in-tree given in Fig. 1.2b is a digraph $G_2 = (V_2, A_2)$ in which $V_2 = \{1, 2, 3\}$ and $A_2 = \{(2, 1), (3, 1)\}$. The out-tree given in Fig. 1.2c is a digraph $G_3 = (V_3, A_3)$ in which $V_3 = \{1, 2, 3\}$ and $A_3 = \{(1, 2), (1, 3)\}$.

The sp-digraph depicted in Fig. 1.2d is a digraph $G_4 = (V_4, A_4)$ in which $V_4 = \{1, 2, 3, 4\}$ and $A_4 = \{(1, 3), (1, 4), (2, 3), (2, 4)\}$. The sp-digraph G_4 is a series composition of sp-digraphs $G_5 = (V_5, A_5)$ and $G_6 = (V_6, A_6)$, where $V_5 = \{1, 2\}$, $V_6 = \{3, 4\}$ and $A_5 = A_6 = \emptyset$. Notice that the sp-digraph G_5, in turn, is a parallel composition of single-vertex sp-digraphs $G_5' = (V_5', A_5')$ and $G_5'' = (V_5'', A_5'')$, where $V_5' = \{1\}$, $V_5'' = \{2\}$ and $A_5' = A_5'' = \emptyset$. Similarly, the sp-digraph G_6 is a parallel composition of single-vertex sp-digraphs $G_6' = (V_6', A_6')$ and $G_6'' = (V_6'', A_6'')$, where $V_6' = \{3\}$, $V_6'' = \{4\}$ and $A_6' = A_6'' = \emptyset$. ◆

Remark 1.13. From Definition 1.9 it follows that every series-parallel digraph $G = (V, A)$ can be represented in a natural way by a binary *decomposition tree* $T(G)$. Each leaf of the tree represents a vertex in G and each internal node is a series (parallel) composition of its successors. Hence we can construct G, starting from the root of the decomposition tree $T(G)$, by successive compositions of the nodes of the tree. For a given series-parallel digraph, its decomposition tree can be constructed in $O(|V| + |A|)$ time (see Valdes et al. [271]). The decomposition tree of the sp-digraph from Fig. 1.2d is given in Fig. 1.3.

Remark 1.14. Throughout the book, the internal nodes of a decomposition tree that correspond to the parallel composition and series composition will be labelled by P and S, respectively.

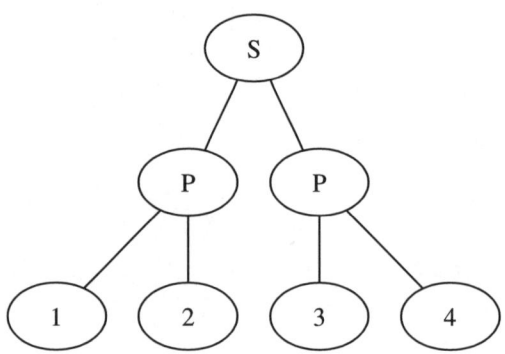

Fig. 1.3: The decomposition tree of the sp-digraph from Fig. 1.2d

Theorem 1.15. (Mean value theorems)
(a) *If functions $f : \langle a, b \rangle \to \mathbb{R}$ and $g : \langle a, b \rangle \to \mathbb{R}$ are differentiable on the interval (a, b) and continuous on the interval $\langle a, b \rangle$, then there exists at least one point $c \in (a, b)$ such that $\frac{f'(c)}{g'(c)} = \frac{f(b)-f(a)}{g(b)-g(a)}$.*
(b) *If function $f : \langle a, b \rangle \to \mathbb{R}$ is differentiable on the interval (a, b) and continuous on the interval $\langle a, b \rangle$, then there exists at least one point $c \in (a, b)$ such that $f'(c) = \frac{f(b)-f(a)}{b-a}$.*

Proof. (a) This is the *generalized mean-value theorem*; see Rudin [248, Chap. 5, Theorem 5.9].   ◇

(b) This is the *mean-value theorem*. Applying Theorem 1.15 (a) for $g(x) = x$, we obtain the result.   ∎

Remark 1.16. The counterparts of mean-value theorems for functions defined in vector spaces are given, e.g., by Maurin [205, Chap. VII].   ◇

Definition 1.17. (A norm)
A norm on a vector space X is the function $\|\cdot\| : X \to \mathbb{R}$ such that for all $x, y \in X$ and any $a \in \mathbb{R}$ the following conditions are satisfied:
(a) $\|x + y\| \leq \|x\| + \|y\|$,
(b) $\|ax\| = |a|\|x\|$,
(c) $\|x\| = 0 \Leftrightarrow x = 0$.
The value $\|x\|$ is called a norm of vector $x \in X$.

Definition 1.18. (Hölder's vector norm l_p)
Given an arbitrary $p \geq 1$, the l_p norm of vector $x \in \mathbb{R}^n$ is defined as follows:

$$\|x\|_p := \begin{cases} \left(\sum_{i=1}^n |x_i|^p\right)^{\frac{1}{p}} & , 1 \leq p < +\infty, \\ \max_{1 \leq i \leq n}\{|x_i|\} & , p = +\infty. \end{cases}$$

Definition 1.19. (A priority-generating function)
Let $\pi' = (\pi^{(1)}, \pi^{(a)}, \pi^{(b)}, \pi^{(2)})$, $\pi'' = (\pi^{(1)}, \pi^{(b)}, \pi^{(a)}, \pi^{(2)}) \in \mathfrak{S}_n$, where $\pi^{(1)}, \pi^{(a)}, \pi^{(b)}, \pi^{(2)} \in \hat{\mathfrak{S}}_n$.
(a) *A function $\mathcal{F} : \mathfrak{S}_n \to \mathbb{R}$ is called a priority-generating function, if there exists a function $\omega : \hat{\mathfrak{S}}_n \to \mathbb{R}$ (called priority function) such that there holds either the implication $\omega(\pi^{(a)}) > \omega(\pi^{(b)}) \Rightarrow \mathcal{F}(\pi') \leq \mathcal{F}(\pi'')$ or the implication $\omega(\pi^{(a)}) = \omega(\pi^{(b)}) \Rightarrow \mathcal{F}(\pi') = \mathcal{F}(\pi'')$.*
(b) *If $\pi^{(a)}, \pi^{(b)} \in \hat{\mathfrak{S}}_1$, then a priority-generating function is called 1-priority-generating function.*

Remark 1.20. Notice that by Definition 1.19, every priority-generating function is a 1-priority-generating function (but not vice versa).

Remark 1.21. Definition 1.19 concerns the priority-generating function of a single variable (see Tanaev et al. [264, Chap. 3]).

Remark 1.22. Priority-generating functions of a single variable are also considered by Gordon et al. [117]. The authors identify several cases in which an objective function for a scheduling problem with some time-dependent job processing times is priority generating (see [117, Sect. 7–9]). They also explore the relationship between the existence of priority functions for different criterion functions for such problems (see [117, Theorems 1–2]).

Remark 1.23. Priority-generating functions of many variables are considered by Janiak et al. [149].

Theorem 1.24. (Tanaev et al. [264]) *If $\mathcal{F} : \mathfrak{S}_n \to \mathbb{R}$ is a 1-priority generating function and $\omega : \hat{\mathfrak{S}}_n \to \mathbb{R}$ is a priority function corresponding to $\mathcal{F}$, then the permutation in which the elements are arranged in the non-increasing order of their priorities minimizes $\mathcal{F}$ over $\mathfrak{S}_n$.*

Proof. See Tanaev et al. [264, Chap. 3, Theorem 7.1]. ◇

Let X denote the set of feasible solutions of a bicriterion optimization problem and let $f : X \to \mathbb{R}^2$, $f = (f_1, f_2)$, be the minimized criterion function, where $f_i : X \to \mathbb{R}$ are single-valued criteria for $i = 1, 2$.

Definition 1.25. (Pareto optimal solutions)
(a) *A solution $x^\star \in X$ is said to be* Pareto optimal, $x^\star \in X_{\mathrm{Par}}$ *in short, if there is no $x \in X$ such that $f(x) \prec f(x^\star)$.*
(b) *A solution $x^\star \in X$ is said to be* weakly Pareto optimal, $x^\star \in X_{\mathrm{w-Par}}$ *in short, if there is no $x \in X$ such that $f_i(x) < f_i(x^\star)$ for $i = 1, 2$.*

The images of sets X_{Par} and $X_{\mathrm{w-Par}}$ under the function $f = (f_1, f_2)$, $f(X_{\mathrm{Par}})$ and $f(X_{\mathrm{w-Par}})$, will be denoted by Y_{eff} and $Y_{\mathrm{w-eff}}$, respectively. (Notice that $X_{\mathrm{Par}} \subset X_{\mathrm{w-Par}}$ and $Y_{\mathrm{eff}} \subset X_{\mathrm{w-eff}}$.)

Example 1.26. (Ehrgott [76]) Consider a set X and a function $f = (f_1, f_2)$, where $X := \{(x_1, x_2) \in \mathbb{R}^2 : 0 < x_1 < 1 \land 0 \le x_2 \le 1\}$, $f_1 := x_1$ and $f_2 := x_2$. Then $Y_{\mathrm{eff}} = \emptyset$ and $Y_{\mathrm{w-eff}} = \{(x_1, x_2) \in X : 0 < x_1 < 1, x_2 = 0\}$.
If we define $X := \{(x_1, x_2) \in \mathbb{R}^2 : 0 \le x_i \le 1$ for $i = 1, 2\}$, then for the function f as above we have $Y_{\mathrm{eff}} = \{(0, 0)\}$ and $Y_{\mathrm{w-eff}} = \{(x_1, x_2) \in X : x_1 = 0 \lor x_2 = 0\}$. ♦

Lemma 1.27. (Scalar optimality vs. Pareto optimality)
If $x^\star$ is an optimal solution with respect to the scalar criterion $\omega \circ f$ for a certain $f = (f_1, f_2)$ and $\omega = (\omega_1, \omega_2)$, then
(a) *if $\omega \in \mathbb{R}^2$, then $x^\star \in X_{\mathrm{w-Par}}$,*
(b) *if $\omega \in \mathrm{int} \mathbb{R}^2$, then $x^\star \in X_{\mathrm{Par}}$.*

Proof. (a), (b) See Ehrgott [76, Proposition 3.7]. ◇

Definition 1.28. (A convex function)
A function f is convex *on the interval $\langle a, b \rangle$ if for any $x_1, x_2 \in \langle a, b \rangle$ and any $\lambda \in \langle 0, 1 \rangle$ there holds the inequality $f(\lambda x_1 + (1-\lambda)x_2) \le \lambda f(x_1) + (1-\lambda)f(x_2)$.*

In other words, a convex function is such a continuous function that the value at any point within every interval in its domain does not exceed the value of a convex combination of its values at the ends of the interval. (A *convex combination* of elements x_1 and x_2 is the element $y := \lambda x_1 + (1-\lambda)x_2$, where $\lambda \in \langle 0, 1 \rangle$ is a given number.)

Remark 1.29. If the symbol '$\le$' is replaced by '$<$' in Definition 1.28, then the function f is *strictly* convex.

Remark 1.30. A function f is (strictly) *concave*, if $-f$ is (strictly) convex.

Definition 1.31. (A convex set)
A set $X \subseteq \mathbb{R}^n$ is convex *if for any $x_1, x_2 \in X$ and any $\lambda \in \langle 0, 1 \rangle$ the convex combination $\lambda x_1 + (1 - \lambda)x_2 \in X$.*

In other words, a set X is convex if the line segment joining any pair of points of X lies entirely in X.

Remark 1.32. Basic facts concerning convex functions and convex sets are given, e.g., by Walk [278, Chap. 1].   ◇

Lemma 1.33. (Pareto optimality vs. scalar optimality)
If X is a convex set and f_1, f_2 are convex functions, then if $x^\star \in X_{\mathrm{w-Par}}$, there exists $\omega \in \mathrm{int}\mathbb{R}^2$ such that $x^\star$ is an optimal solution with respect to the scalar criterion $\omega \circ f$.

Proof. See Ehrgott [76, Proposition 3.8].   ◇

With this lemma, we end the presentation of notation, definitions and auxiliary results used throughout the book. In subsequent chapters, we will introduce basic definitions and results concerning algorithms (Chap. 2), $\mathcal{NP}$-complete problems (Chap. 3), the scheduling theory (Chap. 4) and time-dependent scheduling (Chap. 5).

1.3 Bibliographic notes

A comprehensive presentation of basic mathematical notions and mathematical notation may be found in Rasiowa [241]. Inference rules and proof techniques are discussed in Copi [65].

Bullen et al. [37], Hardy et al. [131] and Mitrinović et al. [212] give a wide range of various inequalities.

Berge [22], Harary [130] and Wilson [294] present the graph theory from different perspectives. Brandstädt et al. [32] study the properties of different classes of graphs. Applications of graphs in computer science and engineering are discussed by Deo [70].

Maurin [205] and Rudin [248] give a concise presentation of calculus and mathematical analysis.

Priority-generating functions are discussed by Tanaev et al. [264, Chap. 3] and by Gordon et al. [117]. The extension of these functions to the multiple criteria case is presented by Janiak et al. [149].

The properties of sp-(di)graphs and applications of these (di)graphs in the scheduling theory are discussed, e.g., by Gordon [116], Gordon et al. [120], Lawler [185], Möhring [221] and Valdes et al. [271].

Ehrgott [76] presents a comprehensive introduction to Pareto optimality.

2

Problems and algorithms

S cheduling problems considered in this book are formulated as either decision or optimization problems. The problems are solved by algorithms of different types. Therefore, in this chapter we present the essential concepts related to decision and optimization problems and to algorithms.

Chapter 2 is composed of three sections. In Sect. 2.1, we recall the notions of decision and optimization problems. In Sect. 2.2, we present the basic concepts related to algorithms. The chapter is completed with bibliographic notes in Sect. 2.3.

2.1 Decision and optimization problems

A *problem* is a general question to be answered. The question concerns a certain mathematical object and it is expressed in terms of a number of *parameters*, whose values are left unspecified. A problem is formulated by giving a general description of all its parameters and a statement concerning properties that the object must have. This object is called a *solution* to the problem.

In this book, we will consider decision and optimization problems. A *decision problem* is a problem of existence of a solution that has properties specified in the formulation of the problem. An *optimization problem* is a problem in which a solution that optimizes (i.e., minimizes or maximizes) a certain *objective (criterion) function* is searched.

Example 2.1. An example of a decision problem is the *SUBSET PRODUCT* problem: given a set of integer numbers and a *threshold* integer value, does there exists a subset of this set such that the product of all elements of this subset is equal to the threshold? ♦

Example 2.2. An example of an optimization problem is the problem of minimizing the value of a function over a set of elements. ♦

Remark 2.3. Formulations of decision problems considered in this book are given in Sect. 3.2.

Remark 2.4. We will come back to optimization problems in Chap. 11, in which we will present several methods of solving such problems.

An optimization problem may have a number of solutions which may be optimal or suboptimal. A solution to an optimization problem is *optimal* if the value of the minimized (maximized) objective function for this solution is minimal (maximal). A solution is *suboptimal* if the value of objective function for this solution is only close to an optimal value. Optimal and suboptimal solutions will be also called *feasible* solutions.

Remark 2.5. A *combinatorial optimization problem* is a special case of an optimization problem in which the set of solutions is finite (see Sect. 11.1 for more details). Scheduling problems considered in this book are examples of combinatorial optimization problems.

The optimization and decision problems are closely related to each other. For example, suppose that in an optimization problem, an objective function is minimized. Then, in a decision counterpart of this problem, we ask if there exists a solution for which the value of the objective function is not greater than a given *threshold* value. (If the objective function is maximized, then we ask for a solution for which the value of the objective function is not lower than a given threshold.) Therefore, though scheduling problems considered in this book are formulated mainly as combinatorial optimization problems, they can be considered either in the optimization or in the decision version.

2.1.1 Encoding schemata

By assigning a specific value to each parameter of a problem, we define an *instance* of the problem. (The set of all instances of a given problem P will be denoted by D_P.) The instance, encoded in a certain format, is the input of any procedure used for finding a solution to this instance. The rules that describe the coding format constitute an *encoding scheme*. The encoding scheme encodes any instance into a sequence of symbols from a certain finite alphabet. The length of this sequence is called the *length of input* for this instance. (The number of symbols needed for encoding a number k and the length of input for an instance I in an encoding scheme e will be denoted by $|k|_e$ and $|I|_e$, respectively.)

Remark 2.6. There exist many different encoding schemata. Each encoding scheme specifies the rules of encoding numbers, sets, graphs and other mathematical objects. The most important rules concern the encoding of numbers. The rules should ensure that the encoding scheme is concise, i.e., it does not cause an exponential growth of the length of input, and that numbers are represented in any positional numeral system.

The simplest scheme is the *unary encoding scheme u*, in which a number n is represented by n 1's, i.e., $|n|_u = n$. The scheme, however, is not concise (cf. Remark 2.6).

Throughout this book, we will use the *binary encoding scheme b*. In this scheme, a number n is represented in the binary system, i.e., $|n|_b = \lceil \log n \rceil = O(\log n)$. This scheme is concise and unlike the unary encoding scheme, it does not cause an exponential growth of the length of input.

2.1.2 Undecidable and decidable problems

A problem is solved if a solution to any instance of the problem may be found. However, the time needed for finding the solution may or may not be finite. Hence, from the point of view of the computation time, all problems are divided into two classes of problems: undecidable and decidable ones.

A problem is said to be *undecidable* (*unsolvable*) if there is no finite procedure that solves the problem. An example of such a problem is the *halting problem*: given a Turing machine and an input, does the machine halt for this input? Another example of an undecidable problem is the *tenth Hilbert's problem*: given an arbitrary *Diophantine equation* (i.e., an equation in which only integer solutions are allowed), does there exist a solution of this equation?

Remark 2.7. Though from the undecidability of a problem there follows a negative answer to the question of existence of a finite procedure solving any instance of the problem, it does not exclude the possibility of finding a solution procedure for a particular type of instances of the problem. For example, there are procedures that determine if a Turing machine halts for a given type of an input and procedures that solve particular types of Diophantine equations.

A problem is said to be *decidable* (*solvable*) if there exists a finite procedure which solves the problem. This means that for any instance of the problem, a solution to the instance can be found in a finite time.

In this book, we will consider only decidable scheduling problems.

2.2 Basic concepts related to algorithms

A finite procedure that finds a solution to an arbitrary instance of a problem is called an *algorithm*. (We say that the algorithm solves the problem.) Generally speaking, an algorithm consists of an *input*, a sequence of *steps* and an *output*. The input describes a specific instance. The sequence of steps must be performed by the algorithm to find a solution to the instance. Each step, in turn, can be decomposed into a finite number of *elementary operations*. Examples of elementary operations are *arithmetical operations* (addition, subtraction, multiplication, division), *logical operations* (negation, conjuction, disjunction), the *assignment statement* (assigning a value to a variable) and *control operations* (conditional jump, call of a function or a procedure). *Complex operations* which can appear in an algorithm (conditional loops, iterative loops, functions and procedures) are sequences of elementary operations. The algorithm, for a given input, performs the steps and produces an output, which is the solution to the input.

Remark 2.8. The solution produced by an algorithm at some step will be called a *partial solution*, contrary to the *complete solution* obtained after the completion of the algorithm.

2.2.1 Time and space complexity of algorithms

For a given problem, there may exist various algorithms which have different efficiency. We will now introduce two basic measures of efficiency of an algorithm: time and space.

The efficiency of an algorithm may be measured by the number of elementary operations which must be performed by the algorithm to find a solution to any instance of a given problem. The number of operations is a function of the input of this algorithm: if the input is longer, the algorithm will perform more operations. The *time complexity* of this algorithm is the function that maps each length of input into the maximal number of elementary operations needed for finding a solution to any instance of that length.

The efficiency of an algorithm may also be measured by the total space needed for the execution of the algorithm. The *space complexity* of this algorithm is the function that maps each length of input into the maximal amount of computer memory cells needed for the execution of the algorithm for any instance of that length.

In this book, we will mainly consider the time complexity of algorithms.

2.2.2 Polynomial-time algorithms

We say that an algorithm is efficient if its time complexity is polynomially bounded with respect to the length of input. This means that the number of elementary operations performed by this algorithm for any instance of a given problem will be not greater than a polynomial of the length of input for this instance. Such an algorithm is called a *polynomial-time (polynomial) algorithm* and is defined as follows.

Definition 2.9. (A polynomial-time algorithm)
An algorithm that solves a problem is said to be a polynomial-time (polynomial) algorithm *if there exists a polynomial q such that for any instance I of the problem the number of elementary operations performed by the algorithm is bounded from above by $q(|I|_b)$.*

In other words, the time complexity of a polynomial-time algorithm for an input of length $|I|_b$ is of order $O(q(|I|_b))$ for a certain polynomial q. Throughout this book, we will say that a problem is *polynomially solvable (computationally tractable)* if there exists a polynomial-time algorithm for this problem.

2.2.3 Exponential-time algorithms

An algorithm which is not a polynomial-time algorithm is called an *exponential-time (exponential) algorithm*. This means that the number of elementary operations performed by such an algorithm cannot be bounded from above by any polynomial of the length of input. The time complexity of an exponential algorithm either is an exponential function or grows at least as quickly as an exponential function. (Notice that the time complexity function of an exponential algorithm need not be an exponential function in the strict sense.) Since exponential functions grow faster than polynomials, the exponential-time algorithms are not efficient. The problems for which only exponential-time algorithms are known will be called *computationally intractable*.

2.2.4 Pseudopolynomial-time algorithms

Algorithms which are polynomial with respect to both the length of input and the maximum value in the input are called *pseudopolynomial-time (pseudopolynomial) algorithms*. Since polynomial and pseudopolynomial algorithms are related in a certain way, we shall make a few remarks concerning the relation between them.

Remark 2.10. By Definition 2.9, any polynomial algorithm is a pseudopolynomial algorithm as well.

Remark 2.11. Pseudopolynomial algorithms are not polynomial, since the maximum value in input exponentially depends on the representation of this value both in the binary encoding scheme and in any other concise encoding scheme (cf. Remark 2.6).

Remark 2.12. Pseudopolynomial algorithms would be polynomial if either the unary encoding scheme was used or the maximum value was bounded from above by a polynomial of the length of input.

Since the existence of a pseudopolynomial algorithm for a computationally intractable problem has important consequences, we will come back to the concept of pseudopolynomial-time algorithm in Sect. 3.1.

2.2.5 Exact algorithms

The solutions generated by an algorithm may or may not be exact. For example, if different schedules for a set of jobs exist, the schedule which meets all requirements and which has the smallest value of criterion function is the *exact (optimal)* solution, while the schedule which meets all requirements but has a cost greater than the cost of the optimal schedule is only a *feasible (suboptimal)* solution. Throughout this book, by an *exact algorithm*, we will understand such an algorithm that finds the exact solution. Polynomial-time algorithms are examples of exact algorithms that find the exact solution in a polynomial-time. For the problems for which no polynomial-time algorithms are known, exact solutions can be found by enumerative, branch-and-bound or dynamic programming algorithms.

2.2.6 Enumerative algorithms

An *enumerative algorithm* directly enumerates all possible solutions. For example, if a problem consists in finding such a permutation of n elements of a set that for this permutation the value of a function is minimal, then a simple enumerative algorithm generates all possible permutations and selects the optimal one. However, since there are $n! > 2^n$ permutations of set I_n, this algorithm runs in exponential time.

2.2.7 Branch-and-bound algorithms

A *branch-and-bound algorithm* finds an optimal solution by indirect enumeration of all possible solutions through examination of smaller and smaller subsets of the set of all solutions. The algorithm consists of two procedures: *branching* and *bounding*.

Branching is a procedure of partitioning a large problem into a number of subproblems. The subproblems, in turn, are partitioned into smaller subproblems and so on. The branching procedure allows to construct a tree, in which a node corresponds to a partial solution (called a *child* solution) to a subproblem. Two nodes in the tree are connected by an edge if the solutions corresponding to these nodes are child solutions of a solution. The leaves of the tree correspond to complete solutions to the problem.

Bounding is a procedure that allows to cut off a certain part of the solution tree. This part includes the partial solutions for which the value of the objective function is not better than the currently best value. The procedure uses an estimation of the optimal value of the objective function, the so-called *lower bound*. (If such an estimation is not known, then the value of the objective function for a known feasible solution is assumed.)

2.2.8 Dynamic programming algorithms

An exact algorithm may also be constructed by *dynamic programming*. In this case, the optimal solution is generated by a multi-stage decision process, which, starting from an *initial state*, constructs subsequent partial solutions from previously generated *states* in a step-by-step manner. The initial state is defined by some *initial conditions*, the subsequent states are defined by a *recursive formula*, and the final state is the *goal* we want to achieve. This final state corresponds to an optimal (exact) solution.

The theoretical foundation of dynamic programming is given by the so-called *Bellman's principle of optimality* (cf. [20, Chap. III, § 3]). The principle says that whatever the initial state and initial decision are, the remaining decisions must constitute an optimal solution with respect to the state resulting from the first decision.

According to this principle, at each stage of the process, the decisions that lead to subsequent partial solutions are ranked in some way and the ones with the highest rank are taken into account in subsequent stages until the final (optimal) solution is achieved.

2.2.9 Approximation algorithms

Not all problems encountered in practice have polynomial-time algorithms. Many problems have been proved to be *computationally intractable*, which means that for such problems polynomial algorithms probably do not exist.

However, even if it is known that a problem is computationally intractable, there still remains the question how to find a solution to the problem. If all known algorithms for the problem are inefficient, we may apply an *approximation algorithm* to solve this problem. The approximation algorithm is a polynomial algorithm that generates an *approximate (suboptimal)* solution that is close (in the sense defined below) to an optimal solution.

Since the solution generated by an approximation algorithm is only a suboptimal solution, it is useful to know how close to the optimal one the solution is. A measure of the closeness is the *worst-case ratio* of the algorithm.

Definition 2.13. (An approximation algorithm and its worst-case ratio)
Let $A(I)$ and $OPT(I)$ denote a solution generated by an algorithm A and an optimal solution to a given instance I of a minimization (maximization) problem, respectively. Let $\epsilon > 0$ and $r = 1 + \epsilon$ ($r = 1 - \epsilon$). An algorithm A is said to be an r-approximation algorithm for a problem P *if for any instance $I \in D_P$, there holds $|A(I) - OPT(I)| \leq \epsilon \cdot OPT(I)$. The value r is called the* worst-case ratio *of the algorithm A.*

From Definition 2.13, it follows that if the algorithm A solves a minimization problem, then $A(I) \leq (1 + \epsilon) \cdot OPT(I) = r \cdot OPT(I)$ and the worst-case ratio $r \in \langle 1, +\infty)$. (If the algorithm A solves a maximization problem, we have $A(I) \geq (1 - \epsilon) \cdot OPT(I) = r \cdot OPT(I)$ and $r \in \langle 0, 1\rangle$.) In other words, an r-approximation algorithm generates a solution which is at most r times worse than the optimal one.

Remark 2.14. If I is an arbitrary instance of a minimization problem and A is an approximation algorithm for the problem, for this instance we can calculate the *absolute ratio* $R_A^a(I)$ of the value of the solution $A(I)$ generated by the algorithm and the value of the optimal solution $OPT(I)$, $R_A^a(I) := \frac{A(I)}{OPT(I)}$. (For a maximization problem we have $R_A^a(I) := \frac{OPT(I)}{A(I)}$.) The worst-case ratio r from Definition 2.13 is an infimum (a supremum for a maximization problem) over the ratios $R_A^a(I)$ over all possible instances I of a given problem.

Remark 2.15. Apart from the absolute ratio $R_A^a(I)$, for an instance I we can calculate the *relative ratio* $R_A^r(I) := \frac{A(I)-OPT(I)}{OPT(I)}$. (For a maximization problem we have $R_A^r(I) := \frac{OPT(I)-A(I)}{A(I)}$.) It is easy to notice that $R_A^r(I) = R_A^a(I) - 1$.

2.2.10 Approximation schemata

For some computationally intractable problems, it is possible to construct a family of approximation algorithms which generate solutions as close to the optimal one as it is desired. Such a family of algorithms is called an *approximation scheme*. There exist two types of approximation schemata: *polynomial-time* and *fully polynomial-time*.

Definition 2.16. (Approximation schemata)
(a) *A family of r-approximation algorithms is called a* polynomial-time approximation scheme *if for an arbitrary $\epsilon > 0$ any algorithm from this family has polynomial time complexity.*
(b) *If a polynomial-time approximation scheme is running in polynomial time with respect to $\frac{1}{\epsilon}$, it is called a* fully polynomial-time approximation scheme.

Throughout this book, a polynomial-time approximation scheme and a fully polynomial-time approximation scheme will be called *PTAS* and *FPTAS*, respectively. In Sect. 3.1 we will specify the conditions which have to be satisfied for a problem to have a PTAS (an FPTAS).

2.2.11 Offline algorithms vs. online algorithms

We have assumed so far that all input data of an algorithm are known at the moment of the start of execution of the algorithm. However, such a complete knowledge is not always possible. Therefore, we can divide all algorithms into *offline* algorithms and *online* algorithms, depending on whether the whole input data are available or not when an algorithm begins its execution. These two classes of algorithms are defined as follows.

Definition 2.17. (Offline algorithm and online algorithm)
(a) *An algorithm is called an* offline algorithm *if it processes its input as one unit and the whole input data are available at the moment of the start of execution of this algorithm.*
(b) *An algorithm is called an* online algorithm *if it processes its input piece by piece and only a part of the input is available at the moment of the start of execution of this algorithm.*

Remark 2.18. Sometimes, even the data concerning a piece of input of an online algorithm may be known only partially. In this case, the online algorithm is called a *semi-online* algorithm.

An online (a semi-online) algorithm can be evaluated by its *competitive ratio*. The ratio is a counterpart of the worst-case ratio for an approximation algorithm (cf. Definition 2.13) and it can be defined as follows.

Definition 2.19. (*c*-competitive algorithm and competitive ratio)
Let $A(I)$ and $OPT(I)$ denote the solutions generated by an online (semi-online) algorithm A and by optimal offline algorithm, respectively, for a given instance I of a minimization problem.
(a) *Algorithm A is called c*-competitive *if there exist constant values c and k such that $A(I) \leq c \cdot OPT(I) + k$ for all I.*
(b) *The constant c defined as above is called the* competitive ratio *of algorithm A.*

Remark 2.20. Definition 2.19 (a) concerns an online minimization algorithm. An online maximization algorithm is *c*-competitive if there exist constant values c and k such that $A(I) \geq c \cdot OPT(I) + k$ for all I.

Remark 2.21. In general, the constant k from Definition 2.19 is a non-zero value. In the case of the problems considered in the book, we have $k = 0$.

2.2.12 Heuristic algorithms

Sometimes, it is difficult to establish the worst-case ratio (the competitive ratio) of an approximation (online) algorithm. An algorithm is called a *heuristic algorithm* (a *heuristic*, in short) if its worst-case ratio is unknown. This means that one cannot predict the behaviour of this algorithm for all instances of the considered problem.

The efficiency of a heuristic algorithm can be evaluated with a *computational experiment*. In the experiment, a set of *test instances* is generated and the solutions obtained by the heuristic under evaluation are compared with optimal solutions found by an exact algorithm.

2.2.13 Greedy algorithms

A huge spectrum of heuristics is known. An example of a simple heuristic is the so-called *greedy algorithm*.

A greedy algorithm repeatedly executes a procedure which tries to construct a solution by choosing a locally best partial solution at each step. In some cases, such a strategy leads to finding optimal solutions. In general, however, greedy algorithms produce only relatively good suboptimal solutions. Therefore, for more complex problems more sophisticated algorithms such as local search algorithms have been proposed.

2.2.14 Local search algorithms

Other examples of heuristics are *local search algorithms*. These algorithms start from an initial solution and iteratively generate a *neighbourhood* of the solution which is currently the best one. The neighbourhood is a set of all solutions that can be obtained from the current solution by feasible *moves*. The moves are performed by different *operators* whose definitions depend on a particular problem. The aim of an operator is to produce a new feasible solution from another feasible solution.

Given a neighbourhood and all feasible moves, a local search algorithm finds a new solution by using a *strategy* of local search of the neighbourhood.

The above described procedure of finding a new solution by a local search algorithm is performed until a *stop condition* is met. In this case, the algorithm stops, since the further decrease (increase) of the minimized (maximized) objective function is very unlikely.

2.2.15 Metaheuristics

An important group of heuristics algorithms is composed of metaheuristics. A *metaheuristic* is a template of a local search algorithm. The template includes a number of control parameters which have an impact on the quality of the solutions that are generated by the given metaheuristic and the conditions that cause the termination of its execution.

Examples of metaheuristics are *simulated annealing*, *tabu search* and *evolutionary algorithms*. In the metaheuristics, different complicated strategies are applied to construct a new solution from a current solution.

2.2.16 The presentation of algorithms

Throughout this book, algorithms will be presented in a pseudo-code similar to Pascal. The formulation of an algorithm in the pseudo-code will start with a header with the name of the algorithm, followed by a description of its input and output. The remaining part of the pseudo-code will be divided into sections which will correspond to particular steps of the algorithm.

In the pseudo-code, we will use standard statements of Pascal such as conditional jump **if .. then .. else**, iterative loop **for .. do**, conditional loops **while .. do** and **repeat .. until**. The **exit** statement will denote an immediate termination of the execution of a loop and passing the control to the first statement after the loop. The assignment operator will be denoted by the symbol '←'. The instruction of printing a message will be denoted by **write**. The **return** statement will denote the end of the execution of the current pseudo-code and the returning of the specified value. The **stop** statement will denote unconditional halting of the execution of a given algorithm.

The level of nesting in complex statements will be denoted by indentation rather than the **begin .. end** statement. The necessary comments will be printed in `teletype` font and preceded by the symbol '▷'. The consecutive operations of an algorithm usually will be followed by a semicolon (';'). The continuation of the previous line of pseudo-code will be denoted by the symbol '↪'. The last statement of an algorithm will be followed by a dot ('.').

As an example of application of the pseudo-code, we present the branch-and-bound algorithm, which is discussed earlier in the section. Let $\mathfrak{F}$ denote the set of all possible solutions to an optimization problem.

Algorithm *BranchAndBound*

Input: a suboptimal solution s_0, criterion f
Output: an optimal solution $s^\star$

▷ Step 1: initialization
$\quad U \leftarrow f(s_0);$
$\quad s^\star \leftarrow s_0;$
$\quad L \leftarrow \mathfrak{F};$
▷ Step 2: the main loop
$\quad$ **while** $(L \neq \emptyset)$ **do**
$\qquad$ Choose a solution $s_{tmp} \in L;$
$\qquad L \leftarrow L \setminus \{s_{tmp}\};$
$\qquad$ Generate all child solutions $s_{n_1}, s_{n_2}, \ldots, s_{n_k}$ of $s_{tmp};$ $\quad$▷ branching
$\qquad$ **for** $i \leftarrow n_1$ **to** n_k **do** ▷ bounding
$\qquad\quad$ Calculate $LB(s_{n_i});$
$\qquad\quad$ **if** $(LB(s_{n_i}) < U)$ **then**
$\qquad\qquad$ **if** $(s_{n_i}$ is a complete solution) **then**
$\qquad\qquad\quad U \leftarrow LB(s_{n_i});$
$\qquad\qquad\quad s^\star \leftarrow s_{n_i}$
$\qquad\qquad$ **else** $L \leftarrow L \cup \{s_{n_i}\};$
▷ Step 3: the final solution
$\quad$ **return** $s^\star.$

Remark 2.22. Throughout this book, pseudo-codes of exact polynomial-time algorithms, enumerative algorithms and heuristic algorithms will be denoted by symbols A_i, E_j and H_k, respectively, where i, j and k will denote the number of the consecutive algorithm.

Remark 2.23. The formulations of algorithms presented in this book slightly differ from the original formulations. The reason for that is the desire to unify the notation and the way of presentation. Therefore, some variable names are changed, added or deleted, and conditional loops are used instead of unconditional jump statements.

With these remarks, we end the presentation of fundamental concepts concerning algorithms. In subsequent chapters, we will introduce basic definitions and results concerning $\mathcal{NP}$-complete problems (Chap. 3), the scheduling theory (Chap. 4) and time-dependent scheduling (Chap. 5).

2.3 Bibliographic notes

The basic concepts related to algorithms are presented by Aho et al. [2], Atallah [8], Cormen et al. [66] and Knuth [167].

The unary and the binary encoding schemata are discussed by Garey and Johnson [85, Chap. 2].

The Turing machine was introduced by Turing [270]. Aho et al. [2], Hopcroft and Ullman [139] and Lewis and Papadimitriou [202] give a detailed description of the Turing machine and its variants.

The undecidability of the halting problem was proved by Turing [270] in 1936. The undecidability of the tenth Hilbert's problem was proved by Matijasevich [204] in 1970.

Davies [67, 68], Hopcroft and Ullman [139], Lewis and Papadimitriou [202], Papadimitriou [233] and Rogers [247] discuss the undecidability and its consequences for computer science.

Different aspects of combinatorial optimization problems are studied by Cook et al. [64], Lawler [183], Nemhauser and Wolsey [225], and Papadimitriou and Steiglitz [234].

Enumerative algorithms are considered by Kohler and Steiglitz [168] and Woeginger [297].

Branch-and-bound algorithms are discussed by Lawler and Woods [187], Papadimitriou and Steiglitz [234] and Walukiewicz [279]. Mitten [211] and Rinnooy Kan [246] give axioms for branch-and-bound algorithms.

Dynamic programming and its applications are studied by Bellman [20], Bellman and Dreyfuss [21] and Lew and Mauch [201].

Graham [124], Graham et al. [125], Hochbaum [136], Klein and Young [166], Schuurman and Woeginger [252], and Vazirani [272] discuss at length different aspects concerning approximation algorithms and approximation schemata.

Online algorithms are considered by Albers [3], Borodin and El-Yaniv [28], Fiat and Woeginger [79], Irani and Karlin [143], Phillips and Westbrook [236], and Pruhs et al. [239].

A greedy algorithm is optimal for a problem if the set of all solutions to the problem constitutes the so-called *matroid*. Matroids are studied in detail by Lawler [183], Oxley [231], Revyakin [244] and Welsh [292].

Applications of matroids in mathematics and computer science are presented by White [293].

Generalizations of matroids, *greedoids*, are studied by Korte et al. [175].

General properties and applications of metaheuristics are discussed by Blum and Roli [25], Gendreau and Potvin [111], Glover and Kochenberger [113], Michiels et al. [210], Osman and Kelly [230], and Voss et al. [274].

Simulated annealing algorithms are considered by Aarts and Korst [1] and Salamon et al. [249].

Tabu search algorithms are studied by Glover and Laguna [114].

Evolutionary algorithms are discussed by Bäck [18], Eiben et al. [77], Fogel [83], Hart et al. [132], Michalewicz [208], Michalewicz and Fogel [209].

$\mathcal{NP}$-complete problems

The theory of $\mathcal{NP}$-completeness has a great impact on the scheduling theory, since the knowledge of the complexity status of a problem allows to facilitate further research of the problem. Therefore, in this chapter, we recall the fundamental concepts related to $\mathcal{NP}$-completeness.

Chapter 3 is composed of three sections. In Sect. 3.1, we recall the basic definitions and results concerning the theory of $\mathcal{NP}$-completeness. In Sect. 3.2, we formulate all $\mathcal{NP}$-complete problems, which appear in $\mathcal{NP}$-completeness proofs presented in this book. The chapter is completed with bibliographic notes in Sect. 3.3.

3.1 Basic definitions and results

Let $\mathcal{P}$ ($\mathcal{NP}$) denote the class of all decision problems solved in polynomial time by a deterministic (non-deterministic) Turing machine.

If a decision problem $P \in \mathcal{P}$, it means that we can solve this problem by a polynomial-time algorithm. If a decision problem $P \in \mathcal{NP}$, it means that we can verify in polynomial time whether a given solution to P has the properties specified in the formulation of this problem. (Notice that if we know how to solve a decision problem in polynomial time, we can also verify in polynomial time any solution to the problem. Hence, $\mathcal{P} \subseteq \mathcal{NP}$.)

Now we introduce the concept of a polynomial-time transformation.

Definition 3.1. (A polynomial-time transformation)
A polynomial-time transformation of a decision problem P' into a decision problem P is a function $f : D_{P'} \rightarrow D_P$ satisfying the following conditions:
(a) the function can be computed in polynomial time;
(b) for all instances $I \in D_{P'}$, there exists a solution to I if and only if there exists a solution to $f(I) \in D_P$.

If there exists a polynomial-time transformation of the problem P' to the problem P (i.e., if P' is *polynomially transformable* to P), we will write $P' \propto P$.

Definition 3.2. (An $\mathcal{NP}$-complete problem)
A decision problem P is said to be $\mathcal{NP}$-complete, if $P \in \mathcal{NP}$ and $P' \propto P$ for any $P' \in \mathcal{NP}$.

In other words, a problem P is $\mathcal{NP}$-complete if any solution to P can be verified in polynomial time and any other problem from the $\mathcal{NP}$ class is polynomially transformable to P.

Since the notion of an $\mathcal{NP}$-complete problem is one of the fundamental notions in the complexity theory, some remarks are necessary.

Remark 3.3. If for a problem P and any $P' \in \mathcal{NP}$ we have $P' \propto P$, the problem P is said to be $\mathcal{NP}$-*hard*.

Remark 3.4. The problems which are $\mathcal{NP}$-complete ($\mathcal{NP}$-hard) with respect to the binary encoding scheme become polynomial with respect to the unary encoding scheme. Therefore, such problems are also called $\mathcal{NP}$-*complete* ($\mathcal{NP}$-*hard*) *in the ordinary sense, ordinary $\mathcal{NP}$-complete* ($\mathcal{NP}$-*hard*) or *binary* $\mathcal{NP}$-*complete* ($\mathcal{NP}$-*hard*) problems.

Remark 3.5. All $\mathcal{NP}$-complete problems are related to each other in the sense that a polynomial-time algorithm which would solve at least one $\mathcal{NP}$-complete problem would solve all $\mathcal{NP}$-complete problems. Therefore, $\mathcal{NP}$-complete problems most probably do not belong to the $\mathcal{P}$ class.

Remark 3.6. Since $\mathcal{P} \subseteq \mathcal{NP}$ and since no polynomial-time algorithm has been found so far for any problem from the $\mathcal{NP}$ class, the question whether $\mathcal{NP} \subseteq \mathcal{P}$ is still open. This fact implies conditional truth of $\mathcal{NP}$-completeness results: they hold, unless $\mathcal{P} = \mathcal{NP}$.

Remark 3.7. $\mathcal{NP}$-completeness of a problem is a very strong argument for the conjecture that this problem cannot be solved by a polynomial-time algorithm, unless $\mathcal{P} = \mathcal{NP}$. (An $\mathcal{NP}$-complete problem, however, may be solved by a pseudopolynomial-time algorithm.)

The class of all $\mathcal{NP}$-complete problems will be denoted by $\mathcal{NPC}$.

Proving the $\mathcal{NP}$-completeness of a decision problem immediately from Definition 3.2 is usually a difficult task, since we have to show that any problem from the $\mathcal{NP}$ class is polynomially transformable to our problem. Hence, in order to prove that a decision problem is $\mathcal{NP}$-complete, the following result is commonly used.

Lemma 3.8. (Basic properties of the $\propto$ relation)
(a) *The relation $\propto$ is transitive, i.e., if $P_1 \propto P_2$ and $P_2 \propto P_3$, then $P_1 \propto P_3$.*
(b) *If $P_1 \in \mathcal{P}$ and $P_1 \propto P_2$, then $P_2 \in \mathcal{P}$.*
(c) *If P_1 and P_2 belong to $\mathcal{NP}$, P_1 is $\mathcal{NP}$-complete and $P_1 \propto P_2$, then P_2 is $\mathcal{NP}$-complete.*

Proof. (a) See Garey and Johnson [85, Chap. 2, Lemma 2.2].
(b) See Garey and Johnson [85, Chap. 2, Lemma 2.1].
(c) See Garey and Johnson [85, Chap. 2, Lemma 2.3].   ◊

Now, by Lemma 3.8 (c), in order to prove that a decision problem P is
$\mathcal{NP}$-complete, it is sufficient to show that $P \in \mathcal{NP}$ and that there exists an
$\mathcal{NP}$-complete problem P' that is transformable to P.

In Sect. 2.2, the so-called pseudopolynomial algorithms were defined.
Pseudopolynomial algorithms exist only for some $\mathcal{NP}$-complete problems.
Hence, there is a need to characterize in more detail those problems from
the $\mathcal{NPC}$ class for which such algorithms exist.

Given a decision problem P, let $Length : D_P \rightarrow \mathbb{Z}_+$ denote the function
returning the number of symbols used to describe any instance $I \in D_P$ and let
$Max : D_P \rightarrow \mathbb{Z}_+$ denote the function returning the magnitude of the largest
number in any instance $I \in D_P$.

For any problem P and any polynomial q over $\mathbb{Z}$, let P_q denote the subproblem of P obtained by restricting P to instances I satisfying the inequality
$Max(I) \leq q(Length(I))$.

Definition 3.9. (A strongly $\mathcal{NP}$-complete problem)
*A decision problem P is said to be strongly $\mathcal{NP}$-complete ($\mathcal{NP}$-complete in
the strong sense) if $P \in \mathcal{NP}$ and if there exists a polynomial q over $\mathbb{Z}$ for
which the problem P_q is $\mathcal{NP}$-complete.*

In other words, a decision problem is $\mathcal{NP}$-complete in the strong sense
if the problem is $\mathcal{NP}$-complete in the ordinary sense even if we restrict it to
these instances in which the maximum value in input is polynomially bounded
with respect to the length of the input.

Since the notion of an $\mathcal{NP}$-complete problem in the strong sense is
as important as the notion of an $\mathcal{NP}$-complete problem, a few remarks
are necessary.

Remark 3.10. If for a problem P there exists a polynomial q over $\mathbb{Z}$ for which
the problem P_q is $\mathcal{NP}$-hard, the problem P is said to be *strongly $\mathcal{NP}$-hard
($\mathcal{NP}$-hard in the strong sense).*

Remark 3.11. Problems which are $\mathcal{NP}$-complete ($\mathcal{NP}$-hard) in the strong
sense with respect to the binary encoding scheme, will remain $\mathcal{NP}$-complete
($\mathcal{NP}$-hard) also with respect to the unary encoding scheme. Hence, the problems are sometimes called *unary $\mathcal{NP}$-complete ($\mathcal{NP}$-hard).*

Remark 3.12. The notion of strong $\mathcal{NP}$-completeness allows to divide all problems from the $\mathcal{NPC}$ class into the problems that can be solved by a pseudopolynomial algorithm and the problems that cannot be solved by such an
algorithm.

Remark 3.13. A decision problem which is $\mathcal{NP}$-complete in the strong sense cannot be solved by a pseudopolynomial-time algorithm unless $\mathcal{P} = \mathcal{NP}$ and unless this problem is a *number problem*. (A problem P is called a *number problem*, if there exists no polynomial p such that $Max(I) \leq p(Length(I))$ for any instance $I \in D_P$. The $\mathcal{NP}$-complete problems given in Sect. 3.2 are examples of number problems.)

Remark 3.14. If a problem is $\mathcal{NP}$-complete in the ordinary sense, then for the problem there is no difference between a pseudopolynomial and a polynomial algorithm. Such a distinction, however, exists for a problem that is $\mathcal{NP}$-complete in the strong sense.

The class of all strongly $\mathcal{NP}$-complete problems will be denoted by $\mathcal{SNPC}$.

Since in order to prove that an optimization problem is ordinary (strongly) $\mathcal{NP}$-hard it is sufficient to show that its decision counterpart is ordinary (strongly) $\mathcal{NP}$-complete, from now on we will mainly consider the problems which are ordinary (strongly) $\mathcal{NP}$-complete.

As in the case of $\mathcal{NP}$-complete problems, it is not easy to prove that a decision problem is $\mathcal{NP}$-complete in the strong sense, using Definition 3.9. The notion of a pseudopolynomial transformation is commonly used instead.

Definition 3.15. (A pseudopolynomial transformation)
A pseudopolynomial transformation from a decision problem P to a decision problem P' is a function $f : D_P \to D_{P'}$ such that
(a) for all instances $I \in D_P$ there exists a solution to I if and only if there exists a solution to $f(I) \in D_{P'}$;
(b) f can be computed in time which is polynomial with respect to $Max(I)$ and $Length(I)$;
(c) there exists a polynomial q_1 such that for all instances $I \in D_P$, there holds the inequality $q_1(Length'(f(I))) \geq Length(I)$;
(d) there exists a two-variable polynomial q_2 such that for all instances $I \in D_P$, there holds the inequality $Max'(f(I)) \leq q_2(Max(I), Length(I))$, where functions $Length()$ and $Max()$ correspond to the problem P and functions $Length()'$ and $Max()'$ correspond to the problem P'.

The application of the pseudopolynomial transformation simplifies proofs of the strong $\mathcal{NP}$-completeness, since there holds the following result.

Lemma 3.16. (Basic properties of the pseudopolynomial transformation)
(a) If $P' \in \mathcal{NP}$, P is $\mathcal{NP}$-complete in the strong sense and if there exists a pseudopolynomial transformation from P to P', then P' is $\mathcal{NP}$-complete in the strong sense.
(b) If $P' \in \mathcal{NP}$, P is $\mathcal{NP}$-complete in the strong sense and if there exists a polynomial transformation from P to P', then P' is $\mathcal{NP}$-complete in the strong sense.

Proof. (a) See Garey and Johnson [85, Chap. 4, Lemma 4.1]. $\diamond$

(b) Note that any polynomial transformation is a pseudopolynomial transformation as well. Applying Lemma 3.16 (a), we obtain the result. $\square$

We complete the section with results concerning the application of approximation algorithms and approximation schemata for $\mathcal{NP}$-complete problems.

Theorem 3.17. *If there exists a polynomial q of two variables such that for any instance $I \in D_P$, there holds the inequality*

$$OPT(I) < q(Length(I), Max(I)),$$

then from the existence of an FPTAS for a problem P there follows the existence of a pseudopolynomial approximation algorithm for this problem.

Proof. See Garey and Johnson [85, Chap. 6].   ◇

Lemma 3.18. *Let P be an optimization problem with integer solutions and let the assumptions of Theorem 3.17 be satisfied. If P is $\mathcal{NP}$-hard in the strong sense, then it cannot be solved by an FPTAS unless $\mathcal{P} = \mathcal{NP}$.*

Proof. The result is a corollary from Theorem 3.17.   □

3.2 Examples of $\mathcal{NP}$-complete problems

The following $\mathcal{NP}$-complete problems will be used in $\mathcal{NP}$-completeness proofs presented in this book.

PARTITION PROBLEM (PP): given $A \in \mathbb{Z}_+$ and a set $X = \{x_1, x_2, \ldots, x_k\}$ of positive integers, $\sum_{i=1}^{k} x_i = 2A$, does there exist a subset $X' \subset X$ such that $\sum_{x_i \in X'} x_i = \sum_{x_i \in X \setminus X'} x_i = A$?

The PP problem is $\mathcal{NP}$-complete in the ordinary sense (see Garey and Johnson [85, Chap. 3, Theorem 3.5]).

SUBSET SUM (SS): given $C \in \mathbb{Z}_+$, a set $R = \{1, 2, \ldots, r\}$ and a value $u_i \in \mathbb{Z}_+$ for each $i \in R$, does there exist a subset $R' \subseteq R$ such that $\sum_{i \in R'} u_i = C$?

The SS problem is $\mathcal{NP}$-complete in the ordinary sense (see Karp [162]).

SUBSET PRODUCT (SP): given $B \in \mathbb{Z}_+$, a set $P = \{1, 2, \ldots, p\}$ and a value $y_i \in \mathbb{Z}_+$ for each $i \in P$, does there exist a subset $P' \subseteq P$ such that $\prod_{i \in P'} y_i = B$?

The SP problem is $\mathcal{NP}$-complete in the ordinary sense (see Johnson [158]).

EQUAL PRODUCTS PROBLEM (EPP): given a set $Q = \{1, 2, \ldots, q\}$ and a value $z_i \in \mathbb{Z}_+$ for each $i \in Q$ such that $\prod_{i \in Q} z_i = E^2$ for a certain $E \in \mathbb{Z}_+$, does there exist a subset $Q' \subset Q$ such that $\prod_{i \in Q'} z_i = \prod_{i \in Q \setminus Q'} z_i = E$?

In order to illustrate the main steps of typical $\mathcal{NP}$-completeness proof, we will show that the EPP problem is computationally intractable.

Lemma 3.19. *The EPP problem is $\mathcal{NP}$-complete in the ordinary sense.*

Proof. We will show that the SP problem is polynomially transformable to the EPP problem and therefore, by Lemma 3.8 (c), the latter problem is $\mathcal{NP}$-complete in the ordinary sense.

Consider the following transformation from the SP problem: $q = p + 2$, $z_i = y_i$ for $1 \leq i \leq p$, $z_{p+1} = 2\frac{Y}{B}$, $z_{p+2} = 2B$, threshold $G = 2Y$, where $Y = \prod_{i \in P} y_i$.

First, note that the above transformation is polynomial. Second, since for a given $Q' \subset Q$ we can check in polynomial time whether $\prod_{i \in Q'} z_i = E$, we have $EPP \in \mathcal{NP}$.

Hence, to end the proof it is sufficient to show that the SP problem has a solution if and only if the EPP problem has a solution.

If the SP problem has a solution, define $Q' := P' \cup \{z_{p+1}\}$. Then it is easy to check that $\prod_{i \in Q'} z_i = 2Y = G$. Hence the EPP problem has a solution.

If the EPP problem has a solution, then there exists a set $Q' \subset Q$ such that $\prod_{i \in Q'} z_i \leq G = 2Y$. Since $\prod_{i \in Q} z_i = 4Y^2$ and $Q' \cap (Q \setminus Q') = \emptyset$, the inequality $\prod_{i \in Q'} z_i < 2Y$ does not hold. Hence it must be $\prod_{i \in Q'} z_i = 2Y$.

Since $z_{p+1} \times z_{p+2} = 4Y$, the elements z_{p+1} and z_{p+2} cannot both belong to set Q'. Assume first that $z_{p+1} \in Q'$. Then $\prod_{i \in Q' \cup \{p+1\}} z_i = \prod_{i \in Q'} z_i \times z_{p+1} = \prod_{i \in Q'} \frac{2Y}{B} \times z_{p+1} = 2Y$. Hence $\prod_{i \in Q'} z_i = \prod_{i \in Q'} y_i = B$ and the SP problem has a solution. If $z_{m+1} \in Q \setminus Q'$, then by similar reasoning as above we have $\prod_{i \in Q \setminus Q' \setminus \{p+1\}} y_i = B$ and the SP problem has a solution as well.   ■

3-PARTITION (3-P): given $K \in \mathbb{Z}_+$ and a set $C = \{c_1, c_2, \ldots, c_{3h}\}$ of $3h$ integers such that $\frac{K}{4} < c_i < \frac{K}{2}$ for $1 \leq i \leq 3h$ and $\sum_{i=1}^{3h} c_i = hK$, can C be partitioned into disjoint sets $C_1, C_2, \ldots, C_h$ such that $\sum_{c_i \in C_j} c_i = K$ for each $1 \leq j \leq h$?

The 3-P problem is $\mathcal{NP}$-complete in the strong sense (see Garey and Johnson [85, Sect. 4.2, Theorem 4.4]).

NON-NUMBER 3-PARTITION (N3P): given $Z \in \mathbb{Z}_+$ and $3w$ positive integers $z_1, z_2, \ldots, z_{3w}$ such that $\sum_{i=1}^{3w} h_i = wZ$, where z_i is bounded by a polynomial of w and $\frac{Z}{4} < z_i < \frac{Z}{2}$ for $1 \leq i \leq 3w$, does there exist a partition of set $\{1, 2, \ldots, 3w\}$ into w disjoint subsets $Z_1, Z_2, \ldots, Z_w$ such that $\sum_{i \in Z_j} z_i = Z$ for $1 \leq j \leq w$?

The N3P problem is $\mathcal{NP}$-complete in the ordinary sense (see Bachman et al. [13]).

4-PRODUCT (4-P): given $D \in \mathbb{Q}_+$, a set $N = \{1, 2, \ldots, 4p\}$ and a value $D^{\frac{1}{5}} < u_i < D^{\frac{1}{3}} \in \mathbb{Q}_+$ for each $i \in N$, $\prod_{i \in N} u_i = D^p$, do there exist disjoint subsets $N_1, N_2, \ldots, N_p$ such that $\bigcup_{i=1}^{p} N_i = N$ and $\prod_{i \in N_j} u_i = D$ for $1 \leq j \leq p$?

The 4-P problem, which is a multiplicative version of the 4-PARTITION problem (see Garey and Johnson [85, Sect. 4.2, Theorem 4.3]), is $\mathcal{NP}$-complete in the strong sense (see Kononov [169]).

KNAPSACK (KP): given $U, W \in \mathbb{Z}_+$, a set $K = \{1, 2, \ldots, k\}$ and values $u_i \in \mathbb{Z}_+$ and $w_i \in \mathbb{Z}_+$ for each $k \in K$, does there exist $K' \subseteq K$ such that $\sum_{k \in K'} u_k \leq U$ and $\sum_{k \in K'} w_k \leq W$?

The KP problem is $\mathcal{NP}$-complete in the ordinary sense (see Karp [162]).

BIN PACKING (BP): given $T, V \in \mathbb{Z}_+$, a set $L = \{1, 2, \ldots, l\}$ and values $u_i \in \mathbb{Z}_+$ for each $l \in L$, does there exist a partition of L into disjoint sets $L_1, L_2, \ldots, L_V$ such that $\sum_{l \in L_k} u_l \leq T$ for $1 \leq k \leq V$?

The BP problem is $\mathcal{NP}$-complete in the strong sense (see Garey and Johnson [85, p. 226]).

With this remark, we end the presentation of basic definitions and results concerning $\mathcal{NP}$-complete problems. In Chap. 4 and Chap. 5, we will introduce the basics of the scheduling theory and time-dependent scheduling, respectively.

3.3 Bibliographic notes

The definitions and notions of the theory of $\mathcal{NP}$-completeness presented in this chapter are expressed in terms of decision problems and transformations between these problems. Alternatively, these definitions and notions can be expressed in terms of *languages* and *reductions* between languages.

Ausiello et al. [9], Bovet and Crescenzi [31], Garey and Johnson [85], Hopcroft and Ullman [139], Papadimitriou [233], Savage [250], Sipser [258], Wagner and Wechsung [276] present the theory of $\mathcal{NP}$-completeness from different perspectives.

From Definition 3.2, it does not follow that $\mathcal{NP}$-complete problems exist at all. The fact that $\mathcal{NPC} \neq \emptyset$ was proved indepedently by Cook [63] in 1971 and Levin [200] in 1973.

The classes $\mathcal{P}$, $\mathcal{NP}$ and $\mathcal{NPC}$ have a fundamental meaning in the complexity theory. Johnson [157] presents a detailed review of other complexity classes.

The list of $\mathcal{NP}$-complete problems, initiated by Cook [63], Karp [162] and Levin [200], contains a great number of problems and is still growing. Ausiello et al. [9], Brucker and Knust [35], Garey and Johnson [85] and Johnson [158] present extensive excerpts from this list, including problems from different areas of computer science and discrete mathematics.

The functions $Length()$ and $Max()$ are discussed in detail by Garey and Johnson [85, Sect. 4.2].

4

Basics of the scheduling theory

Time-dependent scheduling is a branch of the scheduling theory. Therefore, before formally introducing the time-dependent scheduling, we need a precise formulation of fundamentals of the scheduling theory. In this chapter, we recall the basic facts concerning the scheduling theory.

Chapter 4 is composed of five sections. In Sect. 4.1, we define the parameters of problems considered in the framework of the scheduling theory. In Sect. 4.2, we introduce the notion of schedule. In Sect. 4.3, we define different criteria of optimality of schedule. In Sect. 4.4, we introduce the notation $\alpha|\beta|\gamma$, which is used in the book for symbolic description of scheduling problems. The chapter is completed with bibliographic notes in Sect. 4.5.

4.1 Parameters of the scheduling problem

Regardless of its nature, every scheduling problem $\mathcal{S}$ can be formulated as a quadruple, $\mathcal{S} = (\mathcal{J}, \mathcal{M}, \mathcal{R}, \varphi)$, where $\mathcal{J}$ is a set of pieces of work to be executed, $\mathcal{M}$ is a set of entities that will perform the pieces of work, $\mathcal{R}$ is a set of additional entities needed for performing these pieces of work and φ is a function that is used as a measure of quality of solutions to the problem under consideration. We start this section with a brief description of the parameters of the quadruple.

4.1.1 Parameters of the set of jobs

The elements of set $\mathcal{J}$ are called *jobs*. Unless otherwise specified, we will assume that $|\mathcal{J}| = n$, i.e., there are n jobs. We will denote jobs by $J_1, J_2, \ldots, J_n$. The set of indices of jobs from the set $\mathcal{J}$ will be denoted by $N_{\mathcal{J}}$.

Job $J_j, 1 \le j \le n$, consists of n_j *operations*, $O_{1,j}, O_{2,j}, \ldots, O_{n_j,j}$. For each operation, we define the *processing time* of the operation, i.e., the time needed for processing this operation. (If a job consists of one operation only, we will identify the job with the operation.) The processing time of job J_j (operation $O_{i,j}$) will be denoted by p_j ($p_{i,j}$), where $1 \le j \le n$ ($1 \le i \le n_j$ and $1 \le j \le n$).

For job J_j, $1 \le j \le n$, there may be defined a *ready time*, r_j, a *deadline*, d_j, and a *weight*, w_j. The first operation of job J_j cannot be started before the ready time r_j and the last operation of the job cannot be completed after the deadline d_j. We will say that there are no ready times (deadlines) if $r_j = 0$ ($d_j = \infty$) for all j. The weight w_j indicates the importance of job J_j compared to other jobs. We will say that job weights are equal, if $w_j = 1$ for all j.

Throughout the book, unless otherwise stated, we will assume that job (operation) parameters are positive integer numbers, i.e., p_j ($p_{i,j}$), r_j, d_j, $w_j \in \mathbb{Z}^+$ for $1 \le j \le n$ ($1 \le i \le n_j$ and $1 \le j \le n$).

Example 4.1. Let the set $\mathcal{J}$ be composed of 4 jobs, $\mathcal{J} = \{J_1, J_2, J_3, J_4\}$, such that $p_1 = 1$, $p_2 = 2$, $p_3 = 3$ and $p_4 = 4$, with no ready times and deadlines, and with unit job weights. Then $r_j = 0$, $d_j = +\infty$ and $w_j = 1$ for $1 \le j \le 4$.

There may be also defined *precedence constraints* among jobs, which reflect the fact that some jobs have to be executed before others. The precedence constraints correspond to a partial order relation $\prec \subseteq \mathcal{J} \times \mathcal{J}$. We will assume that precedence constraints between jobs can be given in the form of a set of chains, a tree, a forest, a series-parallel or an arbitrary acyclic digraph.

If precedence constraints are defined on the set of jobs, $\prec \ne \emptyset$, we will call the jobs *dependent*; otherwise we will say that the jobs are *independent*.

Example 4.2. The jobs from Example 4.1 are independent, $\prec = \emptyset$. If we assume that job precedence constraints in the set are as in Fig. 1.1b or Fig. 1.2d, then the jobs are dependent.

Jobs can be *preemptable* or *non-preemptable*. If a job is preemptable, then the execution of this job can be interrupted at any time without any cost, and resumed at a later time on the machine on which it was executed before the preemption, or on another one. Otherwise, the job is non-preemptable.

In this book, we will mainly consider scheduling problems with non-preemptable and independent jobs. We will also assume that there are no ready times, no deadlines and all job weights are equal, unless otherwise specified.

4.1.2 Parameters of the set of machines

The jobs are performed by the elements of set $\mathcal{M}$, called *machines*. (Sometimes these elements have other names than 'machines'. For example, in scheduling problems that arise in computer systems, the elements of set $\mathcal{M}$ are called *processors*. Throughout the book, we will use the term 'machine' to denote a single element of set $\mathcal{M}$.) We will asume that $|\mathcal{M}| = m$, i.e., we are given m machines. The machines will be denoted by $M_1, M_2, \ldots, M_m$.

In the simplest case, when $m = 1$, we deal with a *single machine*. Despite its simplicity, this case is worth considering, since it appears in more complex machine environments described below.

If all $m \geq 2$ machines are of the same kind and have the same processing speed, we deal with *parallel identical machines*. In this case, job J_j with the processing time p_j, $1 \leq j \leq n$, can be performed by any machine and its execution on the machine will take p_j units of time.

If among the available machines, there is a slowest machine, M_1, and any other machine, $M_k \neq M_1$, has a speed s_k that is a multiple of the speed s_1 of machine M_1, we deal with *parallel uniform machines*. In this case, the execution of job J_j on machine M_k will take $\frac{p_j}{s_k}$ units of time.

Finally, if the machines differ in speed but the speeds depend on the performed job, we deal with *parallel unrelated machines*. In this case, the symbol $p_{i,j}$ is used for denoting the processing time of job J_j on machine M_i, where $1 \leq i \leq m$ and $1 \leq j \leq n$.

So far, we assumed that all machines in a machine environment perform the same functions, i.e., any job that consists of only one operation can be performed on any machine. If the available machines have different functions, i.e., some of them cannot perform some jobs, we deal with *dedicated machines*. In this case, any job consists of a number of different operations, which are performed by different machines. We will consider three main types of such machine environment: flow shop, open shop and job shop.

A *flow shop* consists of $m \geq 2$ machines, $M_1, M_2, \ldots, M_m$. Each job consists of m operations, $n_j = m$ for $1 \leq j \leq n$. The i-th operation of any job has to be executed by machine M_i, $1 \leq i \leq m$. Moreover, this operation can start only if the previous operation of this particular job has been completed. All jobs follow the same route from the first machine to the last one. (In other words, precedence constraints between operations of any job in a flow shop are in the form of a chain whose length is equal to the number of machines in the flow shop.)

An *open shop* consists of $m \geq 2$ machines. Each job consists of m operations, $n_j = m$ for $1 \leq j \leq n$, but the order of processing of operations can be different for different jobs. This means that each job has to go through all machines but the route can be arbitrary. (In other words, the operations of any job in an open shop are independent and the number of the operations is equal to the number of machines in the open shop.)

A *job shop* consists of $m \geq 2$ machines. Each job can consist of n_j operations, where not necessarily $n_j = m$ for $1 \leq j \leq n$. Moreover, each job has its own route of performing its operations, and it can visit a certain machine more than once or may not visit some machines at all. (In other words, precedence constraints between operations of any job in a job shop are in the form of a chain and the number of operations may be arbitrary.)

Remark 4.3. In some dedicated-machine environments, additional constraints, which restrict the job flow through the machines, may be imposed on available machines. For example, a flow shop may be of the 'no-wait' type. The *no-wait* constraint means that buffers between machines are of zero capacity and a job after completion of its processing on one machine must immediately start on the next (consecutive) machine.

In all the above cases of dedicated machine environments, the symbol $p_{i,j}$ will be used for denoting the processing time of operation $O_{i,j}$ of job J_j, where $1 \leq i \leq n_j = m$ and $1 \leq j \leq n$ for flow shop and open shop problems, and $1 \leq i \leq n_j$ and $1 \leq j \leq n$ for job shop problem.

Remark 4.4. Throughout the book, unless otherwise stated, we will assume that jobs are processed on machines that are *continuously available*. In some applications, however, it is required to consider *machine non-availability periods* in which the machines are not available for processing due to maintenance operations, rest periods or machine breakdowns. We will come back to scheduling problems with machine non-availability periods in Sect. 6.1.

We will not define other types of machine environment, since, in this book, we will consider scheduling problems on a single machine, on parallel machines and on dedicated machines only.

4.1.3 Parameters of the set of resources

In some problems, the execution of jobs requires additional entities other than machines. The entities, elements of set $\mathcal{R}$, are called *resources*. The resources may be continuous or discrete. A resource is *continuous* if it can be allocated to a job in an arbitrary amount. A resource is *discrete* if it can be allocated to a job only in a non-negative integer number of units.

Example 4.5. Energy, gas and power are continuous resources. Tools, robots and automated guided vehicles are discrete resources.   ◆

In real-world applications, the available resources are usually constrained. A resource is *constrained* if it can be allocated only in an amount which is between the minimum and the maximum number of units of the resource; otherwise, it is *unconstrained*. Example 4.5 concerns constrained resources. There also exist applications in which constrained resources are available in a huge number of units. These resources, as a rule, can be considered as unconstrained resources.

Example 4.6. Virtual memory in computer systems, manpower in problems of scheduling very-large-scale projects and money in some finance management problems are unconstrained resources.   ◆

Since, in this book, we consider the scheduling problems in which jobs do not need additional resources for execution, $\mathcal{R} = \emptyset$, we omit a more detailed description of the parameters of set $\mathcal{R}$.

4.2 The notion of schedule

The value of a criterion function may be calculated once a solution to the instance of a particular scheduling problem is known. Before we define possible forms of the criterion function, we describe the solution.

Given a description of sets $\mathcal{J}$ and $\mathcal{M}$ for a scheduling problem, we can start looking for a solution to the problem. Roughly speaking, a solution to a scheduling problem is an assignment of machines to jobs in time that satisfies some (defined below) requirements. The solution will be called a *schedule*. For the purpose of this book, we assume the following definition of the notion.

Definition 4.7. (A schedule)
A schedule is an assignment of machines (and possibly resources) to jobs in time such that the following conditions are satisfied:
(a) at every moment of time, each machine is assigned to at most one job and each job is processed by at most one machine;
(b) job J_j, $1 \leq j \leq n$, is processed in time interval $\langle r_j, +\infty \rangle$;
(c) all jobs are completed;
(d) if there exist precedence constraints for some jobs, then the jobs are executed in the order consistent with these constraints;
(e) if there exist resource contraints, then they are satisfied;
(f) if jobs are non-preemptable, then no job is preempted; otherwise the number of preemptions of each job is finite.

Since the notion of schedule plays a fundamental role in the scheduling theory, we will now add a few remarks to Definition 4.7.

Remark 4.8. An arbitrary schedule specifies two sets of time intervals. The first set consists of the time intervals in which available machines perform some jobs. In every interval of this kind, a job is executed by a machine. If no job was preempted, then only one time interval corresponds to each job; otherwise, a number of intervals correspond to each job. The first set is always non-empty and the intervals are not necessarily disjoint. The second set, which may be empty, consists of the time intervals in which the available machines do not work. These time intervals will be called *idle times* of the machines.

Remark 4.9. In some dedicated machine environments, the available machines may have some limitations which concern idle times. For example, a flow shop may be of the 'no-idle' type. The *no-idle* constraint means that each machine, once it commences its work, must process all operations assigned to it without idle times. (Another constraint concerning the flow shop environment, 'no-wait', is described in Remark 4.3.)

Remark 4.10. An arbitrary schedule is composed of a number of partial schedules that correspond to particular machines. The partial schedules will be called *subschedules*. The number of subschedules of a schedule is equal to the number of machines in the schedule. Note that a schedule for a single machine is identical with its subschedule.

Remark 4.11. In some cases (e.g., no preemptions, no idle times, ready times of all jobs are equal) a schedule may be fully described by the permutations of indices of jobs that are assigned to particular machines in that schedule. The permutation corresponding to such a schedule (subschedule) will be called a *job sequence (subsequence).*

4.2.1 The presentation of schedules

Schedules are usually presented by Gantt charts. A *Gantt chart* is a two-dimensional diagram composed of a number of labelled rectangles and a number of horizontal axes. When the rectangles represent jobs (operations) and the axes correspond to machines, we say that the Gantt chart is *machine-oriented*. When the rectangles represent machines and the axes correspond to jobs (operations), the Gantt chart is *job-oriented*.

Throughout this book, we will use machine-oriented Gantt charts.

Example 4.12. Consider the set of jobs $\mathcal{J}$ defined as in Example 4.1. Since the jobs are independent and there are no ready times and deadlines, any sequence of the jobs corresponds to a schedule. An example schedule for this set of jobs, corresponding to sequence (J_4, J_3, J_1, J_2), is depicted in Fig. 4.1.

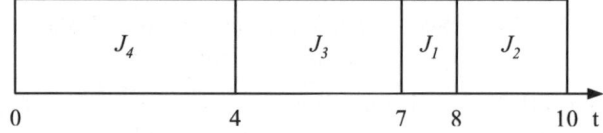

Fig. 4.1: A schedule for Example 4.1

4.2.2 Parameters characterizing a job in schedule

If we know a schedule σ for an instance of a scheduling problem, then for any job J_j, $1 \le j \le n$, we may calculate the values of parameters characterizing this job in schedule σ. Examples of such parameters are *the starting time* $S_j(\sigma)$, *the completion time* $C_j(\sigma) = S_j(\sigma) + p_j$, *the waiting time* $W_j(\sigma) = C_j(\sigma) - r_j - p_j$, *the lateness* $L_j(\sigma) = C_j(\sigma) - d_j$ and *the tardiness* $T_j(\sigma) = \max\{0, L_j(\sigma)\}$ of the job J_j in schedule σ. (If it is clear which schedule we will consider, we will omit the symbol σ and write S_j, C_j, W_j, L_j and T_j, respectively.)

Example 4.13. Consider the Gantt chart given in Fig. 4.1. From the chart we can read, e.g., that $S_4 = 0$ and $C_4 = 4$, while $S_1 = 7$ and $C_1 = 8$.

4.2.3 Types of schedules

As a rule, there can be found different schedules for a given scheduling problem $\mathcal{S}$. The set of all schedules for a given $\mathcal{S}$ will be denoted by $\mathcal{Z}(\mathcal{S})$. The schedules which compose the set $\mathcal{Z}(\mathcal{S})$ can be of different types. Throughout this book, we will distinguish the following types of schedules.

Definition 4.14. (A feasible schedule)
(a) *A schedule is said to be* feasible *if it satisfies all conditions of Definition 4.7, and if other conditions specific for a given problem are satisfied.*
(b) *A feasible schedule is said to be* non-preemptive *if no job has been preempted; otherwise, it is* preemptive.

Example 4.15. The schedule depicted in Fig. 4.1 is a feasible schedule for the set of jobs from Example 4.1. Moreover, any other schedule obtained from the schedule by a rearrangement of jobs is also a feasible schedule.

The set of all feasible schedules for a given scheduling problem $\mathcal{S}$ will be denoted by $\mathcal{Z}_{feas}(\mathcal{S})$.

Definition 4.16. (A semi-active schedule)
A schedule is said to be semi-active *if it is obtained from any feasible schedule by shifting all jobs (operations) to start as early as possible but without changing any job sequence.*

In other words, a schedule is semi-active if jobs (operations) in the schedule cannot be shifted to start earlier without changing the job sequence, violating precedence constraints or ready times.

Example 4.17. The schedule depicted in Fig. 4.1 is a semi-active schedule for the set of jobs from Example 4.1, since no job can be shifted to start earlier without changing the job sequence.

Example 4.18. The schedule depicted in Fig. 4.2 is another feasible schedule for the set of jobs from Example 4.1. The schedule, however, is not a semi-active schedule, since we can shift job J_2 one unit of time to the left.

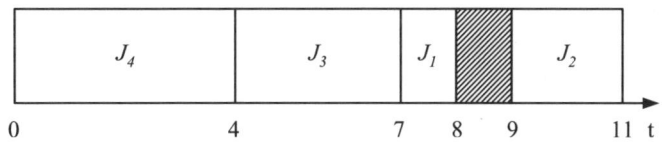

Fig. 4.2: A feasible schedule for Example 4.6

Remark 4.19. Since each job (operation) starts its execution as early as possible in any semi-active schedule, each semi-active schedule is completely characterized by the job sequence (subsequences) corresponding to the schedule (subschedules).

Remark 4.20. A schedule that is completely specified by the job sequence(s) is called a *permutation schedule.*

Remark 4.21. For any job sequence, there exists only one semi-active schedule.

The set of all semi-active schedules for a given scheduling problem S will be denoted by $Z_{s-act}(S)$.

Definition 4.22. (An active schedule)
A schedule is said to be active *if it is obtained from any semi-active schedule by shifting all jobs (operations) to start as early as possible even if the shifting causes a change in some job sequence(s).*

In other words, a schedule is active if jobs (operations) in the schedule cannot be shifted to start earlier without violating precedence constraints or ready times.

Example 4.23. The semi-active schedule from Fig. 4.1 is also an active schedule for the set of jobs from Example 4.1.

The set of all active schedules for a given scheduling problem S will be denoted by $Z_{act}(S)$.

Definition 4.24. (An optimal schedule)
A schedule is said to be optimal *if the value of optimality criterion for the schedule is optimal.*

The set of all optimal schedules for a given scheduling problem S will be denoted by $Z_{opt}(S)$.

Remark 4.25. The criteria of optimality of a schedule will be defined in Definition 4.29.

Remark 4.26. The optimal schedule need not be unique, i.e., $|Z_{opt}(S)| \geq 1$.

Definition 4.27. (A dominant set)
A set of schedules is said to be dominant *if it contains at least one optimal schedule.*

Remark 4.28. Notice that there hold the inclusions $Z_{act}(S) \subseteq Z_{s-act}(S) \subset Z_{feas}(S) \subset Z(S)$ and $Z_{opt}(S) \subset Z_{s-act}(S)$.

In this book, we will consider mainly non-preemptive semi-active schedules.

4.3 The criteria of schedule optimality

In general, the criterion of optimality in a scheduling problem can be an arbitrary function, which has real values and has been defined on the set of all feasible schedules for the problem. In this book, we will consider mainly the following optimality criteria.

Definition 4.29. (Criteria of optimality of a schedule)
Let $C_1, C_2, \ldots, C_n$ be the completion times of jobs in a schedule. The criteria of the maximum completion time ($C_{\max}$), the maximum lateness ($L_{\max}$), the maximum tardiness ($T_{\max}$), the maximum cost ($f_{\max}$), the total completion time ($\sum C_j$), the total weighted completion time ($\sum w_j C_j$), the total machine load ($\sum C_{\max}^{(k)}$), the number of tardy jobs ($\sum U_j$) and the total cost ($\sum f_j$) are defined as follows:

(a) $C_{\max} := \max\limits_{1 \le j \le n} \{C_j\}$;

(b) $L_{\max} := \max\limits_{1 \le j \le n} \{L_j\} := \max\limits_{1 \le j \le n} \{C_j - d_j\}$;

(c) $T_{\max} := \max\limits_{1 \le j \le n} \{T_j\} := \max\limits_{1 \le j \le n} \{\max\{0, L_j\}\}$;

(d) $f_{\max} := \max\limits_{1 \le j \le n} \{f_j(C_j)\}$, *where $f_1, f_2, \ldots, f_n$ are given cost functions;*

(e) $\sum C_j := \sum_{j=1}^{n} C_j$;

(f) $\sum w_j C_j := \sum_{j=1}^{n} w_j C_j$;

(g) $\sum C_{\max}^{(k)} := \sum_{j=1}^{m} C_{\max}^{(k)}$, *where $C_{\max}^{(k)}$ denotes the maximum completion time over all jobs assigned to machine M_k, $1 \le k \le m$;*

(h) $\sum U_j := \sum_{j=1}^{n} U_j$, *where $U_j := 0$ if $L_j \le 0$ and $U_j := 1$ if $L_j > 0$;*

(i) $\sum f_j := \sum_{j=1}^{n} f_j(C_j)$, *where $f_1, f_2, \ldots, f_n$ are given cost functions.*

Example 4.30. The schedule depicted in Fig. 4.2 is an optimal schedule for the set of jobs from Example 4.1 with respect to the $C_{\max}$ criterion. This schedule, however, is not optimal with respect to the $\sum C_j$ citerion.

Since the criteria of optimality have a fundamental meaning to the scheduling theory, we shall make a few remarks concerning the above definition.

Remark 4.31. Some of the above criteria are special cases of the l_p norm (cf. Definition 1.18). For example, if $C = [C_1, C_2, \ldots, C_n]$ is the vector of job completion times in a schedule, then $\|C\|_1 \equiv \sum_{j=1}^{n} C_j$ and $\|C\|_\infty \equiv C_{\max}$. We will come back to this topic in Chap. 12.

Remark 4.32. Some criteria may also be defined in terms of other criteria. For example, $C_{\max} := \max\limits_{1 \le j \le n} \{C_j\} \equiv \max\limits_{1 \le k \le m} \{C_{\max}^{(k)}\}$.

Remark 4.33. The completion time of a job is the basic parameter characterizing the job in a schedule, since the optimality criteria from Definition 4.29 are functions of the job completion times. Given any feasible schedule, the starting time of a job in the schedule is the startpoint of the first time interval corresponding to the job. Similarly, the completion time of a job in the schedule is the endpoint of the last time interval corresponding to the job.

Remark 4.34. The criteria of optimality of a schedule given in Definition 4.29 are single-valued functions. In Chap. 14, we will consider *bicriterion* scheduling problems. In the chapter, we will extend the definition to include schedule optimality criteria composed of two single-valued functions.

As a rule, the applied criterion of optimality of a schedule is minimized. Therefore, the criteria that have properties that allow to find the optimum are of practical interest. Examples of such criteria are *regular criteria*, introduced by Conway et al. [62].

Definition 4.35. (A regular criterion)
Let $C = [C_1, C_2, \ldots, C_n]$ be the vector of job completion times and let $\varphi : C \to \mathbb{R}$ be a criterion function. Criterion φ is said to be a regular criterion if for any other vector of job completion times, $C' = [C'_1, C'_2, \ldots, C'_n]$, the inequality $\varphi(C') \geq \varphi(C)$ holds if and only if there exists an index k, $1 \leq k \leq n$, such that $C'_k \geq C_k$.

In other words, φ is a regular criterion only if it is a non-decreasing function with respect to job completion times.

Example 4.36. The $C_{\max}$ criterion (also called the *schedule length* or *makespan*) and other criteria from Definition 4.29 are regular criteria. ◆

Example 4.37. An example of a *non-regular criterion* is *the total absolute deviation of job completion times*, $\sum |C_j - d_j| := \sum_{j=1}^{n} |C_j - d_j| \equiv \sum_{j=1}^{n} (E_j + T_j)$, where $E_j := \max\{0, d_j - C_j\}$, for $1 \leq j \leq n$, is the earliness of job J_j. ◆

Throughout this book, the value of the criterion function φ for a schedule σ will be denoted by $\varphi(\sigma)$. The optimal schedule and the optimal value of the criterion φ will be denoted by $\sigma^\star$ and $\varphi^\star := \varphi(\sigma^\star)$, respectively.

The following result shows the importance of regular criteria.

Lemma 4.38. (A dominant set for regular criteria)
The set $\mathcal{Z}_{s-act}(\mathcal{S})$ is dominant for regular criteria of optimality of schedule.

Proof. See Baker [14, Theorem 2.1], Conway et al. [62, Sect. 6.5]. ◇

From now on, we will assume that a *scheduling problem* is defined if the sets $\mathcal{J}, \mathcal{M}, \mathcal{R}$ have been described using the parameters given in Sect. 4.1 and if the form of the criterion φ is known. If the parameters have been assigned specific values, we deal with an instance of a given scheduling problem.

An algorithm which solves a scheduling problem will be called a *scheduling algorithm*. Scheduling algorithms can be divided into offline and online algorithms (cf. Definition 2.17). In the case of an offline scheduling algorithm, the input data concerning all jobs to be scheduled are given in advance, and the schedule constructed by the algorithm exploits the knowledge about the whole set of jobs. An online scheduling algorithm generates a schedule in a job-by-job manner using only partial data concerning a single job that is currently processed.

In this book, we will consider offline and online scheduling algorithms.

4.4 Notation of scheduling problems

For simplicity of presentation, we will denote the scheduling problems using the $\alpha|\beta|\gamma$ notation, introduced by Graham et al. [125]. Our description of the notation will be restricted only to the symbols used in this book.

The description of any scheduling problem in the $\alpha|\beta|\gamma$ notation is a complex symbol composed of three fields, separated by the character '|'.

The first field, α, refers to the machine environment and is composed of two symbols, $\alpha = \alpha_1\alpha_2$. Symbol α_1 characterizes the type of machine. If α_1 is an empty symbol, we deal with a single machine case. Otherwise, symbols P, Q, R, F, O and J denote parallel identical machine, parallel uniform machine, parallel unrelated machine, flow shop, open shop and job shop environment, respectively. Symbol α_2 denotes the number of machines. If α_1 is not an empty symbol and $\alpha_2 = m$, we deal with $m \geq 2$ machines. If α_1 is an empty symbol and $\alpha_2 = 1$, we deal with a single machine.

Field β describes the parameters of the set of jobs. In this field, we will use the following symbols (in parentheses we give the meaning of a particular symbol): *pmtn* (job preemption is allowed), *chains, tree, ser-par, prec* (precedence constraints among jobs are in the form of a set of chains, a tree, a series-parallel digraph or an arbitrary acyclic digraph), r_j (a ready time is defined for each job), d_j (a deadline is defined for each job).

If no symbol appears in β, *default values* are assumed: no preemption, arbitrary (but fixed) job processing times, no additional resources, no precedence constraints, no ready times and no deadlines.

Field γ contains the form of the criterion function, expressed in terms of the symbols from Definition 4.29. The dash symbol ('−') in this field means that testing for the existence of a feasible schedule is considered.

Example 4.39.

(a) The symbol $1|prec|\sum w_j C_j$ denotes a single machine scheduling problem with arbitrary job processing times, arbitrary precedence constraints, arbitrary job weights, no ready times, no deadlines and the total weighted completion time criterion.

(b) The symbol $Pm|r_j = r, p_j = 1, tree|C_{\max}$ denotes an m-identical-machine scheduling problem with unit processing time jobs, a common ready time for all jobs, no deadlines, precedence constraints among jobs in the form of a tree and the maximum completion time criterion.

(c) The symbol $F2|no\text{-}wait|\sum C_j$ denotes a two-machine 'no-wait' flow shop problem (see Remark 4.3 for the description of the 'no-wait' constraint), with arbitrary job processing times, no precedence constraints, no ready times, arbitrary deadlines and the total completion time criterion.

(d) The symbol $O3||L_{max}$ denotes a three-machine open shop problem with arbitrary job processing times, no precedence constraints, no ready times, arbitrary deadlines and the maximum lateness criterion. (Note that since deadlines d_j are used in definition of the L_{max} criterion, symbol d_j does not appear in the field β.)

(e) The symbol $Jm||C_{max}$ denotes an m-machine job shop problem with arbitrary job processing times, no precedence constraints, no ready times, no deadlines and the maximum completion time criterion.   ◆

Remark 4.40. In Sect. 5.3, we will extend the $\alpha|\beta|\gamma$ notation to include the symbols describing time-dependent job processing times. In Sect. 6.1.1, we will extend the notation further to include the symbols describing time-dependent scheduling problems on machines with non-availability periods and time-dependent batch scheduling problems.

With this remark, we end the presentation of the basics of the scheduling theory. In Chap. 5, we will introduce time-dependent scheduling.

4.5 Bibliographic notes

Błażewicz et al. [27, Chap. 3], Brucker [34, Chap. 1] and Leung [198, Chap. 1] present general description of scheduling problems.

Brucker [34, Chap. 1] presents a detailed description of different machine environments.

Błażewicz et al. [26] give a comprehensive description of the set $\mathcal{R}$ of resources and problems of scheduling under resource constraints.

Definitions of criteria other than these from Definition 4.29 may be found in Błażewicz et al. [27, Chap. 3] and Brucker [34, Chap. 1].

Numerous books have been published on the scheduling theory, including Baker [14], Błażewicz et al. [26, 27], Brucker [34], Chrétienne et al. [60], Coffman [61], Conway et al. [62], Dempster et al. [69], Elmaghraby [78], French [84], Hartmann [133], Leung [198], Morton and Pentico [214], Muth and Thompson [223], Parker [235], Pinedo [237], Rinnooy Kan [245], Słowiński and Hapke [259], Sule [262], Tanaev et al. [264, 266]. These books cover a huge spectrum of different aspects of scheduling with single-valued criteria and may serve as excellent references on the theory.

Problems of scheduling with multiple criteria are discussed by Dileepan and Sen [71], Hoogeveen [137, 138], Lee and Vairaktarakis [191], Nagar et al. [224] and T'kindt and Billaut [267].

Full explanation of the $\alpha|\beta|\gamma$ notation may be found, e.g., in Błażewicz et al. [27, Chap. 3], Graham et al. [125, Sect. 2] and Lawler et al. [186].

5

Basics of time-dependent scheduling

This chapter completes the first, the introductory, part of the book. In this chapter, we indicate the place of time-dependent scheduling in the general framework of the scheduling theory, formulate the problems of time-dependent scheduling in a more formal way and introduce the terminology used throughout the whole book.

Chapter 5 is composed of five sections. In Sect. 5.1, we present a comparison of the scheduling theory and time-dependent scheduling. In Sect. 5.2, we give the formulation of a generic time-dependent scheduling problem, which is the basis for all time-dependent scheduling problems considered in the book. In Sect. 5.3, we introduce the terminology and notation used in the book for describing time-dependent scheduling problems. In Sect. 5.4, we discuss applications of time-dependent scheduling. The chapter is completed with bibliographic notes in Sect. 5.5.

5.1 The scheduling theory vs. time-dependent scheduling

In Chap. 4, we briefly described the basics of the scheduling theory. The following are the most important assumptions of this theory:

(A1) at every moment of time, each job (operation) can be processed by at most one machine and each machine can process at most one job (operation);
(A2) processing speeds of machines may be different but during the execution of jobs (operations) the speeds do not change in time;
(A3) the processing times of jobs (operations) are fixed and known in advance.

Throughout this book, the scheduling theory with assumptions (A1)–(A3) will be called the *classic scheduling theory*, as opposed to the *non-classic scheduling theory*, where at least one of these assumptions has been changed.

In the period of almost 60 years that elapsed since the classic scheduling theory was formulated, numerous practical problems have appeared, which could not be solved in the framework of this theory. The main reason for that was a certain restrictiveness of assumptions (A1)–(A3). For example, a machine may have a variable speed of processing due to the changing state of this machine, job processing times may increase due to job deterioration, etc. In order to overcome these difficulties and to adapt the theory to cover new problems, assumptions (A1)-(A3) were repeatedly modified. This, in turn, led to new research directions in the scheduling theory, such as scheduling multiprocessor tasks, scheduling on machines with variable processing speed and scheduling jobs with variable processing times. For the completeness of further presentation, we will now shortly describe each of these directions.

5.1.1 Scheduling multiprocessor tasks

In this case, assumption (A1) has been modified: the same operations (called *tasks*) may be performed at the same time by two or more different machines (processors).

The applications of scheduling multiprocessor tasks concern reliable computing in fault-tolerant systems, which are able to detect errors and recover the status of the systems from before an error. Examples of fault-tolerant systems are aircraft control systems, in which the same tasks are executed by two or more machines simultaneously in order to increase the safety of the systems. Other applications of scheduling multiprocessor tasks concern modelling the work of parallel computers, problems of dynamic bandwidth allocation in communication systems and loom scheduling in textile industry.

5.1.2 Scheduling on machines with variable processing speeds

In this case, assumption (A2) has been modified: the machines have variable processing speeds, i.e., the speeds change in time.

There are three main approaches to the phenomenon of the variable processing speeds. In the first approach, it is assumed that the speed is described by a differential equation and depends on a continuous resource. Alternatively, the speed is described by a continuous (the second approach) or a discrete (the third approach) function. In both cases, the speed depends on a resource that is either continuous or discrete.

Scheduling with continuous resources has applications in such production environments in which jobs are executed on machines driven by a common power source, for example, common mixing machines or refueling terminals. Scheduling with discrete resources is applied in modern manufacturing systems, in which jobs to be executed need machines as well as other resources such as robots or automated guided vehicles.

5.1.3 Scheduling jobs with variable processing times

In this case, assumption (A3) has been modified: the processing times of jobs are variable and can change in time.

The variability of job processing times can be modelled in different ways. For example, one can assume that the processing time of a job is a fuzzy number, a function of a continuous resource, a function of the job waiting time, a function of the position of the job in a schedule or is varying in some interval between a certain minimum and maximum value.

Scheduling with variable job processing times has numerous applications, e.g., in the modelling of the forging process in steel plants, manufacturing of preheated parts in plastic molding or in silverware production, finance management and scheduling maintenance or learning activities.

The time-dependent scheduling problems that we will consider in this book are scheduling problems with variable job processing times.

5.2 Formulation of time-dependent scheduling problems

As we said in Sect. 5.1, in time-dependent scheduling problems, the processing time of each job is variable. The general form of the job processing time is as follows.

In parallel-machine time-dependent scheduling problems, the processing time of each job depends on the starting time of the job, i.e.,

$$p_j(S_j) = g_j(S_j), \tag{5.1}$$

where g_j are arbitrary non-negative functions of $S_j \geq 0$ for $1 \leq j \leq n$.

In dedicated-machine time-dependent scheduling problems, the processing time of each operation is in the form of

$$p_{i,j}(S_{i,j}) = g_{i,j}(S_{i,j}), \tag{5.2}$$

where $g_{i,j}$ are arbitrary non-negative functions of $S_{i,j} \geq 0$ for $1 \leq i \leq n_j$ and $1 \leq j \leq n$.

These two forms of presentation, (5.1) and (5.2), are rarely used, since they do not give us any information about the way in which the processing times are changing.

The second way of describing the time-dependent processing time of a job,

$$p_j(S_j) = a_j + f_j(S_j), \tag{5.3}$$

where $a_j \geq 0$ and functions f_j are arbitrary non-negative functions of $S_j \geq 0$ for $1 \leq j \leq n$, is more often encountered. Similarly, the following form of the processing time of an operation,

$$p_{i,j}(S_{i,j}) = a_{i,j} + f_{i,j}(S_{i,j}), \tag{5.4}$$

where $a_{i,j} \geq 0$ and $f_{i,j}$ are arbitrary non-negative functions of $S_{i,j} \geq 0$ for $1 \leq i \leq n_j$ and $1 \leq j \leq n$, is more common than the form (5.2). The main reason for that is the fact that in (5.3) and (5.4), we indicate the constant part a_j $(a_{i,j})$ and the variable part $f_j(S_j)$ $(f_{i,j}(S_{i,j}))$ of the job (operation) processing time.

The constant part of a job (operation) processing time, a_j $(a_{i,j})$, will be called the *basic processing time*.

Remark 5.1. The assumption that functions $g_j(S_j)$ and $f_j(S_j)$ $(g_{i,j}(S_{i,j})$ and $f_{i,j}(S_{i,j}))$ are non-negative for non-negative arguments is essential and from now on, unless otherwise stated, we will consider it to be satisfied.

Remark 5.2. Since the forms (5.3) and (5.4) of job processing times give us more information, in further considerations, we will mainly use the functions $f_j(S_j)$ $(f_{i,j}(S_{i,j}))$.

Remark 5.3. Since the starting time S_j is the variable on which the processing time p_j depends, we will write $p_j(t)$ and $f_j(t)$ instead of $p_j(S_j)$ and $f_j(S_j)$, respectively. Similarly, we will write $p_{i,j}(t)$ and $f_{i,j}(t)$ instead of $p_{i,j}(S_{i,j})$ and $f_{i,j}(S_{i,j})$, respectively.

Remark 5.4. A few authors (Cheng and Sun [45], Lee [192], Lee et al. [195], Toksarı and Güner [268], Wang [281], Wang and Cheng [282, 290]) considered time-dependent scheduling problems with the so-called *learning effect* (cf. Bachman and Janiak [12], Biskup [24]). Since, in this case, job processing times are functions of both the starting time of the job and the job position in the schedule, the problems of this type will be not studied in the book.

Other parameters which describe a time-dependent scheduling problem, such as the parameters of a set of jobs (machines) or the applied optimality criterion, are as those in the classical scheduling (cf. Chap. 4).

Example 5.5. Assume that the set $\mathcal{J}$ is composed of 3 jobs, $\mathcal{J} = \{J_1, J_2, J_3\}$, such that $p_1 = 1 + 3t$, $p_2 = 2 + t$ and $p_3 = 3 + 2t$, there are no ready times and deadlines, and all jobs have unit weights.

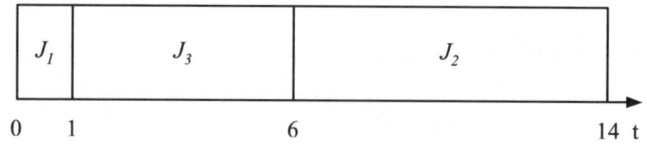

Fig. 5.1: The optimal schedule for Example 5.5

For this set of jobs, there exist the following semi-active schedules (cf. Definition 4.16): $\sigma^1 = (1, 2, 3)$, $\sigma^2 = (1, 3, 2)$, $\sigma^3 = (2, 1, 3)$, $\sigma^4 = (2, 3, 1)$, $\sigma^5 = (3, 1, 2)$ and $\sigma^6 = (3, 2, 1)$. The optimal schedule for the $C_{\max}$ criterion is schedule σ^2, $C_{\max}(\sigma^2) = 14$. The schedule is presented in Fig. 5.1. ◆

Example 5.6. In time-dependent scheduling problems the processing time of the same job may be different in different schedules. For example, consider schedules σ^1 and σ^5 from Example 5.5. The processing time of job J_2 in schedule σ^1 is equal to 4, while in schedule σ^5 it is equal to 41.

Algorithms that solve time-dependent scheduling problems will be called *time-dependent scheduling algorithms.* In this book, we will consider mainly offline time-dependent scheduling algorithms. Examples of online and semi-online time-dependent scheduling algorithms will be given in Chap. 9.

5.3 Terminology and notation

As we said in Sect. 5.2, in any time-dependent scheduling problem, job processing times are described by the functions $f_j(t)$ and $f_{i,j}(t)$, which appear in (5.3) and (5.4), respectively. The form of these functions is related to the problem we consider. For example, if we know nothing about the properties of these functions, then we deal with the *alteration* of job processing time: the processing time of a job varies in time in an unknown way. If we know something more, e.g., whether these functions are monotonic, then two cases are worth considering:

1° $f_j(t)$ and $f_{i,j}(t)$ are increasing (or non-decreasing);

2° $f_j(t)$ and $f_{i,j}(t)$ are decreasing (non-increasing).

The first case is more often encountered in the literature and, as it seems, it is easier to study. The case when job processing times are described by increasing (non-decreasing) functions will be called *deteriorating processing times*: while waiting for processing, the jobs deteriorate and as a result the processing time of each job increases in time.

The second case may cause some problems already at the stage of problem formulation, since we have to make some additional assumptions to avoid the case of negative job processing times. The case when job processing times are described by decreasing (non-increasing) functions will be called *shortening processing times*: unlike the previous case, jobs grow shorter and the processing time of each job is reduced in time.

Remark 5.7. Regardless of the type of functions that we have chosen to describe job processing times in our problem, we still deal with *deterministic* scheduling, since all parameters of the problem are assumed to be known in advance. This objection is important, since *stochastic* scheduling problems with deteriorating jobs are also considered (see, e.g., Glazebrook [112]).

Generally, the time-scheduling problems considered in this book will be denoted using the $\alpha|\beta|\gamma$ notation (see Sect. 4.4 for details). Each problem will be denoted by $\alpha_1\alpha_2|p_j(t) = a_j + f_j(t)|\varphi$ or $\alpha_1\alpha_2|p_{i,j}(t) = a_{i,j} + f_{i,j}(t)|\varphi$, where $\alpha_1\alpha_2$, $f_j(t)$ and φ denote the machine environment, the form of the variable part of job processing time and the criterion function, respectively.

Remark 5.8. We will use the $\alpha|\beta|\gamma$ notation if it will yield a simple notation for the considered scheduling problem. In some cases, however, we will resign from the notation in favour of the description by words if the descriptive approach will be more readable.

The short form of the symbol $\alpha_1\alpha_2|p_j(t) = a_j + f_j(t)|\varphi$ is the symbol $\alpha_1\alpha_2|p_j = a_j + f_j(t)|\varphi$. The short form of the symbol $\alpha_1\alpha_2|p_{i,j}(t) = a_{i,j} + f_{i,j}(t)|\varphi$ is the symbol $\alpha_1\alpha_2|p_{i,j} = a_{i,j} + f_{i,j}(t)|\varphi$. Throughout this book, we will use the short form of the symbols which will denote time-dependent scheduling problems.

If the form of the functions $f_j(t)$ $(f_{i,j}(t))$ is known, we will call the processing times $p_j = a_j + f_j(t)$ $(p_{i,j} = a_{i,j} + f_{i,j}(t))$ by the name of the function. For example, if the functions $f_j(t)$ $(f_{i,j}(t))$ are proportional (linear, polynomial, etc.), the processing times will be called *proportional (linear, polynomial,* etc.) processing times. If the functions are non-negative (non-positive), the processing times will be called *deteriorating (shortening)* processing times.

In a similar way, we will call the processing times of jobs, if non-linear forms of job deterioration are considered. For example, if the functions $f_j(t)$ are step functions or piecewise proportional-step functions, the processing times will be called *step* and *proportional-step* processing times, respectively.

If the same function $f(t)$ is used for all jobs, $f_j(t) = f(t)$ for $1 \le j \le n$ or $f_{i,j}(t) = f(t)$ for $1 \le i \le n_j$ and $1 \le j \le n$, we will speak about *simple deterioration (shortening)* of job processing times. In the opposite case, we will speak about *general deterioration (shortening)* of job processing times.

Example 5.9.

(a) The symbol $1|p_j = b_j t|C_{\max}$ will denote a single machine scheduling problem with proportional job processing times and the $C_{\max}$ criterion.

(b) The symbol $Pm|p_j = a_j + f(t)| \sum C_j$ will denote a multiple identical machine scheduling problem with simple general deterioration of jobs and the $\sum C_j$ criterion.

(c) The symbol $F2|p_{i,j} = a_{i,j} + b_{i,j}t|L_{\max}$ will denote a two-machine flow shop problem with linear job processing times and the $L_{\max}$ criterion.

(d) The symbol $O3|p_{i,j} = b_{i,j}t, b_{i,3} = b|C_{\max}$ will denote a three-machine open shop problem with proportional job processing times such that all job processing times on machine M_3 are equal to each other, and with the $C_{\max}$ criterion.

(e) The symbol $J2|p_{i,j} = b_{i,j}t|C_{\max}$ will denote a two-machine job shop problem with proportional job processing times and the $C_{\max}$ criterion. ◆

Remark 5.10. In Sect. 6.1, we will extend the $\alpha|\beta|\gamma$ notation to include the symbols describing time-dependent scheduling problems in batch environments and on machines with non-availability periods.

5.4 Applications of time-dependent scheduling

The motivation for research into time-dependent scheduling follows from the existence of many real-life problems which can be formulated in terms of scheduling jobs with time-dependent processing times. Such problems appear in all cases in which any delay in processing causes an increase (a decrease) of the processing times of executed jobs. If job processing times increase, we deal with deteriorating job processing times; if they decrease, we deal with shortening job processing times. In this section, we give a few examples of problems which can be modelled in time-dependent scheduling.

5.4.1 Scheduling problems with deteriorating job processing times

Gupta et al. [129] consider the problem of the *repayment of multiple loans*. We have to repay n loans, $L_1, L_2, \ldots, L_n$. A loan may represent an amount of borrowed cash or a payment to be made for a credit purchase. Loan L_k qualifies for a discount u_k if it is paid on or before a specified time b_k. A penalty at the rate v_k per day is imposed if the loan is not paid by due date $d_k, 1 \leq k \leq n$. The debtor earmarks a constant amount of q dollars per day, $q < v_k$, for repayment of the loans. Cash flows are continuously discounted with discount factor $(1+r)^{-1}$. The aim is to find an optimal repayment schedule that minimizes the present value PV of all cash outflows, $PV := \sum_{k=1}^{n} \frac{A_k}{(1+r)^{T_k}}$, where A_k and T_k denote, respectively, the actual amount paid for loan L_k and the time at which the loan L_k is repaid, $1 \leq k \leq n$. This problem can be modelled as a single-machine scheduling problem with time-dependent job processing times and the PV criterion.

Mosheiov [217] considers the following problem of *scheduling maintenance procedures*. A set of n maintenance procedures P_k, $1 \leq k \leq n$, has to be executed by $m \geq 1$ machines. A maintenance procedure P_k has to take place before a specified deadline d_k. The procedure consists of a series of actions, which last altogether p_k^1 time units. If the procedure does not complete by the deadline, several additional actions are required. The new processing time of procedure P_k is $p_k^2 > p_k^1$ time units. The aim is to find an order of execution of maintenance procedures $P_1, P_2, \ldots, P_n$, which minimizes the maximum completion time of the last executed procedure. This problem can be modelled as a single- or multiple-machine scheduling problem with two-step deteriorating job processing times.

Gawiejnowicz et al. [103] consider the following problem of *scheduling derusting operations*. We are given n items (e.g., parts of devices), which are subject to maintenance (e.g., they should be cleared from rust). This maintenance is performed by a single worker, who can himself determine the sequence of maintenance procedures. All procedures are non-preemptable, i.e., no maintenance procedure can be interrupted once it has started. At the moment $t = 0$, all items need the same amount of time for maintenance, e.g., one unit of time. As time elapses, each item corrodes at a rate that depends on the kind of the material from which the particular item is made. The rate of corrosion for the j-th item is equal to b_j, $1 \leq j \leq n$, and the time needed for the maintenance of each item grows proportionally to the time that elapsed from the moment $t = 0$. The problem is to choose such a sequence of the maintenance procedures that minimizes the total completion time of maintenance of all items. This problem can be modelled as the single-machine time-dependent scheduling problem $1|p_j = 1 + b_j t| \sum C_j$.

Rachaniotis and Pappis [240] consider the problem of *scheduling a single fire-fighting resource* in the case when there are several fires to be controlled. The aim is to find such order of supressing n existing fires that the total damage caused by the fires is minimized. The problem can be modelled as a single machine scheduling problem with time-dependent processing times and the total cost minimization criterion.

5.4.2 Scheduling problems with shortening job processing times

Ho et al. [135] consider the problem of *recognizing aerial threats*. A radar station recognizes some aerial threats approaching the station. The time required to recognize the threats decreases as they get closer. The aim is to find an optimal order of recognizing the threats which minimizes the maximum completion time. This problem can be modelled as a single-machine scheduling problem with shortening job processing times and the $C_{\max}$ criterion.

Kunnathur and Gupta [178] and Ng et al. [226] consider the problem of *producing ingots in a steel mill*. A set of ingots has to be produced in a steel mill. After being heated in a blast furnace, hot liquid metal is poured into steel ladles and next into ingot moulds, where it solidifies. Next, after the ingot stripper process, the ingots are segregated into batches and transported to the soaking pits, where they are preheated up to a certain temperature. Finally, the ingots are hot-rolled on the blooming mill. If the temperature of an ingot, while waiting in a buffer between the furnace and the rolling machine, has dropped below a certain value, then the ingot needs to be reheated to the temperature required for rolling. The reheating time depends on the time spent by the ingot in the buffer. The problem is to find a sequence of preheating the ingots which minimizes the maximum completion time of the last ingot produced. This problem can be modelled as a single machine scheduling problem with shortening job processing times and the $C_{\max}$ criterion.

5.4.3 Other examples of time-dependent scheduling problems

Shakeri and Logendran [254] consider the following problem of *maximizing satisfaction level* in a multitasking environment. Several plates are spinning on vertical poles. An operator has to ensure all plates spin as smoothly as possible. A value, called the *satisfaction level*, can be assigned to each plate's spinning state. The satisfaction level of a plate is ranging from 0% (i.e., the plate is not spinning) up to 100% (the plate is spinning perfectly). The objective is to maximize the average satisfaction level of all plates over time.

The above problem is applicable to multitasking environments in which we cannot easily determine the completion time of any job. Examples of such environments are the environments of the control of a plane flight parameters, monitoring air traffic or the work of nuclear power plants. A special case of the problem, when a 100% satisfaction level is equivalent to the completion of a job, is a single-machine time-dependent scheduling problem.

Other examples of practical problems which can be modelled in terms of time-dependent scheduling include the control of queues in communication systems in which jobs deteriorate as they wait for processing (Browne and Yechiali [33]), search for an object in worsening weather or growing darkness, performance of medical procedures under deterioration of the patient conditions and repair of machines or vehicles under deteriorating mechanical conditions (Mosheiov [216]).

We refer the reader to the literature (see Alidaee and Womer [6] and Cheng et al. [55]) for more examples of time-dependent scheduling applications.

5.4.4 Scheduling problems with time-dependent parameters

The time dependence may concern not only job processing times but also other parameters of a scheduling problem. For example, Cai et al. [38] consider the following *crackdown scheduling problem*. There are n illicit drug markets, all of which need to be brought down to a negligible level of activity. Each market is eliminated by a procedure consisting in a crackdown phase and a maintenance phase. The crackdown phase utilizes all the available resources until the market is brought down to the desired level. The maintenance phase, which follows after the crackdown phase and uses a significantly smaller amount of resources, maintains the market at this level. The aim is to find an order of elimination of the drug markets that minimizes the total time spent in eliminating all drug markets. The problem can be modelled as a single-machine scheduling problem of minimizing the total cost $\sum f_j$, where f_j are monotonically increasing time-dependent cost functions.

Other examples of scheduling problems in which some parameters are time dependent include multiprocessor tasks scheduling (Bampis and Kononov [16]), scheduling in a contaminated area (Janiak and Kovalyov [147, 148]), multicriteria project sequencing (Klamroth and Wiecek [165]), selection problems (Seegmuller et al. [253]) and scheduling jobs with deteriorating job values (Voutsinas and Pappis [275]).

With these remarks, we end the presentation of the basics of time-dependent scheduling. This chapter also ends the first part of the book. In the next part, we will consider the complexity of time-dependent scheduling problems.

5.5 Bibliographic notes

The problems of scheduling multiprocessor tasks are reviewed in detail by Drozdowski [73, 74] and Lee et al. [190].

The problems of scheduling with continuous resources are discussed by Błażewicz et al. [27, Chap. 12] and Gawiejnowicz [87].

The problems of scheduling on machines with variable speed are considered, e.g., by Dror et al. [72], Gawiejnowicz [88, 89, 90], Meilijson and Tamir [206] and Trick [269].

The variability of the processing time of a job can be modelled in many different ways. The job processing time can be, e.g., a function of the job waiting time (see, e.g., Barketau et al. [17], Finke and Jiang [80], Finke et al. [81], Finke and Oulamara [82], Leung et al. [199], Lin and Cheng [203], Sriskandarajah and Goyal [261]), a function of a continuous resource (see, e.g., Janiak [144, 145]) or a fuzzy number (see, e.g., Słowiński and Hapke [259]).

The processing time of a job may also depend on the position of the job in a schedule (Bachman and Janiak [12], Biskup [24]), the length of machine non-availability period (Lahlou and Dauzère-Pérès [181]) or varies in some interval between a certain minimum and maximum value (see, e.g., Nowicki and Zdrzałka [227], Shakhlevich and Strusevich [255], Vickson [273]).

The problems of time-dependent scheduling are reviewed by Alidaee and Womer [6] and Cheng et al. [55].

Gawiejnowicz [87] discusses time-dependent scheduling problems in the framework of scheduling with discrete and continuous resources.

Part II

COMPLEXITY

6

Single-machine time-dependent scheduling

The knowledge of the complexity of a scheduling problem is essential in further research of the problem. Therefore, the second part of the book is devoted to the complexity of time-dependent scheduling problems. To give the reader full insight into the subject, we include proofs or sketches of proofs of the greater part of discussed results. We also present the pseudo-codes of formulations of exact polynomial-time algorithms.

This part is composed of three chapters. In Chap. 6, we present the complexity results concerning time-dependent scheduling on a single machine. The complexity of the problems of time-dependent scheduling on parallel and dedicated machines is considered in Chaps. 7 and 8, respectively.

Chapter 6 is composed of five sections. In Sect. 6.1, we present the results concerning a single machine and minimization of the $C_{\max}$ criterion. In Sect. 6.2, we present the results concerning a single machine and minimization of the $\sum C_j$ criterion. In Sect. 6.3, we present the results concerning a single machine and minimization of the $L_{\max}$ criterion. In Sect. 6.4, we present the results concerning a single machine and minimization of criteria other than $C_{\max}$, $\sum C_j$ and $L_{\max}$. The chapter is completed with Sect. 6.5 including the summary and tables.

6.1 Minimizing the maximum completion time

In this section, we consider the results concerning the criterion $C_{\max}$.

6.1.1 Proportional deterioration

This is the simplest form of job deterioration. In this case, we assume that

$$p_j = b_j t, \tag{6.1}$$

where $b_j > 0$ for $1 \leq j \leq n$ and $S_1 \equiv t_0 > 0$. This form of job deterioration was introduced by Mosheiov [216]. Number b_j will be called the *deterioration rate* of job J_j, $1 \leq j \leq n$.

Equal ready times and deadlines

First, we consider the single machine time-dependent scheduling problems with proportional job processing times, in which neither non-zero ready times nor finite deadlines have been defined, i.e., we will assume that $r_j = 0$ and $d_j = \infty$ for $1 \leq j \leq n$.

Theorem 6.1. (Mosheiov [216]) *The problem $1|p_j = b_j t|C_{\max}$ is solvable in $O(n)$ time, and the maximum completion time does not depend on the schedule of jobs.*

Proof. Let σ be an arbitrary schedule and let $[j]$ denote the index of the j-th job in σ. Since

$$C_{[j]}(\sigma) = S_1 \prod_{i=1}^{j}(1 + b_{[i]}) = t_0 \prod_{i=1}^{j}(1 + b_{[i]}), \qquad (6.2)$$

we have $C_{\max} = C_{[n]}(\sigma) = t_0 \prod_{i=1}^{n}(1 + b_{[i]})$. Since the product $\prod_{i=1}^{n}(1 + b_{[i]})$ can be calculated in $O(n)$ time and it is independent of the schedule, the result follows.

An alternative proof uses pairwise job interchange argument. We consider schedule σ' in which job $J_{[i]}$ is immediately followed by job $J_{[j]}$, and schedule σ'' in which the jobs are in the reverse order. Since we have $C_{[j]}(\sigma') - C_{[i]}(\sigma'') = 0$, the result follows. □

The problem $1|p_j = b_j t|C_{\max}$ can be a basis for more general problems. Below, we consider two of them.

Cheng and Sun [44] reformulated the problem $1|p_j = b_j t|C_{\max}$ into the following *time-dependent batch scheduling problem.*

We are given n jobs $J_1, J_2, \ldots, J_n$, which are available starting from time $t_0 = 0$. The jobs are classified into m groups $G_1, G_2, \ldots, G_m$. Group G_i is composed of k_i jobs, where $1 \leq i \leq m$ and $\sum_{i=1}^{m} k_i = n$. Jobs in the same group G_i, $1 \leq i \leq m$, are processed consecutively and without idle times. The setup time θ_i precedes the processing of the group G_i, $1 \leq i \leq m$. The processing time of the j-th job in group G_i is in the form of $p_{i,j} = b_{i,j} t$, where $b_{i,j} > 0$ for $1 \leq i \leq m$ and $1 \leq j \leq k_i$. The aim is to find the sequence of groups and the sequence of jobs in each group, which together minimize the $C_{\max}$ criterion.

Remark 6.2. The assumption that jobs in the same group are processed consecutively and without idle times is called *group technology*; see Potts and Kovalyov [238], Tanaev et al. [265] for more details.

Remark 6.3. If we put symbols GT and θ_i in field β of a symbol in the $\alpha|\beta|\gamma$ notation (cf. Sect. 4.4), the whole symbol will denote a batch scheduling problem with group technology and setup times. For example, the symbol $1|p_{i,j} = b_{i,j} t, \theta_i, GT|C_{\max}$ will denote the above described time-dependent batch scheduling problem.

For the problem $1|p_{i,j} = b_{i,j}t, \theta_i, GT|C_{\max}$, Cheng and Sun [44] proposed the following algorithm.

Algorithm A_1 for the problem $1|p_{i,j} = b_{i,j}t, \theta_i, GT|C_{\max}$ ([44])

Input: sequences $(\theta_1, \theta_2, \ldots, \theta_m)$, $(b_{i,j})$ for $1 \leq i \leq m$ and $1 \leq j \leq k_i$
Output: an optimal schedule

▷ Step 1:
   for $i \leftarrow 1$ **to** m **do** $B_i \leftarrow \prod_{j=1}^{k_i}(1 + b_{i,j})$;

▷ Step 2:
   Schedule groups of jobs in the non-decreasing order of $\frac{\theta_i B_i}{B_i - 1}$ ratios;

▷ Step 3:
   for $i \leftarrow 1$ **to** m **do** Schedule jobs in group G_i in an arbitrary order.

Theorem 6.4. (Cheng and Sun [44]) *The problem* $1|p_{i,j} = b_{i,j}t, \theta_i, GT|C_{\max}$ *is solvable in* $O(n \log n)$ *time by algorithm* A_1.

Proof. Let G_i, $1 \leq i \leq m$, denote the i-th group of jobs. Note that the completion time C_{i,k_i} of the last job in group G_i is given by the equation

$$C_{i,k_i} = (C_{i-1,k_{i-1}} + \theta_i) \prod_{j=1}^{k_i}(1 + b_{i,j}),$$

where $1 \leq i \leq m$ and $C_{0,0} := 0$.
   Let $\sigma^1 = (G_1, G_2, \ldots, G_{i-1}, G_i, G_{i+1}, \ldots, G_m)$ be a schedule such that

$$\frac{\theta_i \prod_{j=1}^{k_i}(1 + b_{i,j})}{\prod_{j=1}^{k_i}(1 + b_{i,j}) - 1} \geq \frac{\theta_{i+1} \prod_{j=1}^{k_{i+1}}(1 + b_{i+1,j})}{\prod_{j=1}^{k_{i+1}}(1 + b_{i+1,j}) - 1} \tag{6.3}$$

for some $1 \leq i \leq m - 1$. Let $\sigma^2 = (G_1, G_2, \ldots, G_{i-1}, G_{i+1}, G_i, \ldots, G_m)$ be the schedule obtained from σ^1 by mutual exchange of groups G_i and G_{i+1}. Since

$$C_{i+1,k_{i+1}}(\sigma^1) - C_{i,k_i}(\sigma^2) = \left(\prod_{j=1}^{k_{i+1}}(1 + b_{i+1,j}) - 1\right)\left(\prod_{j=1}^{k_i}(1 + b_{i,j}) - 1\right) \times$$

$$\times \left(\frac{\theta_i \prod_{j=1}^{k_i}(1+b_{i,j})}{\prod_{j=1}^{k_i}(1+b_{i,j})-1} - \frac{\theta_{i+1}\prod_{j=1}^{k_{i+1}}(1+b_{i,j})}{\prod_{j=1}^{k_{i+1}}(1+b_{i,j})-1}\right),$$

by (6.3) the difference is non-positive. Hence, σ^2 is better than σ^1.

Repeating, if necessary, the above described exchange, we obtain a schedule in which all groups of jobs are in the non-decreasing order of $\dfrac{\theta_i \prod\limits_{j=1}^{k_i} b_{i,j}}{\prod\limits_{j=1}^{k_i} b_{i,j}-1}$ ratios.

To complete the proof, it is sufficient to note that by Theorem 6.1, the sequence of jobs in each group is immaterial. The overall time complexity of algorithm A_1 is $O(n \log n)$, since Step 1 needs $O(m \log m)$ time, Step 2 needs $O(n)$ time and $m = O(n)$. □

Theorem 6.4 has been generalized by Wu et al. [300], who considered time-dependent setup times, $\theta_i = \delta_i t$, where $\delta_i > 0$ for $1 \leq i \leq m$.

Theorem 6.5. (Wu et al. [300]) *The problem $1|GT, p_{i,j} = b_{i,j}t, \theta_i = \delta_i t|C_{\max}$ is solvable in $O(n)$ time, and the maximum completion time does not depend either on the schedule of jobs in the group or on the order of groups.*

Proof. Consider an arbitrary schedule σ for the problem $1|GT, p_{i,j} = b_{i,j}t$, $\theta_i = \delta_i|C_{\max}$. Without loss of generality, we can assume that σ is in the form of $(\sigma_{1,[1]}, \sigma_{1,[2]}, \ldots, \sigma_{1,[k_1]}, \sigma_{2,[1]}, \sigma_{2,[2]}, \ldots, \sigma_{2,[k_2]}, \ldots, \sigma_{m,[1]}, \sigma_{m,[2]}, \ldots, \sigma_{m,[k_m]})$, where $\sum_{i=1}^{m} k_i = n$. Since, by Theorem 6.1,

$$C_{\max}(\sigma) = t_0 \prod_{i=1}^{m}(1+\delta_i) \prod_{i=1}^{m}\prod_{j=1}^{k_i}(1+b_{i,[j]}) \tag{6.4}$$

and since the value of the right side of (6.4) does not depend on the order of jobs, the result follows. □

Remark 6.6. Scheduling deteriorating jobs with setup times and batch scheduling of deteriorating jobs are new topics in time-dependent scheduling. In the classic scheduling (cf. Sect. 5.1), both these topics have been studied since early 1960s and have an extensive literature. We refer the reader to the reviews by Allahverdi et al. [7], Potts and Kovalyov [238] and Webster and Baker [290] and to the book by Tanaev et al. [265] for details.

Remark 6.7. Batch scheduling problems with time-dependent job processing times are also considered by Barketau et al. [17] and Leung et al. [199]. In these papers, however, the processing time of a job depends on the *waiting time* of the job.

Another generalization of the problem $1|p_j = b_j t|C_{\max}$ is the problem of scheduling proportionally deteriorating jobs on a single machine with non-availability periods (cf. Remark 4.4).

Assume that the used machine is not continuously available and there are given k disjoint *periods of machine non-availability*. These periods are described by time intervals $\langle W_{i,1}, W_{i,2} \rangle$, where $W_{1,1} > t_0$ and $W_{i,1} < W_{i,2}$ for $1 \leq i \leq k < n$. Since any non-availability period can interrupt the processing of a job, we have to decide what to do in the case when the job has been interrupted by the start time of a non-availability period (cf. Lee [188]).

Remark 6.8. If in field α of the $\alpha|\beta|\gamma$ notation appears the symbol h_{ik}, where $1 \leq i \leq m$ is the number of machines and k is the number of non-availability periods, it will denote the number of non-availability periods. For example, the symbols $1, h_{11}|p_j = b_j t|C_{\max}$ and $1, h_{1k}|p_j = b_j t|C_{\max}$ will denote the problem $1|p_j = b_j t|C_{\max}$ with a single non-availability period and an arbitrary number of non-availability periods, respectively.

Definition 6.9. (Job preemption vs. machine non-availability periods)
(a) *A job is said to be* non-resumable *if in the case when the job has been interrupted by the start time of a non-availability period this job has to be restarted after the machine becomes available again.*
(b) *A job is said to be* resumable *if in the case when the job has been interrupted by the start time of a non-availability period this job does not need to be restarted and can be completed after the machine becomes available again.*

Remark 6.10. The fact that jobs are non-resumable (resumable) will be denoted in field β of the $\alpha|\beta|\gamma$ notation by symbol *nres* (*res*). For example, the symbols $1, h_{11}|p_j = b_j t, nres|C_{\max}$ and $1, h_{11}|p_j = b_j t, res|C_{\max}$ will denote the problem $1, h_{11}|p_j = b_j t|C_{\max}$ with non-resumable jobs and resumable jobs, respectively.

The problem of scheduling proportional jobs on a machine with a single non-availability period was introduced, for the case of resumable jobs, by Wu and Lee [298]. Gawiejnowicz [91] and Ji et al. [155] proved that scheduling non-resumable jobs with proportional processing times and one period of machine non-availability is a computationally intractable problem.

Theorem 6.11. (Gawiejnowicz [91], Ji et al. [155]) *The decision version of the problem* $1, h_{11}|p_j = b_j t, nres|C_{\max}$ *is $\mathcal{NP}$-complete in the ordinary sense.*

Proof. Gawiejnowicz [91] uses the following transformation from the SP problem (cf. Sect. 3.2): $n = p$, $t_0 = 1$, $\alpha_j = y_j - 1$ for $1 \leq j \leq n$, $k = 1$, $W_{1,1} = B, W_{1,2} = 2B$ and threshold $G = 2Y$, where $Y = \prod_{j=1}^{p} y_j$.
    Note that by Lemma 6.1, we can check in polynomial time whether $C_{\max}(\sigma) \leq G$ for a given schedule σ for the above instance of the problem $1|p_j = b_j t|C_{\max}$ with a single non-availability period. Therefore, the decision version of this problem is in the $\mathcal{NP}$ class.
    Since the above transformation can be done in polynomial time, in order to complete the proof, it is sufficient to show that the SP problem has a solution if and only if there exists a feasible schedule σ for the above instance of the problem $1, h_{11}|p_j = b_j t, nres|C_{\max}$ with the non-availability period $\langle W_{1,1}, W_{1,2} \rangle$ such that $C_{\max}(\sigma) \leq G$ (see Fig. 6.1 and Remark 6.12).
    Ji et al. [155] use the following transformation from the SP problem: $n = p$, t_0 arbitrary, $W_{1,1} = t_0 B$, $W_{1,2} > W_{1,1}$ arbitrary, $b_j = y_j - 1$ for $1 \leq j \leq n$ and threshold $G = W_{1,2}\frac{Y}{B}$. The rest of the proof is as above.   □

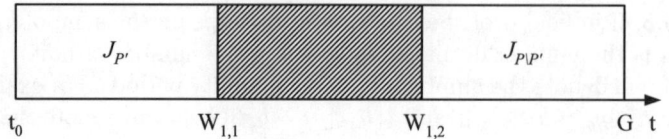

Fig. 6.1: Example schedule in the proof of Theorem 6.11

Remark 6.12. In figures included in some $\mathcal{N}\mathcal{P}$-completeness proofs, by J_X we will denote the set of jobs with deterioration rates corresponding to the elements of set X. For example, in Fig. 6.1, the symbols $J_{P'}$ and $J_{P \setminus P'}$ denote the set of jobs with deterioration rates corresponding to the elements of set P' and $P \setminus P'$, respectively.

Remark 6.13. Ji et al. [155], using a dynamic programming approach, formulated for the problem $1, h_{11}|p_j = b_j t, nres|C_{\max}$ a pseudopolynomial algorithm that runs in $O(n(W_{1,1} - t_0))$ time; see [155, Sect. 2.1] for details. Hence, by Lemma 3.18, the problem cannot be $\mathcal{N}\mathcal{P}$-hard in the strong sense.

Remark 6.14. Since in the book we will consider mainly such problems for which it is easy to show that the decision versions of the problems are in the $\mathcal{N}\mathcal{P}$ class or that the applied transformations are polynomial (pseudopolynomial), in most cases we will omit these parts of $\mathcal{N}\mathcal{P}$-completeness proofs.

The case of an arbitrary number of non-availability periods has been considered by Gawiejnowicz.

Theorem 6.15. (Gawiejnowicz [91]) *The decision version of the problem* $1, h_{1k}|p_j = b_j t, nres|C_{\max}$ *is $\mathcal{N}\mathcal{P}$-complete in the strong sense.*

Proof. The transformation from the 4-P problem (cf. Sect. 3.2) is as follows: $n = 4p$, $t_0 = 1$, $b_j = u_j - 1$ for $1 \leq j \leq n$, $k = p$, $W_{i,1} = \sum_{j=1}^{i} D^j$ and $W_{i,2} = \sum_{j=0}^{i} D^j$ for $1 \leq i \leq k$, and threshold $G = \sum_{j=1}^{p} D^j$.
To complete the proof, it is sufficient to show that the 4-P problem has a solution if and only if there exists a feasible schedule σ for the above instance of the problem $1, h_{1k}|p_j = b_j t, nres|C_{\max}$ with non-availability periods $\langle W_{i,1}, W_{i,2} \rangle$, $1 \leq i \leq p$, such that $C_{\max}(\sigma) \leq G$.   □

Gawiejnowicz and Kononov proved the the problem with a single non-availability period remains computationally intractable for resumable jobs.

Theorem 6.16. (Gawiejnowicz and Kononov [92]) *The decision version of the problem* $1, h_{11}|p_j = b_j t, res|C_{\max}$ *is $\mathcal{N}\mathcal{P}$-complete in the ordinary sense.*

Proof. The transformation from the SP problem (cf. Sect. 3.2) is as follows: $n = p + 1$, $t_0 = 1$, $\alpha_j = y_j - 1$ for $1 \leq j \leq p$, $\alpha_{p+1} = B - 1$, $k = 1$, $W_{1,1} = B+1$, $W_{1,2} = 2B+1$ and threshold $G = (B+1)Y$, where $Y = \prod_{j=1}^{p} y_j$.

In order to complete the proof, it is sufficient to show that the SP problem has a solution if and only if there exists a feasible schedule σ for the above instance of the problem $1, h_{11}|p_j = b_j t, res|C_{\max}$ with the non-availability period $\langle W_{1,1}, W_{1,2} \rangle$ such that $C_{\max}(\sigma) \leq G$.   □

Remark 6.17. Scheduling deteriorating jobs on machines with non-availability periods is a new topic in time-dependent scheduling. In the classic scheduling (cf. Sect. 5.1), however, the matter has been studied since the early 1980s and it has an extensive literature; see the reviews by Lee [188, 189], Lee et al. [190] and Schmidt [251].

Distinct ready times and deadlines

Now, we pass to the single-machine time-dependent scheduling problems with proportional job processing times in which jobs have distinct ready times and (or) distinct deadlines.

Theorem 6.18. (Gawiejnowicz [91]) *The decision version of the problem* $1|p_j = b_j t, r_j, d_j|C_{\max}$ *with two distinct ready times and two distinct deadlines is $\mathcal{NP}$-complete in the ordinary sense.*

Proof. We use the following transformation from the SP problem (cf. Sect. 3.2): $n = p+1$, $t_0 = 1$, $b_j = y_j - 1$, $r_j = 1$ and $d_j = BY$ for $1 \leq j \leq p$, $b_{p+1} = B - 1$, $r_{p+1} = B$, $d_{p+1} = B^2$ and threshold $G = BY$, where $Y = \prod_{j=1}^{p} y_j$.

Notice that the completion time of the j-th job in any feasible schedule for the problem $1|p_j = b_j t, r_j, d_j|C_{\max}$ is equal to $C_{[j]} = S_{[j]}(1 + b_{[j]}) = \max\{C_{[j-1]}, r_{[j]}\}(1 + b_{[j]})$, where $1 \leq j \leq n$ and $C_{[0]} := t_0$. Hence, the decision version of the problem $1|p_j = b_j t, r_j, d_j|C_{\max}$ is in the $\mathcal{NP}$ class.

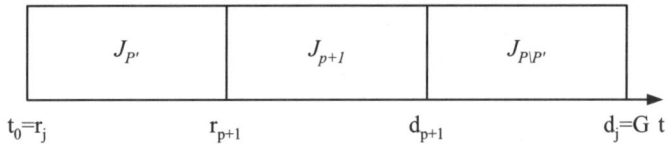

$$t_0 = r_j \qquad\qquad r_{p+1} \qquad\qquad d_{p+1} \qquad\qquad d_j = G\ t$$

Fig. 6.2: Example schedule in the proof of Theorem 6.18

In order to complete the proof it is sufficient to show that the SP problem has a solution if and only if there exists a feasible schedule σ for the above instance of the problem $1|p_j = b_j t, r_j, d_j|C_{\max}$ (see Fig. 6.2 and Remark 6.12) such that $C_{\max}(\sigma) \leq G$.   □

Theorem 6.19. (Gawiejnowicz [91]) *The decision version of the problem* $1|p_j = b_j t, r_j, d_j|C_{\max}$ *with an arbitrary number of distinct ready times and an arbitrary number of deadlines is $\mathcal{NP}$-complete in the strong sense.*

Proof. We use the following transformation from the 4-P problem (cf. Sect. 3.2). Let $n = 5p$ and $t_0 = 1$. Job deterioration rates, ready times and deadlines are defined as follows: $b_j = u_j - 1, r_j = t_0$ and $d_j = G$ for $1 \le j \le 4p$, $b_{4p+k} = D - 1$, $r_{4p+k} = D^{2k-1}$ and $d_{4p+k} = D^{2k}$ for $1 \le k \le p$, where the threshold $G = D^{2p}$.

In order to complete the proof, it is sufficient to show that the 4-P problem has a solution if and only if for the above instance of the problem $1|p_j = b_j t, r_j, d_j|C_{\max}$ there exists a feasible schedule σ such that $C_{\max}(\sigma) \le G$.

$\square$

6.1.2 Proportional-linear deterioration

Equal ready times and deadlines

Theorem 6.1 was generalized by Kononov to the case of *proportional-linear* job processing times:

$$p_j = b_j(A + Bt), \tag{6.5}$$

where $1 \le j \le n$ and $S_1 \equiv t_0 \ge 0$.

Theorem 6.20. (Kononov [173]) *If there hold inequalities*

$$Bt_0 + A > 0 \tag{6.6}$$

and

$$1 + b_j B > 0 \quad \text{for} \quad 1 \le j \le n, \tag{6.7}$$

then the problem $1|p_j = b_j(A + Bt)|C_{\max}$ *is solvable in $O(n)$ time, the maximum completion time*

$$C_{\max} = \begin{cases} t_0 + A \sum\limits_{j=1}^{n} b_j, & \text{if } B = 0, \\ (t_0 + \frac{A}{B}) \prod\limits_{j=1}^{n} (b_j B + 1) - \frac{A}{B}, & \text{if } B \ne 0, \end{cases} \tag{6.8}$$

and it does not depend on the schedule of jobs.

Proof. We proceed by induction with respect to the number of jobs. For $n = 1$, the equality (6.8) is satisfied. Assume that it is satisfied for $n = k-1$. If $B = 0$, then we have $C_k = C_{k-1} + p_k = t_0 + A\sum_{j=1}^{k-1} b_j + Ab_k = t_0 + A\sum_{j=1}^{k} b_j$.

If $B \ne 0$, then we have $C_k = C_{k-1} + p_k = C_{k-1} + b_k(A + BC_{k-1}) = (t_0 + \frac{A}{B}) \prod_{j=1}^{k-1}(b_j B + 1) - \frac{A}{B} + b_k B((t_0 + \frac{A}{B})\prod_{j=1}^{k-1}(b_j B + 1) - \frac{A}{B}) + b_k A = (b_k B + 1)(t_0 + \frac{A}{B})\prod_{j=1}^{k-1}(b_j B + 1) + b_k A - b_k A - \frac{A}{B} = (t_0 + \frac{A}{B}) \prod_{j=1}^{k}(b_j B + 1) - \frac{A}{B}$.

■

Remark 6.21. A slightly different form of Theorem 6.20, without conditions (6.6) and (6.7) but with assumptions $A > 0, B > 0, b_j > 0$ for $1 \leq j \leq n$, was given by Zhao et al. [304, Theorem 1].

Guo and Wang [127] generalized Theorem 6.4 to the case of proportional-linear job processing times. For the problem $1|p_{i,j} = b_{i,j}(A+Bt), \theta_i, GT|C_{\max}$ the authors proposed the following algorithm.

Algorithm A_2 for the problem $1|p_{i,j} = b_{i,j}(A + Bt), \theta_i, GT|C_{\max}$ ([127])

   Input: sequences $(\theta_1, \theta_2, \ldots, \theta_m)$, $(b_{i,j})$ for $1 \leq i \leq m$ and $1 \leq j \leq k_i$,
            numbers A, B
   Output: an optimal schedule
▷ **Step 1:**

   Schedule groups of jobs in the non-decreasing order of $\dfrac{\theta_i \prod\limits_{j=1}^{k_i} b_{i,j}}{\prod\limits_{j=1}^{k_i} b_{i,j} - 1}$ ratios;

▷ **Step 2:**
   for $i \leftarrow 1$ **to** m **do** Schedule jobs in group G_i in an arbitrary order.

Theorem 6.22. (Guo and Wang [127]) *The problem $1|p_{i,j} = b_{i,j}(A+Bt), \theta_i, GT|C_{\max}$ is solvable in $O(n \log n)$ time by algorithm A_2.*

Proof. Similar to the proof of Theorem 6.4.   □

Theorem 6.5 and Theorem 6.22 have been generalized by Wang et al. [284]. The authors assumed that job processing times are in the form of (6.5) and setup times are proportional-linear, i.e., for $1 \leq i \leq m$, we have

$$\theta_i = \delta_i(A + Bt). \tag{6.9}$$

Theorem 6.23. (Wang et al. [284]) *The problem $1|p_{i,j} = b_{i,j}(A + Bt), \theta_i = \delta_i(A + Bt), GT|C_{\max}$ is solvable in $O(n)$ time, and the maximum completion time does not depend either on the schedule of jobs in the group or on the order of groups.*

Proof. Similar to the proof of Theorem 6.5; see [284, Theorem 1].   ◇

6.1.3 Linear deterioration

This is the next form of job deterioration. In this case, the job processing time is a linear function of time,

$$p_j = a_j + b_j t, \tag{6.10}$$

where $S_1 \equiv t_0 = 0$, $a_j > 0$ and $b_j > 0$ for $1 \leq j \leq n$. This form of job deterioration was introduced by Tanaev et al. [264]. Numbers a_j and b_j will be called, respectively, the *basic processing time* and the *deterioration rate* of job J_j, $1 \leq j \leq n$.

Equal ready times and deadlines

In this subsection, we consider the single-machine time-dependent scheduling problems with linear job processing times in which all ready times and deadlines are equal.

The following result has been obtained independently by a number of authors. Since the authors used different proof techniques, we will shortly describe the approaches which have been applied in order to prove the result.

Theorem 6.24. (Gawiejnowicz and Pankowska [109]; Gupta and Gupta [128]; Tanaev et al. [264]; Wajs [277]) *The problem* $1|p_j = a_j + b_j t|C_{\max}$ *is solvable in* $O(n \log n)$ *time by scheduling jobs in the non-increasing order of* $\frac{b_j}{a_j}$ *ratios.*

Proof. Notice that for a given schedule σ there holds the equality

$$C_j(\sigma) = \sum_{i=1}^{j} a_{\sigma_i} \prod_{k=i+1}^{j} (1 + b_{\sigma_k}), \tag{6.11}$$

where $1 \leq j \leq n$. (The equality can be proved by induction with respect to j.)

The first and simplest proof uses the pairwise job interchange argument: we assume that in schedule σ' job J_i precedes job J_j and in schedule σ'' job J_i follows job J_j. Next, we calculate the difference between $C_{\max}(\sigma') \equiv C_n(\sigma')$ and $C_{\max}(\sigma'') \equiv C_n(\sigma'')$. Finally, we show that the difference does not depend on time. (The approach has been used by Gupta and Gupta [128], Wajs [277].)

The second proof is more complicated and uses a priority-generating function (see Definition 1.19). Let $C_{\max}(\sigma)$ denote the length of a schedule for a given job sequence σ. It has been proved (see [264, Chap. 3]) that function $\omega(\sigma) = \frac{\Psi(\sigma)}{C_{\max}(\sigma)}$, where

$$\Psi(\sigma) = \sum_{j=1}^{n} b_{\pi_j}(1 + x_{\sigma_j}), x_{\sigma_1} = 0, x_{\sigma_j} = \sum_{i=1}^{j-1} b_{\sigma_i}(1 + x_{\sigma_i}),$$

is a priority function and $C_{\max}$ is a priority generating function for the problem $1|p_j = a_j + b_j t|C_{\max}$. Thus, by Remark 1.20 and Theorem 1.24, the optimal schedule for the problem can be obtained in $O(n \log n)$ time by scheduling jobs in the non-increasing order of their priorities. (This approach to the proof has been used by Tanaev et al. [264].)

The third way of proving the result uses the following idea. Denote by $F = \{f_1, f_2, \ldots, f_n\}$, where $f_j = a_j + b_j t$ for $1 \leq j \leq n$, the set of linear functions, which describe jobs processing times. Let $\sigma = (\sigma_1, \sigma_2, \ldots, \sigma_n)$ and $\pi = (\pi_1, \pi_2, \ldots, \pi_n)$, $\sigma \neq \pi$, be permutations of elements of set I_n, and let $\preceq$ be an ordering relation on set $N_{\mathcal{J}}$ such that $i \preceq j \Leftrightarrow a_i b_j - a_j b_i \leq 0$. Let sequence $p_{\pi_1}, p_{\pi_2}, \ldots, p_{\pi_n}$ be defined in the following way: $p_{\pi_1} = f_{\pi_1}(0)$, $p_{\pi_2} = f_{\pi_2}(p_{\pi_1}), \ldots, p_{\pi_n} = f_{\pi_n}(\sum_{j=1}^{n-1} p_{\pi_j})$, where $f_{\pi_j} \in F$ for $1 \leq j \leq n$. Then

$$\sum_{j=1}^{n} p_{\sigma_j} = \min_{\pi \in \mathfrak{S}_n} \sum_{j=1}^{n} p_{\pi_j} \quad \Leftrightarrow \quad \sigma_1 \preceq \sigma_2 \preceq \ldots \preceq \sigma_n. \qquad (6.12)$$

In other words, the optimal schedule for the problem $1|p_j = a_j + b_j t|C_{\max}$ (equivalently, the permutation of job indices) is generated by non-decreasing sorting of job indices according to the $\preceq$ relation. The rest of this proof is technical: the main idea is to consider an expanded form of the formulae, which describe the length of a schedule. (This approach, exploiting an ordering relation in the set of functions which describe processing times, has been applied by Gawiejnowicz and Pankowska [109].) $\square$

Remark 6.25. A probabilistic counterpart of Theorem 6.24 is known. Namely, Browne and Yechiali [33] have used Lemma 1.2 (a) for deriving the expected value and the variance of a single machine schedule length for linearly deteriorating jobs. The authors proved that scheduling jobs in the non-decreasing order of $\frac{E(a_j)}{b_j}$ ratios minimizes the expected maximum completion time and scheduling jobs in the non-decreasing order of $\frac{Var(a_j)}{(1+b_j)^2-1}$ ratios minimizes the variance of the maximum completion time, where $E(a_j)$ and $Var(a_j)$ are the expected maximum completion time and the variance of the maximum completion time for $1 \le j \le n$, respectively; see [33, Sect. 1] for details. $\diamond$

Remark 6.26. Note that by formula (6.11), we can easily prove that sequence $(C_j)_{j=0}^{n}$ is non-decreasing, since $C_{\sigma_j} - C_{\sigma_{j-1}} = a_{\sigma_j} + (1+b_{\sigma_j})C_{\sigma_{j-1}} - C_{\sigma_{j-1}} = a_{\sigma_j} + b_{\sigma_j}C_{\sigma_{j-1}} > 0$ for $j = 1, 2, \ldots, n$.

Remark 6.27. Formula (6.11) is an extension of formula (6.2) to the case when $a_{\sigma_j} \ne 0$ for $1 \le j \le n$.

Remark 6.28. Formula (6.11), in turn, is a special case of the formula

$$C_j(\sigma) = \sum_{i=1}^{j} a_{\sigma_i} \prod_{k=i+1}^{j} (1 + b_{\sigma_k}) + t_0 \prod_{i=1}^{j}(1 + b_{\sigma_i}), \qquad (6.13)$$

which describes $C_j(\sigma)$ in the case when $t_0 \ne 0$. Some authors give special cases of (6.13); see, e.g., Zhao et al. [305, Lemma 2].

By Theorem 6.24, we can construct the following scheduling algorithm.

Algorithm A_3
for the problem $1|p_j = a_j + b_j t|C_{\max}$ ([33, 128, 129, 264, 277])

Input: sequence $((a_1, b_1), (a_2, b_2), \ldots, (a_n, b_n))$
Output: an optimal schedule

Schedule jobs in the non-increasing order of $\frac{b_j}{a_j}$ ratios.

Remark 6.29. Note that algorithm A_3 is equivalent to sorting a sequence of numbers. For simplicity of further presentation, for algorithms similar to A_3, which for a given input number sequence (x_j) schedule jobs according to an order of a number sequence (y_j), we will use the notation $Q : (x_j) \mapsto (y_j)$, where Q is one of symbols introduced in Remark 2.22. In the notation, algorithm A_3 can be denoted as '$A_3 : (a_j|b_j) \mapsto (\frac{b_j}{a_j} \searrow)$'.

We can, however, consider a more general case than (6.10). The next algorithm, proposed by Gawiejnowicz and Pankowska [109], additionally covers the case when in (6.10) for $1 \leq j \leq n$, we have either $a_j = 0$ or $b_j = 0$.

Algorithm A_4 for the problem $1|p_j = a_j + b_j t|C_{\max}$ ([109])

Input: sequences $(a_1, a_2, \ldots, a_n), (b_1, b_2, \ldots, b_n)$
Output: an optimal schedule

▷ **Step 1:**
 Schedule jobs with $b_j = 0$ in an arbitrary order;
▷ **Step 2:**
 Schedule jobs with $a_j, b_j > 0$ in the non-increasing order of $\frac{b_j}{a_j}$ ratios;
▷ **Step 3:**
 Schedule jobs with $a_j = 0$ in an arbitrary order.

There exists yet another algorithm for the problem $1|p_j = a_j + b_j t|C_{\max}$, also proposed by Gawiejnowicz and Pankowska [108]. The algorithm is based on equivalence (6.12) and hence it does not use division operation. This is important in the case when some a_j are very small numbers, since then the calculation of the quotients $\frac{b_j}{a_j}$ may lead to numerical errors.

The algorithm uses two matrices, A and B. Matrix A contains all products in the form of $a_i * b_j$, $1 \leq i, j \leq n$. Matrix B is a $\{0, 1\}$-matrix in which $B[i, j] = 0$ if $A[i, j] = a_i * b_j < a_j * b_i = A[j, i]$ and $B[i, j] = 1$ otherwise.

Though the algorithm runs in $O(n^2)$ time, it needs only $O(k \log k)$ time in the case of adding k new jobs to the set $\mathcal{J}$, while algorithms A_3 and A_4 need $O((n + k) \log(n + k))$ time. (This is caused by the fact that we do not need to fill the whole matrices A and B again but only their new parts.) The pseudo-code of the algorithm can be formulated as follows.

Algorithm A_5 for the problem $1|p_j = a_j + b_j t|C_{\max}$ ([108])

Input: sequences $(a_1, a_2, \ldots, a_n), (b_1, b_2, \ldots, b_n)$
Output: an optimal schedule σ^*

▷ **Step 1:**
 for $i \leftarrow 1$ **to** n **do**
  for $j \leftarrow 1$ **to** n **do**
   $A_{i,j} \leftarrow a_i * b_j;$

▷ Step 2:
  for $i \leftarrow 1$ to n do
    for $j \leftarrow 1$ to n do
      if $(i \neq j)$ then $B_{i,j} \leftarrow 0$
      else $B_{i,j} \leftarrow 1$;
▷ Step 3:
  for $i \leftarrow 1$ to n do
    for $j \leftarrow 1$ to n do
      if $(A_{i,j} \neq A_{j,i})$ then
        if $(A_{i,j} < A_{j,i})$ then $B_{j,i} \leftarrow 0$
        else $B_{i,j} \leftarrow 1$;
▷ Step 4:
  for $i \leftarrow 1$ to n do
    $\sigma_i^\star \leftarrow 0$;
    for $j \leftarrow 1$ to n do
      $\sigma_i^\star \leftarrow \sigma_i^\star + B_{i,j}$.

Example 6.30. Let jobs J_1, J_2, J_3 have the processing times in the form of $p_1 = 1 + 2t$, $p_2 = 2 + t$, $p_3 = 5$. Then

$$A = \begin{bmatrix} 2 & 1 & 0 \\ 4 & 2 & 0 \\ 10 & 5 & 0 \end{bmatrix} \quad \text{and} \quad B = \begin{bmatrix} 1 & 0 & 0 \\ 1 & 1 & 0 \\ 1 & 1 & 1 \end{bmatrix}.$$

Hence, $\sigma^\star = (1, 2, 3)$ is an optimal schedule, with $C_{\max}(\sigma^\star) = 9$. The schedule is depicted in Fig. 6.3. ◆

Fig. 6.3: Gantt chart for schedule $\sigma^\star$ in Example 6.30

Example 6.31. Let jobs J_1, J_2, J_3 now have the processing times $p_1 = 1 + 2t$, $p_2 = 2 + t$, $p_3 = 4 + 2t$. In this case,

$$A = \begin{bmatrix} 2 & 1 & 2 \\ 4 & 2 & 4 \\ 8 & 4 & 8 \end{bmatrix} \quad \text{and} \quad B = \begin{bmatrix} 1 & 0 & 0 \\ 1 & 1 & 0 \\ 1 & 0 & 1 \end{bmatrix}.$$

The sums of the second and the third rows of B are the same and equal to 2. This means that in the optimal schedule, the order of jobs J_2 and J_3 is immaterial. Therefore, there are two optimal schedules, $\sigma^{\star'} = (1, 2, 3)$ and $\sigma^{\star''} = (1, 3, 2)$, both of which have the schedule length equal to 16. The schedules are depicted in Fig. 6.4a and Fig. 6.4b, respectively.   ◆

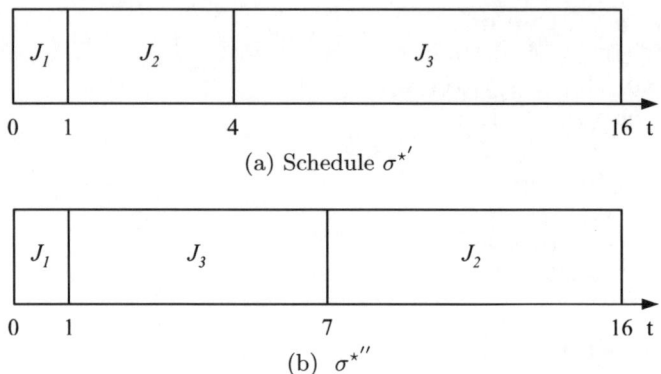

Fig. 6.4: Optimal schedules in Example 6.31

Remark 6.32. Algorithm A_5 is a time-dependent scheduling algorithm that uses matrices. Other examples of applications of matrices in time-dependent scheduling will be given in Chap. 12.

Distinct ready times and deadlines

Now we pass to single machine time-dependent scheduling problems with linear job processing times (6.10), distinct ready times and distinct deadlines.

The general problem, with an arbitrary number of distinct ready times and distinct deadlines, is computationally intractable.

Theorem 6.33. *The decision version of the problem $1|p_j = a_j + b_j t, r_j, d_j|C_{\max}$ is $\mathcal{NP}$-complete in the strong sense.*

Proof. It is sufficient to note that assuming $a_j = 0$ for $1 \leq j \leq n$, we obtain the problem $1|p_j = b_j t, r_j, d_j|C_{\max}$. The decision version of the problem, by Theorem 6.19, is $\mathcal{NP}$-complete in the strong sense.

Alternatively, assuming $b_j = 0$ for $1 \leq j \leq n$, we obtain the problem $1|r_j, d_j|C_{\max}$. The decision version of the problem is $\mathcal{NP}$-complete in the strong sense as well (cf. Lenstra et al. [197]).   ■

The problem remains computationally intractable if all ready times are equal to zero, i.e., if $r_j = 0$ for $1 \leq j \leq n$.

Theorem 6.34. (Cheng and Ding [52]) *The decision version of the problem* $1|p_j = a_j + b_j t, d_j|C_{\max}$ *is* $\mathcal{NP}$-*complete in the strong sense.*

Proof. The transformation from the 3-P problem (cf. Sect. 3.2) is as follows. Let v denote an integer larger than $2^8 h^3 K^3$. Define $n = 4h$ jobs with $a_j = vc_j$ and $b_j = \frac{c_j}{v}$ for $1 \le j \le 3h$, $a_{3h+i} = v$ and $b_{3h+i} = 0$ for $1 \le i \le h$. Deadlines $d_j = D = vh(K+1) + \sum_{i=1}^{n} \sum_{j=1}^{i-1} c_i c_j + \frac{1}{2}h(h-1)K + 1$ for $1 \le j \le 3h$ and $d_{3h+i} = iv + (i-1)(vK + 4hK^2)$ for $1 \le i \le h$. The threshold $G = D$.

To complete the proof, it is sufficient to show that the 3-P problem has a solution if and only if for the instance of the problem $1|p_j = a_j + b_j t, d_j|C_{\max}$ there exists a schedule σ such that $C_{\max}(\sigma) \le G$; see [52, Lemmata 6–7].   □

If we simplify the problem, some cases are polynomially solvable.

Theorem 6.35. (Cheng and Ding [53]) *The problem* $1|p_j = a + b_j t, d_j,$ $b_j \in \{B_1, B_2\}|C_{\max}$ *is solvable in* $O(n \log n)$ *time by a version of algorithm* A_{14}.

Proof. Similar to the proof of Theorem 6.92.   □

The problem $1|p_j = a_j + b_j t, r_j, d_j|C_{\max}$ still remains computationally intractable even if all basic processing times and all ready times are equal, i.e., if $a_j = 1$ and $r_j = 0$ for $1 \le j \le n$.

Theorem 6.36. (Cheng and Ding [53]) *The problem of whether there exists a feasible schedule for the problem* $1|p_j = 1 + b_j t, d_j|C_{\max}$ *is* $\mathcal{NP}$-*complete in the strong sense.*

Proof. Given an instance of the 3-P problem (cf. Sect. 3.2), construct an instance of the problem $1|p_j = 1 + b_j t, d_j|C_{\max}$ as follows.

The set of jobs $J = V \cup R \cup Q_1 \cup \ldots \cup Q_{m-1}$, where $q = 32h^2 K$, $v = 16h^2 qK$, $V = \{J_{0,1}, J_{0,2}, \ldots, J_{0,v}\}$, $R = \{J_1, J_2, \ldots, J_{3h}\}$ and $Q_i = \{J_{i,1}, J_{i,2}, \ldots, J_{i,q}\}$ for $1 \le i \le h-1$. Define $n = v + 3h + (h-1)q$, $E = 4hnK$ and $A = 32n^3 E^2$.

The job deterioration rates and deadlines are the following: $b_{0,i} = 0$ and $d_{0,i} = v$ for $1 \le i \le v$, $b_{i,j} = 0$ and $d_{i,j} = D_i$ for $1 \le i \le h-1$ and $1 \le j \le q$, $b_i = \frac{E+c_i}{A}$ and $d_i = G$ for $1 \le i \le 3h$, where the threshold

$$G = n + \sum_{k=0}^{h-1} \frac{3E(v + qk + 3k + 1)}{A} + \sum_{k=0}^{h-1} \frac{K(v + qk + 3k + 1)}{A} + \frac{2hK}{A}$$

and the constants $D_i = v + qi + 3i + \sum_{k=0}^{i-1} \frac{3E(v+qk+3k+1)}{A} + \sum_{k=0}^{i-1} \frac{K(v+qk+3k+1)}{A} + \frac{2hK}{A}$ for $1 \le i \le h-1$.

By showing that the 3-P problem has a solution if and only if for the above instance of the problem $1|p_j = 1 + b_j t, d_j|C_{\max}$, there exists a feasible schedule σ such that $C_{\max}(\sigma) \le G$, we obtain the result.   □

The restricted version of the problem $1|p_j = 1 + b_j t, d_j|C_{\max}$, with only two distinct deadlines, is also computationally intractable.

Theorem 6.37. (Cheng and Ding [53]) *The problem whether there exists a feasible schedule for the problem* $1|p_j = 1 + b_j t, d_j \in \{D_1, D_2\}|C_{\max}$ *is* $\mathcal{NP}$-*complete in the ordinary sense.*

Proof. We use the following transformation from the PP problem (cf. Sect. 3.2). Define $n = (k+1)(k+2)$, $E = n^2 2^{2k} k^{2k} A$ and $B = 16n^3 E^2$. Job deterioration rates are as follows: $b_{0,0} = b_{0,1} = \frac{2E}{B}$, $b_{0,j} = 0$ for $2 \leq j \leq k+1$, $b_{i,0} = \frac{E + 2^{2k-2i+2}k^{2k-2i+2}A + x_j}{(i+1)B}$, $b_{i,j} = \frac{x_i}{(i+1)A}$ for $1 \leq i \leq k$ and $1 \leq j \leq k+1$.

The deadlines are the following: $d_{0,j} = D_1$ and $d_{i,j} = D_2$ for $1 \leq i \leq k$ and $0 \leq j \leq k+1$, where $D_1 = 2k + 2 + \frac{4E - 2A + 1}{2B} + \sum_{i=1}^{k}(i+1)b_{i,0}$ and $D_2 = n + \frac{4E + 2kA + 1}{2B} + \sum_{i=1}^{k}(i+1)b_{i,0} \sum_{i=1}^{k} \sum_{j=0}^{k} ((i+1)(k+1) + j) b_{i,j+1}$. The threshold $G = D_2$.

To complete the proof, it is sufficient to construct a schedule for the above instance of the problem $1|p_j = 1 + b_j t, d_j \in \{D_1, D_2\}|C_{\max}$ and to show that the PP problem has a solution if and only if this schedule is feasible.   □

Simplifying the problem $1|p_j = a_j + b_j t, d_j|C_{\max}$ further, we can obtain polynomially solvable cases. Let $b_j = b$ for $1 \leq j \leq n$, i.e.,

$$p_j = a_j + bt, \tag{6.14}$$

where $b > 0$ and $a_j > 0$ for $1 \leq j \leq n$. This problem was considered for the first time by Cheng and Ding [52]. The authors proposed the following algorithm for job processing times given by (6.14) and for distinct deadlines. For a given $\sigma^i \in \tilde{\mathfrak{S}}_n$, let $\mathcal{J}(\sigma^i)$ denote the set of jobs with indices from σ^i.

Algorithm A_6 for the problem $1|p_j = a_j + bt, d_j|C_{\max}$ ([52])

Input: sequences $(a_1, a_2, \ldots, a_n)$, $(d_1, d_2, \ldots, d_n)$, number b
Output: an optimal schedule

▷ Step 1:
  Arrange jobs in the non-decreasing order of a_j values;
  $\sigma^1 \leftarrow ([1], [2], \ldots, [n])$;
  $\sigma^2 \leftarrow (\phi)$;
  $C_{[0]} \leftarrow 0$;
  $C \leftarrow 0$;
▷ Step 2:
  for $i \leftarrow 1$ **to** n **do**
    $C_{[i]} \leftarrow (1+b)C_{[i-1]} + a_{[i]}$;
▷ Step 3:
  while $(C \neq C_{[n]})$ **do**
    $s \leftarrow C_{[n]}$;
    for $i \leftarrow n$ **downto** 1 **do**
      $t \leftarrow \max\{C_{[i]}, s\}$;
      Find job $J_{(i)} \in \mathcal{J}(\sigma^1)$ with maximal $a_{(i)}$ and $d_{(i)} \geq t$;

if (there exists no such $J_{(i)}$) **then**
    write 'There exists no feasible schedule';
    stop
    else $s \leftarrow \frac{t - a_{(i)}}{1+b}$;
        $\sigma^1 \leftarrow \sigma^1 \setminus \{(i)\}$;
        $\sigma^2 \leftarrow \sigma^2 \cup \{(i)\}$;
$C \leftarrow C_{[n]}$;
$\sigma^1 \leftarrow \sigma^2$;
for $i \leftarrow 1$ **to** n **do**
    $C_{[i]} \leftarrow (1+b)C_{[i-1]} + a_{[i]}$.

Theorem 6.38. (Cheng and Ding [52]) *The problem* $1|p_j = a_j + bt, d_j|C_{\max}$ *is solvable in* $O(n^5)$ *time by algorithm* A_6.

Proof. The optimality of the schedule generated by algorithm A_6 follows from some results concerning a special form of schedules for the problem, which are very similar to *canonical* schedules (cf. Definition 6.89) considered in Sect. 6.1.7. We refer the reader to [53, Sect. 3] for more details. ◇

The time complexity of algorithm A_6 can be established as follows. Step 1 and Step 2 can be completed in $O(n \log n)$ and $O(n)$ time, respectively. The 'while' loop in Step 3 needs $O(n^2)$ time, while the loop '**for**' in this step needs $O(n^2)$ time. Finding job $J_{(i)} \in \mathcal{J}(\sigma^1)$ with maximal $a_{(i)}$ and $d_{(i)} \geq t$ needs $O(n)$ time. Therefore, the overall time complexity of A_6 is $O(n^5)$. □

The problems of scheduling linearly deteriorating jobs with non-zero ready times are considered in Sect. 6.1.7, since they are closely related to problems of scheduling with linearly shortening job processing times.

6.1.4 Simple non-linear deterioration

Equal ready times and deadlines

Now, we pass to more general forms of job deterioration than the linear one. The first result from the area concerns the case of *simple general non-linear deterioration*, when job processing times are in the form of

$$p_j = a_j + f(t), \tag{6.15}$$

where $a_j > 0$ for $1 \leq j \leq n$ and $f(t)$ is an arbitrary function such that

$$f(t) \geq 0 \quad \text{for} \quad t \geq 0. \tag{6.16}$$

This form of job deterioration was introduced by Melnikov and Shafransky[207].

From the point of view of applications, the most interesting case is when the function $f(t)$ is a non-decreasing function, i.e., when

$$\text{if } t_1 \leq t_2, \text{ then } f(t_1) \leq f(t_2). \tag{6.17}$$

In such a case, there holds the following result.

Theorem 6.39. (Melnikov and Shafransky [207]) *If $f(t)$ is an arbitrary function satisfying conditions* (6.16)–(6.17), *then the problem* $1|p_j = a_j + f(t)|C_{\max}$ *is solvable in* $O(n \log n)$ *time by scheduling jobs in the non-decreasing order of a_j values.*

Proof. We apply the pairwise job interchange argument. Let the schedule σ',

$$\sigma' = (\sigma_1, \ldots, \sigma_{i-1}, \sigma_i, \sigma_{i+1}, \sigma_{i+2}, \ldots, \sigma_n),$$

start from time $t_0 > 0$ and let two jobs, J_{σ_i} and $J_{\sigma_{i+1}}$, be such that $a_{\sigma_i} \geq a_{\sigma_{i+1}}$.
    Consider now schedule σ'',

$$\sigma'' = (\sigma_1, \ldots, \sigma_{i-1}, \sigma_{i+1}, \sigma_i, \sigma_{i+2}, \ldots, \sigma_n),$$

differing from σ' only in the order of jobs J_{σ_i} and $J_{\sigma_{i+1}}$. We will show that $C_{\max}(\sigma'') \leq C_{\max}(\sigma')$.
    Note that since the first $i-1$ jobs in schedule σ' are the same as the first $i-1$ jobs in schedule σ'', job J_{σ_i} in schedule σ' and job $J_{\sigma_{i+1}}$ in schedule σ'' start at the same time t_i. Calculate the completion times $C_{\sigma_i}(\sigma')$ and $C_{\sigma_{i+1}}(\sigma'')$. We then have $C_{\sigma_i}(\sigma') = t_i + p_{\sigma_i}(t_i) = t_i + a_{\sigma_i} + f(t_i)$ and $C_{\sigma_{i+1}}(\sigma') = C_{\sigma_i}(\sigma') + p_{\sigma_{i+1}}(C_{\sigma_i}(\sigma')) = t_i + a_{\sigma_i} + f(t_i) + a_{\sigma_{i+1}} + f(t_i + a_{\sigma_i} + f(t_i))$.
    Next, we have $C_{\sigma_{i+1}}(\sigma'') = t_i + p_{\sigma_{i+1}}(t_i) = t_i + a_{\sigma_{i+1}} + f(t_i)$ and $C_{\sigma_i}(\sigma'') = C_{\sigma_{i+1}}(\sigma'') + p_{\sigma_i}(C_{\sigma_{i+1}}(\sigma'')) = t_i + a_{\sigma_{i+1}} + f(t_i) + a_{\sigma_i} + f(t_i + a_{\sigma_{i+1}} + f(t_i))$. Therefore,

$$C_{\sigma_i}(\sigma'') - C_{\sigma_{i+1}}(\sigma') = f(t_i + a_{\sigma_{i+1}} + f(t_i)) - f(t_i + a_{\sigma_i} + f(t_i)) \leq 0,$$

since $f(t)$ is an increasing function and $a_{\sigma_i} \geq a_{\sigma_{i+1}}$ by assumption. From that it follows that $C_{\max}(\sigma'') - C_{\max}(\sigma') \leq 0$, since the $C_{\max}$ criterion is regular. Repeating, if necessary, the above mutual exchange for other pairs of jobs, we will obtain an optimal schedule in which all jobs are scheduled in the non-decreasing order of a_j values. ∎

By Theorem 6.39, the problem $1|p_j = a_j + f(t), f \nearrow |C_{\max}$ is solved by the algorithm $A_7 : (a_j) \mapsto (a_j \nearrow)$.

Remark 6.40. Kuo and Yang [179, Proposition 1] considered a special case of Theorem 6.39, with $f(t) := \sum_{i=1}^{m} \lambda_i t^{r_i}$ and $r_i \in [0, +\infty)$ for $1 \leq i \leq n$.

If we assume that $f(t)$ is an arbitrary non-increasing function, i.e.,

$$\text{if } t_1 \leq t_2, \text{ then } f(t_1) \geq f(t_2), \tag{6.18}$$

$f(t)$ is differentiable and its first derivative is bounded,

$$\left| \frac{df}{dt} \right| \leq 1, \tag{6.19}$$

then the following result holds.

Theorem 6.41. (Melnikov and Shafransky [207]) *If $f(t)$ is an arbitrary differentiable function satisfying conditions* (6.16), (6.18) *and* (6.19), *then the problem* $1|p_j = a_j + f(t)|C_{\max}$ *is solvable in $O(n \log n)$ time by scheduling jobs in the non-increasing order of a_j values.*

Proof. Since there holds condition (6.19), for any $t \geq 0$ there holds inequality $|f(t + \Delta t) - f(t)| \leq \Delta t$. Repeating the reasoning from the proof of Theorem 6.39, we obtain the result. $\square$

By Theorem 6.41, the problem $1|p_j = a_j + f(t), f \searrow, |\frac{df}{dt}| \leq 1|C_{\max}$ is solved by the algorithm $A_8 : (a_j) \mapsto (a_j \searrow)$.

Remark 6.42. Kuo and Yang [179, Proposition 2] considered a special case of Theorem 6.41, with $f(t) := \sum_{i=1}^{m} \lambda_i t^{r_i}$ and $r_i \in (-\infty, 0]$ for $1 \leq i \leq n$.

Another form of simple general non-linear deterioration is the one in which job processing times proportionally deteriorate according to a certain function,

$$p_j = b_j h(t), \tag{6.20}$$

where $h(t)$ is a convex (concave) function for $t \geq t_0$. (Convex and concave functions were introduced in Definition 1.28 and Remark 1.30, respectively.) This form of job deterioration was introduced by Kononov [173].

Theorem 6.43. (Kononov [173]) *If $h(t)$ is a convex (concave) function for $t \geq 0$ and there hold conditions*

$$h(t_0) > 0 \tag{6.21}$$

and

$$t_1 + b_j h(t_1) \leq t_2 + b_j(t_2) \text{ for all } t_2 > t_1 \geq t_0 \text{ and all } J_j \in \mathcal{J}, \tag{6.22}$$

then the problem $1|p_j = b_j h(t)|C_{\max}$ is solvable in $O(n \log n)$ time by scheduling jobs in the non-decreasing (non-increasing) order of b_j values.

Proof. The main idea is to prove that the criterion $C_{\max}$ is a 1-priority-generating function with priority function $\omega_i = -b_i$ ($\omega_i = b_i$). Then, by Theorem 1.24, the result follows. $\square$

Remark 6.44. The 1-priority-generating functions and priority functions were introduced in Definition 1.19.

By Theorem 6.43, if $h(t)$ is a convex function and there hold conditions (6.21) and (6.22), the problem $1|p_j = b_j h(t)|C_{\max}$ is solved by the algorithm $A_9 : (b_j) \mapsto (b_j \nearrow)$.

Remark 6.45. Kuo and Yang [179, Propositions 3–4] considered special cases of Theorem 6.43, with $f(t) := 1 + \sum_{i=1}^{m} \lambda_i t^{r_i}$ and $r_i \in [1, +\infty)$ or $r_i \in (0, 1)$ for $1 \leq i \leq n$.

Now we pass to the problems with general non-linear processing times.

6.1.5 General non-linear deterioration

In this subsection, we consider the forms of non-linear deterioration in which distinct job processing times are described by distinct functions.

In this case, job processing times are in the form of

$$p_j = g_j(t), \tag{6.23}$$

where $g_j(t)$ are arbitrary non-negative functions for $1 \le j \le n$. This type of job deterioration was introduced by Gupta and Gupta [128].

Equal ready times and deadlines

Gupta and Gupta [128] introduced the *polynomial* job deterioration in which the processing times of jobs are in the form of

$$p_j = a_{0,j} + a_{1,j}t + a_{2,j}t^2 + \ldots + a_{m,j}t^m = \sum_{i=0}^{m} a_{i,j}t^i, \tag{6.24}$$

where $a_{i,j}$ are positive constants for $0 \le i \le m$, $1 \le j \le n$ and integer $m \ge 1$.

Theorem 6.46. (Alidaee [4], Gupta and Gupta [128]) *Let $\sigma^{(1)}$ and $\sigma^{(2)}$ be two different partial schedules for the problem $1|p_j = \sum_{i=0}^{m} a_{i,j}t^i|C_{\max}$, in which the same subset of set $\mathcal{J}$ is scheduled and let $\tau^{(r)}$ be a schedule of the set of remaining jobs. If $C_{\max}(\sigma^{(1)}) \le C_{\max}(\sigma^{(2)})$, then $C_{\max}(\sigma^{(1)}, \tau^{(r)}) \le C_{\max}(\sigma^{(2)}, \tau^{(r)})$.*

Proof. Alidaee [4] gives the proof by pairwise job interchange argument. Gupta and Gupta [128] state the result without a proof.   ⋄

By Theorem 6.46, Gupta and Gupta [128] proposed an exact algorithm for the problem $1|p_j = \sum_{i=0}^{m} a_{i,j}t^i|C_{\max}$. The authors also proposed two heuristics and reported the results of their experimental analysis. We will consider these heuristics in Chap. 9.

Alidaee [4] considered the problem with general processing times in the case when $g_j(t), 1 \le j \le n$, are differentiable and non-decreasing functions.

Theorem 6.47. (Alidaee [4]) *Let σ' be a sequence of jobs, $\sigma'' = \sigma'(i \leftrightarrow i+1)$ be the sequence σ' with mutually exchanged positions i and $i + 1$, and let T denote the maximum completion time for the first $i - 1$ jobs in the sequence. Then there exist real numbers $z_{j,i} \in [T, T + p_{j,i+1}(T)]$ and $z_{j,i+1} \in [T, T + p_{j,i}(T)]$ such that if there holds inequality*

$$\frac{g_{j,i}(z_{j,i})}{g_{j,i}(T)} \ge \frac{g_{j,i+1}(z_{j,i+1})}{g_{j,i+1}(T)}, \tag{6.25}$$

then $C_{\max}(\sigma') \le C_{\max}(\sigma'')$.

Proof. Consider sequences of jobs σ' and σ'', where $\sigma'' = \sigma'(i \leftrightarrow i+1)$. Let T denote the maximum completion time for the first $i-1$ jobs in a sequence and let $\Delta(\sigma'', \sigma') = C_{\max}(\sigma'') - C_{\max}(\sigma')$ be the difference between criterion values for schedules σ'' and σ'. Then, if there holds inequality

$$g_{j,i}(T + g_{j,i+1}(T)) - g_{j,i}(T) \geq g_{j,i+1}(T + g_{j,i}(T)) - g_{j,i+1}(T),$$

then $\Delta(\sigma'', \sigma') \geq 0$.

By Theorem 1.15 (b), there exist real numbers $z_{j,i}$ and $z_{j,i+1}$ such that $z_{j,i} \in [T, T+g_{j,i+1}(T)]$, $z_{j,i+1} \in [T, T+g_{j,i}(T)]$ and such that (6.25) is satisfied. $\square$

For the problem $1|p_j = a_j + b_j t + \ldots + m_j t^m|C_{\max}$, Alidaee [4] also proposed a heuristic algorithm, constructed on the basis of Theorem 6.47. We will consider this algorithm in Chap. 9.

A special case of non-linear job deterioration is an exponential deterioration. The *exponential* job deterioration can have a few distinct forms and one of them is the following one:

$$p_j = e^{b_j t}, \tag{6.26}$$

where $0 < b_j < 1$ for $1 \leq j \leq n$. This form of job deterioration was considered for the first time by Alidaee [4] as an example of general non-linear deterioration (6.23). Hsieh [140] proposed a heuristic algorithm for a single-machine time-dependent scheduling problem with exponentially deteriorating processing times (6.26). We will consider this algorithm in Chap. 9.

Distinct ready times and deadlines

We start this subsection with *step* deterioration, in which job processing times are described by step functions.

Mosheiov [217] introduced *step* deteriorating job processing times

$$p_j = \begin{cases} a_j, & \text{if } t \leq d_j, \\ b_j, & \text{if } t > d_j, \end{cases} \tag{6.27}$$

where

$$b_j \geq a_j \text{ for } 1 \leq j \leq n. \tag{6.28}$$

Remark 6.48. Without loss of generality, we can also assume that in this case $d_1 \leq d_2 \leq \ldots \leq d_n$.

Remark 6.49. Non-linear job processing times (6.27) will be denoted in short as $p_j \in \{a_j, b_j : a_j \leq b_j\}$.

Theorem 6.50. (Cheng and Ding [51], Mosheiov [217]) *If there hold inequalities (6.28), the decision version of the problem $1|p_j \in \{a_j, b_j : a_j \leq b_j\}|C_{\max}$ is $\mathcal{NP}$-complete in the ordinary sense, even if $d_j = D$ for $1 \leq j \leq n$.*

Proof. Mosheiov [217] transformed an integer programming formulation of the considered problem to the KP problem (cf. Sect. 3.2) in the following way. Introduce 0-1 variables x_j defined as follows: $x_j = 1$ if job J_j starts not later than time $t = d_j$ and $x_j = 0$ otherwise. Then, the maximum completion time for a given schedule is equal to

$$\sum_{j=1}^{n} x_j a_j + \sum_{j=1}^{n} (1 - x_j) b_j = \sum_{j=1}^{n} b_j - \sum_{j=1}^{n} x_j (b_j - a_j). \qquad (6.29)$$

Since the value $\sum_{j=1}^{n} b_j$ is a constant in (6.29), the problem of minimizing the maximum completion time is equal to the problem of maximizing the sum $\sum_{j=1}^{n} x_j(b_j - a_j)$. Therefore, we can reformulate the problem of minimizing the $C_{\max}$ criterion to the following problem (P^1):

$$\max \sum_{j=1}^{n} x_j(b_j - a_j)$$

subject to

$$\sum_{j=1}^{i-1} x_j a_j \leq d_i + (1 - x_i)L, \quad i = 1, 2, \ldots, n, \qquad (6.30)$$

$$x_i \in \{0, 1\} \quad \text{for} \quad i = 1, 2, \ldots, n,$$

where $L \geq \max\{d_n, \sum_{j=1}^{n} a_j\}$ is sufficiently large.

Consider a new problem (P^2) obtained by ignoring constraints (6.30) for $i = 1, 2, \ldots, n - 1$. The problem (P^2) is as follows:

$$\max \sum_{j=1}^{n} x_j u_j$$

subject to

$$\sum_{j=1}^{i-1} x_j v_j \leq D,$$

$$x_i \in \{0, 1\} \quad \text{for} \quad i = 1, 2, \ldots, n,$$

where $u_j = b_j - a_j$ for $1 \leq j \leq n$, $c_j = a_j$ for $1 \leq j \leq n - 1$, and $D = d_n + L$.

Since the problem (P^2) is equivalent to the KP problem, the result follows.

Cheng and Ding [51] used the following transformation from the PP problem (cf. Sect. 3.2). Let a_0 and b_0 be two numbers larger than A. Define $n = k$, $a_j = b_j = x_j$ for $1 \leq j \leq k$, $D = A$ and $G = a_0 + 3A$.

If the PP problem has a solution, then there exist disjoint subsets of X, X_1 and X_1, such that $X_1 \cup X_2 = X$ and $\sum_{x_i \in X_1} x_i = \sum_{x_i \in X_2} x_i = A$. Construct schedule σ in which jobs corresponding to elements of X_1 are scheduled first, the job corresponding to a_0 and b_0 is scheduled next and is followed by jobs corresponding to elements of X_2 (see Fig. 6.5 and Remark 6.12). Then, we have $C_{\max}(\sigma) = \sum_{x_j \in X_1} a_j + a_0 + \sum_{x_j \in X_2}(a_j + b_j) = G$. Hence, the considered problem has a solution.

Assume that there exists a schedule $\sigma := (R_1, J_0, R_2)$ for the problem $1|p_j \in \{a_j, b_j : a_j \le b_j\}, d_j = D|C_{\max}$ such that $C_{\max} \le G$ and R_1 (R_2) is the set of jobs scheduled before (after) job J_0. By contradiction one can show that neither $C_{\max}(R_1) > A$ nor $C_{\max}(R_1) < A$. Therefore, it must be $C_{\max}(R_1) = A$. By the selection of elements of X, which correspond to jobs of R_1, we obtain a solution of the PP problem.   □

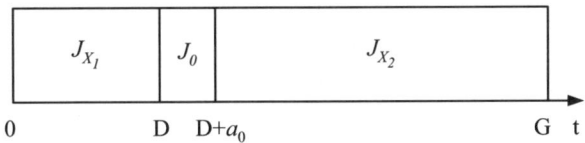

Fig. 6.5: Example schedule in the proof of Theorem 6.50

For the problem $1|p_j \in \{a_j, b_j : a_j \le b_j\}|C_{\max}$, Mosheiov [217] proposed a heuristic algorithm. We will consider this algorithm in Chap. 9.

For the case when $d_j = D$ for $1 \le j \le n$, Cheng and Ding [51] proposed to apply the enumerative algorithm E_2. (Since the algorithm was proposed for the total completion time criterion, it is presented in Sect. 6.2.5; now, we describe only the general idea behind it.) Roughly speaking, this algorithm divides the set of jobs $\mathcal{J}$ into a number of chains, using the approach applied for the problems $1|p_j \in \{a_j, b_j : a_j \le b_j\}, d_j = D|\sum C_j$ (cf. Sect. 6.2.5) and $1|p_j \in \{a_j, b_j : a_j \le b_j\}, d_j = D|\sum w_j C_j$ (cf. Sect. 6.4.5). The modified algorithm E_2 will be denoted by E_3.

Theorem 6.51. (Cheng and Ding [51]) *If an instance of the problem* $1|p_j \in \{a_j, b_j : a_j \le b_j\}, d_j = D|C_{\max}$ *has a fixed number of chains, then algorithm E_3 is a polynomial-time algorithm for the instance.*

Proof. Similar to the proof of Theorem 6.154.   □

Remark 6.52. Since in the problem $1|p_j \in \{a_j, b_j : a_j \le b_j\}, d_j = D|C_{\max}$ jobs are independent, the chains from Theorem 6.51 are not related to job precedence constraints. The problems of time-dependent scheduling with dependent jobs will be considered in Chap. 13.

For the general problem, $1|p_j \in \{a_j, b_j : a_j \le b_j\}, d_j|C_{\max}$, Jeng and Lin [151] proposed a pseudopolynomial-time exact algorithm, based on dynamic programming.

Lemma 6.53. (Jeng and Lin [151]) *There exists an optimal schedule for the problem $1|p_j \in \{a_j, b_j : a_j \leq b_j\}, d_j|C_{\max}$, in which early jobs precede tardy jobs.*

Proof. Assume that in an optimal schedule a tardy job precedes an early job. If we schedule the tardy job as the last one, we do not increase the $C_{\max}$ value, since the processing time of the tardy job will not change. Therefore, by repeating such a movement for other tardy jobs, we obtain an optimal schedule in which early jobs precede tardy jobs. □

Lemma 6.54. (Jeng and Lin [151]) *There exists an optimal schedule for the problem $1|p_j \in \{a_j, b_j : a_j \leq b_j\}, d_j|C_{\max}$, in which early jobs are scheduled in the non-decreasing order of $a_j + b_j$ values.*

Proof. By pairwise job interchange argument. □

The dynamic programming algorithm proposed by Jeng and Lin [151] exploits Lemmata 6.53–6.54 and is formulated as follows:

$$\textbf{Initial conditions}: \ F(j,t) := \begin{cases} 0, & \text{if } j = 0 \land t = 0, \\ +\infty, & \text{otherwise}; \end{cases} \tag{6.31}$$

Recursive formula for $1 \leq j \leq n, 0 \leq t \leq a_{\max} + d_{\max}$:

$$F(j,t) := \begin{cases} \min \left\{ \begin{array}{l} F(j-1, t-a_j) + a_j \\ F(j-1, t) + a_j + b_j \end{array} \right\}, & \text{if } t \geq a_j, d_j \geq t - a_j, \\ F(j-1, t) + a_j + b_j, & \text{otherwise}; \end{cases} \tag{6.32}$$

$$\textbf{Goal}: \ \min\{F(n,t) : 0 \leq t \leq a_{\max} + d_{\max}\}, \tag{6.33}$$

where $a_{\max} := \max_{1 \leq j \leq n}\{a_j\}$ and $d_{\max} := \max_{1 \leq j \leq n}\{d_j\}$.

The time complexity of the dynamic programming algorithm (6.31)–(6.33) is $O(n(a_{\max} + d_{\max}))$.

For the problem $1|p_j \in \{a_j, b_j : a_j \leq b_j\}, d_j|C_{\max}$, Jeng and Lin [151] also proposed a branch-and-bound algorithm. Computational experiments have shown that the algorithm is able to solve instances with $n = 80$ jobs in 10 minutes (see [151, Sect. 'Computational experiments']).

Mosheiov [217] extended the step deterioration (6.27), introducing *multi-step* deterioration. In this case, we have processing times in the form of

$$p_j^k = \begin{cases} a_j^k, & \text{if } d_j^{k-1} < S_j \leq d_j^k, \\ a_j^{m+1}, & \text{if } S_j > d_j^m, \end{cases} \tag{6.34}$$

where $d_j^0 = 0$,

$$a_j^1 < a_j^2 < \ldots < a_j^{m+1} \tag{6.35}$$

and

$$d_j^1 < d_j^2 < \ldots < d_j^{m+1} \tag{6.36}$$

for $1 \leq j \leq n$. For the problem, Mosheiov [217] proposed a heuristic algorithm. We will consider this algorithm in Chap. 9.

Kunnathur and Gupta [178] introduced *unbounded step-linear* processing times

$$p_j = \begin{cases} a_j, & \text{if } t \leq d_j, \\ a_j + b_j(t - d_j), & \text{if } t > d_j, \end{cases} \tag{6.37}$$

where $1 \leq j \leq n$. In this case, job processing times can grow up to infinity after some prespecified time has elapsed.

Remark 6.55. The equivalent form of job processing times (6.37) is $p_j = a_j + \max\{0, b_j(t - d_j)\}$ for all j. We will use both these forms interchangeably.

Kunnathur and Gupta proved a few properties of the single-machine time-dependent scheduling problem with job processing times given by (6.37).

Property 6.56. (Kunnathur and Gupta [178]) Let T be the earliest possible starting time for jobs in the set of unscheduled jobs, U, for the problem $1|p_j = a_j + \max\{0, b_j(t - d_j)\}|C_{\max}$. If $d_j < T$ for all $J_k \in U$, then scheduling jobs from U in the non-decreasing order of $\frac{a_j - b_j d_j}{b_j}$ ratios minimizes $C_{\max}(U)$.

Proof. By pairwise job interchange argument. □

Theorem 6.57. (Kunnathur and Gupta [178]) *The problem* $1|p_j = a_j + \max\{0, b(t - d)\}|C_{\max}$ *is solvable in* $O(n \log n)$ *time by scheduling jobs in the non-decreasing order of* a_j *values.*

Proof. The result is a corollary from Property 6.56 for $b_j = b$ and $d_j = d$ for $1 \leq j \leq n$. □

By Theorem 6.57, the problem $1|p_j = a_j + \max\{0, b(t - d)\}|C_{\max}$ is solved by the algorithm $A_7 : (a_j) \mapsto (a_j \nearrow)$.

Property 6.58. (Kunnathur and Gupta [178]) Let S be the set of scheduled jobs, $U := \mathcal{J} \setminus S$ and $T := C_{\max}(S)$. Then the maximum completion time $C_{\max}(S|U)$ for the problem $1|p_j = a_j + \max\{0, b_j(t - d_j)\}|C_{\max}$ is not less than $T + \sum_{J_j \in S}(a_j + \max\{0, b_j(T - d_j)\})$.

Proof. The result follows from the fact that for any U the schedule $(S|U)$ starts with subschedule S, and from the definition of job processing times (6.37). □

Property 6.59. (Kunnathur and Gupta [178]) Let S'' be the set of jobs scheduled after jobs from the set $S' := \mathcal{J} \setminus S''$. Then, the maximum completion time for the problem $1|p_j = a_j + \max\{0, b_j(t - d_j)\}|C_{\max}$ for sequences in the form of $(S'|S'')$ is not less than $\sum_{J_j \in S'} a_j + \sum_{J_j \in S''}(a_j + \max\{0, b_j(t - d_j)\})$.

Proof. Similar to the proof of Property 6.58. □

Property 6.60. (Kunnathur and Gupta [178]) Let S be the set of jobs for the problem $1|p_j = a_j + \max\{0, b_j(t - d_j)\}|C_{\max}$, scheduled in the non-decreasing order of $a_j + d_j$ values. If $S_j > d_j$ for any job $J_j \in S$, then there does not exist a schedule σ such that $C_{\max}(\sigma) = \sum_{J_j \in \mathcal{J}} a_j$.

Proof. Since $C_{\max}(\sigma) = \sum_{J_j \in \mathcal{J}} a_j$ holds only for the schedules in which no job is tardy, the result follows.   □

Property 6.61. (Kunnathur and Gupta [178]) Let S be the set of jobs for the problem $1|p_j = a_j + \max\{0, b_j(t - d_j)\}|C_{\max}$, scheduled in the non-decreasing order of $a_j + d_j$ values, let S_j be computed assuming that $b_j = 0$ for all $J_j \in S$ and let S' be the set S with job J_j such that $b_j > 0$. Then, $C_{\max}(S')$ is not less than $a_j + \max_{J_j \in S}\{0, (S_j - d_j)\} \min_{J_j \in \mathcal{J}}\{b_j\}$.

Proof. If $b_j = 0$ for all $J_j \in S$, then scheduling jobs in the non-decreasing order of d_j values minimizes the maximum tardiness. Since $\max_{J_j \in S}\{0, (S_j - d_j)\}$ equals the minimum value of the maximum tardiness and, simultaneously, equals the minimum value of the maximum delay in the starting time of any job in the case when $b_j > 0$ for a job $J_j \in \mathcal{J}$, the result follows.   □

Remark 6.62. From now on, if jobs are scheduled in the non-decreasing order of d_j values, we will say that the jobs are in the *EDD order*.

The time complexity of the problem was established by Kononov.

Theorem 6.63. (Kononov [171]) *The decision version of the problem* $1|p_j = a_j + \max\{0, b_j(t - d_j)\}|C_{\max}$ *is*
(a) $\mathcal{NP}$-*complete in the strong sense, if deterioration rates* b_j *are arbitrary,*
(b) $\mathcal{NP}$-*complete in the ordinary sense, if* $b_j = B$ *for* $1 \le j \le n$ *and*
(c) $\mathcal{NP}$-*complete in the ordinary sense, if* $d_j = D$ *for* $1 \le j \le n$.

Proof. (a) The idea is to use a transformation from the strongly $\mathcal{NP}$-hard problem $1|| \sum w_j T_j$ and to show that given an input for this problem and an arbitrary $\epsilon > 0$ one can construct such an input for the problem $1|p_j = a_j + \max\{0, b_j(t - d_j)\}|C_{\max}$ that the solving of the first problem reduces to the solving of the second one, and that an optimal schedule for the second problem is an optimal schedule for the first one for sufficiently small ϵ; see [171, Theorem 1].

(b) By applying a similar transformation from the ordinary $\mathcal{NP}$-hard problem $1|| \sum T_j$, the result follows; see [171, Theorem 2].

(c) Given an instance of the SS problem, (cf. Sect. 3.2), define $u_{\max} := \max_{i \in R}\{u_i\}$, $U := \sum_{i \in R} u_i$, $\epsilon := \frac{1}{u_{\max}^2 n^2}$ and $\mu := \frac{\epsilon}{28Un}$. Construct an instance of the problem $1|p_j = a_j + \max\{0, b_j(t - d_j)\}|C_{\max}$ as follows: $n = r+1$, $a_1 = 1$, $b_1 = \frac{1}{\mu}$, $a_i = \mu u_i$ and $b_i = \epsilon u_i$ for $2 \le i \le r+1$ and $D = \mu C$. By applying the reasoning similar to (a), the result follows; see [171, Theorem 7].   ◇

Remark 6.64. The ordinary $\mathcal{NP}$-hardness of the problem $1||\sum T_j$ was proved by Du and Leung [75]. The strong $\mathcal{NP}$-hardness of the problem $1||\sum w_j C_j$ was proved by Lawler [184].

Remark 6.65. Since Theorem 6.63 was originally formulated in an optimization form, the transformations in its proof were made from optimization versions of the problems used in these transformations. The decision formulation has been used for compatibility with other results presented in the book.

For the problem $1|p_j = a_j + \max\{0, b_j(t - d_j)\}|C_{\max}$, Kunnathur and Gupta [178] proposed two exact algorithms.

The first exact algorithm is based on dynamic programming. Let S' and S'' be defined as in Lemma 6.59. The algorithm starts with $S'' := \emptyset$ and adds to S'' one job from S' at a time in such a way that the schedule corresponding to the ordered set S'' is always optimal. Let $r_j := \max\{0, b_j(t^{\star} - d_j)\}$, where $t^{\star} = \min_{S'' \setminus \{J_j\}} \left\{ \sum_{J_k \in S'' \setminus \{J_j\}} a_k + \max\{0, b_k(t - d_k)\} \right\}$. The optimal schedule can be found by the recursive equation $C_{\max}(S'') = C_{\max}(S'' \setminus \{J_j\}) + r_j$, with $C_{\max}(\emptyset) := 0$. The time complexity of the dynamic programming algorithm is $O(n2^n)$.

The second exact algorithm proposed by Kunnathur and Gupta [178] is a branch-and-bound algorithm. In the algorithm, the lower bound is computed using Lemma 6.59. The upper bound is obtained by one of the heuristic algorithms and, if possible, improved by pairwise job interchange. Branching is performed by the standard depth-first search procedure. The branch-and-bound algorithm was tested on instances with up to $n = 15$ jobs. We refer the reader to [178, Sect. 3] for more details on the exact algorithms and to [178, Sect. 6] for the results of a computational experiment, conducted in order to evaluate the quality of schedules generated by the algorithms.

Kunnathur and Gupta [178] also proposed five heuristic algorithms to solve the problem. We will consider these heuristics in Chap. 9.

Cai et al. [39] considered deteriorating job processing times

$$p_j = a_j + b_j f(t, t_0),\qquad(6.38)$$

where $f(t, t_0) = 0$ for $t \leq t_0$ and $f(t, t_0) \nearrow$ for $t > t_0$. The first result proved by the authors concerns the case when $f(t, t_0) := 1_X$ for a set X. (Definition of the function 1_X was given in Sect. 1.1.)

Theorem 6.66. (Cai et al. [39]) *If* $X := \{t : t - t_0 > 0\}$, $f(t, t_0) := 1_X$ *and* $\sum_{j=1}^{n} a_j > t_0$, *then the decision version of the problem* $1|p_j = a_j + b_j f(t, t_0)|C_{\max}$ *is* $\mathcal{NP}$-*complete in the ordinary sense.*

Proof. Let $X := \{t : t - t_0 > 0\}$ and $f(t, t_0) := 1_X$. Then, the problem $1|p_j = a_j + b_j f(t, t_0)|C_{\max}$ is equivalent to a version of the KP problem (cf. Sect. 3.2). The version of the KP problem can be formulated as follows. Given

$t_0 > 0$ and $k - 1$ pairs of positive integers $\{(a_1, b_1), (a_2, b_2), \ldots, (a_{k-1}, b_{k-1})\}$, find a subset $K \subseteq \{1, 2, \ldots, k - 1\}$ which maximizes $\sum_{j \in K} b_j$ subject to $\sum_{j \in K} a_j \leq t_0$. By letting $a_k > t_0$ and $b_k > \max_{1 \leq j \leq k-1}\{b_j\}$, we obtain an instance of the problem $1|p_j = a_j + b_j f(t, t_0)|C_{\max}$. Since the latter problem has an optimal schedule if and only if the KP problem has an optimal solution, and since the KP problem is $\mathcal{NP}$-complete in the ordinary sense, the result follows. $\qquad\square$

Cai et al. also considered the case when $f(t, t_0) := \max\{t - t_0, 0\}$.

Lemma 6.67. (Cai et al. [39]) *For a given schedule* $\sigma = (\sigma_1, \sigma_2, \ldots, \sigma_n)$ *for the problem* $1|p_j = a_j + b_j \max\{t - t_0, 0\}|C_{\max}$, *the maximum completion time is equal to*

$$C_{\max}(\sigma) = t_0 + \left(\sum_{j=1}^{k} a_{\sigma_j} - t_0\right) \prod_{r=k+1}^{n} (1 + b_{\sigma_r}) + \sum_{j=k+1}^{n} a_{\sigma_j} \prod_{r=j+1}^{n} (1 + b_{\sigma_r}).$$

Proof. By induction with respect to n. $\qquad\square$

Theorem 6.68. (Cai et al. [39]) *Given the set of jobs that start at* $t \leq t_0$, *the maximum completion time for the problem* $1|p_j = a_j + b_j \max\{t - t_0, 0\}|C_{\max}$ *is minimized by scheduling jobs that start after time* t_0 *in the non-decreasing order of* $\frac{a_j}{b_j}$ *ratios.*

Proof. The result follows from Lemma 6.67 and Lemma 1.2 (a). $\qquad\square$

By Theorem 6.68, given the set of jobs that start at $t \leq t_0$, the problem $1|p_j = a_j + b_j \max\{t - t_0, 0\}|C_{\max}$ is solved by the algorithm $A_{10} : (a_j|b_j) \mapsto (\frac{a_j}{b_j} \nearrow)$.

Theorem 6.69. (Cai et al. [39])
(a) *The problem* $1|p_j = a + b_j \max\{t - t_0, 0\}|C_{\max}$ *is solvable in* $O(n \log n)$ *time by scheduling jobs in the non-increasing order of* b_j *values.*
(b) *The problem* $1|p_j = a_j + b \max\{t - t_0, 0\}|C_{\max}$ *is solvable in* $O(n \log n)$ *time by scheduling jobs in the non-decreasing order of* a_j *values.*
(c) *If* $a_j = k b_j$ *for* $1 \leq j \leq n$ *and a constant* $k > 0$, *then the problem* $1|p_j = a_j + b_j \max\{t - t_0, 0\}|C_{\max}$ *is solvable in* $O(n \log n)$ *time by scheduling jobs in the non-decreasing order of* b_j *values.*

Proof. (a),(b) The results are corollaries from Theorem 6.68. $\qquad\square$
(c) See [39, Theorem 4]. $\qquad\diamond$

By Theorem 6.69 (a), the problem $1|p_j = a + b_j \max\{t - t_0, 0\}|C_{\max}$ is solved by the algorithm $A_{11} : (b_j) \mapsto (b_j \searrow)$.

By Theorem 6.69 (b), the problem $1|p_j = a_j + b \max\{t - t_0, 0\}|C_{\max}$ is solved by the algorithm $A_7 : (a_j) \mapsto (a_j \nearrow)$.

By Theorem 6.69 (c), the problem $1|p_j = k b_j + b_j \max\{t - t_0, 0\}|C_{\max}$ is solved by the algorithm $A_9 : (b_j) \mapsto (b_j \nearrow)$.

Remark 6.70. Cai et al. also proved that a restricted version of the problem $1|p_j = a_j + b_j \max\{t - t_0, 0\}|C_{\max}$ is still $\mathcal{NP}$-complete in the ordinary sense; see [39, Sect. 4.1]. $\diamond$

Now we pass to *bounded step-linear* deterioration, in which the processing time of each job can grow only up to some limit value. Formally,

$$
p_j = \begin{cases} a_j, & \text{if } t \leq d, \\ a_j + b_j(t - D), & \text{if } d < t < D, \\ a_j + b_j(D - d), & \text{if } t \geq D, \end{cases} \tag{6.39}
$$

where d and D, $d < D$, are *common critical date* and *common maximum deterioration date*, respectively. We will also assume that $\sum_{j=1}^{n} a_j > d$, since otherwise the problem is trivial (all jobs can start by time d). This form of job deterioration was introduced by Kubiak and van de Velde [177].

Remark 6.71. In the case of job processing times given by (6.39), the deterioration will be called *bounded* if $D < \infty$ and it will be called *unbounded* if $D = \infty$.

Theorem 6.72. (Kubiak and van de Velde [177]) *If $d > 0$ and $D = \infty$, then the decision version of the problem of minimizing the maximum completion time for a single machine and for job processing times in the form of (6.39) is $\mathcal{NP}$-complete in the ordinary sense.*

Proof. We use the following transformation from the PP problem (cf. Sect. 3.2), provided that $|X| = k = 2l$ for some $l \in \mathbb{N}$. Let $n = k + 1 = 2l + 1$, $H_1 = A^2$, $d = lH_1 + a$, $H_2 = A^5$, $H_3 = A(H_2^{l-1}(H_2 + A + 1) + 1)$ and $H_4 = H_3 \sum_{i=0}^{l} H_2^{l-i} A^i + l^2 H_2^{l-1} d$.

Job processing times are defined as follows: $a_j = x_j + H_1$ and $b_j = H_2 + x_j - 1$ for $1 \leq j \leq k$, $a_{k+1} = H_3$ and $b_{k+1} = H_4 + 1$. The threshold $G = H_4$.

To complete the proof it is sufficient to show that the PP problem has a solution if and only if for the above instance there exists a feasible schedule σ such that $C_{\max}(\sigma) \leq G$ (see [177, Lemmata 2-4]). $\square$

Kubiak and van de Velde established a few properties of the single machine problem with job processing times in the form of (6.39) and $C_{\max}$ criterion. Before we formulate the properties, we introduce a definition (cf. [177]).

Definition 6.73. (Job types in the problem $1|p_j \equiv (6.39)|C_{\max}$)
(a) *The job that starts by time d and completes after d is called* pivotal.
(b) *The job that starts before d and completes by d is called* early.
(c) *The job that starts after d but completes by D is called* tardy.
(d) *The job that starts after D is called* suspended.

Now, we can formulate the above mentioned properties. (We omit simple proofs by pairwise job interchange argument.)

Property 6.74. (Kubiak and van de Velde [177]) The sequence of the early jobs is immaterial.

Property 6.75. (Kubiak and van de Velde [177]) The tardy jobs are sequenced in the non-increasing order of $\frac{b_j}{a_j}$ ratios.

Property 6.76. (Kubiak and van de Velde [177]) The sequence of the suspended jobs is immaterial.

Property 6.77. (Kubiak and van de Velde [177]) The pivotal job has processing time not smaller than any of the early jobs.

Property 6.78. (Kubiak and van de Velde [177]) If $a_k \leq a_l$ and $b_k \geq b_l$, then job J_k precedes job J_l.

Based on Properties 6.74–6.78, Kubiak and van de Velde proposed a branch-and-bound algorithm. It is reported (see [177, Sect. 5]) that the algorithm solves instances with $n = 50$ job within 1 second on a PC, while most of instances with $n = 100$ jobs are solved within 7 minutes.

Kubiak and van de Velde proposed also three pseudopolynomial-time algorithms for the problem. The first algorithm, designed for the case of unbounded deterioration, runs in $O(nd\sum_{j=1}^{n} a_j)$ time and $O(nd)$ space. The second algorithm, designed for the case of bounded deterioration, requires $O(n^2 d(D - d)\sum_{j=1}^{n} a_j)$ time and $O(nd(D - d))$ space. The third algorithm, also designed for the case of bounded deterioration, requires $O(nd\sum_{j=1}^{n} b_j(\sum_{j=1}^{n} a_j)^2)$ time and $O(nd\sum_{j=1}^{n} b_j \sum_{j=1}^{n} a_j)$ space; see [177, Sect. 4] for more details.

Janiak and Kovalyov [146] introduced exponential deterioration of job processing times in the form of

$$p_j = a_j 2^{b_j(t-r_j)}, \qquad (6.40)$$

where $a_j, b_j, r_j \geq 0$ for $1 \leq j \leq n$.

Theorem 6.79. (Janiak and Kovalyov [146]) *The decision version of the problem* $1|p_j = a_j 2^{b_j(t-r_j)}|C_{\max}$ *is* $\mathcal{NP}$*-complete in the strong sense.*

Proof. The transformation from the 3-P problem (cf. Sect. 3.2) is as follows. There are $n = 4h$ jobs, $r_j = 0, a_j = c_j, b_j = 0$ for $1 \leq j \leq 3h$, $r_{3h+i} = iK + i - 1, a_{3h+i} = 1, b_{3h+i} = K$ for $1 \leq i \leq h$. The threshold $G = hK + h$.

In order to complete the proof it is sufficient to show that the 3-P problem has a solution if and only if there exists a schedule σ for the above instance of the problem $1|p_j = a_j 2^{b_j(t-r_j)}|C_{\max}$, such that $C_{\max}(\sigma) \leq G$. □

If there are only two distinct ready times, the problem remains computationally intractable.

Theorem 6.80. (Janiak and Kovalyov [146]) *The decision version of the problem* $1|p_j = a_j 2^{b_j(t-r_j)}, r_j \in \{0, R\}|C_{\max}$ *is $\mathcal{NP}$-complete in the ordinary sense.*

Proof. The transformation from the PP problem (cf. Sect. 3.2) is as follows. There are $n = k + 1$ jobs, $r_j = 0, a_j = x_j, b_j = 0$ for $1 \le j \le k$, $r_{k+1} = A$, $a_{k+1} = 1$, $b_{k+1} = 1$. The threshold $G = 2A + 1$.

In order to complete the proof, it is sufficient to show that the PP problem has a solution if and only if there exists a schedule σ for the above instance of the problem $1|p_j = a_j 2^{b_j(t-r_j)}, r_j \in \{0, R\}|C_{\max}$, such that $C_{\max}(\sigma) \le G$ (see Fig. 6.6 and Remark 6.12).   □

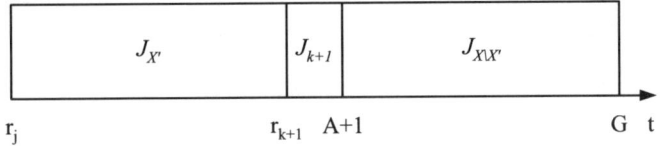

Fig. 6.6: Example schedule in the proof of Theorem 6.80

Remark 6.81. Janiak and Kovalyov state Theorem 6.80 (see [146, Theorem 2]) without proof. The above reduction comes from the present author.

6.1.6 Proportional-linear shortening

A separate group of results concerns job *shortening*, i.e., the case when job processing times are non-increasing linear functions of job starting times.

The simplest case of the shortening is *proportional-linear shortening* in which job processing times are in the form of

$$p_j = a_j(A - Bt), \tag{6.41}$$

for $1 \le j \le n$, the *shortening rates* b_j are rational and satisfy condition

$$0 < a_j B < 1 \tag{6.42}$$

and condition

$$B\left(\sum_{i=1}^{n} a_i - a_j\right) < 1 \tag{6.43}$$

hold for $1 \le j \le n$. Conditions (6.42) and (6.43) eliminate some trivial cases and assure that the constructed instances of scheduling problems with job processing times (6.41) make sense from the practical point of view. This form of job processing times has been introduced by Wang and Xia [289].

Equal ready times and deadlines

Wang and Xia [289] considered job processing times in the form of (6.41) with $A = 1$ and $B := k$,

$$p_j = a_j(1 - kt) \tag{6.44}$$

where $1 \leq j \leq n$ and $k = const > 0$. In this case, condition (6.43) takes the form of

$$k \left(\sum_{j=1}^{n} a_j - a_{\min} \right) < 1, \tag{6.45}$$

where $a_{\min} := \min_{1 \leq j \leq n}\{a_j\}$.

Theorem 6.82. (Wang and Xia [289]) *The problem* $1|p_j = a_j(1 - kt), k > 0,$ $k \left(\sum_{j=1}^{n} a_j - a_{\min} \right) < 1|C_{\max}$ *is solvable in* $O(n)$ *time and the maximum completion time* $C_{\max}(\sigma) = \frac{1}{k} \left(1 - \prod_{j=1}^{n}(1 - ka_{\sigma_j}) \right)$ *does not depend on the schedule of jobs.*

Proof. By induction with respect to n.   □

6.1.7 Linear shortening

The next type of job shortening is *linear shortening* in which the processing times are non-increasing linear functions of job starting times.

In linear shortening, the processing times of jobs are in the form of

$$p_j = a_j - b_j t, \tag{6.46}$$

for $1 \leq j \leq n$, the *shortening rates* b_j are rational and conditions

$$0 < b_j < 1 \tag{6.47}$$

and

$$b_j \left(\sum_{i=1}^{n} a_i - a_j \right) < a_j \tag{6.48}$$

hold for $1 \leq j \leq n$. This form of job processing times (with an additional assumption, see (6.51) below) has been introduced by Ho et al. [135].

Remark 6.83. Conditions (6.47) and (6.48) are counterparts of conditions (6.42) and (6.43), respectively. Moreover, in conjuction with condition (6.51) they eliminate some trivial cases and assure that the constructed instances of scheduling problems with job processing times (6.46) make sense from the practical point of view.

Equal ready times and deadlines

Wang et al. [283] considered the problem of single-machine batch scheduling (cf. Remarks 6.2, 6.3 and 6.6) with job processing times $p_{i,j} = a_{i,j} - b_{i,j}t$, where

$$0 < b_{i,j} < 1 \tag{6.49}$$

and

$$b_{i,j}\left(\sum_{i=1}^{m}\sum_{j=1}^{k_i}(\theta_i + a_{i,j}) - a_{i,j}\right) < a_{i,j} \tag{6.50}$$

for $1 \le i \le m$ and $1 \le j \le k_i$, $\sum_{i=1}^{m} k_i = n$.

Remark 6.84. Conditions (6.50) and (6.50) for time-dependent batch scheduling problems with group technology (cf. Remark 6.2) are counterparts of conditions (6.47) and (6.48) for time-dependent scheduling problems without batching, respectively. Both groups of conditions have the same aim (see Remark 6.83 for details).

For the problem $1|p_{i,j} = a_{i,j} - b_{i,j}t, \theta_i, GT|C_{\max}$, Wang et al. [283] proposed the following algorithm.

Algorithm A_{12}
for the problem $1|p_{i,j} = a_{i,j} - b_{i,j}t, \theta_i, GT|C_{\max}([283])$

Input: sequences $(a_{i,j})$, $(b_{i,j})$, (θ_i) for $1 \le i \le m$ and $1 \le j \le k_i$,
Output: an optimal schedule $\sigma^\star$

▷ Step 1:
    for $i \leftarrow 1$ **to** m **do**
        Arrange jobs in group G_i in the non-increasing order of the $\frac{b_{i,j}}{a_{i,j}}$ ratios;
        Call the sequence $\sigma^{(i)}$;
▷ Step 2:
    for $i \leftarrow 1$ **to** m **do**
        Calculate $\rho(G_i) := \dfrac{\sum_{j=1}^{k_i} a_{i,j} \prod_{k=j+1}^{k_i}(1-b_{i,k}) + \theta_i \prod_{j=1}^{k_i}(1-b_{i,j})}{1 - \prod_{j=1}^{k_i}(1-b_{i,j})}$;
▷ Step 3:
    Schedule groups in the non-increasing order of $\rho(G_i)$ values;
▷ Step 4:
    $\sigma^\star \leftarrow (\sigma^{([1])}|\sigma^{([2])}|\ldots|\sigma^{([m])})$;
    return $\sigma^\star$.

Theorem 6.85. (Wang et al. [283]) *The problem $1|p_{i,j} = a_{i,j} - b_{i,j}t, \theta_i, GT|C_{\max}$ is solvable by algorithm A_{12} in $O(n \log n)$ time.*

Proof. By pairwise job interchange; see [283, Theorems 1–2] for details.   ◇

Remark 6.86. A special case of the Theorem 6.85, with $a_{i,j} = 1$ for $1 \leq i \leq m$ and $1 \leq j \leq k_i$, where $\sum_{i=1}^{m} k_i = n$, has been considered by Cheng and Sun [44, Algorithm C, Theorem 9].

Distinct ready times and deadlines

In this case, processing times given by (6.46) are such that condition (6.47) and

$$b_j d_j < a_j \leq d_j. \tag{6.51}$$

hold for $1 \leq j \leq n$. All jobs have ready times equal to zero. This form of shortening job processing times has been introduced by Ho et al. [135].

Theorem 6.87. (Ho et al. [135]) *If a_j, b_j and d_j satisfy conditions (6.47) and (6.51), then the problem $1|p_j = a_j - b_j t, d_j|-$ is*
(a) $\mathcal{NP}$-complete in the strong sense, if there is an arbitrary number of deadlines;
(b) $\mathcal{NP}$-complete in the ordinary sense, if there are only two distinct deadlines;
(c) solvable in $O(n \log n)$ time by scheduling jobs in the non-increasing order of $\frac{a_j}{b_j}$ ratios if all deadlines are identical.

Proof. (a) The transformation from the 3-P problem (cf. Sect. 3.2) is as follows:

$n = 4h - 1$,

$$d_i = \begin{cases} i(K+1) & \text{for } 1 \leq i \leq h-1, \\ hK + h - 1 & \text{for } h \leq i \leq 4h-1, \end{cases}$$

$$a_i = \begin{cases} iK - 1 & \text{for } 1 \leq i \leq h-1, \\ c_{i-h+1} & \text{for } h \leq i \leq 4h-1, \end{cases}$$

$$b_i = \begin{cases} \frac{iK+i-1}{iK+i-1} & \text{for } 1 \leq i \leq h-1, \\ \frac{1}{16h^6 K^6}, & \text{for } h \leq i \leq 4h-1. \end{cases}$$

To complete the proof, it is sufficient to show that the 3-P problem has a solution if and only if for the above instance of the problem $1|p_j = a_j - b_j t, d_j|C_{\max}$, there exists a non-preemptive feasible schedule.

   (b) The transformation from the PP problem (cf. Sect. 3.2) is as follows:

$n = k + 1$,

$$d_i = \begin{cases} A+1 & \text{for } i = 1, \\ 2A+1 & \text{for } 2 \leq i \leq k+1, \end{cases}$$

$$a_i = \begin{cases} A & \text{for } i = 1, \\ x_{i-1} & \text{for } 2 \leq i \leq k+1. \end{cases}$$

$$b_i = \begin{cases} \frac{A-1}{4} & \text{for } i = 1, \\ \frac{1}{3k^2 A}, & \text{for } 2 \leq i \leq k+1. \end{cases}$$

To complete the proof, it is sufficient to show that the PP problem has a solution if and only if for the above instance of the problem $1|p_j = a_j - b_j t, d_j|C_{\max}$, there exists a non-preemptive feasible schedule.

(c) Since $d_i = D$ for $1 \leq i \leq n$, it is sufficient to start scheduling jobs from time $t = 0$, to apply Theorem 6.24 to construct a schedule and to check if the maximum completion time for the schedule does not exceed time $t = D$. $\square$

By Theorem 6.87 (c), if a_j, b_j and d_j satisfy conditions (6.47) and (6.51), the problem $1|p_j = a_j - b_j t, d_j = D|C_{\max}$ is solved by the algorithm $A_{13} : (a_j|b_j) \mapsto (\frac{a_j}{b_j} \searrow)$.

For the case when job processing times given by (6.46), all shortening rates are equal,

$$p_j = a_j - bt \tag{6.52}$$

for $1 \leq j \leq n$, a number of results are known.

The problem of the complexity of minimizing the maximum completion time in the case when there are arbitrary deadlines and equal shortening rates, stated by Ho et al. [135], has been solved by Cheng and Ding.

Theorem 6.88. (Cheng and Ding [50]) *The decision version of the problem* $1|p_j = a_j - bt, d_j|C_{\max}$ *is $\mathcal{NP}$-complete in the strong sense.*

Proof. The transformation is from the 3-P problem (cf. Sect. 3.2). Define $q = 2^6 h^3 K^2$, the identical decreasing procesing rate $b = \frac{1}{2^3 q^3 hK}$ and the number of jobs $n = 3h + (h-1)q$. Define $D_j = 1 - b((q-1)(jK+j-1) + \frac{1}{q}\sum_{k=1}^{q-1} k)$ for $1 \leq j \leq h - 1$. The deadlines are as follows: $d_i = d^0 = hK + \sum_{j=1}^{h-1} D_j$ for $1 \leq i \leq 3h$, $d_k^j = d^j = jK + j$ for $1 \leq j \leq h - 1$ and $1 \leq k \leq q$. The basic processing times are the following: $a_i = c_i$ for $1 \leq i \leq 3h$, and $a_k^j = a = \frac{1}{q}$ for $1 \leq j \leq h - 1$ and $1 \leq k \leq q$. The threshold value is d^0.

To complete the proof, it is sufficient to show that an instance of the 3-P problem has a solution if and only if the above constructed instance of the problem $1|p_j = a_j - bt, d_j|C_{\max}$ has a solution. The idea which simplifies the proof is introducing a special form of a feasible schedule for the problem $1|p_j = a_j - bt, d_j|C_{\max}$. Due to regularity of the schedule, further calculations are easier (see [50, Lemmata 1–4]). $\square$

If we simplify the problem $1|p_j = a_j - bt, d_j|C_{\max}$, assuming that $a_j = a$ and $b_j \in \{B_1, B_2\}$ for $1 \leq j \leq n$, then the new problem can be solved in polynomial time. Before we state the result, we introduce a definition proposed by Cheng and Ding [53].

Definition 6.89. (A canonical schedule)
Given an instance of the problem $1|p_j = a_j - bt, d_j, b_j \in \{B_1, B_2\}|C_{\max}$ with m distinct deadlines $D_1 < D_2 < \ldots < D_m$, a schedule σ is called canonical if the jobs with the same b_j are scheduled in σ in the EDD order and the jobs in sets $R_j := \{J_j \in \mathcal{J} : D_{j-1} < C_j \leq D_j\}$, where $1 \leq j \leq m$ and $D_0 := 0$, are scheduled in the non-decreasing order of b_j values.

Lemma 6.90. (Cheng and Ding [53]) *If there exists a feasible (optimal) schedule for an instance of the problem $1|p_j = a_j - bt, d_j, b_j \in \{B_1, B_2\}|C_{\max}$, then there exists a canonical feasible (optimal) schedule for the instance.*

Proof. See [53, Lemma 1]. ◇

Now, we briefly describe the construction of a schedule for the problem $1|p_j = a_j - bt, d_j, b_j \in \{B_1, B_2\}|C_{\max}$, [53]. The construction is composed of the following three steps.

In the first step, we schedule jobs with same b_j value in the EDD order, obtaining two chains of jobs, $\mathcal{C}_1$ and $\mathcal{C}_2$.

In the second step, we construct the *middle* schedule $M(\mathcal{C}_1)$. To this end, we insert jobs from $\mathcal{C}_2$ into the time interval $\langle 0, D_m \rangle$ backwards as follows. We assign the jobs with the largest deadline D_m to a subschedule $\sigma^{1,m}$ of $M(\mathcal{C}_1)$ in such a way that the jobs are started at some time T_m^1 and completed at D_m. Next, we assign the jobs with the deadline D_{m-1} to a subschedule $\sigma^{1,m-1}$ in such a way that the jobs are started at some time T_{m-1}^1 and completed at $\min\{D_{m-1}, T_m^1\}$. We continue until all jobs have been assigned to subschedules $\sigma^{1,j}$ for $1 \leq j \leq k \leq m$. (Note that it must be $T_j^1 \geq D_{j-1}$, where $1 \leq j \leq k$ and $D_0 := 0$; hence, if some jobs cannot be put in $\sigma^{1,j}$, we move them to the set of late jobs $L(\mathcal{C}_1)$.) After the completion of this step, we obtain the schedule $M(\mathcal{C}_1) := (\sigma^{1,1}|\sigma^{1,2}|\ldots|\sigma^{1,k})$ in the form of $(\mathcal{I}_1|\mathcal{W}_1|\mathcal{I}_2|\mathcal{W}_2|\ldots|\mathcal{I}_k|\mathcal{W}_k)$, $k \leq m$, where $\mathcal{I}_j$ and $\mathcal{W}_j$ denote, respectively, the j-th idle time and the j-th period of continuous machine working time, $1 \leq j \leq k$.

In the third step, we construct the *final* schedule $F(\mathcal{C}_1, \mathcal{C}_2)$. To this end we insert jobs from $\mathcal{C}_1$ into the idle times of $M(\mathcal{C}_1)$ as follows. Starting from the beginning of $\mathcal{I}_1$, we assign jobs to $\mathcal{I}_1$ until an assigned job cannot meet its deadline. This job is moved to the set of late jobs $L(\mathcal{C}_2)$. We continue the procedure with the remaining jobs until the end of $\mathcal{I}_1$. Then we shift $\sigma^{1,1}$ forward to connect it with $\sigma^{2,1}$, where $\sigma^{2,1}$ denotes the schedule of jobs assigned to $\mathcal{I}_1$. We continue until all jobs have been assigned to idle times $\mathcal{I}_j$, $2 \leq j \leq k$. After the completion of this step, we obtain the schedule $F(\mathcal{C}_1, \mathcal{C}_2) := (\sigma^{2,1}|\sigma^{1,1}|\sigma^{2,2}|\sigma^{1,2}|\cdots|\sigma^{2,k}|\sigma^{1,k})$ and the set of late jobs $L(\mathcal{C}_1, \mathcal{C}_2) := L(\mathcal{C}_1) \cup L(\mathcal{C}_2)$.

Lemma 6.91. (Cheng and Ding [53]) *Let I be an instance of the problem $1|p_j = a_j - b_j t, d_j, b_j \in \{B_1, B_2\}|C_{\max}$. Then*
(a) if $L(I) = \emptyset$, then the final schedule is optimal;
(b) if $L(I) \neq \emptyset$, then for I there does not exists a feasible schedule.

Proof. (a) See [53, Lemma 5].
     (b) See [53, Lemma 6]. ◇

On the basis of Lemma 6.91, Cheng and Ding proposed for the problem $1|p_j = a - b_j t, d_j, b_j \in \{B_1, B_2\}|C_{\max}$ an optimal algorithm. Given an instance I of the problem, the algorithm first constructs described above schedules $M(\mathcal{C}_1)$, $F(\mathcal{C}_1, \mathcal{C}_2)$ and the set $L(\mathcal{C}_1, \mathcal{C}_2)$. Next, it verifies whether $L \neq \emptyset$. If $L = \emptyset$, the final schedule is optimal by Lemma 6.91; otherwise, for I there does not exist a feasible schedule.

The pseudo-code of the algorithm can be formulated as follows.

Algorithm A_{14}
for the problem $1|p_j = a - b_j t, d_j, b_j \in \{B_1, B_2\}|C_{\max}$ ([53])

Input: numbers a, B_1, B_2, sequence $(d_1, d_2, \ldots, d_n)$
Output: an optimal schedule $\sigma^\star$

▷ Step 1:
  $\mathcal{C}_1 \leftarrow \{J_j \in \mathcal{J} : b_j = B_1\}$;
  $\mathcal{C}_2 \leftarrow \{J_j \in \mathcal{J} : b_j = B_2\}$;
▷ Step 2:
  Construct schedules $M(\mathcal{C}_1)$, $F(\mathcal{C}_1, \mathcal{C}_2)$ and the set $L(\mathcal{C}_1, \mathcal{C}_2)$;
  $\sigma^\star \leftarrow (F|L)$;
▷ Step 3:
  if $(L \neq \emptyset)$ then write 'there exists no feasible schedule';
    stop
  else return $\sigma^\star$.

Theorem 6.92. (Cheng and Ding [53]) *The problem $1|p_j = a - b_j t, d_j, b_j \in \{B_1, B_2\}|C_{\max}$ is solvable in $O(n \log n)$ time by algorithm A_{14}.*

Proof. The correctness of algorithm A_{14} follows from Lemma 6.91. Since no step needs more than $O(n \log n)$ time, the overall time complexity of the algorithm is $O(n \log n)$. ☐

Theorem 6.93. (Cheng and Ding [53]) *The problem of whether there exists a feasible schedule for the problem $1|p_j = 1 - b_j t, d_j|C_{\max}$ is $\mathcal{NP}$-complete in the strong sense.*

Proof. The transformation from the 3-P problem (cf. Sect. 3.2) is as follows. Define $q = 2hK$, $v = 32h^2 qK$, $n = v + 3h + (h-1)q$, $A_1 = 3n^3$ and $A_2 = A_3 = 2nhK$.

The deterioration rates and deadlines are the following: $b_{0,i} = 0, d_{0,i} = v$ for $1 \leq i \leq v$, $b_{i,j} = \frac{1}{A_1 A_3}, d_{i,j} = D_i$ for $1 \leq i \leq h-1$ and $1 \leq j \leq q$, $b_i = \frac{c_i}{A_1 A_2 A_3}, d_i = G$ for $1 \leq i \leq 3h$, where the deadlines

$$D_i = v + qi + 3i - \sum_{k=1}^{i-1} \sum_{l=1}^{q} \frac{v + qk + 3k - l}{A_1 A_3} - \sum_{k=0}^{i-1} \frac{(v + qk + 3k)K}{A_1 A_2 A_3}$$

for $1 \leq i \leq h-1$ and

$$G = n - \sum_{k=1}^{h-1} \sum_{l=1}^{q} \frac{v + qk + 3k - l}{A_1 A_3} - \sum_{k=0}^{h-1} \frac{(v + qk + 3k)K}{A_1 A_2 A_3}.$$

The set of jobs in the above instance of the problem $1|p_j = 1 - b_j t, d_j|C_{\max}$ is divided into sets of jobs $V = \{J_{0,1}, J_{0,2}, \ldots, J_{0,v}\}$, $R = \{J_1, J_2, \ldots, J_{3h}\}$ and $Q_i = \{J_{i,1}, J_{i,2}, \ldots, J_{i,q}\}$ for $1 \leq i \leq h-1$. Construct for this instance a schedule in the form of $(V, R_1, Q_1, R_2, \ldots, Q_{h-1}, R_h)$, where the job order in any of these sets is arbitrary. By showing that the 3-P problem has a solution if and only if the schedule is feasible for the problem $1|p_j = 1 - b_j t, d_j|C_{\max}$ (see [53, Lemmata 1–3]), we obtain the result. □

The restricted version of the above problem, when there are only two distinct deadlines, is also computationally intractable.

Theorem 6.94. (Cheng and Ding [53]) *The problem of whether there exists a feasible schedule for the problem* $1|p_j = 1 - b_j t, d_j \in \{D_1, D_2\}|C_{\max}$ *is $\mathcal{NP}$-complete in the ordinary sense.*

Proof. The transformation is from the PP problem (cf. Sect. 3.2). Define $n = (k+1)(k+2)$, $A_1 = 4n^3$ and $A_2 = A_3 = 2^{k+1}k^k n^2 A$.

The job deterioration rates are the following: $b_{0,0} = b_{0,1} = 0$, $b_{0,j} = \frac{1}{A_2 A_3}$ for $2 \le j \le k+1$, $b_{i,0} = \frac{2^i k^i A - x_i}{(i+1)A_1 A_2 A_3}$, $b_{i,j} = \frac{2^i k^i A}{(i+1)A_1 A_2 A_3}$ for $1 \le i \le k$ and $1 \le j \le k+1$.

The deadlines are the following: $d_{0,j} = D_1$ and $d_{i,j} = D_2$ for $1 \le i \le k$ and $0 \le j \le k+1$, where $D_1 = 2k + 2 - \sum_{i=1}^{k}(i+1)b_{i,1} - \sum_{j=2}^{k+1}(k+j)b_{0,j} + \frac{A}{A_1 A_2 A_3} +$

$\frac{1}{2A_1 A_2 A_3}$, $D_2 = n - \sum_{i=1}^{k}(i+1)b_{i,1} - \sum_{j=2}^{k+1}(k+j)b_{0,j} - \sum_{i=1}^{k}(i+1)(k+1)b_{i,0} -$

$\sum_{i=1}^{k}\sum_{j=1}^{k}((i+1)(k+1)+j)b_{i,j+1} - \frac{kA}{A_1 A_2 A_3} + \frac{1}{2A_1 A_2 A_3}$. The threshold $G = D_2$.

In order to complete the proof, it is sufficient to show that the PP problem has a solution if and only if for the above instance of the problem $1|p_j = 1 - b_j t$, $d_j \in \{D_1, D_2\}|C_{\max}$ there exists a feasible schedule. □

Another restricted version of the problem, when job basic processing times are distinct, all deterioration rates are equal and there are only two distinct deadlines, is computationally intractable as well.

Theorem 6.95. (Cheng and Ding [48]) *The decision version of the problem* $1|p_j = a_j - bt, d_j \in \{D_1, D_2\}|C_{\max}$ *is $\mathcal{NP}$-complete in the ordinary sense.*

Proof. The transformation from the PP problem (cf. Sect. 3.2) is as follows. Let $B = 2^{n+3}n^2 A$ and $v = 2^6 n^3 B$. Define $n = 2k + 1$ shortening jobs, where $a_0 = v$, $a_{1,i} = v(B + 2^{n-i+}A + x_i)$ and $a_{2,i} = v(B + 2^{n-i+1}A)$ for $1 \le i \le n$, $b = \frac{2}{v}$, $d_0 = v(nB + 2^{n+1}A - A + 1)$ and $d_{1,i} = d_{2,i} = G$ for $1 \le i \le n$, where the threshold $G = \sum_{i=0}^{2n+1}(x_i - b(E - (n+1)Av) + 1)$, with constant $E = \sum_{i=1}^{n}(2n+1)a_{2,i} + na_0 + \sum_{i=1}^{n-1}(n-i)a_{1,i}$.

In order to complete the proof, it is sufficient to show that the PP problem has a solution if and only if for the above instance of the problem $1|p_j = a_j - bt$, $d_j \in \{D_1, D_2\}|C_{\max}$ there exists a feasible schedule σ such that $C_{\max}(\sigma) \le G$ (see [48, Lemmata 1–4]). □

The problem when $d_j = \infty$ and $r_j \neq 0$ for $1 \leq j \leq n$ has been considered by Cheng and Ding. The authors proposed the following algorithm.

Algorithm A_{15} for the problem $1|p_j = a_j - bt, r_j|C_{\max}$ ([49])

Input: sequences $(a_1, a_2, \ldots, a_n)$, $(r_1, r_2, \ldots, r_n)$, number $b := \frac{v}{u}$
Output: an optimal schedule

▷ Step 1:
$$B_1 \leftarrow \max_{1 \leq j \leq n} \{r_j\};$$

$$B_2 \leftarrow u^{n-1} \left(\sum_{j=1}^{n} a_j + B_1 \right);$$

▷ Step 2:
while $(B_2 - B_1 > 1)$ do
$\qquad G' \leftarrow \lceil \frac{B_1 + B_2}{2} \rceil$; $G \leftarrow \frac{G'}{u^{n-1}}$;
$\qquad$ Use Lemma 6.97 to construct instance I of $1|p_j = a_j + bt, d_j|C_{\max}$;
$\qquad$ Apply Algorithm A_6 to instance I to find an optimal schedule;
$\qquad$ **if** (there exists an optimal schedule for I) **then** $B_1 \leftarrow G'$
$\qquad$ **else** $G \leftarrow \frac{B_2}{u^{n-1}}$.

Theorem 6.96. (Cheng and Ding [49]) *The problem $1|p_j = a_j - bt, r_j|C_{\max}$ is solvable in $O(n^6 \log n)$ time by algorithm A_{15}.*

Proof. Consider any instance I of the problem $1|p_j = a_j - bt, r_j|C_{\max}$. By Lemma 6.97, for this instance there exists an instance I' of the problem $1|p_j = a_j + bt, d_j|C_{\max}$. Changing iteratively the threshold value G and applying algorithm A_6 to I', we can check whether there exists an optimal schedule σ' for I'. Due to the symmetry between problems $1|p_j = a_j - bt, r_j|C_{\max}$ and $1|p_j = a_j + bt, d_j|C_{\max}$, from σ' we can construct an optimal schedule σ for I.

The value of G can be determined in at most $\log(B_2 - B_1) \equiv O(n \log n)$ time. In each iteration of the loop **while** at most $O(n^5)$ time is needed for execution of algorithm A_6. Therefore, the overall time complexity of algorithm A_{15} is $O(n^6 \log n)$. □

Linear shortening vs. linear deterioration

There exists a kind of symmetry between single-machine time-dependent scheduling problems with shortening job processing times and ready times and their counterparts with deteriorating job processing times and deadlines. In this subsection, we will present a few results that illustrate the symmetry.

By the *symmetry reduction*, we will mean that we are able to obtain a schedule for the problem with deadlines using a schedule for its counterpart with ready times, and vice versa. For example, given a schedule for an instance of the $1|p_j = a_j - b_j t, r_j|C_{\max}$ problem, we can take it as a schedule for a corresponding instance of the $1|p_j = a_j + b_j t, d_j|C_{\max}$ problem, viewed from the reverse direction.

Lemma 6.97. (Cheng and Ding [49]) *There exists a symmetry reduction between the problem* $1|p_j = a_j - b_j t, r_j|C_{\max}$ *and problem* $1|p'_j = a'_j + b'_j t, d_j|C_{\max}$ *such that any schedule for the first problem defines a schedule for the second one and vice versa.*

Proof. Given an instance I of the problem $1|p_j = a_j - b_j t, r_j|C_{\max}$ and threshold $G > \max_{1 \le j \le n}\{r_j\}$, an instance II of the problem $1|p_j = a_j + b_j t, d_j|C_{\max}$ can be constructed in the following way: $a'_j = \frac{a_j - b_j G}{1 - b_j}$, $b'_j = \frac{b_j}{1 - b_j}$ and $d_j = G - r_j$ for $1 \le j \le n$.

Since it is sufficient to consider only schedules without idle times and since the above reduction can be done in polynomial time, the result follows. $\square$

By Lemma 6.97, we obtain the following result.

Lemma 6.98. (Cheng and Ding [49]) *There exists a symmetry reduction between the following pairs of problems:*
(a) $1|p_j = a_j + b_j t, r_j|C_{\max}$ *and* $1|p'_j = a'_j - b'_j t, d_j|C_{\max}$,
(b) $1|p_j = a_j + bt, r_j|C_{\max}$ *and* $1|p'_j = a'_j - b'_j t, d_j|C_{\max}$,
(c) $1|p_j = a_j - bt, r_j|C_{\max}$ *and* $1|p'_j = a'_j + b'_j t, d_j|C_{\max}$,
(d) $1|p_j = a_j + bt, r_j \in \{0, R\}|C_{\max}$ *and* $1|p'_j = a'_j - b't, d_j \in \{D_1, D_2\}|C_{\max}$,
(e) $1|p_j = a_j - bt, r_j \in \{0, R\}|C_{\max}$ *and* $1|p'_j = a'_j + b't, d_j \in \{D_1, D_2\}|C_{\max}$,
such that any schedule for the first problem from a pair defines a schedule for the second one and vice versa.

Remark 6.99. In Sect. 12.4, we will consider the so-called *equivalent* problems, which are similar, in some sense, to the problems from Lemma 6.98.

Theorem 6.100. (Cheng and Ding [49]) *The decision versions of the problems* $1|p_j = a_j + b_j t, r_j|C_{\max}$ *and* $1|p_j = a_j - b_j t, d_j|C_{\max}$ *are* $\mathcal{NP}$-*complete in the strong sense.*

Proof. By Theorem 6.88 and Lemma 6.97, the results follow. $\square$

Theorem 6.101. (Cheng and Ding [49]) *The decision versions of the problems* $1|p_j = a_j + b_j t, r_j \in \{0, R\}|C_{\max}$ *and* $1|p_j = a_j - b_j t, d_j \in \{D_1, D_2\}|C_{\max}$ *are* $\mathcal{NP}$-*complete in the ordinary sense.*

Proof. By Theorem 6.95 and Lemma 6.97, the results follow. $\square$

Bosio and Righini [29] proposed an exact dynamic programming algorithm for the problem $1|p_j = a_j + b_j t, r_j|C_{\max}$. The algorithm starts from an empty schedule and iteratively adds an unscheduled job to the partial schedule constructed so far. In order to speed up the process of finding the final schedule, the authors used some upper and lower bounds on the optimal value of $C_{\max}$.

Lee et al. [194] considered the above problem in the case when $b_j = b$ for $1 \le j \le n$. The authors established a few properties of an optimal schedule for the latter problem.

Let J_i and J_j, $1 \le i, j \le n$, be two adjacent jobs in a schedule for the problem $1|p_j = a_j + bt, r_j|C_{\max}$.

Property 6.102. (Lee et al. [194]) If $S_i < r_i$ and $(1+b)r_i + a_i < r_j$, then there exists an optimal schedule in which job J_i immediately precedes job J_j.

Proof. Let σ be any feasible schedule for the problem $1|p_j = a_j + bt, r_j|C_{\max}$ in which job J_i immediately precedes job J_j. Since $S_i < r_i$ and $(1+b)r_i + a_i = C_i(\sigma) < r_j$, we have $C_j(\sigma) = (1+b)r_j + a_j$.

Let now σ' denote the schedule σ in which job J_i and J_j have been mutually replaced. Since $S_i < r_i < (1+b)r_i + a_i < r_j$, we have $S_i < r_j$ and $r_i < r_j$. Therefore, $C_j(\sigma') = (1+b)r_j + a_j$ and $C_i(\sigma') = (1+b)^2 r_j + (1+b)a_j + a_i$. Since $C_j(\sigma) < C_i(\sigma')$, we have $C_{\max}(\sigma) < C_{\max}(\sigma')$. ∎

Applying similar reasoning as above, one can prove the following properties.

Property 6.103. (Lee et al. [194]) If $S_i \geq \max\{r_i, r_j\}$ and $a_j > a_i$, then there exists an optimal schedule in which job J_i immediately precedes job J_j.

Property 6.104. (Lee et al. [194]) If $r_i \leq S_i \leq r_j, (1+b)S_i + a_i \geq r_j$ and $a_j > a_i$, then there exists an optimal schedule in which job J_i immediately precedes job J_j.

Property 6.105. (Lee et al. [194]) If $r_i \leq S_i$ and $(1+b)S_i + a_i < r_j$, then there exists an optimal schedule in which job J_i immediately precedes job J_j.

Property 6.106. (Lee et al. [194]) If $r_j \leq S_i < r_i, (1+b)S_i + a_j \geq r_i$ and $b(a_j - a_i) + (1+b)^2(S_i - r_i) > 0$, then there exists an optimal schedule in which job J_i immediately precedes job J_j.

Property 6.107. (Lee et al. [194]) If $r_j > r_i > S_i, (1+b)r_i + a_i \geq r_j$ and $a_j > a_i$, then there exists an optimal schedule in which job J_i immediately precedes job J_j.

The authors also proposed the following lower bound on the optimal value of the schedule length for the problem.

Lemma 6.108. (Lee et al. [194]) *Let $\sigma^{(k)}$ be a partial schedule for the problem $1|p_j = a_j + bt, r_j|C_{\max}$ such that $|\sigma^{(k)}| = k$ and let σ be a complete schedule obtained from $\sigma^{(k)}$. Then $C_{\max}(\sigma) \geq \max\{LB_1, LB_2\}$, where*

$$LB_1 = (1+b)^l C_{[k]}(\sigma) + \sum_{j=1}^{l}(1+b)^{l-j} a_{(k+j)},$$

$$LB_2 = \max\{(1+b)r_{k+j} + a_{k+j} : 1 \leq j \leq l\}$$

and $a_{(k+1)} \leq a_{(k+2)} \leq \ldots \leq a_{(k+l)}$.

Proof. By a direct computation, see [194, Sect. 3.2] for details. ◇

Based on Properties 6.102–6.107 and Lemma 6.108, Lee et al. [194] proposed a branch-and-bound algorithm for the problem $1|p_j = a_j + bt, r_j|C_{\max}$. The algorithm has been tested on instances with $12 \leq n \leq 28$ jobs; see [194, Sect. 5] for details.

The authors also proposed two heuristic algorithms for the above problem. We will consider these algorithms in Chap. 9.

6.1.8 Non-linear shortening

Distinct ready times and deadlines

Now, we pass to single-machine time-dependent scheduling problems with shortening job processing times other than linear ones.

Cheng et al. [56] introduced *decreasing step* job processing times

$$p_{i,j} = \begin{cases} a_j, & \text{if } t < D, \\ a_j - b_j, & \text{if } t \geq D, \end{cases} \qquad (6.53)$$

where $\sum_{j=1}^{n} a_j \geq D$ and $0 \leq b_j \leq a_j$ for $1 \leq j \leq n$.

Theorem 6.109. (Cheng et al. [56]) *The decision version of the problem* $1|p_j \in \{a_j, a_j - b_j : 0 \leq b_j \leq a_j\}|C_{\max}$ *is $\mathcal{NP}$-complete in the ordinary sense.*

Proof. The transformation from the PP problem (cf. Sect. 3.2) is as follows. Let $n = k$, $a_j = 2x_j$ and $b_j = x_j$ for $1 \leq j \leq n$, $D = 2A$ and the threshold $G = 3A$.

To complete the proof, it is sufficient to show that the PP problem has a solution if and only if for the above instance of the problem $1|p_j \in \{a_j, a_j - b_j : 0 \leq b_j \leq a_j\}|C_{\max}$ there exists a schedule σ such that $C_{\max}(\sigma) \leq G$.   □

There exists a relationship between the KP problem and the problem $1|p_j \in \{a_j, a_j - b_j : 0 \leq b_j \leq a_j\}|C_{\max}$.

Lemma 6.110. (Cheng et al. [56]) *A version of the KP problem is equivalent to the problem* $1|p_j \in \{a_j, a_j - b_j : 0 \leq b_j \leq a_j\}|C_{\max}$.

Proof. Consider an optimal schedule σ for the above scheduling problem. Let $\mathcal{J}_E := \{J_k \in \mathcal{J} : S_k < D\}$ and $\mathcal{J}_T := \mathcal{J} \setminus \mathcal{J}_E = \{J_k \in \mathcal{J} : S_k \geq D\}$. Let E and T denote sets of indices of jobs from the set $\mathcal{J}_E$ and $\mathcal{J}_T$, respectively. Only two cases are possible: either $\sum_{j\in E} a_j \leq D - 1$ or $\sum_{j\in E} a_j \geq D$.

In the first case, we have $C_{\max}(\sigma) = D + \sum_{j\in T}(a_j - b_j)$. This, in turn, corresponds to the solution of the following KP problem: $\min \sum_{j\in T}(a_j - b_j)$ subject to $\sum_{j\in T} a_j \geq \sum_{j=1}^{n} a_j - D + 1$ for $T \subseteq \{1, 2, \ldots, n\}$.

In the second case, we have $C_{\max}(\sigma) = \sum_{j\in E}(a_j) + \sum_{j\in T}(a_j - b_j) = \sum_{j\in E} b_j + \sum_{j\in E}(a_j - b_j) + \sum_{j\in T}(a_j - b_j) = \sum_{j=1}^{n}(a_j - b_j) + \sum_{j\in E} b_j$. This, in turn, corresponds to the solution of the following KP problem: $\min \sum_{j\in T} b_j$ subject to $\sum_{j\in E} a_j \geq D$ for $E \subseteq \{1, 2, \ldots, n\}$.

The optimal $C_{\max}$ value equals $\min\{D + z_1, \sum_{j=1}^{n}(a_j - b_j) + z_2\}$, where z_1 (z_2) is the solution of the first (the second) KP problem.   □

By Lemma 6.110, we obtain the following result.

Theorem 6.111. (Cheng et al. [56])
The problem $1|p_j \in \{a_j, a_j - b_j : 0 \leq b_j \leq a_j\}|C_{\max}$ *is solvable in* $O(n \sum_{j=1}^{n} a_j)$ *time by a pseudopolynomial algorithm.*

Proof. Since, by Lemma 6.110, a version of the KP problem is equivalent to the problem $1|p_j \in \{a_j, a_j - b_j : 0 \le b_j \le a_j\}|C_{\max}$, and since the KP problem can be solved in $O(n\sum_{j=1}^{n} u_j)$ time by a pseudopolynomial algorithm (see, e.g., Kellerer et al. [163]), the result follows.   □

Cheng et al. [56] and Ji et al. [156] proposed approximation algorithms for the problem $1|p_j \in \{a_j, a_j - b_j : 0 \le b_j \le a_j\}|C_{\max}$. We will consider these algorithms in Chap. 9.

Cheng et al. [54] introduced the *decreasing step-linear* job processing times. In this case, the processing time $p_{i,j}$ of the job J_i scheduled on machine M_j, $1 \le i \le n$ and $1 \le j \le m$, is as follows:

$$p_{i,j} = \begin{cases} a_{i,j}, & \text{if } t \le y, \\ a_{i,j} - b_{i,j}(t-y), & \text{if } y < t < Y, \\ a_{i,j} - b_{i,j}(Y-y), & \text{if } t \ge Y, \end{cases} \tag{6.54}$$

where $a_{i,j} > 0, b_{i,j} > 0, y \ge 0$ and $Y \ge y$ are the *basic* processing time, the *shortening rate*, the *common initial shortening date* and the *common final shortening date*, respectively. It is also assumed that

$$0 < b_{i,j} < 1 \tag{6.55}$$

and

$$a_{i,j} > b_{i,j}\left(\min\{\sum_{k=1}^{n} a_{i,k} - a_{i,j}, Y\} - y\right) \tag{6.56}$$

for $1 \le j \le n$ and $1 \le i \le m$.

Remark 6.112. The decreasing step-linear processing times given by (6.54) are counterparts of the increasing step-linear processing times given by (6.39).

Remark 6.113. Conditions (6.55) and (6.56) are generalizations of conditions (6.47) and (6.48) for the case of parallel machines.

Cheng et al. [54] considered the case of job processing times (6.54) for a single machine, i.e., when $m = 1$.

Theorem 6.114. (Cheng et al. [54]) *The decision version of the problem* $1|p_j = a_j - b_j(t-y), y = 0, 0 < b_j < 1, Y < \infty|C_{\max}$ *is $\mathcal{NP}$-complete in the ordinary sense.*

Proof. The transformation from the PP problem (cf. Sect. 3.2) is as follows. Let $V = (k!)(2k)^{3k+6}A^2$, $B = V^4$, $\alpha = \frac{1}{V^{20}}$ and $\beta = \frac{1}{V^{22}}$. Define $2k$ jobs with the following job processing times: $a_{1,j} = B + 2^j A + x_j$ and $a_{2,j} = B + 2^j A$ for $1 \le j \le k$, $b_{1,j} = \alpha a_{1,j} - \beta(2k)^j A - \frac{x_j}{k-j+1}$, $b_{2,j} = \alpha a_{2,j} - \beta(2k)^j A$ for $1 \le j \le k$.

The common initial shortening date is $y = 0$ and the common final shortening date $Y = kB + A(2^{k+1} - 1)$. The threshold $G = 2Y - \alpha E + \beta BF + 2\alpha V$, where the constants E and F are defined as follows: $E = \frac{3Y^2 - kB^2}{2} - BA(2^{k+1} - 1)$ and $F = \sum_{j=1}^{k} \left((k + j - 1)(2k)^{k+j-1} A + \frac{kx_j}{k-j+1} \right) - A$.

To complete the proof, it is sufficient to show that the PP problem has a solution if and only if for the above instance of the problem $1|p_j = a_j - b_j(t-y)$, $y = 0, 0 < b_j < 1, Y < \infty|C_{\max}$ there exists a feasible schedule σ such that $C_{\max}(\sigma) \leq G$ (see [54, Lemmata 1–3]).   □

Cheng et al. [54] also proved a few properties of optimal schedules of the problem $1|p_j = a_j - b_j(t - y), y > 0, 0 < b_j < 1, Y < \infty|C_{\max}$, where $\sum_{j=1}^{n} a_j > y$. The terminology used in formulations of the properties is equivalent to the terminology introduced in Definition 6.73 for bounded step-linear processing times (6.39), provided that $y \equiv d$ and $Y \equiv D$.

Property 6.115. (Cheng et al. [54]) The order of the early jobs and the order of the suspended jobs are immaterial.

Proof. Since both the early and suspended jobs have fixed processing times and the completion time of the last early (suspended) job does not depend on the order of the early (suspended) jobs, the result follows.   □

Property 6.116. (Cheng et al. [54]) The tardy jobs are sequenced in the non-increasing order of $\frac{a_j}{b_j}$ ratios.

Proof. The result is a corollary from Theorem 6.87 (c).   □

Property 6.117. (Cheng et al. [54]) The pivotal job has a processing time not larger than that of any of the early jobs.

Proof. By pairwise job interchange argument.   □

Property 6.118. (Cheng et al. [54]) If $a_i \geq a_j$ and $b_i \leq b_j$, then job J_i precedes job J_j.

Proof. By pairwise job interchange argument.   □

Theorem 6.119. (Cheng et al. [54])
(a) *The problem* $1|p_j = a_j - b(t - y), y > 0, 0 < b_j < 1, Y < \infty|C_{\max}$ *is solvable in* $O(n \log n)$ *time by scheduling jobs in the non-increasing order of* a_j *values.*
(b) *The problem* $1|p_j = a - b_j(t - y), y > 0, 0 < b_j < 1, Y < \infty|C_{\max}$ *is solvable in* $O(n \log n)$ *time by scheduling jobs in the non-decreasing order of* b_j *values.*

Proof. (a), (b) The results are corollaries from Property 6.118.   □

By Theorem 6.119, the problems $1|p_j = a_j - b(t - y), y > 0, 0 < b_j < 1$, $Y < \infty|C_{\max}$ and $1|p_j = a - b_j(t - y), y > 0, 0 < b_j < 1, Y < \infty|C_{\max}$

are solved by the algorithms $A_8 : (a_j) \mapsto (a_j \searrow)$ and $A_9 : (a_j) \mapsto (b_j \nearrow)$, respectively.

For the problem $1|p_j = a_j - b_j(t - y), y = 0, 0 < b_j < 1, Y < \infty|C_{\max}$ Cheng et al. proposed a pseudopolynomial-time algorithm. The algorithm runs in $O(n^2 \sum_{j=1}^n a_j \sum_{j=1}^n v_j)$ time, where $v_j = b_j L$, $L = const$; see [54, Sect. 2.3] for details.

Cheng et al. [54] also proposed three heuristic algorithms for the problem $1|p_j = a_j - b_j(t - y), y = 0, 0 < b_j < 1, Y < \infty|C_{\max}$. We will present these algorithms in Chap. 9.

6.2 Minimizing the total completion time

In this section, we consider single-machine time-dependent scheduling problems with the $\sum C_j$ criterion.

6.2.1 Proportional deterioration

Equal ready times and deadlines

The case of jobs with proportional processing times (6.1) is polynomially solvable.

Theorem 6.120. (Mosheiov [216]) *The problem* $1|p_j = b_j t| \sum C_j$ *is solvable in $O(n \log n)$ time by scheduling jobs in the non-decreasing order of b_j values.*

Proof. First, note that by summing $C_j \equiv C_{[j]}(\sigma)$ from (6.2) for $1 \le j \le n$ we obtain the formula for the total completion time:

$$\sum_{j=1}^n C_j = S_1 \sum_{j=1}^n \prod_{k=1}^j (1 + b_{[k]}) = t_0 \sum_{j=1}^n \prod_{k=1}^j (1 + b_{[k]}). \tag{6.57}$$

Now we can prove that the right side of formula (6.57) is minimized by sequencing b_j in the non-decreasing order in two different ways. The first way is to apply pairwise job interchange argument. To this end, we calculate the total completion time for schedule σ' in which job J_i is followed by job J_j and $a_i > a_j$. Next, we show that schedule σ'', which is obtained from σ' by changing the order of jobs J_i and J_j, has lower total completion time than σ'. Repeating the above reasoning for all such pairs of jobs, we obtain the result. The second way of proof is to apply Lemma 1.2 (a) to formula (6.57). □

By Theorem 6.120, the problem $1|p_j = b_j t| \sum C_j$ is solved by the algorithm $A_9 : (b_j) \mapsto (b_j \nearrow)$.

The problem $1|p_j = b_j t| \sum C_j$ was also considered in more general settings. Wu et al. [300] reformulated the problem $1|p_j = b_j t| \sum C_j$ into a batch scheduling problem (cf. Theorem 6.5 and Remark 6.6).

Theorem 6.121. (Wu et al. [300]) *In the optimal schedule for the problem* $1|p_{i,j} = b_{i,j}t, \theta_i = \delta_i, GT| \sum C_j$ *in each group jobs are scheduled in the non-decreasing order of* $b_{i,j}$ *values and groups are scheduled in the non-decreasing order of* $\frac{(1+\delta_i)A_{k_i}-1}{(1+\delta_i)B_{k_i}}$ *ratios, where* $A_{k_i} := \prod_{j=1}^{k_i}(1 + b_{i,j})$ *and* $B_{k_i} := \sum_{k=1}^{k_i} \prod_{j=1}^{k}(1 + b_{i,j})$ *for* $1 \leq i \leq m$.

Proof. The first part of the theorem is a corollary from Theorem 6.120. The second part can be proved by contradiction, see [300, Theorem 2].   ◇

On the basis of Theorem 6.121 Wu et al. proposed the following algorithm.

Algorithm A_{16} for the problem $1|p_{i,j} = b_{i,j}t, \theta_i = \delta_i, GT| \sum C_j$ ([300])

Input: sequences $(\delta_1, \delta_2, \ldots, \delta_n), (b_{i,j})$ for $1 \leq i \leq m$ and $1 \leq j \leq k_i$
Output: an optimal schedule σ^*

▷ Step 1:
   for $i \leftarrow 1$ **to** m **do**
   Arrange jobs in group G_i in the non-decreasing order of $b_{i,j}$ values;
▷ Step 2:
   Arrange groups $G_1, G_2, \ldots, G_m$ in the non-decreasing order of $\frac{(1+\delta_i)A_{k_i}-1}{(1+\delta_i)B_{k_i}}$
   ↪ ratios, where $A_{k_i} := \prod_{j=1}^{k_i}(1 + b_{i,j})$ and $B_{k_i} := \sum_{k=1}^{k_i} \prod_{j=1}^{k}(1 + b_{i,j})$
   ↪ for $1 \leq i \leq m$;
▷ Step 3:
   $\sigma^* \leftarrow$ the schedule obtained in Step 2;
   return σ^*.

Theorem 6.122. (Wu et al. [300]) *The problem* $1|p_{i,j} = b_{i,j}t, \theta_i = \delta_i, GT| \sum C_j$ *is solvable in* $O(n \log n)$ *time by algorithm* A_{16}.

Proof. The result is a consequence of Theorem 6.121 and the fact that Step 1 and Step 2 need $O(n \log n)$ and $O(m \log m)$ time, respectively.   □

The problem $1|p_j = b_j t| \sum C_j$ becomes computationally intractable, if the applied machine is not continuously available (cf. Remark 4.4) and if jobs are non-resumable (cf. Definition 6.9).

Theorem 6.123. (Ji et al. [155]) *The decision version of the problem* $1, h_{11}|p_j = b_j t, nres| \sum C_j$ *is $\mathcal{NP}$-complete in the ordinary sense.*

Proof. Ji et al. [155] use the following transformation from the SP problem (cf. Sect. 3.2): $n = p + 4$, arbitrary t_0, $W_{1,1} = t_0 B^5$, arbitrary $W_{1,2} > W_{1,1}$, $b_j = y_j - 1$ for $1 \leq j \leq p$, $b_{p+1} = YB - 1$, $b_{p+2} = \frac{Y^2}{B} - 1$, $b_{p+3} = b_{p+4} = Y^3 - 1$ and threshold $G = (p + 2)W_{1,2}B^2 + (t_0 + W_{1,2})B^5$, where $Y = \prod_{j=1}^{p} y_j$.
   To complete the proof, it is sufficient to show that the SP problem has a solution if and only if there exists a feasible schedule σ for the above instance of the problem $1|p_j = b_j t| \sum C_j$ with the non-availability period $\langle W_{1,1}, W_{1,2} \rangle$ such that $\sum C_j(\sigma) \leq G$.   □

Remark 6.124. For the problem $1, h_{11}|p_j = b_j t, nres| \sum C_j$ Ji et al. [155] proposed a pseudopolynomial algorithm which runs in $O(n(W_{1,1} - t_0)W$, where $W := \prod_{j=1}^{n}(1 + b_j)$. Hence, by Lemma 3.18, this problem cannot be $\mathcal{NP}$-hard in the strong sense.

Distinct ready times and deadlines

If all jobs have distinct deadlines, the problem $1|p_j = b_j t, d_j| \sum C_j$ is polynomially solvable. Provided that there exists a schedule in which all jobs are completed before their deadlines, the following algorithm solves the problem.

Algorithm A_{17} for the problem $1|p_j = b_j t, d_j| \sum C_j$

Input: sequences $(b_1, b_2, \ldots, b_n), (d_1, d_2, \ldots, d_n)$
Output: an optimal schedule $\sigma^\star$

▷ Step 1:
$$B \leftarrow \prod_{j=1}^{n}(b_j + 1);$$
$N_{\mathcal{J}} \leftarrow \{1, 2, \ldots, n\};$
$k \leftarrow n;$
$\sigma^\star \leftarrow (\phi);$
▷ Step 2:
while $(N_{\mathcal{J}} \neq \emptyset)$ **do**
$\quad \mathcal{J}_B \leftarrow \{J_i \in \mathcal{J} : d_i \geq B\};$
$\quad$ Choose job $J_j \in \mathcal{J}_B$ with maximal b_j;
$\quad$ Schedule J_j in $\sigma^\star$ in position k;
$\quad k \leftarrow k - 1;$
$\quad N_{\mathcal{J}} \leftarrow N_{\mathcal{J}} \setminus \{j\};$
$\quad B \leftarrow \frac{B}{b_j + 1};$
▷ Step 3:
$\quad$ **return** $\sigma^\star$.

Theorem 6.125. (Cheng et al. [55]) *The problem $1|p_j = b_j t, d_j| \sum C_j$ is solvable in $O(n \log n)$ time by algorithm A_{17}.*

Proof. The result follows from the fact that algorithm A_{17} is an adaptation of *Smith's backward scheduling rule* for the problem $1|d_j| \sum C_j$ (Smith [260]) to the problem $1|p_j = b_j t, d_j| \sum C_j$. $\qquad \square$

Remark 6.126. Cheng et al. [55] give the result without a proof. The formulation of algorithm A_{17} comes from the present author.

6.2.2 Proportional-linear deterioration

The problem of minimizing the $\sum C_j$ criterion with job processing times given by (6.5) is polynomially solvable.

Theorem 6.127. *The problem* $1|p_j = b_j(A+Bt)|\sum C_j$ *is solvable in* $O(n \log n)$ *time by scheduling jobs in the non-decreasing order of* $\frac{b_j}{1+Bb_j}$ *ratios.*

Proof. The result is a corollary from Theorem 6.197 for $w_j = 1$ for $1 \le j \le n$. ∎

By Theorem 6.127, the problem $1|p_j = b_j(A + Bt)|\sum C_j$ is solved by the algorithm $A_{18} : (b_j|B) \mapsto (\frac{b_j}{1+Bb_j} \nearrow)$.

6.2.3 Linear deterioration

Now, we pass to single-machine time-dependent scheduling problems with job processing times given by (6.10).

Equal ready times and deadlines

If we consider the problem of single-machine scheduling jobs with linear processing times, the situation is unclear. On the one hand, it seems that the problem is not difficult since by summing the right side of (6.13) for $1 \le j \le n$ we obtain the formula for the total completion time:

$$\sum_{j=1}^{n} C_j(\sigma) = \sum_{i=1}^{n}\sum_{j=1}^{i} a_{\sigma_j} \prod_{k=j+1}^{i}(1 + b_{\sigma_k}) + t_0 \sum_{i=1}^{n}\prod_{j=1}^{i}(1 + b_{\sigma_j}), \qquad (6.58)$$

from which we can obtain formulae for special cases, e.g., when $a_{\sigma_j} = 0$ for $1 \le j \le n$ or $t_0 = 0$. Therefore, for a given schedule we can easily calculate job completion times.

On the other hand, the complexity of the problem $1|p_j = a_j + b_j t|\sum C_j$ is still unknown, even if $a_j = 1$ for $1 \le j \le n$. In this case, the following result is known.

Property 6.128. (Mosheiov [215]) If $k = \arg \max_{1 \le j \le n} \{b_j\}$, then in an optimal schedule for the problem $1|p_j = 1 + b_j t|\sum C_j$ job J_k is scheduled as the first one.

Proof. Consider job J_k with the greatest deterioration rate, $k = \arg \max\{b_j\}$. (If there are several jobs that have the same rate, choose any of them.) Since the completion time of job J_k is $C_k = 1 + (1 + b_k)S_k$, this completion time (and thus the total completion time) will be the smallest if $S_k = 0$. ∎

Notice that by Property 6.128, we may reduce the original problem to the problem with only $n - 1$ jobs, since the greatest job (i.e., the one with the greatest deterioration rate) has to be scheduled as the first one. Moreover, since in this case this job contributes only one unit of time to the total completion time, we can also reformulate the criterion function. Namely, for any schedule σ for the problem $1|p_j = 1 + b_j t| \sum C_j$ there holds the equality $\sum C_j(\sigma) = g(\sigma) + n$. The function g is defined in the next property.

Property 6.129. (Mosheiov [215]) Let $g(\sigma) := \sum_{i=1}^{n} \sum_{k=1}^{i} \prod_{j=k}^{i} b_{[j]}$ and let σ and $\bar{\sigma}$ be a schedule for the problem $1|p_j = 1 + b_j t| \sum C_j$ and the reverse schedule to σ, respectively. Then $g(\sigma) = g(\bar{\sigma})$.

Proof. Since $\sum_{i=1}^{n} \sum_{j=1}^{i} \prod_{k=j+1}^{i}(1 + b_k) = \sum_{i=1}^{n} \sum_{j=1}^{i} \prod_{k=j+1}^{i}(1 + b_{n-k+1})$, the result follows. ∎

By Property 6.129, we obtain the following *symmetry property* for the $\sum C_j$ criterion in the problem $1|p_j = 1 + b_j t| \sum C_j$.

Property 6.130. (Mosheiov [215]) If σ and $\bar{\sigma}$ are defined as in Property 6.129, then $\sum C_j(\sigma) = \sum C_j(\bar{\sigma})$.

Proof. We have $\sum C_j(\sigma) = g(\sigma) + n$. But, by Property 6.129, $g(\sigma) = g(\bar{\sigma})$. Hence $\sum C_j(\sigma) = g(\sigma) + n = g(\bar{\sigma}) + n = \sum C_j(\bar{\sigma})$. ∎

From Property 6.129 follows the next property.

Property 6.131. (Mosheiov [215]) Let $k = \arg \min_{1 \leq j \leq n} \{b_j\}$. Then, in the optimal schedule for the problem $1|p_j = 1 + b_j t| \sum C_j$, job J_k is scheduled neither as the first nor as the last one.

Proof. Let $\sigma = (k, \sigma_2, \dots, \sigma_n)$ be any schedule in which job J_k is scheduled as the first one. Consider schedule $\sigma' = (\sigma_2, k, \dots, \sigma_n)$. Then we have $g(\sigma') - g(\sigma) = (b_k - b_2) \sum_{i=3}^{n} \prod_{j=3}^{i}(1 + b_j) \leq 0$, since $b_k - b_2 \leq 0$ by assumption. Hence, schedule σ' is better than schedule σ.

By Property 6.129, job J_k cannot be scheduled as the last one either. □

Property 6.131 allows us to prove the following result.

Property 6.132. (Mosheiov [215]) Let J_{i-1}, J_i and J_{i+1} be three consecutive jobs in a schedule for the problem $1|p_j = 1 + b_j t| \sum C_j$. If $b_i > b_{i-1}$ and $b_i > b_{i+1}$, then this schedule cannot be optimal.

Proof. The main idea of the proof is to show that the exchange of jobs J_{i-1} and J_i or of jobs J_i and J_{i+1} leads to a better schedule.

Let σ, σ' and σ'' denote schedules in which jobs J_{i-1}, J_i, J_{i+1} are scheduled in the order $(\sigma^{(a)}, i-1, i, i+1, \sigma^{(b)})$, $(\sigma^{(a)}, i, i-1, i+1, \sigma^{(b)})$ and $(\sigma^{(a)}, i-1, i+1, i, \sigma^{(b)})$, respectively, where $\sigma^{(a)}$ ($\sigma^{(b)}$) denotes the part of schedule σ before (after) the jobs J_{i-1}, J_i, J_{i+1}.

Since

$$g(\sigma') - g(\sigma) = (b_i - b_{i-1})\sum_{k=1}^{i-2}\prod_{j=k}^{i-2}(1+b_j) + (b_{i-1} - b_i)\sum_{k=i+1}^{n}\prod_{j=i+1}^{k}(1+b_j)$$

and

$$g(\sigma'') - g(\sigma) = (b_{i+1} - b_i)\sum_{k=1}^{i-1}\prod_{j=k}^{i-1}(1+b_j) + (b_i - b_{i+1})\sum_{k=i+2}^{n}\prod_{j=i+2}^{k}(1+b_j),$$

it can be shown (see [215, Lemma 2]) that the two differences cannot be both positive. Hence, either σ' or σ'' are better schedules than schedule σ. □

From Properties 6.131–6.132, there follows the result describing the so-called *V-shape property* for the problem $1|p_j = 1 + b_j t|\sum C_j$.

Theorem 6.133. (Mosheiov [215]) *The optimal schedule for the problem $1|p_j = 1 + b_j t|\sum C_j$ is V-shaped with respect to deterioration rates b_j.*

Proof. The result is a consequence of Property 6.131 and Property 6.132. □

Remark 6.134. V-shaped sequences were introduced in Definition 1.3.

Theorem 6.133 allows to decrease the number of possible schedules from $n!$ to $2^{n-3} - 1$. In some cases, we can obtain a V-shaped sequence that is optimal.

Definition 6.135. (Mosheiov [215]) *A sequence (x_j), $1 \leq j \leq n$, is said to be perfectly symmetric V-shaped (to have a perfect V-shape), if it is V-shaped and $x_i = x_{n-i+2}$, $2 \leq i \leq n$.*

The following result shows the importance of perfectly symmetric V-shaped sequences for the problem $1|p_j = 1 + b_j t|\sum C_j$.

Theorem 6.136. (Mosheiov [215]) *If a perfectly symmetric V-shaped sequence can be constructed from the sequence of deterioration rates of an instance of the problem $1|p_j = 1 + b_j t|\sum C_j$, then the sequence is optimal.*

Proof. See Mosheiov [215, Proposition 3]. ◇

For the problem $1|p_j = 1 + b_j t|\sum C_j$, Mosheiov [215] proposed two heuristic algorithms, both running in $O(n \log n)$ time. We will consider these heuristics in Chap. 9.

Now we consider some special cases of the problem $1|p_j = 1 + b_j t|\sum C_j$. Assume that in the set $\{b_1, b_2, \ldots, b_n\}$ there exist only k different values, i.e.,

$$b_j \in \{B_1, B_2, \ldots, B_k\}, \tag{6.59}$$

where $1 \leq j \leq n$ and $k < n$ is fixed. Without loss of generality, we can assume that

$$B_1 \geq B_2 \geq \ldots B_{k-1} \geq B_k.$$

Lemma 6.137. (Gawiejnowicz et al. [106]) *If condition (6.59) is satisfied, then in the optimal schedule for the problem $1|p_j = 1 + b_j t|\sum C_j$ the jobs with the smallest deterioration rate b_k are scheduled as one group without inserting between them jobs with deterioration rates greater than b_k.*

Proof. The result is a corollary from Theorem 6.133.   □

For purposes of the next result, we will say that a *deterioration rate b is of type b_k*, if $b = B_k$.

Theorem 6.138. (Gawiejnowicz et al. [106]) *If condition (6.59) is satisfied, then the optimal schedule for the problem $1|p_j = 1 + b_j t|\sum C_j$ is V-shaped with respect to types of jobs.*

Proof. By mathematical induction with respect to the number of types k and by Lemma 6.137, the result follows.   □

On the basis of Theorem 6.138, we can construct the following algorithm.

Algorithm E_1
for the problem $1|p_j = 1 + b_j t, b_j \in \{B_1, B_2, \ldots, B_k\}|\sum C_j$ ([106])

Input: sequence $(b_1, b_2, \ldots, b_n)$
Output: an optimal schedule $\sigma^\star$

▷ **Step 1:**
Construct set $\mathcal{V}$ of the schedules which have the V-shape property
$\hookrightarrow$ described in Theorem 6.138;
▷ **Step 2:**
for all $\nu \in \mathcal{V}$ **do** calculate $\sum C_j(\nu)$;
▷ **Step 3:**
$\sigma^\star \leftarrow \arg\min\{\sum C_j(\tau) : \tau \in \mathcal{V}\}$;
return $\sigma^\star$.

Theorem 6.139. (Gawiejnowicz et al. [106]) *If $b_j \in \{B_1, B_2, \ldots, B_k\}$, the problem $1|p_j = 1 + b_j t|\sum C_j$ is solvable by algorithm E_1 in $O(n^{k+1})$ time.*

Proof. Since Step 1 and Step 3 need $O(n^k)$ time and calculation of $\sum C_j(\sigma)$ for a given $\sigma \in \mathcal{V}$ needs $O(n)$ time, the result follows.   ∎

Another polynomial case of the problem $1|p_j = 1 + b_j t|\sum C_j$ has been considered by Ocetkiewicz [228]. The author has shown that if for any i and j such that $1 \leq i \neq j \leq n$ there holds the implication

$$b_i > b_j \implies b_i \geq \frac{b_{\min} + 1}{b_{\min}} + \frac{1}{b_{\min}}, \tag{6.60}$$

with $b_{\min} := \min_{1 \leq j \leq n}\{b_j\}$, the optimal V-shaped job sequence for the problem (cf. Definition 1.3) can be constructed in polynomial time. This is caused by the fact that implication (6.60) allows to indicate to which branch of the constructed V-shaped sequence a given ratio b_i should be added.

The pseudo-code of the algorithm can be formulated as follows.

Algorithm A_{19} for the problem $1|p_j = 1 + b_j t| \sum C_j$ ([228])

Input: sequence $(b_1, b_2, \ldots, b_n)$
Output: an optimal schedule $\sigma^\star$

▷ Step 1:
Arrange jobs in the non-decreasing order of b_j values;
$L \leftarrow (1 + b_{[n-1]})$;
$\sigma^{(L)} \leftarrow (b_{[n-1]})$;
$R \leftarrow 0$;
$\sigma^{(R)} \leftarrow (\phi)$;
▷ Step 2:
for $i \leftarrow n - 2$ **downto** 2 **do**
    if $(L > R)$ **then** $\sigma^{(R)} \leftarrow (b_{[i]}|\sigma^{(R)})$;
                    $R \leftarrow (R + 1)(1 + b_{[i]})$
    else $\sigma^{(L)} \leftarrow (\sigma^{(L)}|b_{[i]})$;
           $L \leftarrow (L + 1)(1 + b_{[i]})$;
▷ Step 3:
$\sigma^\star \leftarrow (b_{[n]}|\sigma^{(L)}|b_{[1]}|\sigma^{(R)})$;
return $\sigma^\star$.

Theorem 6.140. (Ocetkiewicz [228]) *If all jobs have distinct deterioration rates and for any $1 \leq i \neq j \leq n$ there holds implication (6.60), then the problem $1|p_j = 1 + b_j t| \sum C_j$ is solvable by algorithm A_{19} in $O(n \log n)$ time.*

Proof. By direct calculation, see [228, Sect. 2]   ◇

Unlike the case of proportional job processing times (6.1), the problem of minimization of the total completion time for a set of jobs with linear processing times (6.10) is computationally intractable even if all deadlines are equal.

Theorem 6.141. (Cheng et al. [55]) *The decision version of the problem $1|p_j = a_j + b_j t, d_j = D| \sum C_j$ is $\mathcal{NP}$-complete in the ordinary sense.*

Proof. The authors state the result without proof, see [55, Theorem 2].   ◇

Distinct ready times and deadlines

If deadlines are arbitrary, then the problem of minimizing $\sum C_j$ is also computationally intractable.

Theorem 6.142. (Cheng and Ding [52]) *The problem* $1|p_j = a_j + b_j t, d_j| \sum C_j$ *is* $\mathcal{NP}$-*complete in the strong sense.*

Proof. The authors give only a sketch of proof, see [52, Theorem 6].   ◇

If $b_j = b$ for $1 \leq j \leq n$, then there holds the following result.

Lemma 6.143. (Cheng and Ding [52]) *The problem* $1p_j = a_j + bt, d_j|C_{\max}$ *is equivalent to the problem* $1|p_j = a_j + bt, d_j| \sum C_j$.

Proof. Let σ be an arbitrary schedule for the problem $1p_j = a_j + bt, d_j|C_{\max}$. Then

$$C_i(\sigma) = a_i + Ba_{i-1} + \ldots + B^{i-1}a_1 = \sum_{k=1}^{i} B^{i-k} a_k,$$

where $B = 1 + b$.

Since $C_i(\sigma) = \sum_{j=1}^{i} p_j$ and $p_j = a_j + bS_j$, we have

$$C_i(\sigma) = \sum_{j=1}^{i} (a_j + bS_j) = \sum_{j=1}^{i} a_j + b \sum_{j=1}^{i} S_j = \sum_{j=1}^{i} a_j + b \sum_{j=1}^{i} \frac{1}{B}(C_j(\sigma) - a_j) =$$

$$\frac{1}{B} \sum_{j=1}^{i} a_j + \frac{b}{B} \sum_{j=1}^{i} C_j.$$

Therefore, the problem of minimizing the maximum completion time is equivalent to the problem of minimizing the total completion time. ∎

Remark 6.144. Lemma 6.143 gives an example of two time-dependent scheduling problems that are equivalent in some sense. In Sect. 12.4, we will consider other forms of equivalent time-dependent scheduling problems.

Theorem 6.145. (Cheng and Ding [48]) *The problem* $1|p_j = a_j + bt, d_j| \sum C_j$ *is solvable in* $O(n^5)$ *time by algorithm* A_6.

Proof. By Lemma 6.143 and Theorem 6.38, the result follows. □

6.2.4 Simple non-linear deterioration

Equal ready times and deadlines

A single-machine time-dependent scheduling problem with simple non-linear job processing times given by (6.15) is polynomially solvable.

Theorem 6.146. (Gawiejnowicz [89]) *The problem* $1|p_j = a_j + f(t)| \sum C_j$, *where* $f(t)$ *is an arbitrary increasing function, is optimally solved in* $O(n \log n)$ *time by scheduling jobs in the non-decreasing order of* a_j *values.*

Proof. Since $\sum_{j=1}^{n} C_j = \sum_{j=1}^{n} (n-j+1)(a_j + f(S_j))$, and sequence $(n-j+1)$ is decreasing, the result follows from Lemma 1.2 (a). ∎

By Theorem 6.146, the problem $1|p_j = a_j + f(t), f \nearrow| \sum C_j$ is solved by the algorithm $A_7 : (a_j) \mapsto (a_j \nearrow)$.

6.2.5 General non-linear deterioration

Equal ready times and deadlines

For job processing times given by (6.27), the following result is known.

Theorem 6.147. (Cheng and Ding [51]) *The decision version of the problem* $1|p_j \in \{a_j, b_j : a_j \leq b_j\}| \sum C_j$ *is $\mathcal{NP}$-complete in the ordinary sense even if* $d_j = D$ *for* $1 \leq j \leq n$.

Proof. Given an instance of the PP problem (cf. Sect. 3.2), construct an instance of the problem $1|p_j \in \{a_j, b_j : a_j \leq b_j\}, d_j = D| \sum C_j$ as follows. Let $n = 2k+1$ and $B = 2^{k+1}A$. Define $a_0 = (k+1)B$, $a_{2j-1} = B + 2^j A + x_j$ and $a_{2k} = B + 2^j A$ for $1 \leq j \leq k$. Let $C = 4kA$, $E = \sum_{j=1}^{k}(2k + 2 - j)a_{2j-1} + (k+1)a_0 + (A+1)C$ and $F = E + E\sum_{j=1}^{k}(k+1-j)k^j$. Define $b_0 = F^2$, $b_{2j-1} = F - a_{2j} + k^j E + \frac{Cx_j}{k+1-j}$ and $b_{2j} = F - a_{2j} + k^j E$ for $1 \leq j \leq k$. The identical deteriorating rate $D = kB + 2^{k+1}A - A$. The threshold $G = F + F\sum_{j=1}^{k}(k+1-j)$.

To complete the proof, it is suficient to show that the PP problem has a solution if and only if for the above instance of the considered problem there exists a schedule σ such that $C_{\max}(\sigma) \leq G$ (see [51, Appendix A]). □

For the case when $d_j = D$ for $1 \leq j \leq n$, Jeng and Lin [152] formulated a branch-and-bound algorithm. Before we formulate the next result concerning this case, we define a new type of a schedule.

Definition 6.148. (A normal schedule)
A schedule for the problem $1|p_j \in \{a_j, b_j : a_j \leq b_j\}, d_j = D| \sum C_j$ *is called normal, if the last job in the set of early jobs is started by the common due date D and finished after D.*

The given below result allows to estimate the number of early jobs in a normal schedule.

Lemma 6.149. Jeng and Lin [152]) *Let $E := \{J_k \in \mathcal{J} : C_k \leq d\}$ and let U (L) be the smallest (largest) integer satisfying the inequality $\sum_{j=1}^{U} a_i > D$ ($\sum_{j=n-U}^{U} a_i > D$). Then for an arbitrary normal schedule for the problem* $1|p_j \in \{a_j, b_j : a_j \leq b_j\}, d_j = D| \sum C_j$ *we have $U \leq |E| \leq L$.*

Proof. Note that without loss of generality, we can consider only the schedules in which jobs are indexed in the non-decreasing order of a_j values. Since time interval $\langle 0, d \rangle$ can be considered as a one-dimensional bin, and job processing times a_j as the items to pack into the bin, the result follows. □

Remark 6.150. The terminology used in the proof of Lemma 6.149 is related to the BP problem (cf. Sect. 3.2).

Lemma 6.151. (Jeng and Lin [152]) *Let $b_{(j)}$ be the j-th smallest deterioration rate in an instance of the problem $1|p_j \in \{a_j, b_j : a_j \leq b_j\}, d_j = D| \sum C_j$. Then for any schedule σ for the problem there holds the inequality $\sum C_j(\sigma) \geq \sum_{j=1}^{n} (n - j + 1)a_j + \sum_{j=1}^{n-U} (n - U - j + 1)b_{(j)}$.*

Proof. Let $E := \{J_k \in \mathcal{J} : C_k \leq D\}$. By definition of the problem $1|p_j \in \{a_j, b_j : a_j \leq b_j\}, d_j = D| \sum C_j$, for any schedule σ we have $\sum C_j(\sigma) = \sum_{j=1}^{n} (n - j + 1)a_{[j]} + \sum_{j=1}^{n-|E|} (n - |E| - j + 1)b_{[j]}$. Since jobs are indexed in the non-decreasing order of a_j values, $\sum C_j(\sigma) = \sum_{j=1}^{n} (n - j + 1)a_{[j]} \geq \sum_{j=1}^{n} (n - j + 1)a_j$. By Lemma 6.149, $|E| \leq U$ and hence $|L| = |\mathcal{J} \setminus E| \geq n - U$. Since $b_{(j)} \leq b_{[j]}$ for any $1 \leq j \leq n$, the result follows.   □

By Lemma 6.151, the value $\sum_{j=1}^{n} (n - j + 1)a_j + \sum_{j=1}^{n-U} (n - U - j + 1)b_{(j)}$ may be used as an initial lower bound of the $\sum C_j$ criterion.

Jeng and Lin also obtained a few results, concerning the possible dominance relationships between jobs. One of the results is the following.

Lemma 6.152. (Jeng and Lin [152]) *Let σ and E be an arbitrary subschedule for the problem $1|p_j \in \{a_j, b_j : a_j \leq b_j\}, d_j = D| \sum C_j$ and the set of early jobs in the subschedule, respectively. If there exists a job J_j not belonging to σ and such that $\sum_{j \in E} a_j + \sum_{i=j}^{n} a_i \leq d$, then each subtree rooted at $E \cup \{k\}$, $j \leq k \leq n$, can be eliminated.*

Proof. See [152, Lemma 4].   ◇

Based on Lemmata 6.149–6.152 and some other results (see [152, Sect. 4]) Jeng and Lin constructed the mentioned above branch-and-bound algorithm for the problem $1|p_j \in \{a_j, b_j : a_j \leq b_j\}, d_j = D| \sum C_j$. The reported results of conducted computational experiments (see [152, Sect. 5]) suggest that the algorithm is quite effective, since it can solve most instances with $n = 100$ jobs in time not longer than 3 minutes.

Cheng and Ding identified some polynomially solvable cases of the problem $1|p_j \in \{a_j, b_j : a_j \leq b_j\}, d_j = D| \sum C_j$. Assume that the conjunction $a_i \neq a_j \wedge b_i \neq b_j$ holds for any $1 \leq i, j \leq n$. Let $E := \{J_k \in \mathcal{J} : C_k \leq d\}$ and $L := \mathcal{J} \setminus E$.

Lemma 6.153. (Cheng and Ding [51]) *If $a_i \leq a_j$ and $a_i + b_i \leq a_j + b_j$, then job J_i precedes job J_j in any optimal schedule for the problem $1|p_j \in \{a_j, b_j : a_j \leq b_j\}| \sum C_j$.*

Proof. By pairwise job interchange argument.   □

By Lemma 6.153, we can divide the set of jobs $\mathcal{J}$ into a number of *chains* as follows. Renumber jobs in the non-decreasing order of a_j values, ties in the non-increasing order of $a_j + b_j$ values. Assume that a job is the head of a chain. Put the next job at the end of the chain, if it has agreeable parameters with the last job in the chain. Repeat the procedure until all jobs are checked. Create other chains from the remaining jobs in the same way.

Notice that knowing all chains, we know which jobs are early and tardy in each chain. Moreover, since the early jobs have known processing times and in this case the total completion time is minimized by scheduling the jobs in the non-decreasing order with respect to a_j values, we obtain a local optimal schedule. If we enumerate all locally optimal schedules, we find an optimal schedule.

Based on the above reasoning, Cheng and Ding [51] proposed the following exact algorithm.

Algorithm E_2
for the problem $1|p_j \in \{a_j, b_j : a_j \le b_j\}, d_j = D|\sum C_j$ ([51])

Input: sequences $(a_1, a_2, \dots, a_n)$, $(b_1, b_2, \dots, b_n)$, number D
Output: an optimal schedule $\sigma^\star$

▷ Step 1:
Construct all chains of jobs $\mathcal{C}_1, \mathcal{C}_2, \dots, \mathcal{C}_k$;
▷ Step 2:
  for all possible $(e_1, e_2, \dots, e_k)$ such that $e_i \le |\mathcal{C}_i|$ for $1 \le i \le k$ **do**
    for $\mathcal{C} \leftarrow \mathcal{C}_1$ **to** $\mathcal{C}_k$ **do**
      Set the number of elements of $\{J_k \in \mathcal{C} : C_k \le D\}$ to $e_{\mathcal{C}}$;
    $E \leftarrow \{J_k \in \mathcal{J} : C_k \le D\}; L \leftarrow \mathcal{J} \setminus E$;
    Schedule jobs in E in the non-decreasing order of a_j values;
    Schedule jobs in L in the non-increasing order of $a_j + b_j$ values;
    if $(C_{\max}(E) \le D)$ **then** compute $\sum C_j(E)$;
▷ Step 3:
  $\sigma^\star \leftarrow$ the best schedule among all schedules generated in Step 2;
  return $\sigma^\star$.

Algorithm E_2 runs in $O(n^k \log n)$ time, where k is the number of chains created in Step 1. If k is a fixed number, then there holds the following result.

Theorem 6.154. (Cheng and Ding [51]) *If for an instance of the problem* $1|p_j \in \{a_j, b_j : a_j \le b_j\}, d_j = D|\sum C_j$ *there is a fixed number of chains* k, *then algorithm* E_2 *is a polynomial-time algorithm for this instance.*

Proof. For a fixed k, the total running time of algorithm E_2 becomes polynomial with respect to n. □

6.2.6 Linear shortening

We will end this section with the results concerning single-machine time-dependent scheduling problems with shortening job processing times.

Equal ready times and deadlines

First we consider shortening job processing times given by (6.46), where for $1 \leq j \leq n$ there hold conditions (6.47) and (6.48). The assumptions eliminate some trivial cases and assure that the constructed instances make sense from the practical point of view.

Property 6.155. (Ng et al. [226]) The problem $1|p_j = a_j - bt| \sum C_j$ is solvable in $O(n \log n)$ time by scheduling jobs in the non-decreasing order of a_j values.

Proof. By pairwise job interchange argument. □

Property 6.156. (Ng et al. [226]) The problem $1|p_j = a_j(1 - kt)| \sum C_j$ is solvable in $O(n \log n)$ time by scheduling jobs in the non-decreasing order of a_j values.

Proof. By pairwise job interchange argument. □

By Properties 6.155–6.156, the problems $1|p_j = a_j - bt, 0 < b < 1| \sum C_j$ and $1|p_j = a_j - ka_j t| \sum C_j$ are both solved by the algorithm $A_7 : (a_j) \mapsto (a_j \nearrow)$.

Property 6.157. (Ng et al. [226]) Any optimal schedule for the problem $1|p_j = a - b_j t| \sum C_j$ is Λ-shaped with respect to job shortening rates b_j.

Proof. Let J_{i-1}, J_i, J_{i+1} be three consecutive jobs such that $b_i < b_{i-1}$ and $b_i < b_{i+1}$. Assume that job sequence σ^* is optimal, where $\sigma^* = (1, 2, \ldots, i-2, i-1, i, i+1, i+2, \ldots, n)$. Consider two job sequences: $\sigma' = (1, 2, \ldots, i-2, i, i-1, i+1, i+2, \ldots, n)$ and $\sigma'' = (1, 2, \ldots, i-2, i-1, i+1, i, i+2, \ldots, n)$. Since the differences $\sum w_j C_j(\sigma^*) - \sum w_j C_j(\sigma')$ and $\sum w_j C_j(\sigma^*) - \sum w_j C_j(\sigma'')$ cannot be both negative (see [226, Property 3]), either σ' or σ'' is a better sequence than σ^*. A contradiction. □

Property 6.158. (Ng et al. [226]) If σ is a schedule for the problem $1|p_j = a - b_j t| \sum C_j$ and $\bar{\sigma}$ is a schedule reverse to σ, then $\sum C_j(\sigma) = \sum C_j(\bar{\sigma})$.

Proof. By direct calculation, see [226, Property 4]. ◇

Property 6.159. (Ng et al. [226]) In any optimal schedule for the problem $1|p_j = a - b_j t| \sum C_j$, the job with the smallest deterioration rate is scheduled in the first position.

Proof. Assuming that the job with the smallest deterioration rate is not scheduled in the first position, by Property 6.158, we obtain a contradiction, see [226, Property 5]. ◇

Remark 6.160. For the problem $1|p_j = a - b_j t| \sum C_j$, Ng et al. also proposed a pseudopolynomial-time dynamic programming algorithm with running time $O(n^3 h^2)$, where h is the product of denominators of all shortening rates b_j, $1 \leq j \leq n$; see [226, Sect. 4] for more details. ◇

Theorem 6.161. (Cheng and Ding [55]) *The decision version of the problem* $1|p_j = a_j - bt, d_j = D| \sum C_j$ *is* $\mathcal{NP}$-*complete in the ordinary sense.*

Proof. The authors only give the idea of the proof; see [55, Theorem 1].   ◇

Wang et al. [283] considered the problem of single-machine batch scheduling (cf. Remarks 6.2, 6.3 and 6.6) with job processing times $p_{i,j} = a_{i,j} - b_{i,j}t$, where

$$0 < b_{i,j} < 1 \tag{6.61}$$

and

$$b_{i,j} \left(\sum_{i=1}^{m} \sum_{j=1}^{k_i} (\theta_i + a_{i,j}) - a_{i,j} \right) < a_{i,j} \tag{6.62}$$

for $1 \le i \le m$ and $1 \le j \le k_i$, $\sum_{i=1}^{m} k_i = n$.

Remark 6.162. Conditions (6.61) and (6.62) for time-dependent scheduling problems with group technology (cf. Remark 6.2) are counterparts of conditions (6.47) and (6.48), respectively, for time-dependent scheduling problems without batching (cf. Remark 6.83).

For the problem $1|p_{i,j} = a_{i,j} - b_{i,j}t, \theta_i, GT| \sum C_j$, with $b_{i,j} = b_i$ for all j, the authors proposed the following algorithm.

Algorithm A_{20}
for the problem $1|p_{i,j} = a_{i,j} - b_it, \theta_i, GT| \sum C_j([283])$

Input: sequences $(a_{i,j})$, (b_i), (θ_i) for $1 \le i \le m$ and $1 \le j \le k_i$
Output: an optimal schedule $\sigma^\star$

▷ Step 1:
    for $i \leftarrow 1$ **to** m **do**
        Arrange jobs in group G_i in the non-increasing order of the $a_{i,j}$ values;
        Call the sequence $\sigma^{(i)}$;
▷ Step 2:
    for $i \leftarrow 1$ **to** m **do**
        Calculate $\rho(G_i) := \dfrac{\theta_i(1-b_i)^{k_i} + \sum_{j=1}^{k_i} a_{i,j}(1-b_i)^{k_i-j}}{\sum_{j=1}^{k_i} a_{i,j}(1-b_i)^j}$;
▷ Step 3:
    Schedule groups in the non-decreasing order of $\rho(G_i)$ values;
▷ Step 4:
    $\sigma^\star \leftarrow (\sigma^{([1])}|\sigma^{([2])}| \dots |\sigma^{([m])})$;
    return $\sigma^\star$.

Theorem 6.163. (Wang et al. [283]) *The problem* $1|p_{i,j} = a_{i,j} - b_it, \theta_i, GT| \sum C_j$ *is solvable by algorithm* A_{20} *in* $O(n \log n)$ *time.*

Proof. By pairwise job interchange; see [283, Theorems 4–5] for details.   ◇

Remark 6.164. Wang et al. [283] also considered the case when $b_{i,j} = Ba_{i,j}$ for all i, j. This case is solved by an algorithm that is similar to A_{20}; see[283, Algorithm 3, Theorems 7–8] for details.   ◇

Now we pass to the results concerning job processing times given by (6.54).

Theorem 6.165. (Cheng et al. [54]) *The decision version of the problem* $1|p_j = a_j - b_j(t - y), y = 0, 0 < b_j < 1, Y < \infty| \sum C_j$ *is $\mathcal{NP}$-complete in the ordinary sense.*

Proof. The transformation from the PP problem (cf. Sect. 3.2) is as follows. Let $V = (2kA)^6$ and $B = V^3$. Define $2k + 1$ jobs with shortening processing times as follows: $a_0 = 4k^2B, b_0 = 1, a_{1,j} = jB + x_j(\frac{1}{2} + (2k - 3j + 2))$ and $a_{2,j} = jB$ for $1 \leq j \leq k$, and $b_{1,j} = 0, b_{2,j} = \frac{x_j}{jB}$ for $1 \leq j \leq k$.

The common initial shortening date is $y = 0$ and the common final shortening date $Y = \sum_{j=1}^k (a_{1,j} + a_{2,j}) - \sum_{j=1}^k (j - 1)x_j - A$. The threshold $G = E + a_0 - F + \frac{H}{2} + \frac{1}{V}$, where constants E and F are defined as follows: $E = \sum_{j=1}^k \left((4k - 4j + 3)jB + (2k - 2j - 1)(\frac{1}{2} + (2k - 3j + 2))x_j\right)$ and $F = 2\sum_{j=1}^k (k - j + 1)(j - 1)x_j$.

To complete the proof, it is sufficient to show that the PP problem has a solution if and only if for the above instance of the problem $1|p_j = a_j - b_jt, 0 < b_j < 1, 0 \leq t \leq Y| \sum C_j$ there exists a schedule σ such that $\sum C_j(\sigma) \leq G$. ☐

Cheng et al. [54] also established two properties of an optimal schedule for the problem $1|p_j = a_j - b_j(t - y), y > 0, 0 < b_j < 1, Y < \infty| \sum C_j$.

Property 6.166. (Cheng et al. [54]) The early jobs are sequenced in the non-decreasing order of a_j values.

Proof. By pairwise job interchange argument.   ☐

Property 6.167. (Cheng et al. [54]) The suspended jobs are sequenced in the non-decreasing order of $a_j - b_j(Y - y)$ values.

Proof. By pairwise job interchange argument.   ☐

The authors also proposed four heuristic algorithms for the latter problem. We will consider these algorithms in Chap. 9.

Distinct ready times and deadlines

The problem of minimizing the total completion time for a set of jobs that have the same shortening rate, $b_j = b$, and only two distinct deadlines, $d_j \in \{D_1, D_2\}$, is computationally intractable.

Theorem 6.168. (Cheng and Ding [48]) *The decision version of the problem* $1|p_j = a_j - bt, d_j \in \{D_1, D_2\}| \sum C_j$ *is $\mathcal{NP}$-complete in the ordinary sense.*

Proof. The result is a corollary from Theorem 6.95.   ☐

6.3 Minimizing the maximum lateness

In this section, we consider single-machine time-dependent scheduling problems with the $L_{\max}$ criterion.

6.3.1 Proportional deterioration

Equal ready times and deadlines

The single-machine time-dependent scheduling problem with proportional job processing times given by (6.1) is easy to solve.

Theorem 6.169. (Mosheiov [216]) *The problem* $1|p_j = b_j t|L_{\max}$ *is solvable in* $O(n \log n)$ *time by scheduling jobs in the non-decreasing order of* d_j *values.*

Proof. The first proof is by pairwise job interchange argument. Consider schedule σ' in which job J_i is followed by job J_j and $d_i > d_j$. Then $L_i(\sigma') = (1 + b_i)S_i - d_i$ and $L_j(\sigma') = (1 + b_j)(1 + b_i)S_i - d_j$.

Consider now schedule σ'', obtained from σ' by interchanging jobs J_i and J_j. Then $L_j(\sigma'') = (1 + b_j)S_i - d_j$ and $L_i(\sigma'') = (1 + b_i)(1 + b_j)S_i - d_i$.

Since $L_j(\sigma') > L_j(\sigma'')$ and $L_j(\sigma') > L_i(\sigma'')$, schedule σ'' is better than schedule σ'.

Repeating this reasoning for all other pairs of jobs that are scheduled in the non-increasing order of d_j, we obtain an optimal schedule in which all jobs are scheduled in the non-decreasing order of deadlines. □

By Theorem 6.169, the problem $1|p_j = b_j t|L_{\max}$ is solved by the algorithm $A_{21} : (b_j|d_j) \mapsto (d_j \nearrow)$. In the final schedule, the jobs are in the EDD order.

6.3.2 Proportional-linear deterioration

Theorem 6.169 was generalized by Kononov for proportional-linear job processing times given by (6.5).

Theorem 6.170. (Kononov [173]) *If there hold inequalities* (6.6) *and* (6.7), *then the problem* $1|p_j = b_j(A + Bt)|L_{\max}$ *is solvable in* $O(n \log n)$ *time by scheduling jobs in the non-increasing order of* d_j *values.*

Proof. By pairwise job interchange argument. □

By Theorem 6.170, the problem $1|p_j = b_j(A + Bt)|L_{\max}$ is solved by the algorithm $A_{21} : (b_j|d_j) \mapsto (d_j \nearrow)$.

Remark 6.171. A version of Theorem 6.170, without conditions (6.6) and (6.7) but with assumptions $A > 0, B > 0, b_j > 0$ for $1 \leq j \leq n$, was given by Zhao et al., see [304, Theorem 4].

Now we pass to the presentation of the results concerning more general job processing times than proportional or proportional-linear ones.

6.3.3 Linear deterioration

Distinct ready time and deadlines

The general problem, with linearly deteriorating processing times (6.10), is computationally intractable.

Theorem 6.172. (Kononov [172]) *The decision version of the problem* $1|p_j = a_j + b_j t|L_{\max}$ *is $\mathcal{NP}$-complete in the ordinary sense even if only one* $a_k \neq 0$ *for some* $1 \leq k \leq n$, *and* $d_j = D$ *for jobs with* $a_j = 0$.

Proof. The transformation from the SP problem (cf. Sect. 3.2) is as follows. We are given $n = p + 1$ jobs, where $a_0 = 1$, $b_0 = 0$, $d_0 = B + 1$ and $a_j = 0$, $b_j = y_j - 1$, $d_j = \frac{Y(B+1)}{B}$ for $1 \leq j \leq p$, with $Y = \prod_{j=1}^{p} y_j$. The threshold $G = 0$. To prove the result, it is sufficient to apply (6.2) and to show that the SP problem has a solution if and only if for the above instance of the problem $1|p_j = a_j + b_j t|L_{\max}$ there exists a schedule σ such that $L_{\max}(\sigma) \leq G$. □

The problem $1|p_j = a_j + b_j t|L_{\max}$ was also studied by other authors.

Theorem 6.173. (Bachman and Janiak [11]) *The decision version of the problem* $1|p_j = a_j + b_j t|L_{\max}$ *is $\mathcal{NP}$-complete in the ordinary sense even if there are only two distinct deadlines.*

Proof. The transformation from the PP problem (cf. Sect. 3.2) is as follows. We have $n = k + 1$ jobs, $d_i = \left(k^{q+2}A + kA + k + A + 1 + \frac{1}{k^q} + \frac{1}{k^{q-1}}\right) \times \left(1 + \frac{2}{2k^q-1}\right) - k^q A$, $a_i = x_i$ and $b_i = \frac{x_i}{k^q A}$ for $1 \leq i \leq k$, and $d_{k+1} = k^{q+2}A + kA + A + k + 1 + \frac{(A+2)(k+1)}{2k^q-1}$, $a_{k+1} = k^{q+2}A$, $b_{k+1} = k$, where $q = \lceil \frac{\ln(A+1)-\ln(2)}{\ln k} \rceil + 3$. The threshold $G = 0$. In order to complete the proof, it is sufficient to show that the PP problem has a solution if and only if for the above instance of the problem $1|p_j = a_j + b_j t|L_{\max}$ there exists a schedule σ such that $L_{\max}(\sigma) \leq G$. □

Theorem 6.174. (Cheng and Ding [52]) *The decision version of the problem* $1|p_j = a_j + b_j t|L_{\max}$ *is $\mathcal{NP}$-complete in the strong sense.*

Proof. The authors give only a sketch of proof, see [52, Theorem 6]. ◇

Bachman and Janiak [11] also proposed two heuristic algorithms for the problem. We will present these algorithms in Chap. 9.

Hsu and Lin [142] proposed a branch-and-bound algorithm for deriving exact solutions for the problem $1|p_j = a_j + b_j t|L_{\max}$. The algorithm exploits several properties concerning dominance relations among different schedules.

Property 6.175. (Hsu and Lin [142]) Let $\sigma^{(a)}$ and $\sigma^{(b)}$ be two schedules of a given subset of jobs. If $C_{\max}(\sigma^{(a)}) \leq C_{\max}(\sigma^{(b)})$ and $L_{\max}(\sigma^{(a)}) \leq L_{\max}(\sigma^{(b)})$, then the subtree rooted at $\sigma^{(b)}$ can be eliminated.

Proof. The result follows from the regularity of the $C_{\max}$ and $L_{\max}$ criteria.
□

Property 6.176. (Hsu and Lin [142]) Let $J_i, J_j \in \mathcal{J}$ be any two jobs scheduled consecutively. If $\frac{a_i}{b_i} \leq \frac{a_j}{b_j}$ and $d_i \leq d_j$, then there exists an optimal schedule in which job J_i is an immediate predecessor of job J_j.

Proof. Let σ' (σ'') be a schedule in which job J_i (job J_j) precedes job J_j (job J_i). First, note that if $\frac{a_i}{b_i} \leq \frac{a_j}{b_j}$ and $d_i \leq d_j$, then by Theorem 6.24 we have $C_{\max}(\sigma') \leq C_{\max}(\sigma'')$. Second, since then $L_i(\sigma') \leq L_i(\sigma'')$ and $L_j(\sigma'') = C_j(\sigma') - d_j \leq L_i(\sigma'')$, we have $\max\{L_i(\sigma'), L_j(\sigma'')\} \leq \max\{L_i(\sigma''), L_j(\sigma'')\}$. Hence, $L_{\max}(\sigma') \leq L_{\max}(\sigma'')$.
□

Let $(\sigma^{(a)}|\tau^{(a)})$ denote a schedule composed of partial schedules $\sigma^{(a)}$ and $\tau^{(a)}$, where $|\tau^{(a)}| \geq 0$.

Property 6.177. (Hsu and Lin [142]) Given a partial schedule $(\sigma^{(a)}|j)$ and an unscheduled job J_i, if $\frac{a_i}{b_i} \leq \frac{a_j}{b_j}$ and $C_{\max}(\sigma^{(a)}|i|j) - d_j \leq L_{\max}(\sigma^{(a)}|j)$, then the subtree rooted at $(\sigma^{(a)}|j|i)$ can be eliminated.

Proof. First, inequality $\frac{a_i}{b_i} \leq \frac{a_j}{b_j}$ implies that $C_{\max}(\sigma^{(a)}|i|j) \leq C_{\max}(\sigma^{(a)}|j|i)$. Second, $L_i(\sigma^{(a)}|i|j) \leq L_i(\sigma^{(a)}|j|i)$. Finally, $C_{\max}(\sigma^{(a)}|i|j) - d_j \leq L_{\max}(\sigma^{(a)}|j)$ by assumption. Hence, $L_{\max}(\sigma^{(a)}|i|j) \leq \max\{C_{\max}(\sigma^{(a)}|j|i) - d_i, L_{\max}(\sigma^{(a)}|j)\}$ $= L_{\max}(\sigma^{(a)}|j|i)$.
□

Properties 6.175–6.177 allow to cut off some subtrees during the process of searching for an optimal schedule in the tree of all possible schedules. In order to estimate the lateness of an optimal schedule from below, we need a lower bound on the value of $L_{\max}$ for the optimal schedule.

Property 6.178. (Hsu and Lin [142]) Let $\sigma^{(a)}$ and $\tau^{(a)}$ denote, respectively, a schedule of a subset of jobs and a schedule with the remaining jobs arranged in the non-decreasing order of $\frac{a_j}{b_j}$ ratios. Then $L_{\max}(\sigma^{(a)}|\tau^{(a)\prime}) \leq C_{\max}(\sigma^{(a)}|\tau^{(a)}) - \max\{d_j : j \in N_{\mathcal{J}} \setminus N(\sigma^{(a)})\}$, where $\tau^{(a)\prime} \neq \tau^{(a)}$ and $N(\sigma^{(a)})$ denotes the set of indices of jobs from the subschedule $\sigma^{(a)}$.

Proof. See [142, Lemma 4].
◇

Another lower bound is obtained by a transformation of the initial set of jobs into a new one, called an *ideal* set (see [142, Lemma 5]).

The branch-and-bound algorithm, obtained by implementation of the above properties and using the above lower bounds, appears to be quite effective, since it is reported (see [142, Sect. 5]) that problems of no more than 100 jobs can be solved, on average, within 1 minute.

If we assume that all jobs deteriorate at the same rate, the problem is polynomially solvable.

Algorithm A_{22} for the problem $1|p_j = a_j + bt|L_{\max}$ ([52])

Input: sequence $(a_1, a_2, \ldots, a_n)$, $(d_1, d_2, \ldots, d_n)$, number $b := \frac{v}{u}$
Output: the minimal value of $L_{\max}$

▷ Step 1:

$$B_1 \leftarrow u^{n-1}\left(\min_{1 \leq j \leq n}\{d_j - a_j\}\right);$$

$$B_2 \leftarrow (u+v)^{n-1}\sum_{j=1}^{n} a_j;$$

▷ Step 2:

while $(B_2 - B_1 > 1)$ do
  $L' \leftarrow \lceil\frac{B_1 + B_2}{2}\rceil$; $L \leftarrow \frac{L'}{u^{n-1}}$;
  for $i \leftarrow 1$ to n do $d_i' \leftarrow d_i + L$;
  Apply Algorithm A_6 to the modified instance I;
  if (there exists an optimal schedule for I) then $B_1 \leftarrow L'$
    else $B_2 \leftarrow L'$;
$L \leftarrow \frac{B_1}{u^{n-1}}$;
return L.

Theorem 6.179. (Cheng and Ding [52]) *The problem $1|p_j = a_j + bt|L_{\max}$ is solvable in $O(n^6 \log n)$ time by algorithm A_{22}.*

Proof. The optimality of the schedule generated by algorithm A_{22} follows from the relation betweeen the problem $1|p_j = a_j + bt|L_{\max}$ and the problem $1|p_j = a_j + bt, d_j|C_{\max}$; see [52, Sect. 5].

The time complexity of algorithm A_{22} follows from the fact that the 'while' loop in Step 2 is executed at most $O(n \log n)$ times and each iteration of the loop needs $O(n^5)$ time due to the execution of algorithm A_6. □

6.3.4 Simple non-linear deterioration

Kononov proved that a simple non-linear deterioration is polynomially solvable for convex (concave) functions.

Theorem 6.180. (Kononov [173]) *If $h(t)$ is a convex (concave) function for $t \geq 0$ and there hold conditions (6.21) and (6.22), then the problem $1|p_j = b_j h(t)|L_{\max}$ is solvable in $O(n \log n)$ time by scheduling jobs in the non-decreasing (non-increasing) order of $b_j + d_j$ values.*

Proof. The main idea is to prove that the criterion $L_{\max}$ is a 1-priority-generating function (cf. Definition 1.19) with priority function $w_i = -b_i - d_i$ ($w_i = b_i + d_i$). Then, by Theorem 1.24, the result follows. □

By Theorem 6.180, if $h(t)$ is a convex or concave function for $t \geq 0$ and there hold conditions (6.21) and (6.22), the problem $1|p_j = b_j h(t)|L_{\max}$ is solved, respectively, by the algorithm $A_{23} : (b_j|d_j) \mapsto (b_j + d_j \searrow)$ or by the algorithm $A_{24} : (b_j|d_j) \mapsto (b_j + d_j \nearrow)$.

6.3.5 General non-linear deterioration

Distinct ready times and deadlines

In this subsection, we consider results concerning an exponential deterioration. In this case, the processing time of a job is in the form of

$$p_j = a_j 2^{-b_j t}, \tag{6.63}$$

where $a_j > 0, b_j \geq 0$ for $1 \leq j \leq n$. This form of job deterioration was introduced by Janiak and Kovalyov [146].

Theorem 6.181. (Janiak and Kovalyov [146]) *The decision version of the problem* $1|p_j = a_j 2^{-b_j t}|L_{\max}$ *is* $\mathcal{NP}$*-complete in the strong sense.*

Proof. The transformation from the 3-P problem (cf. Sect. 3.2) is as follows. There are $n = 4h$ jobs, $a_j = c_j, b_j = 0, d_j = hK + \frac{h-1}{2}$ for $1 \leq j \leq 3h$, $a_{3h+i} = 1, b_{3h+i} = (iK + \frac{i-1}{2})^{-1}$ for $1 \leq i \leq h$. The threshold $G = 0$.

To complete the proof, it is sufficient to show that the 3-P problem has a solution if and only if for the above instance of the $1|p_j = a_j 2^{-b_j t}|L_{\max}$ problem there exists a schedule σ such that $L_{\max}(\sigma) \leq G$. □

The restricted version of the above problem, with only two distinct deadlines, is computationally intractable as well.

Theorem 6.182. (Janiak and Kovalyov [146]) *The decision version of the problem* $1|p_j = a_j 2^{-b_j t}, d_j \in \{d, D\}|L_{\max}$ *is* $\mathcal{NP}$*-complete in the ordinary sense.*

Proof. The transformation from the PP problem (cf. Sect. 3.2) is as follows. There are $n = k + 1$ jobs, $a_j = x_j, b_j = 0, d_j = 2A+1$ for $1 \leq j \leq k$, $a_{k+1} = 1$, $b_{k+1} = 0$ and $d_{k+1} = A + 1$. The threshold $G = 0$.

To complete the proof, it is sufficient to show that the PP problem has a solution if and only if for the above instance of the $1|p_j = a_j 2^{-b_j t}$, $d_j \in \{d, D\}|L_{\max}$ problem there exists a schedule σ such that $L_{\max}(\sigma) \leq G$. □

Remark 6.183. Janiak and Kovalyov state Theorem 6.182 (see [146, Theorem 4]) without proof. The above reduction comes from the present author.

6.3.6 Linear shortening

Distinct ready times and deadlines

The problem of minimizing the maximum lateness for a set of jobs which have the same shortening rate, $b_j = b$, and only two distinct deadlines, $d_j \in \{D_1, D_2\}$, is computationally intractable.

Theorem 6.184. (Cheng and Ding [48]) *The decision version of the problem* $1|p_j = a_j - bt, d_j \in \{D_1, D_2\}|L_{\max}$ *is* $\mathcal{NP}$*-complete in the ordinary sense.*

Proof. The result is a corollary from Theorem 6.95. □

6.4 Other criteria

In this section, we consider results concerning the problems of single-machine time-dependent scheduling with criteria other than $C_{\max}$, $\sum C_j$ or $L_{\max}$.

6.4.1 Proportional deterioration

Equal ready times and deadlines

The $\sum w_j C_j$ criterion. For proportional job processing times given by (6.1), the following result is known.

Theorem 6.185. (Mosheiov [216]) *The problem* $1|p_j = b_j t|\sum w_j C_j$ *is solvable in* $O(n \log n)$ *time by scheduling jobs in the non-decreasing order of* $\frac{b_j}{(1+b_j)w_j}$ *ratios.*

Proof. By pairwise job interchange argument.   □

By Theorem 6.185, the problem $1|p_j = b_j t|\sum w_j C_j$ is solved by the algorithm $A_{25} : (b_j|w_j) \mapsto (\frac{b_j}{(1+b_j)w_j} \nearrow)$.

Remark 6.186. Note that if $w_j = 1$ for $1 \le j \le n$, the scheduling rule from Theorem 6.185 is reduced to the rule given in Theorem 6.57.

The $f_{\max}$ criterion. The problem of minimizing the maximum cost for proportionally deteriorating jobs is polynomially solvable as well.

Theorem 6.187. *The problem* $1|p_j = b_j t|f_{\max}$ *is solvable in* $O(n^2)$ *time by algorithm* A_{29}.

Proof. The result is a corollary from Theorem 6.204 for $A = 0$ and $B = 1$. ■

The $\sum(C_i - C_j)$ criterion. Oron [229] considered the total deviation of job completion times criterion, $\sum(C_i - C_j) := \sum_{i=1}^{n} \sum_{k=i+1}^{n} (C_{[k]} - C_{[i]})$.

Notice that $\sum(C_i - C_j) \equiv \sum_{i=1}^{n} \sum_{k=i+1}^{n} C_{[k]} - \sum_{i=1}^{n} \sum_{k=i+1}^{n} C_{[i]} = \sum_{i=1}^{n} (i-1)C_{[i]} - \sum_{i=1}^{n} (n-i)C_{[i]} = \sum_{i=1}^{n} (2i - n - 1)C_{[i]}$. Hence, by (6.2), $\sum(C_i - C_j) \equiv S_1 \sum_{i=1}^{n} (2i - n - 1) \prod_{j=1}^{i} (1 + b_{[j]})$.

Oron [229] proved a few properties of an optimal schedule for the problem $1|p_j = b_j t|\sum(C_i - C_j)$.

Property 6.188. (Oron [229]) If $n \ge 2$, then there exists an optimal schedule for the problem $1|p_j = b_j t|\sum(C_i - C_j)$ in which the job with the smallest deterioration rate is not scheduled as the first one.

Proof. Let $b_1 := \min_{1 \le j \le n}\{b_j\}$, $\sigma^1 := (1, [2], \dots, [n])$ and $\sigma^2 := ([2], 1, \dots, [n])$. Since $\sum(C_i - C_j)(\sigma^1) - \sum(C_i - C_j)(\sigma^2) = (n-1)(b_{[2]} - b_1) \ge 0$, the result follows. ■

Property 6.189. (Oron [229]) If $n \geq 3$, then there exists an optimal schedule for the problem $1|p_j = b_j t| \sum(C_i - C_j)$ in which the job with the smallest deterioration rate is not scheduled as the last one.

Proof. Similar to the proof of Property 6.188, see [229, Proposition 2]. ◇

Property 6.190. (Oron [229]) Let $J_{[i-1]}, J_{[i]}$ and $J_{[i+1]}$ be three consecutive jobs in a schedule for the problem $1|p_j = b_j t| \sum(C_i - C_j)$. If $b_{[i]} > b_{[i-1]}$ and $b_{[i]} > b_{[i+1]}$, then this schedule cannot be optimal.

Proof. Similar to the proof of Property 6.132, see [229, Proposition 3]. ◇

Property 6.191. (Oron [229]) Let $b_{[k]} := \min\{b_j : 1 \leq j \leq n\}$ and $b_{[l]} := \min\{b_j : 1 \leq j \neq k \leq n\}$ be two smallest job deterioration rates.
(a) If n is even, then in optimal schedule for the problem $1|p_j = b_j t| \sum(C_i - C_j)$ the job $J_{[k]}$ is scheduled in the $\frac{n}{2} + 1$ position.
(b) If n is odd, then in optimal schedule for the problem $1|p_j = b_j t| \sum(C_i - C_j)$ the jobs $J_{[k]}$ and $J_{[l]}$ are scheduled in positions $\frac{n+1}{2}$ and $\frac{n+3}{2}$, respectively.

Proof. By direct calculation, see [229, Propositions 4–5]. ◇

Property 6.192. (Oron [229]) The optimal value of the total deviation of job completion times for the problem $1|p_j = b_j t| \sum(C_i - C_j)$ is not less than

$$\sum_{i=1}^{\frac{n}{2}} (2i - n - 1) \prod_{j=1}^{i} (1 + b_{[n+2-2j]}) + \sum_{i=\frac{n}{2}+1}^{n} (2i - n - 1) \prod_{j=1}^{i} (1 + b_{[j]}).$$

Proof. By direct computations, see [229, Proposition 8]. ◇

Oron proved also that for the problem $1|p_j = b_j t| \sum(C_i - C_j)$ there holds the following counterpart of Theorem 6.133.

Theorem 6.193. (Oron [229]) *The optimal schedule for the problem $1|p_j = b_j t| \sum(C_i - C_j)$ is V-shaped with respect to deterioration rates b_j.*

Proof. The result is a consequence of Properties 6.188–6.190. □

For the problem $1|p_j = b_j t| \sum(C_i - C_j)$, Oron [229] proposed two heuristic algorithms. We will consider these heuristics in Chap. 9.

Distinct ready times and deadlines

The criteria $\sum L_j$ and $T_{\max}$. For proportional job processing times given by (6.1), the following results are known.

Theorem 6.194. (Mosheiov [216]) *The problem* $1|p_j = b_j t|\varphi$ *is solvable in* $O(n \log n)$ *time by scheduling jobs*
(a) *in the non-decreasing order of* b_j *values, if* φ *is the total lateness criterion* ($\varphi \equiv \sum L_j$);
(b) *in the non-decreasing order of* d_j *values, if* φ *is the maximum tardiness criterion* ($\varphi \equiv T_{\max}$).

Proof. (a) Since the total lateness $\sum_{j=1}^{n} L_j = \sum_{j=1}^{n} (C_j - d_j) = \sum_{j=1}^{n} C_j - \sum_{j=1}^{n} d_j$, and the sum $\sum_{j=1}^{n} d_j$ is a constant value, the problem of minimizing the sum $\sum_{j=1}^{n} L_j$ is equivalent to the problem of minimizing the sum $\sum_{j=1}^{n} C_j$. The latter problem, by Theorem 6.57, is solvable in $O(n \log n)$ time by scheduling jobs in the non-decreasing order of b_j values.

(b) Since for any schedule σ, we have $T_i(\sigma) = \max\{0, L_i(\sigma)\}$, $1 \leq i \leq n$, the result follows by the reasoning from the proof of Theorem 6.169. $\square$

By Theorem 6.194, problems $1|p_j = b_j t|\sum L_j$ and $1|p_j = b_j t|T_{\max}$ are solved by algorithms $A_9 : (b_j|d_j) \mapsto (b_j \nearrow)$ and $A_{17} : (b_j|d_j) \mapsto (d_j \nearrow)$, respectively.

The $\sum U_j$ criterion. The problem of minimizing the number of tardy jobs, which proportionally deteriorate, is optimally solved by the following algorithm, which is an adaptation of Moore–Hodgson's algorithm for the problem $1||\sum U_j$ (Moore [213]).

Algorithm A_{26} for the problem $1|p_j = b_j t|\sum U_j$ ([216])

Input: sequences $(b_1, b_2, \ldots, b_n)$, $(d_1, d_2, \ldots, d_n)$
Output: an optimal schedule $\sigma^\star$

▷ Step 1:
  Arrange jobs in the non-decreasing order of d_j values;
  Call the sequence $\sigma^\star$;
▷ Step 2:
  while $(TRUE)$ **do**
    if (no jobs in sequence $\sigma^\star$ are late) **then exit**
    else find in $\sigma^\star$ the first late job, $J_{[m]}$;
    Find a job, $J_{[k]}$, such that $b_{[k]} = \max\limits_{1 \leq i \leq m} \{b_{[i]}\}$;
    Move job $J_{[k]}$ to the end of $\sigma^\star$;
▷ Step 3:
  return $\sigma^\star$.

Remark 6.195. By the constant $TRUE$ we will denote the logical truth. Similarly, by $FALSE$ we will denote the logical false.

Theorem 6.196. (Mosheiov [216]) *The problem* $1|p_j = b_j t|\sum U_j$ *is solvable in* $O(n \log n)$ *time by algorithm* A_{26}.

Proof. The proof of optimality of algorithm A_{26} is similar to the original proof of optimality of Moore–Hodgson's algorithm (cf. [213]) and consists of the following two steps.

In the first step, we prove that if there exists a schedule with no late jobs, then there are no late jobs in the schedule obtained by arranging jobs in the EDD order.

In the second step, by mathematical induction, we prove that if algorithm A_{26} generates a schedule σ that has k late jobs, then there does not exist another schedule, σ', with only $k - 1$ late jobs.   □

6.4.2 Proportional-linear deterioration

The $\sum w_j C_j$ criterion. Theorem 6.185 was generalized by Kononov for job processing times given by (6.5).

Theorem 6.197. (Kononov [173]) *If there hold inequalities (6.6) and (6.7), then the problem $1|p_j = b_j(A + Bt)| \sum w_j C_j$ is solvable in $O(n \log n)$ time by scheduling jobs in the non-increasing order of $w_i(b_i^{-1} + A)$ values.*

Proof. By pairwise job interchange argument.   □

By Theorem 6.197, if there hold inequalities (6.6) and (6.7), the problem $1|p_j = b_j(A + Bt)| \sum w_j C_j$ is solved by the algorithm $A_{27} : (b_j|w_j|A|B) \mapsto (w_j(b_j^{-1} + A) \nearrow)$.

Remark 6.198. A version of Theorem 6.197, without conditions (6.6) and (6.7) but with assumptions $A > 0, B > 0, b_j > 0$ for $1 \leq j \leq n$, was given by Zhao et al., see [304, Theorem 2].

Wang et al. [284] considered the problem of single-machine batch scheduling with proportional-linear job processing times and setup times. The authors proposed the following algorithm for this problem.

Algorithm A_{28}
for the problem $1|p_{i,j} = b_{i,j}(A + Bt), \theta_i = \delta_i(A + Bt), GT| \sum w_j C_j$ ([284])

Input: sequences $(\delta_1, \delta_2, \ldots, \delta_m)$, $(b_{i,j})$, $(w_{i,j})$ for $1 \leq i \leq m$ and
$\qquad\quad$ $1 \leq j \leq k_i$, numbers A, B
Output: an optimal schedule $\sigma^\star$

▷ Step 1:
$\quad$ **for** $i \leftarrow 1$ **to** m **do**
$\qquad$ Arrange jobs in group G_i in the non-decreasing order of
$\qquad \hookrightarrow$ the $\frac{b_{i,j}}{w_{i,j}(1+b_{i,j})}$ values;
$\qquad$ Call the sequence $\sigma^{(i)}$;

▷ **Step 2:**

    for $i \leftarrow 1$ **to** m **do** Calculate $\rho(G_i) := \dfrac{(1+B\delta_i)\prod_{j=1}^{k_i}(1+Bb_{i,j})-1}{(1+B\delta_i)\sum_{j=1}^{k_j} w_{i,j}\prod_{l=1}^{j}(1+Bb_{i,l})}$;

▷ **Step 3:**

Schedule groups in the non-decreasing order of $\rho(G_i)$ values;

▷ **Step 4:**

    $\sigma^\star \leftarrow (\sigma^{([1])}|\sigma^{([2])}|\dots|\sigma^{([m])})$;

    return $\sigma^\star$.

Theorem 6.199. (Wang et al. [284]) *The problem* $1|GT, p_{i,j} = b_{i,j}(A + Bt),$ $\theta_i = \delta_i(A + Bt)|\sum w_j C_j$ *is solvable by algorithm* A_{28} *in* $O(n\log n)$ *time.*

Proof. By pairwise job interchange; see [284, Theorem 2] for details.   ◇

Remark 6.200. A special case of Theorem 6.199, with $A = 0$, $B = 1$ and $\delta_i = 0$, was given by Cheng and Sun [44, Theorem 5].

Remark 6.201. A special case of Theorem 6.199, with $\theta_i = const$, was given by Xu et al. [302, Theorems 1–2].

The $\sum U_j$ criterion. Theorem 6.196 was generalized by Kononov for job processing times given by (6.5).

Theorem 6.202. (Kononov [173]) *If there hold inequalities* (6.6) *and* (6.7), *then the problem* $1|p_j = b_j(A + Bt)|\sum U_j$ *is solvable in* $O(n\log n)$ *time by algorithm* A_{26}.

Proof. Similar to the proof of Theorem 6.196.   □

The $f_{\max}$ criterion. The problem of minimizing the maximum cost for jobs with proportional-linear processing times is solved by the following algorithm.

Algorithm A_{29} for the problem $1|p_j = b_j(A + Bt)|f_{\max}$ ([169])

Input: sequences $(b_1, b_2, \dots, b_n)$, $(f_1, f_2, \dots, f_n)$, numbers A, B

Output: an optimal schedule $\sigma^\star$

▷ **Step 1:**

    $\sigma^\star \leftarrow (\phi)$;

    $N_\mathcal{J} \leftarrow \{1, 2, \dots, n\}$;

    $T \leftarrow (t_0 + \frac{A}{B}) \prod_{j=1}^{n} (1 + Bb_j)$;

▷ **Step 2:**

    while $(N_\mathcal{J} \neq \emptyset)$ **do**

        Find job J_k such that $f_k(T) = \min\{f_j(T) : j \in N_\mathcal{J}\}$;

        $\sigma^\star \leftarrow (\sigma^\star|k)$;

        $T \leftarrow \frac{T - Ab_k}{1 + Bb_k}$;

        $N_\mathcal{J} \leftarrow N_\mathcal{J} \setminus \{k\}$;

▷ **Step 3:**

    return $\sigma^\star$.

Remark 6.203. Algorithm A_{29} is an adaptation of Lawler's algorithm for the problem $1|prec|f_{\max}$ (Lawler [182]). Since, so far, we assumed that between jobs there are no precedence constraints, A_{29} is a simplified version of Lawler's algorithm. The full version of the algorithm, for dependent jobs with proportional processing times, will be presented in Chap. 13.

Theorem 6.204. (Kononov [173]) *If there hold inequalities (6.6) and (6.7), the problem $1|p_j = b_j(A+Bt)|f_{\max}$ is solvable in $O(n^2)$ time by algorithm A_{29}.*

Proof. Similar to the proof of Theorem 13.35. $\square$

Remark 6.205. Kononov proved Theorem 6.204 in a more general form, admitting arbitrary job precedence constraints in the problem. For simplicity of presentation (cf. Remark 6.203), we assumed no precedence constraints.

Remark 6.206. A version of Theorem 6.204, without job precedence constraints, without conditions (6.6) and (6.7) but with assumptions $A > 0$, $B > 0$, $b_j > 0$ for $1 \leq j \leq n$, was given by Zhao et al. [304, Theorem 3].

6.4.3 Linear deterioration

Equal ready times and deadlines

The $\sum w_j C_j$ criterion. The problem of minimizing the total weighted completion time, $\sum w_j C_j$, for a single machine and linear deterioration given by (6.10) was considered for the first time by Mosheiov [218].

Remark 6.207. Browne and Yechiali [33] studied the problem earlier, but they considered the *expected* total weighted completion time. Namely, if $\frac{E(a_1)}{b_1} < \frac{E(a_1)}{b_1} < \dots < \frac{E(a_n)}{b_n}$ and $\frac{b_1}{w_1(1+a_1)} < \frac{b_2}{w_2(1+a_2)} < \dots < \frac{b_n}{w_n(1+a_n)}$, then schedule $(1, 2, \dots, n)$ minimizes the expected value of $\sum w_j C_j$; see [33, Proposition 2]. $\diamond$

Mosheiov [218] considered the weights of jobs which are proportional to the basic job processing times, i.e., $w_j = \delta a_j$ for a given constant $\delta > 0$ and $b_j = b$ for $1 \leq j \leq n$. The criterion function is in the form of

$$\sum w_j C_j \equiv \sum_{j=1}^{n} w_j C_j = \delta \sum_{j=1}^{n} a_j \sum_{k=1}^{j} a_k (1+b)^{j-k}.$$

The following properties of the problem $1|p_j = a_j + bt|\sum w_j C_j$, with $w_j = \delta a_j$, are known. First, there holds the symmetry property similar to Property 6.129.

Property 6.208. For any job sequence σ, let $\bar{\sigma}$ denote the sequence reverse to σ. Then, there holds the equality $\sum w_j C_j(\sigma) = \sum w_j C_j(\bar{\sigma})$.

Proof. By direct calculation, see [218, Proposition 1].   ◇

Property 6.209. (Mosheiov [218]) If $w_j = \delta a_j$ for $1 \le j \le n$, then the optimal schedule for the problem $1|p_j = a_j + bt| \sum w_j C_j$, where $w_j = \delta a_j$, is Λ-shaped.

Proof. Let J_{i-1}, J_i, J_{i+1} be three consecutive jobs such that $a_i < a_{i-1}$ and $a_i < a_{i+1}$. Assume that job sequence $\sigma^* = (1, 2, \ldots, i-2, i-1, i, i+1, i+2, \ldots, n)$ is optimal. Consider job sequences $\sigma' = (1, 2, \ldots, i-2, i, i-1, i+1, i+2, \ldots, n)$ and $\sigma'' = (1, 2, \ldots, i-2, i-1, i+1, i, i+2, \ldots, n)$. Calculate differences $v_1 = \sum w_j C_j(\sigma^*) - \sum w_j C_j(\sigma')$ and $v_2 = \sum w_j C_j(\sigma^*) - \sum w_j C_j(\sigma'')$.

Since v_1 and v_2 cannot both be negative (see [218, Proposition 2]), either σ' or σ'' is a better sequence than σ^*. A contradiction.   □

By Property 6.209, the dominant set (cf. Definition 4.27) for the problem $1|p_j = a_j + bt| \sum w_j C_j$, where $w_j = \delta a_j$, is composed of Λ-shaped schedules. Since there exist $O(2^n)$ Λ-shaped sequences for a given sequence $a = (a_1, a_2, \ldots, a_n)$, the problem seems to be computationally intractable. However, there holds the following result.

Property 6.210. (Mosheiov [218]) If $w_j = \delta a_j$ for $1 \le j \le n$ and if in an instance of the problem $1|p_j = a_j + bt| \sum w_j C_j$ jobs are numbered in the nondecreasing order of a_j values, then the optimal permutation for the problem is in the form of $\sigma^1 = (1, 3, \ldots, n-2, n, n-1, n-3, \ldots, 4, 2)$ if n is odd, and it is in the form of $\sigma^2 = (1, 3, \ldots, n-1, n, n-2, \ldots, 4, 2)$ if n is even.

Proof. By direct calculation, see [218, Proposition 3].   ◇

By Property 6.210, the problem $1|p_j = a_j + b_j t| \sum w_j C_j$ with equal deterioration rates and weights proportional to basic job processing times is solvable in $O(n)$ time.

Mosheiov [218] stated also the conjecture that the problem $1|p_j = a_j + bt| \sum w_j C_j$ with arbitrary weigths is $\mathcal{NP}$-hard and proposed a heuristic algorithm for the problem. We will consider this algorithm in Chap. 9.

Bachman et al. [13] proved that the problem with arbitrary deterioration rates and arbitrary weights is computationally intractable.

Theorem 6.211. (Bachman et al. [13]) *If $S_1 = 1$, then the decision version of the problem $1|p_j = a_j + b_j t| \sum w_j C_j$ is $\mathcal{NP}$-complete in the ordinary sense.*

Proof. Assume that jobs start at time $S_1 = 1$. The transformation from the N3P problem (cf. Sect. 3.2) is as follows. Let $n = 4w$, $a_i = 0$, $b_i = D^{z_i} - 1$ and $w_i = 1$ for $1 \le i \le 3w$ and $a_{3w+i} = D^{iZ}$, $b_{3w+i} = 0$, $w_{3w+i} = D^{(w+1-i)Z}$ for $1 \le i \le w$, where $D = 2w^2 + 1$. The threshold $G = 2w^2 D^{(w+1)Z}$.

To complete the proof, it is sufficient to show that the N3P problem has a solution if and only if for the above instance of the problem $1|p_j = a_j + b_j t| \sum w_j C_j$ there exists a schedule σ such that $\sum w_j C_j(\sigma) \le G$.   □

Remark 6.212. Notice that the assumption $S_1 = 1$ is not essential. If we assume that $S_1 = 0$ and add to the instance described above an additional job J_0 with parameters $a_0 = 1$, $b_0 = 0$, $w_0 = G+1$, and if we change the threshold value to $2G + 1$, then it can be shown that the result of Theorem 6.211 also holds in this case; see [13, Sect. 2].   ◇

Wu et al. proved a few properties of the problem $1|p_j = a_j + b_j t| \sum w_j C_j$.

Property 6.213. (Wu et al. [301]) Let $J_i, J_k \in \mathcal{J}$ be any two jobs. If $a_i < a_k$, $b_i = b_k$ and $w_i \geq w_k$, then for the problem $1|p_j = a_j + b_j t| \sum w_j C_j$ there exists an optimal schedule in which job J_i precedes job J_k.

Proof. By direct calculation, see [301, Property 1].   ◇

Property 6.214. (Wu et al. [301]) Let $J_i, J_k \in \mathcal{J}$ be any two jobs to be scheduled consecutively. If $a_i = a_k$, $b_i > b_k$ and $\frac{b_i}{b_k} \leq \frac{w_i}{w_k}$, then for the problem $1|p_j = a_j + b_j t| \sum w_j C_j$ there exists an optimal schedule in which job J_i immediately precedes job J_k.

Proof. By direct calculation, see [301, Property 2].   ◇

The next three properties are similar to Property 6.214, Wu et al. state them without proofs.

Property 6.215. (Wu et al. [301]) Let $J_i, J_k \in \mathcal{J}$ be any jobs to be scheduled consecutively and let t_0 be the completion time of the last job scheduled before these two jobs. If $\frac{a_i}{b_i} = \frac{a_k}{b_k}$, $w_i \geq w_k$ and $\frac{a_i + b_i t_0}{w_i} < \frac{a_k + b_k t_0}{w_k}$, then for the problem $1|p_j = a_j + b_j t| \sum w_j C_j$ there exists an optimal schedule in which job J_i immediately precedes job J_k.

Property 6.216. (Wu et al. [301]) Let $J_i, J_k \in \mathcal{J}$ be any jobs to be scheduled consecutively and let t_0 be the completion time of the last job scheduled before these two jobs. If $\frac{a_i}{b_i} \leq \frac{a_k}{b_k}$, $w_i = w_k$ and $a_i + b_i t_0 < a_k + b_k t_0$, then for the problem $1|p_j = a_j + b_j t| \sum w_j C_j$ there exists an optimal schedule in which job J_i immediately precedes job J_k.

Property 6.217. (Wu et al. [301]) Let $J_i, J_k \in \mathcal{J}$ be any jobs to be scheduled consecutively and let t_0 be the completion time of the last job scheduled before these two jobs. If $\frac{a_i}{b_i} \leq \frac{a_k}{b_k}$, $\frac{w_k}{w_i} < \frac{1+b_i}{1+b_k} \min\{\frac{a_k}{a_i}, \frac{b_k}{b_i}\}$, then for the problem $1|p_j = a_j + b_j t| \sum w_j C_j$ there exists an optimal schedule in which job J_i immediately precedes job J_k.

Wu et al. proposed also a lower bound for the considered problem.

Theorem 6.218. (Wu et al. [301]) *Let $\sigma = (\sigma^{(1)}, \sigma^{(2)})$ be a schedule for the problem $1|p_j = a_j + b_j t| \sum w_j C_j$, where $\sigma^{(1)}$ ($\sigma^{(2)}$) denotes the sequence of scheduled (unscheduled) jobs, $|\sigma^{(1)}| = m$ and $|\sigma^{(2)}| = r = n - m$. Then the optimal weighted completion time $\sum w_j C_j(\sigma)$ is not less than $\max\{LB_1, LB_2\}$,* *where*

$$LB_1 = \sum_{k=1}^{m} w_{[k]} C_{[k]}(\sigma) + C_{[m]}(\sigma) \sum_{k=1}^{r} w_{(m+r+1-k)} \prod_{i=1}^{k} (1 + b_{(m+i)}) +$$
$$a_{(m+1)} \sum_{k=1}^{r-1} (\prod_{j=1}^{k} (1 + b_{(m+j)}) (\sum_{i=1}^{r-k} w_{(m+i)}) + a_{(m+1)} \sum_{k=1}^{r} w_{m+k},$$

and

$$LB_2 = \sum_{k=1}^{m} w_{[k]} C_{[k]}(\sigma) + C_{[m]}(\sigma) \sum_{k=1}^{r} w_{(m+r+1-k)} \prod_{i=1}^{k} (1 + b_{(m+1)})^k +$$
$$\sum_{k=1}^{r-1} a_{(m+k)} \sum_{i=k}^{r} w_{(m+r+1-i)} (1 + b_{(m+1)})^i,$$

and all parameters of unscheduled jobs are in non-decreasing order, i.e., $a_{m+1)} \leq a_{(m+2)} \leq \cdots \leq a_{(m+r)}, \; b_{m+1} \leq b_{(m+2)} \leq \cdots \leq b_{(m+r)}$ *and* $w_{m+1)} \leq w_{(m+2)} \leq \cdots \leq w_{(m+r)}.$

Proof. By direct calculation, see [301, Sect. 3.1]. ◇

Based on Properties 6.213–6.217 and Theorem 6.218, Wu et al. [301] proposed a branch-and-bound algorithm for the problem $1|p_j = a_j + b_j t| \sum w_j C_j$. Computational experiments have shown that the algorithm can solve instances with $n = 16$ jobs in time no longer than 3 hours (see [301, Sect. 5]).

For the problem $1|p_j = a_j + b_j t| \sum w_j C_j$, Wu et al. [301] proposed three heuristic algorithms. We will consider these heuristics in Chap. 9.

The $P_{\max}$ criterion. Alidaee and Landram [5] considered linear job processing times and the criterion of minimizing the *maximum processing time*, $P_{\max}$. The problem is to find such a schedule $\sigma \in \mathfrak{S}_n$ that minimizes $\max_{1 \leq j \leq n}\{p_{[j]}(\sigma)\}$. The following example shows that the problem of minimizing $P_{\max}$ is not equivalent to the problem of minimizing $C_{\max}$.

Example 6.219. (Alidaee and Landram [5]) Let $p_1 = 100 + \frac{1}{5}t$, $p_2 = 2 + \frac{2}{9}t$ and $p_3 = 70 + \frac{3}{10}t$. Then schedule $(1, 3, 2)$ is optimal for the $P_{\max}$ criterion, while schedule $(2, 3, 1)$ is optimal for the $C_{\max}$ criterion. ♦

For the $P_{\max}$ criterion, the following results are known.

Property 6.220. (Alidaee and Landram [5]) If $a_j > 0$ and $b_j \geq 1$ for $1 \leq j \leq n$, then for any sequence of jobs with processing times in the form of $p_j = a_j + b_j t$ the maximum processing time occurs for the last job.

Proof. The result follows from the fact that all processing times are described by increasing functions and the jobs start their execution at increasingly ordered starting times. □

Remark 6.221. By Property 6.220, the problem of minimizing the maximum processing time is equivalent to the problem of minimizing the processing time of the last job in a schedule.

Alidaee and Landram [5] also proposed an $O(n^2)$ heuristic algorithm H_{43} for the problem $1p_j = a_j + b_j t | P_{\max}$. (We will consider algorithm H_{43} in Chap. 9.) This algorithm is optimal under some additional assumptions.

Theorem 6.222. (Alidaee and Landram [5]) *If $a_j > 0$ and $b_j \geq 1$ for $1 \leq j \leq n$, then the problem of minimizing the processing time of the last job in a single-machine schedule of a set of linearly deteriorating jobs is solvable in $O(n \log n)$ time by algorithm H_{43}.*

Proof. See [5, Proposition 2]. ◇

Another polynomially solvable case is when $p_j = a_j + bt$.

Theorem 6.223. (Alidaee and Landram [5]) *If $a_j > 0$ and $b_j = b$ for $1 \leq j \leq n$, then the problem of minimizing the maximum processing time is optimally solved by scheduling jobs in the non-decreasing order of a_j values.*

Proof. By pairwise job interchange argument; see [5, Proposition 3]. ◇

By Theorem 6.223, if $a_j > 0$ for $1 \leq j \leq n$, the problem $1|p_j = a_j + bt|P_{\max}$ is solved by the algorithm $A_7 : (a_j) \mapsto (a_j \nearrow)$.

The $\sum U_j$ criterion. Chakaravarthy et al. [41, Theorem 4.2] considered the problem $1|p_j = a_j + b_j t, d_j = D| \sum U_j$. For the problem, the authors proposed a dynamic programming algorithm. We will call this algorithm A_{30}.

Theorem 6.224. (Chakaravarthy et al. [41]) *The problem $1|p_j = a_j + b_j t, d_j = D| \sum U_j$ is solved in $O(n^2)$ time by algorithm A_{30}.*

Proof. Algorithm A_{30} uses a dynamic programming approach as follows. Arrange jobs in the non-increasing order of $\frac{b_j}{a_j}$ ratios. By Theorem 6.24, the schedule corresponding to this order is optimal for the $C_{\max}$ criterion. Let $T(i, j)$ denote the minimum schedule length for a subset of j jobs from the first i jobs arranged in the $\frac{b_j}{a_j} \searrow$ order. The values of $T(i, j)$ can be calculated by using the formula

$$T(i, j) := \min \{T(i - 1, j), T(i - 1, j - 1) + a_i \times T(i - 1, j - 1) + b_i\}$$

if $j \leq i$ (they are not defined if $j > i$). By insertion of the values of $T(i, j)$ in an $n \times n$ table and calculating them row by row, the result follows. □

Remark 6.225. Chakaravarthy et al. [41] also considered the above problem with $\sum w_j U_j$ criterion. Since the problem $1|p_j = a_j + b_j t, d_j = D| \sum w_j U_j$ is a generalization of the KP problem (cf. Sect. 3.2), it is computationally intractable. Applying a dynamic programming approach, the authors have shown (cf. [41, Theorem 4.3]) that for the problem there exists an FPTAS.

The $\sum(\alpha E_j + \beta T_j + \gamma d)$ criterion. Cheng et al. [57] considered the problem of minimizing the sum of earliness, tardiness and due-date penalties and common due-date assignment.

Job processing times are in the form of

$$p_j = a_j + bt, \tag{6.64}$$

where $b > 0$ and $a_j > 0$ for $1 \le j \le n$.

The authors proved the following properties of the problem.

Property 6.226. (Cheng et al. [57]) For any schedule σ for the problem $1|p_j = a_j + bt| \sum(\alpha E_j + \beta T_j + \gamma d)$, there exists an optimal due-date $d^\star = C_{[k]}$ such that $k = \lceil \frac{n\beta - n\gamma}{\alpha + \beta} \rceil$ and exactly k jobs are non-tardy.

Proof. It is an adaptation of the proof of Panwalkar et al. [232, Lemma 1]. ◇

Property 6.227. (Cheng et al. [57]) If k is defined as in Property 6.226, then in any optimal schedule $\sigma^\star$ for the problem $1|p_j = a_j + bt| \sum(\alpha E_j + \beta T_j + \gamma d)$ there hold inequalities $a_{[k+1]} \le a_{[k+2]} \le \ldots \le a_{[n]}$.

Proof. By pairwise job interchange argument. □

Before we formulate the next property, we will introduce new notation. Let

$$m_i := \begin{cases} b \sum_{j=1}^{k} (\alpha(j-1) + n\gamma)(1+b)^{j-1} + \\ +b \sum_{j=k+1}^{n} \beta(n+1-j)(1+b)^{j-1} \text{ for } 2 \le i \le k, \\ b \sum_{j=i}^{n} \beta(n+1-j)(1+b)^{j-i} \text{ for } k+1 \le i \le n. \end{cases} \tag{6.65}$$

Define also the following two functions:

$$g(i) := \alpha(i-1) + n\gamma + m_{i+1} \quad \text{for} \quad 1 \le i \le k \tag{6.66}$$

and

$$f(b) := (\alpha + n\gamma)b - \alpha + bm_3. \tag{6.67}$$

Based on definitions (6.65)–(6.67), Cheng et al. proved a few properties concerning possible relations between $f(b)$, $g(i)$ and m_i.

Property 6.228. (Cheng et al. [57])
(a) If $f(b) = (\alpha + n\gamma)b - \alpha + bm_3 > 0$, then $(\alpha i + n\gamma)b - \alpha + bm_{i+2} > 0$ for $1 \le i \le k - 1$.
(b) (c) The implication (a) in which the symbol '>' has been replaced by the symbol '<' and '=', respectively.

Proof. (a),(b),(c) By induction with respect to i. □

Since the next property is similar to Property 6.228, we do not give its formulation (see [57, Property 5]).

Property 6.229. (Cheng et al. [57]) If k is defined as in Property 6.226 and
(a) $f(b) \geq 0$, then for the problem $1|p_j = a_j + bt| \sum (\alpha E_j + \beta T_j + \gamma d)$ there exists an optimal schedule σ^* such that $a_{\sigma_1^*} \leq a_{\sigma_2^*} \leq \ldots \leq a_{\sigma_k^*}$;
(b) $f(b) < 0$, then for the problem $1|p_j = a_j + bt| \sum (\alpha E_j + \beta T_j + \gamma d)$ in any optimal schedule σ^* there hold inequalities $a_{\sigma_1^*} \geq a_{\sigma_2^*} \geq \ldots \geq a_{\sigma_k^*}$.

Proof. (a), (b) By pairwise job interchange argument.   □

Based on the above properties, the authors proved the following result.

Theorem 6.230. (Cheng et al. [57]) *If k is defined as in Property 6.226 and*
(a) $f(b) \geq 0$, then for the problem $1|p_j = a_j + bt| \sum (\alpha E_j + \beta T_j + \gamma d)$ there exists an optimal schedule σ^* in which $a_{\sigma_1^*} \leq a_{\sigma_2^*} \leq \ldots \leq a_{\sigma_k^*}$ and $a_{\sigma_{k+1}^*} \leq \ldots \leq a_{\sigma_n^*}$;
(b) $f(b) < 0$, then for the problem $1|p_j = a_j + bt| \sum (\alpha E_j + \beta T_j + \gamma d)$ there exists an optimal schedule σ^* which is V-shaped with respect to a_j values and such that $a_{\sigma_k^*} = \min_{1 \leq j \leq n} \{a_j\}$.

Proof. (a) (b) The results follow from Properties 6.228–6.229.   □

Based on the proved properties, the authors proposed also an algorithm for the problem. We will call the algorithm A_{31}. Since algorithm A_{31} is rather complicated (see [57, Algorithm 1] for details), we present only the following result.

Theorem 6.231. (Cheng et al. [57]) *The problem $1|p_j = a_j + bt| \sum (\alpha E_j + \beta T_j + \gamma d)$ is solvable in $O(n \log n)$ time by algorithm A_{31}.*

Proof. See [57, Properties 9–11, Theorem 12].   ◇

For the same problem, $1|p_j = a_j + bt| \sum (\alpha E_j + \beta T_j + \gamma d)$, Kuo and Yang [180] proposed another $O(n \log n)$ algorithm simpler than A_{31}. We will call the new algorithm A_{32}.

Algorithm A_{32} is based on Lemma 1.2 (a) and on the following observation. Given a due-date d and a job sequence σ, we have

$$\sum (\alpha E_j + \beta T_j + \gamma d) = \sum_{j=1}^{n} (\alpha E_{[j]}(\sigma) + \beta T_{[j]}(\sigma) + \gamma d) = \sum_{j=1}^{n} W_j a_{[j]},$$

where the coefficients W_j, $1 \leq j \leq n$, called *positional weights*, are as follows:

$$\begin{aligned}
W_1 &= w_1 + w_2 b + w_3 b(1+b) + \ldots + w_n b(1+b)^{n-2}, \\
W_2 &= w_2 + w_3 b + w_4 b(1+b) + \ldots + w_n b(1+b)^{n-3}, \\
&\ldots, \\
W_{n-1} &= w_{n-1} + w_n b, \\
W_n &= w_n
\end{aligned}$$

(see [180, Sect. 'Preliminary results'] for details). Algorithm A_{32} calculates the positional weights W_j for $1 \leq j \leq n$ and assigns jobs to the weights in an apropriate way. The pseudo-code of the algorithm is as follows.

Algorithm A_{32} for the problem $1|p_j = a_j + bt| \sum (\alpha E_j + \beta T_j + \gamma d)$ ([180])

Input: sequence $(a_1, a_2, \ldots, a_n)$, numbers b, α, β, γ
Output: an optimal due-date $d^\star$, an optimal schedule $\sigma^\star$

▷ Step 1:
   $N_{\mathcal{J}} \leftarrow \{1, 2, \ldots, n\};$
   $W \leftarrow \{1, 2, \ldots, n\};$
   $k \leftarrow \left\lceil \frac{n(\beta - \gamma)}{\alpha + \beta} \right\rceil;$
   Assign to the due-date $d^\star$ the completion time of the k-th job;
▷ Step 2:
   for $j \leftarrow 1$ **to** n **do** Calculate $W_j;$
▷ Step 3:
   Arrange jobs in the non-decreasing order of a_j values;
   while $(N_{\mathcal{J}} \neq \emptyset)$ **do**
      Assign job J_k such that $a_k = \max\{a_j : j \in N_{\mathcal{J}}\}$ to the r-th position
      $\hookrightarrow$ in $\sigma^\star$, where r is such that $W_r = \min\{W_j : j \in N_{\mathcal{J}}\};$
      $N_{\mathcal{J}} \leftarrow N_{\mathcal{J}} \setminus \{k\};$
      $W \leftarrow W \setminus \{r\};$
▷ Step 4:
   return $\sigma^\star$.

Remark 6.232. Scheduling deteriorating jobs with earliness and tardiness penalties and common due-date assignment are new topics in time-dependent scheduling. In the classic scheduling (cf. Sect. 5.1), however, both these topics have been studied since early 1970s; see the reviews by Baker and Scudder [15] and Gordon et al. [118, 119].

6.4.4 Simple non-linear deterioration

The $\sum w_j C_j$ criterion. Kononov [173] proved that some problems with simple non-linear job deterioration, described by convex functions, are polynomially solvable.

Theorem 6.233. (Kononov [173]) *If $h(t)$ is a convex function for $t \geq 0$ and there hold conditions (6.21) and (6.22), then*
(a) *if $h(t) \geq 0$ for all t, $\lim_{t \to \infty} \frac{dh(t)}{dt} = \infty$ and $w_i \geq w_l$ for all $J_i, J_l \in \mathcal{J}$ such that $b_i < b_l$, then the problem $1|p_j = b_j h(t)| \sum w_j C_j$ is solvable in $O(n \log n)$ time by scheduling jobs in the non-decreasing order of $b_j - w_j$ values;*

(b) *if $h(t) \geq 0$ for all t, $\lim_{t \to \infty} \frac{dh(t)}{dt} = H$ and $w_i(b_i^{-1} + H) \geq w_l(b_l^{-1} + H)$ for all $J_i, J_l \in \mathcal{J}$ such that $b_i < b_l$, then the problem $1|p_j = b_j h(t)|\sum w_j C_j$ is solvable in $O(n \log n)$ time by scheduling jobs in the non-decreasing order of $b_j - w_j(b_j^{-1} + H)$ values;*
(c) *if there exists $T > t_0$ such that $h(T) = 0$ and $h(t) > 0$ for any $t \in \langle t_0, T \rangle$, $\lim_{t \to \infty} \frac{dh(t)}{dt} = H$ and $w_i(b_i^{-1} + H) \geq w_l(b_l^{-1} + H)$ for all $J_i, J_l \in \mathcal{J}$ such that $b_i < b_l$, then the problem $1|p_j = b_j h(t)|\sum w_j C_j$ is solvable in $O(n \log n)$ time by scheduling jobs in the non-decreasing order of $b_j - w_j(b_j^{-1} + H)$ values.*

Proof. See [173, Theorem 8].   ◇

By Theorem 6.233, if $h(t)$ is a convex function for $t \geq 0$, there hold conditions (6.21) and (6.22) and according to other assumptions specified in the theorem, the problem $1|p_j = b_j h(t)|\sum w_j C_j$ is solved by the algorithm $A_{33} : (b_j|w_j) \mapsto ((b_j - w_j) \nearrow)$ or by the algorithm $A_{34} : (b_j|w_j|H) \mapsto ((b_j - w_j(b_j^{-1} + H)) \nearrow)$.

Remark 6.234. Kononov proved also a similar result for concave functions; see [173, Theorem 10].   ◇

The $\sum C_j^k$ criterion. Kuo and Yang [179] considered single-machine time-dependent scheduling problems with non-linear job processing times and with the criterion $\sum C_j^k$, where k is a given positive integer.

Theorem 6.235. (Kuo and Yang [179])
(a) *If $f(t) := \sum_{i=1}^{m} \lambda_i t^{r_i}$ and $r_i \in \langle 0, +\infty)$ for $1 \leq i \leq m$, then there exists an optimal schedule for the problem $1|p_j = a_j + f(t)|\sum C_j^k$ in which jobs are in the non-decreasing order of a_j values;*
(b) *If $f(t) := 1 + \sum_{i=1}^{m} \lambda_i t^{r_i}$ and $r_i \in \langle 1, +\infty)$ for $1 \leq i \leq m$, then there exists an optimal schedule for the problem $1|p_j = a_j f(t)|\sum C_j^k$ in which jobs are in the non-decreasing order of a_j values;*
(c) *If $f(t) := 1 + \sum_{i=1}^{m} \lambda_i t^{r_i}$ and $r_i \in (-\infty, 0\rangle$ for $1 \leq i \leq m$, then there exists an optimal schedule for the problem $1|p_j = a_j f(t)|\sum C_j^k$ in which all jobs except the first one are in the non-decreasing order of a_j values.*

Proof. (a) By pairwise job interchange argument; see [179, Proposition 6].
(b) By pairwise job interchange argument; see [179, Proposition 7].
(c) By pairwise job interchange argument; see [179, Proposition 8].   ◇

By Theorem 6.235, if $f(t) := \sum_{i=1}^{m} \lambda_i t^{r_i}$ and $r_i \in \langle 0, +\infty)$ or $r_i \in \langle 1, +\infty)$ for $1 \leq i \leq m$, then the problem $1|p_j = a_j + f(t)|\sum C_j^k$ is solved by algorithm $A_7 : (a_j|\lambda_j|r_i) \mapsto (a_j \nearrow)$.

Remark 6.236. The criterion $\sum C_j^k$ is nothing else than the k-th power of the l_p norm (cf. Definition 1.18), where $p := k$. In Chap. 12, we will consider time-dependent scheduling problems with the l_p norm as optimality criterion.

6.4.5 General non-linear deterioration

Distinct ready times and deadlines

We begin this subsection with results concerning *step deterioration*. In this case, the processing time of each job is described by a step function.

There exist a few forms of the step deterioration. In the simplest case

$$p_j = \begin{cases} a, & \text{if } t \le D, \\ a + b_j, & \text{if } t > D, \end{cases} \tag{6.68}$$

where $a > 0$, $b_j > 0$ for $1 \le j \le n$ and $D > 0$ is the *common critical start time* for all jobs in a schedule. This form of job deterioration was introduced by Sundararaghavan and Kunnathur [263].

The $\sum w_j C_j$ criterion. For job processing times given by (6.68), a few results are known. Let $\mathcal{J}$ denote the set of jobs and let the jobs have only two distinct weights, $w_j \in \{w_1, w_2\}$ for $1 \le j \le n$. Let $k := \lfloor \frac{D}{a} \rfloor + 1$ denote the maximum number of jobs that can be scheduled without job processing time increase.

Lemma 6.237. (Sundararaghavan and Kunnathur [263]) *For a given instance of the problem of minimizing the criterion $\sum w_j C_j$ for a single machine and job processing times in the form of (6.68), let $w_j \in \{w_1, w_2 : w_1 > w_2\}$. Let $E := \{J_k \in \mathcal{J} : C_k \le D\}$, $L := \mathcal{J} - E$, $W_i := \{J_j \in \mathcal{J} : w_j = w_i\}$, $1 \le i \le 2$, and let $J_{[r]}$ denote the job with the greatest starting time among the jobs in $E \cap W_1$. Then the following conditions are necessary for optimality of a schedule for the problem:*
(a) $w_{[i]} \ge w_{[i+1]}$ *for* $1 \le i \le k - 1$ *and* $J_{[i]} \in E$,
(b) $\frac{a+b_{[j]}}{w_{[j]}} \le \frac{a+b_{[j+1]}}{w_{[j+1]}}$ *for* $k + 1 \le j \le n - 1$ *and* $J_{[j]} \in L$,
(c) $b_{[r]} \ge b_i$ *for* $i \in L \cap W_1$, $J_{[r]} \in E$,
(d) $b_{[r+1]} \ge b_i$ *for* $i \in L \cap W_2$, $J_{[r+1]} \in E$.

Proof. The result follows directly from the properties of an optimal schedule for this problem.   □

Remark 6.238. By Lemma 6.237, we know that in an optimal schedule for the problem with two distinct weights there are k jobs in set E, which are arranged in the non-increasing order of b_j values, and r jobs in set L, which are arranged in the non-decreasing order of $\frac{a+b_j}{w_j}$ ratios.

For a given schedule σ, let $J_i \leftrightarrow J_j$ denote mutual exchange of job $J_i \in E$ with job $J_j \in L$, i.e., rearrangement of the jobs in the set $\{J_j\} \cup E \setminus \{J_i\}$ in the non-increasing order of w_j values and rearrangement of the jobs in the set $\{J_i\} \cup L \setminus \{J_j\}$ in the non-decreasing order of $\frac{a+b_j}{w_j}$ ratios. For a given $J_i \leftrightarrow J_j$, let $\Delta(J_i \leftrightarrow J_j)$ denote the difference between total weighted completion time for the schedule before the exchange J_i with J_j and after it.

Algorithm A_{35} for the problem $1|p_j \equiv (6.68)| \sum w_j C_j$ ([263])

Input: $(b_1, b_2, \ldots, b_n)$, numbers a, D, w_1, w_2
Output: an optimal schedule $\sigma^\star$

▷ Step 1:
    $W_1 \leftarrow \{J_k : w_k = w_1\}$;
    $W_2 \leftarrow \{J_k : w_k = w_2\}$;
    $k \leftarrow \lfloor \frac{D}{a} \rfloor + 1$;

▷ Step 2:
    $E \leftarrow W_1$;
    Arrange E in the non-increasing order of b_j values;
    if $(|E| < k)$ **then** $E' \leftarrow \{J_{[k]} \in W_2' \subseteq W_2 : |W_2'| = |W_1| - k\}$;
                     Arrange E' in the non-increasing order of b_j values;
                     $E \leftarrow E \cup E'$;
                     $L \leftarrow \{1, 2, \ldots, n\} \setminus E$;
                     $\sigma^\star \leftarrow (E, L)$;

▷ Step 3:
    repeat
        if $(J_{[i]} \in E \cap W_1 \wedge J_{[j]} \in L \cap W_2 \wedge \Delta(J_{[i]} \leftrightarrow J_{[j]}) > 0)$ **then**
        $\hookrightarrow$ exchange $J_{[i]}$ with $J_{[j]}$;
        Call the obtained schedule σ';
        $\sigma^\star \leftarrow \sigma'$;
    until (no more exchange $J_{[i]} \leftrightarrow J_{[j]}$ exists for $J_{[i]} \in E \cap W_1 \wedge J_{[j]} \in L \cap W_2$);
▷ Step 4:
    repeat
        if $(J_{[i]} \in E \cap W_2 \wedge J_{[j]} \in L \cap W_1 \wedge \Delta(J_{[i]} \leftrightarrow J_{[j]}) > 0)$ **then**
        $\hookrightarrow$ exchange $J_{[i]}$ with $J_{[j]}$;
        Call the obtained schedule σ';
        $\sigma^\star \leftarrow \sigma'$;
    until (no more exchange $J_{[i]} \leftrightarrow J_{[j]}$ exists for $J_{[i]} \in E \cap W_2 \wedge J_{[j]} \in L \cap W_1$);
▷ Step 5:
    return $\sigma^\star$.

Theorem 6.239. (Sundararaghavan and Kunnathur [263]) *The problem of minimizing the $\sum w_j C_j$ criterion for a single machine, job processing times in the form of (6.68) and two distinct weights is solvable in $O(n \log n)$ time by algorithm A_{35}.*

Proof. Let $\sigma^\star$ be the final schedule generated by algorithm A_{35}, let $J_{[r]}$ be the job with the greatest starting time among the jobs in set $E \cap W_1$ and $k := \lfloor \frac{D}{a} \rfloor + 1$. Then, by Lemma 6.237, $b_{[1]} \geq b_{[2]} \geq \ldots \geq b_{[r]}$ and $b_{[r+1]} \leq b_{[r+2]} \leq \ldots \leq b_{[k]}$. Since in $\sigma^\star$ there do not exist $J_{[i]} \in E$ and $J_{[j]} \in L$ such that $\Delta(J_{[i]} \leftrightarrow J_{[j]}) < 0$ (otherwise $\sigma^\star$ would not be a final schedule), for any $J_{[r]} \in E \cap W_1$ and any $J_{[i]} \in L \cap W_2$ there holds the inequality $\Delta(J_{[i]} \leftrightarrow J_{[j]}) \geq 0$.

Since $J_{[r]}$ has the lowest deterioration rate of all jobs in $E \cap W_1$, it follows that the exchange of any subset of jobs in $E \cap W_1$ with any subset of jobs in $L \cap W_2$ cannot improve the schedule $\sigma^\star$. To complete the proof, it is sufficient to show that also other possible exchanges cannot improve $\sigma^\star$, and hence it is optimal.

Since in the exchanges that are performed in either Step 3 or Step 4 a job is exchanged exactly once, and since Step 1 and Step 2 are performed in at most $O(n \log n)$ time, algorithm A_{35} runs in $O(n \log n)$ time.   □

Now let us assume that there are only two distinct deterioration rates, $b_j \in \{b_1, b_2\}$, $b_1 > b_2$. Let $B_1 := \{J_j \in \mathcal{J} : b_j = b_1\}$ and $B_2 := \{J_j \in \mathcal{J} : b_j = b_2\}$. Let σ, E and L be defined as in Lemma 6.237.

Lemma 6.240. (Sundararaghavan and Kunnathur [263]) *For a given instance of the problem of minimizing the criterion $\sum w_j C_j$ for a single machine and job processing times in the form of (6.68), let $b_j \in \{b_1, b_2\}$, where $b_1 > b_2$. Let E denote the set of jobs which can be scheduled before time D, $L := \mathcal{J} - E$, $B_i := \{J_j \in \mathcal{J} : b_j = b_i\}$, $1 \le i \le 2$, and let $J_{[r]}$ denote the job with the greatest starting time among the jobs in $E \cap B_1$. Then the following conditions are necessary for optimality of a schedule for the problem:*
(a) if $J_i \in E \cap B_1$ and $J_j \in L \cap B_1$, then $w_i \ge w_j$,
(b) if $J_i \in E \cap B_2$ and $J_j \in L \cap B_2$, then $w_i \ge w_j$,
(c) if $J_i \in E \cap B_1$ and $J_j \in L \cap B_1$, then $w_i > w_j$.

Proof. The result follows directly from the properties of an optimal schedule for this problem.   □

For the problem with only two deterioration rates, Sundararaghavan and Kunnathur propose to use algorithm A_{35} in which $(b_1, b_2, \ldots, b_n)$ and (w_1, w_2) are replaced with (b_1, b_2) and $(w_1, w_2, \ldots, w_n)$ in the input, respectively, and W_i is replaced with B_i, $1 \le i \le 2$, in each step of the algorithm. Let us call the modified algorithm A_{36}.

Theorem 6.241. (Sundararaghavan and Kunnathur [263]) *The problem of minimizing the $\sum w_j C_j$ criterion for a single machine, job processing times in the form of (6.68) and two distinct deterioration rates is solvable in $O(n \log n)$ time by algorithm A_{36}.*

Proof. Similar to the proof of Theorem 6.239.   □

Sundararaghavan and Kunnathur considered also the case of *agreeable* job weights and deterioration rates, when

$$w_i \ge w_j \Rightarrow b_i \ge b_j \tag{6.69}$$

for any $1 \le i, j \le n$.

Lemma 6.242. (Sundararaghavan and Kunnathur [263]) *For a given instance of the problem of minimizing the criterion* $\sum w_j C_j$ *for a single machine and job processing times in the form of* (6.68), *let the inequality* $w_i \geq w_j$ *imply the inequality* $b_i \geq b_j$ *for any* $1 \leq i, j \leq n$. *Let* E *denote the set of jobs which can be scheduled before time* D *and* $L := \mathcal{J} - E$. *Then the condition*

$$if \ \ w_i < w_j \ \ and \ \ J_i \in E, \ \ then \ \ J_j \in E$$

is a necessary condition for optimality of a schedule for the problem.

Proof. Consider two jobs, J_i and J_j, such that (6.69) is satisfied. Assume that schedule $\sigma^\star$, in which job J_i is followed by J_j, $C_i < D$ and $C_j > D$, is optimal. Consider schedule σ' in which jobs J_i and J_j have been mutually exchanged. Let $R \subseteq L$ denote the set of job started after J_j in $\sigma^\star$ (started after J_i in σ'). Since

$$C_j(\sigma') = C_i(\sigma^\star) \tag{6.70}$$

and

$$C_i(\sigma') = C_j(\sigma^\star) - (b_j - b_i), \tag{6.71}$$

we have

$$\sum w_j C_j(\sigma') = \sum w_j C_j(\sigma^\star) - w_i C_i(\sigma^\star) - w_j C_j(\sigma^\star) + w_i C_i(\sigma') + w_j C_j(\sigma') + r,$$

where r is the change in the deterioration of jobs in R. By (6.69), we have $r < 0$. From that and from (6.70) and (6.71), we have $\sum w_j C_j(\sigma') - \sum w_j C_j(\sigma^\star) < 0$. A contradiction. □

For the problem with agreeable parameters (6.69), Sundararaghavan and Kunnathur [263] proposed the following algorithm.

Algorithm A_{37} for the problem $1|p_j \equiv (6.68) \wedge (6.69)| \sum w_j C_j$ ([263])

Input: sequences $(b_1, b_2, \ldots, b_n)$, $(w_1, w_2, \ldots, w_n)$, numbers a, D
Output: an optimal schedule

▷ Step 1:
   $k \leftarrow \lfloor \frac{D}{a} \rfloor + 1$;
   Arrange k jobs in the non-increasing order of $\frac{1}{w_j}$ ratios;

▷ If necessary, ties are broken in favour of jobs with larger b_j;
▷ Step 2:
   Arrange the remaining $n-k$ jobs in the non-decreasing order of $\frac{a+b_j}{w_j}$ ratios.

   ▷ Ties are broken arbitrarily

Theorem 6.243. (Sundararaghavan and Kunnathur [263]) *The problem of minimizing the* $\sum w_j C_j$ *criterion for a single machine, job processing times given by* (6.68) *and agreeable parameters is solvable in* $O(n \log n)$ *time by algorithm* A_{37}.

Proof. The proof is based on Lemma 6.242 and a lemma concerning the case when $w_i = w_j$ for some $1 \leq i, j \leq n$; see [263, Lemmata 3–5].   ◇

The authors also proposed an algorithm for the problem $1|p_j| \sum w_j C_j$, where job processing times p_j are given by (6.68), and stated the conjecture that it is optimal. We will show in Chap. 9 that the conjecture is false.

Cheng and Ding [51] proposed for the general problem an enumerative algorithm, which under some assumptions allows to solve the problem in polynomial time. First, note that there holds the result similar to Lemma 6.153.

Lemma 6.244. (Cheng and Ding [51]) *If* $a_i \leq a_j$, $\frac{a_i}{w_i} \leq \frac{a_j}{w_j}$ *and* $\frac{a_i + b_i}{w_i} \leq \frac{a_j + b_j}{w_j}$, *then job* J_i *precedes job* J_j *in any optimal schedule for the problem* $1|p_j \in \{a_j, b_j : a_j \leq b_j\}| \sum C_j$.

Proof. By pairwise job interchange argument.   □

By Lemma 6.244, we can divide the set of jobs $\mathcal{J}$ into a number of chains, using the approach applied in algorithm E_2 for the problem $1|p_j \in \{a_j, b_j : a_j \leq b_j\}, d_j = D| \sum C_j$. In order to do that, it is sufficient to renumber jobs in the non-decreasing order of a_j values, breaking ties in the non-decreasing order of $\frac{a_j}{w_j}$ ratios, if jobs have the same values of a_j, and breaking ties in the non-decreasing order of $\frac{a_j + b_j}{w_j}$ ratios, if jobs have the same values of $\frac{a_j}{w_j}$ ratios. Let us call the modified algorithm E_2 as E_4.

Theorem 6.245. (Cheng and Ding [51]) *If an instance of the problem* $1|p_j \in \{a_j, b_j : a_j \leq b_j\}, d_j = D| \sum w_j C_j$ *has a fixed number of chains, then algorithm* E_4 *is a polynomial-time algorithm for the instance.*

Proof. Similar to the proof of Theorem 6.154.   □

We end this subsection with a single result concerning the exponential deterioration given by (6.63).

Theorem 6.246. (Janiak and Kovalyov [146]) *The decision version of the problem* $1|p_j = a_j 2^{-b_j t}| \sum w_j C_j$ *is* $\mathcal{NP}$-complete in the ordinary sense.

Proof. The transformation from the PP problem (cf. Sect. 3.2) is as follows. Define $n = k + 1$ jobs, where $w_j = x_j$, $a_j = x_j$, $b_j = 0$ for $1 \leq j \leq k$, $w_{k+1} = A$, $a_{k+1} = 2A(2 \ln 2 + 1)$, $b_{k+1} = \frac{1}{A}$. The threshold $G = \frac{1}{2} \sum_{j=1}^{k} x_j^2 + A^2 \left(3 + \frac{2}{2 \ln 2 + 1}\right)$. To complete the proof, it is sufficient to show that the PP problem has a solution if and only if for the above instance of the problem $1|p_j = a_j 2^{-b_j t}| \sum w_j C_j$ there exists a schedule σ such $\sum w_j C_j(\sigma) \leq G$.   □

6.4.6 Proportional-linear shortening

Distinct ready times and deadlines

The $\sum U_j$ criterion. Wang and Xia considered job processing times given by (6.44) and proved the following result.

Theorem 6.247. (Wang and Xia [289]) *The problem* $1|p_j = a_j(1-kt), k > 0,$
$$k\left(\sum_{j=1}^{n} a_j - a_{\min}\right) < 1|\sum U_j \text{ is solvable in } O(n\log n) \text{ time by algorithm } A_{26}.$$

Proof. Similar to the proof of Theorem 6.196.   □

The $f_{\max}$ criterion. Wang and Xia considered job processing times given by (6.44) and proved the following result.

Theorem 6.248. (Wang and Xia [289]) *The problem* $1|p_j = a_j(1-kt), k > 0,$
$$k\left(\sum_{j=1}^{n} a_j - a_{\min}\right) < 1|f_{\max} \text{ is solvable in } O(n^2) \text{ time by algorithm } A_{29}.$$

Proof. Similar to the proof of Theorem 6.204.   □

6.4.7 Linear shortening

Equal ready times and deadlines

The $\sum w_j C_j$ criterion. For proportional job processing times given by (6.46), the following results are known.

Theorem 6.249. (Bachman et al. [10]) *The decision version of the problem* $1|p_j = a_j - b_j t, 0 \leq b_j < 1, b_i(\sum_{j=1}^{n} a_j - a_i) < a_i|\sum w_j C_j$ *is $\mathcal{NP}$-complete in the ordinary sense, even if there exists only one non-zero shortening rate.*

Proof. The transformation from the PP problem (cf. Sect. 3.2) is as follows. We have $n = k + 1$ jobs with the following parameters: $a_i = x_i$, $b_i = 0$ and $w_i = x_i$ for $1 \leq i \leq k$, and $a_{k+1} = 2A$, $b_{k+1} = 1 - \frac{1}{A}$, and $w_{k+1} = 2A^2$. The threshold $G = \frac{1}{2}\sum_{i=1}^{k} x_i^2 + A^2 + A(2A+1)^2$.

To complete the proof, it is sufficient to show that the PP problem has a solution if and only if for the above instance of the problem $1|p_j = a_j - b_j t,$ $0 \leq b_j < 1, b_i(\sum_{j=1}^{n} a_j - a_i) < a_i|\sum w_j C_j$ there exists a schedule σ such that the total weighted completion time $\sum w_j C_j(\sigma) \leq G$. The equivalence can be proved using the equalities $\frac{1}{2}\sum_{i=1}^{k} x_i^2 = \frac{1}{2}(\sum_{i=1}^{k} x_i)^2 - \sum_{1 \leq i < j \leq k} x_i x_j = A^2 - \sum_{1 \leq i < j \leq m} x_i x_j$, (see [10, Theorem 1]).   □

The authors also proved a few properties of special cases of the problem $1|p_j = a_j - b_j t, 0 \leq b_j < 1, b_i(\sum_{j=1}^{n} a_j - a_i) < a_i|\sum w_j C_j$.

Property 6.250. (Bachman et al. [10]) The problem $1|p_j = a - bt| \sum w_j C_j$ is solvable in $O(n \log n)$ time by scheduling jobs in the non-increasing order of w_j values.

Proof. Since all jobs have the same form of job processing times, the result follows from Lemma 1.2 (b). ∎

By Property 6.250, if $0 \le b_j < 1$ and $b_i(\sum_{j=1}^{n} a_j - a_i) < a_i$ for any $1 \le i, j \le n$, the problem $1|p_j = a - bt| \sum w_j C_j$ is solved by the algorithm $A_{38} : (a|b|w_j) \mapsto (w_j \searrow)$.

Property 6.251. (Bachman et al. [10]) The problem $1|p_j = a_j - ka_j t| \sum w_j C_j$, where k is a given constant, $k(\sum_{j=1}^{n} a_j - a_{\min}) < 1$ and $a_{\min} = \min_{1 \le j \le n}\{a_j\}$, is solvable in $O(n \log n)$ time by scheduling jobs in the non-decreasing order of $\frac{a_j}{w_j(1-ka_j)}$ ratios.

Proof. By pairwise job interchange argument, see [10, Property 2]. ◇

By Property 6.250, if $k(\sum_{j=1}^{n} a_j - a_{\min}) < 1$ for $1 \le i \le n$, $a_{\min} := \min_{1 \le j \le n}\{a_j\}$ and a given constant k, the problem $1|p_j = a - ka_j t| \sum w_j C_j$ is solved by the algorithm $A_{39} : (a|k|a_j|w_j) \mapsto (\frac{a_j}{w_j(1-ka_j)} \nearrow)$.

Property 6.252. (Bachman et al. [10]) Let $\sigma = (\sigma_1, \sigma_2, \dots, \sigma_n)$ be a schedule for the problem $1|p_j = a_j - bt| \sum w_j C_j$, where $w_j = ka_j$ for $1 \le j \le n$, and let $\bar{\sigma} = (\sigma_n, \sigma_{n-1}, \dots, \sigma_1)$. Then $\sum w_j C_j(\sigma) = \sum w_j C_j(\bar{\sigma})$.

Proof. By direct calculation, see [10, Property 3]. ◇

Property 6.253. (Bachman et al. [10]) For the problem $1|p_j = a_j - bt| \sum w_j C_j$, where $w_j = ka_j$ for $1 \le j \le n$, there exists an optimal schedule that is V-shaped with respect to a_j values.

Proof. Consider three schedules $\sigma^\star$, σ' and σ'', differing only in the order of jobs $J_{\sigma_{i-1}}$, J_{σ_i} and $J_{\sigma_{i+1}}$. Assume that $a_{\sigma_i} - a_{\sigma_{i-1}} > 0$, $a_{\sigma_i} - a_{\sigma_{i+1}} > 0$ and that $\sigma^\star$ is optimal. Calculate differences $\sum w_j C_j(\sigma^\star) - \sum w_j C_j(\sigma')$ and $\sum w_j C_j(\sigma^\star) - \sum w_j C_j(\sigma'')$. Since $\sigma^\star$ is optimal, both these differences are positive. This, in turn, leads to a contradiction (see [10, Property 4]). □

Before we state the next property, we will introduce a new notion (cf. [10]).

Definition 6.254. (An even-odd V-shaped sequence)
A sequence (x_j) is said to be even-odd V-shaped (has an even-odd V-shape), if non-increasingly ordered numbers with even (odd) indices are followed by non-decreasingly ordered numbers with odd (even) indices.

Property 6.255. (Bachman et al. [10]) If in an instance of the problem $1|_j = a_j - bt| \sum ka_j C_j$, the jobs are numbered in the non-decreasing order of a_j values, then an even-odd V-shaped schedule is not optimal.

Proof. By direct calculation, see [10, Property 5].   ◇

For the problem $1|p_j = a_j - b_j t, 0 \leq b_j < 1, b_i(\sum_{j=1}^{n} a_j - a_i) < a_i| \sum w_j C_j$ Bachman et al. [10] proposed two heuristic algorithms. We will consider this algorithms in Chap. 9.

Distinct ready times and deadlines

The $\sum U_j$ criterion. Theorem 6.87 implies that if in the set of jobs there are identical ready times and deadlines, the problem of minimizing the number of tardy jobs is $\mathcal{NP}$-complete. Ho et al. [135] stated the problem of minimizing the number of tardy jobs for a set of jobs with a common deadline, D, as an open problem. The problem has been solved independently by two authors: Chen [42] and Woeginger [295]. First, we consider the result obtained by Chen.

Lemma 6.256. (Chen [42]) *There exists an optimal schedule where the on-time jobs are scheduled in the non-increasing order of $\frac{a_j}{b_j}$ ratios, and the late jobs are scheduled in an arbitrary order.*

Proof. First, note that without loss of generality, we can consider only schedules without idle times. Second, since $d_i = D$, each on-time job is processed before all late jobs. Finally, since by Theorem 6.24 scheduling jobs in the non-increasing order of $\frac{b_j}{a_j}$ ratios minimizes $C_{\max}$, the result follows.   □

Chen [42] proposed to construct a schedule for the problem $1|r_j, p_j = a_j - b_j t| \sum U_j$, where $r_j = R$, $0 < b_j < 1, b_j d_j < a_j \leq d_j$ and $d_j = D$, by assigning an unscheduled job with the largest ratio $\frac{a_j}{b_j}$ either to the position immediately following the current last on-time job or to the position following the current last late job. Some recursive relations, Lemma 6.256 and a dynamic programming aproach allowed Chen to obtain an $O(n^2)$ time algorithm.

Before we formulate the algorithm proposed by Chen, we need some new notation. Let $C(j, k)$ and $S(j, k)$ denote the minimum completion time of the on-time jobs in a partial schedule containing the first j jobs, $1 \leq j \leq n$, among which there are exactly $k \leq j$ on-time jobs, and the set of on-time jobs in the schedule corresponding to $C(j, k)$, respectively.

Define function $F(j, k)$ as follows:

$$F(j, k) := \begin{cases} a_j + (1 - b_j)C(j - 1, k - 1), & \text{if } a_j + (1 - b_j)C(j - 1, k - 1) \leq D, \\ \infty, & \text{otherwise.} \end{cases}$$

Define the set $S(j, k)$ as follows:

$$S(j, k) := \begin{cases} \{j\} \cup S(j - 1, k - 1), & \text{if } C(j, k) = F(j, k) \leq D, \\ S(j - 1, k), & \text{if } C(j, k) = C(j - 1, k) \leq D, \\ \emptyset, & \text{otherwise.} \end{cases}$$

Applying the above notation, the algorithm proposed for the problem $1|r_j = R, p_j = a_j - b_j t, a_j > 0, 0 < b_j < 1, b_j d_j < a_j \leq d_j, d_j = D| \sum U_j$ by Chen [42] can be formulated as follows.

Algorithm A_{40} for the problem
$1|r_j = R, p_j = a_j - b_j t, a_j > 0, 0 < b_j < 1, b_j d_j < a_j \leq d_j, d_j = D| \sum U_j$ ([42])

Input: sequences $(a_1, a_2), \ldots, a_n), (b_1, b_2, \ldots, b_n)$ numbers R, D
Output: the minimum number of late jobs

▷ **Step 1:**
   Arrange jobs according to the non-increasing order of $\frac{a_j}{b_j}$ ratios;
$$C(1,1) \leftarrow \begin{cases} a_{[1]}, \text{ if } & a_{[1]} \leq D, \\ \infty, & \text{otherwise}; \end{cases}$$
▷ **Step 2:**
   for $j \leftarrow 1$ **to** n **do**
      $C(j, 0) \leftarrow 0$;
      $S(j, 0) \leftarrow \emptyset$;
      $S(1,1) \leftarrow \begin{cases} \{1\}, \text{ if } & C(1,1) = a_{[1]}, \\ \emptyset, & \text{otherwise}; \end{cases}$
▷ **Step 3:**
   for $j \leftarrow 2$ **to** n **do**
      for $k \leftarrow 1$ **to** $j - 1$ **do**
         $C(j, k) \leftarrow \min\{C(j - 1, k), F(j, k)\}$;
         $x \leftarrow a_{[j]} + (1 - b_{[j]})C(j - 1, j - 1)$;
         $C(j, j) \leftarrow \begin{cases} x, \text{ if } & x \leq D, \\ \infty, & \text{otherwise}, \end{cases}$
         $S(j, j) \leftarrow \begin{cases} \{j\} \cup S(j - 1, j - 1), \text{ if } & C(j, j) \leq D, \\ \emptyset, & \text{otherwise}; \end{cases}$
▷ **Step 4:**
   Calculate $k^\star \leftarrow \arg\max\{k : C(n, k) \leq D\}$;
▷ **Step 5:**
   return $n - k^\star$.

Theorem 6.257. (Chen [42]) *The problem $1|r_j, p_j = a_j - b_j t| \sum U_j$, where $r_j = R, 0 < b_j < 1, b_j d_j < a_j \leq d_j$ and $d_j = D$ is solvable in $O(n^2)$ time by algorithm A_{40}.*

Proof. Knowing the minimum number of late jobs, we can construct an optimal schedule by scheduling jobs from the set $S(n, k^\star)$ in the order of their rearranged indices, and then scheduling the remaining jobs in an arbitrary order. Since in algorithm A_{40} for each $1 \leq j \leq n$ there are j possible states, the overall time complexity of the algorithm is $O(n^2)$. □

The problem $1|r_j = R, p_j = a_j - b_j t, a_j > 0, 0 < b_j < 1, b_j d_j < a_j \leq d_j, d_j = D| \sum U_j$ has also been considered by Woeginger [295].

Lemma 6.258. (Woeginger [295]) *A subset of jobs, $j_1, j_2, \ldots, j_m$, can be executed during the time interval $\langle t_1, t_2 \rangle$ if and only if the jobs are processed in the non-decreasing order of $\frac{b_j}{a_j}$ ratios.*

Proof. See [295, Lemma 2]. $\diamond$

Using Lemma 6.258, Woeginger proposed to find the solution using dynamic programming in $O(n^3)$ time (see [295, Theorem 3] for details).

The $\sum g(E_j)$ criterion. Zhao and Tang [303] considered job processing times given by (6.52). Let g be a non-decreasing function.

Theorem 6.259. (Zhao and Tang [303]) *The problem $1|p_j = a_j - bt| \sum g(E_j)$ is solvable in $O(n \log n)$ time by scheduling jobs in the non-increasing order of a_j values.*

Proof. By pairwise job interchange argument. $\square$

By Theorem 6.259, the problem $1|p_j = a_j - bt| \sum g(E_j)$ is solved by the algorithm $A_8 : (a_j|b|g) \mapsto (a_j \searrow)$.

The criterion $\sum(\alpha E_j + \beta T_j + \gamma d)$. Cheng et al. [59] considered the problem of minimizing the sum of earlines, tardiness and due-date penalties and common due-date assignment for job processing times in the form of

$$p_j = a_j - bt, \tag{6.72}$$

where

$$0 < b < 1 \tag{6.73}$$

and

$$b \left(\sum_{j=1}^{n} a_j - a_i \right) < a_i \tag{6.74}$$

for $1 \leq i \leq n$.

Remark 6.260. Job processing times (6.72) and conditions (6.73) and (6.74) are special cases, respectively, of job processing times (6.46) and conditions (6.47) and (6.48) for $b_j = b$ and $1 \leq j \leq n$.

Cheng et al. [59], applying the approach used previously for the problem $1|p_j = a_j + bt| \sum(\alpha E_j + \beta T_j + \gamma d)$ (cf. Sect. 6.4.3), proposed for the problem $1|p_j = a_j - bt, 0 < b < 1, b(\sum_{j=1}^{n} a_j - a_i) < a_i| \sum(\alpha E_j + \beta T_j + \gamma d)$ an algorithm (see [59, Algorithm A]). Let us call the algorithm A_{41}. Since the properties (see [59, Properties 1–8]) on which algorithm A_{41} is based are counterparts of the properties from Sect. 6.4.3, we do not present their formulations, giving only the formulation of the main result.

Theorem 6.261. (Cheng et al. [59]) *The problem* $1|p_j = a_j - bt, 0 < b < 1,$ $b(\sum_{j=1}^n a_j - a_i) < a_i| \sum(\alpha E_j + \beta T_j + \gamma d)$ *is solvable in* $O(n \log n)$ *time by algorithm* A_{41}.

Proof. See [59, Properties 10–13, Theorem 14].   ◇

With this theorem we end the review of single-machine time-dependent scheduling. In Chap. 7 and Chap. 8 we will consider the results concerning time-dependent scheduling on parallel and dedicated machines, respectively.

6.5 Summary and tables

In this chapter, we reviewed single-machine time-dependent scheduling problems. We considered different deteriorating or shortening job processing times. The criteria of schedule optimality include the most popular criteria such as the $C_{\max}$, $\sum C_j$ or $L_{\max}$, as well as less popular criteria such as the $\sum w_j C_j$, $\sum U_j$ or $\sum(\alpha E_j + \beta T_j + \gamma d)$.

Approximately two-thirds of the problems presented in the chapter are polynomially solvable. For these problems, we presented pseudo-codes of optimal algorithms, including different variants of these algorithms if they exist. For the remaining problems that are ordinary or strongly $\mathcal{NP}$-complete, we presented $\mathcal{NP}$-completeness proofs or sketches of such proofs.

Below, we classify in the tabular form the single-machine time-dependent scheduling problems and polynomial algorithms considered in the chapter. The problems and algorithms are divided into groups with respect to the applied optimality criterion. The problems with a particular criterion are divided into tractable and intractable problems.

Tables 6.1 and 6.2 present the problems concerning, respectively, tractable and intractable single-machine time-dependent scheduling problems with the $C_{\max}$ criterion.

Tables 6.3 and 6.4 present the problems concerning, respectively, tractable and intractable single-machine time-dependent scheduling problems with the $\sum C_j$ criterion.

Tables 6.5 and 6.6 present the problems concerning, respectively, tractable and intractable single-machine time-dependent scheduling problems with the $L_{\max}$ criterion.

Tables 6.7 and 6.8 present the problems concerning, respectively, tractable and intractable single-machine time-dependent scheduling problems with criteria other than $C_{\max}$, $\sum C_j$ or $L_{\max}$.

Tables 6.9, 6.10 and 6.11 present polynomial algorithms for the $C_{\max}$, $\sum C_j$ and $L_{\max}$ criteria, respectively.

Table 6.12 presents polynomial algorithms for other criteria than $C_{\max}$, $\sum C_j$ or $L_{\max}$.

Table 6.13 presents exact algorithms for single-machine time-dependent scheduling problems.

Table 6.1: Tractable Single-Machine Time-Dependent Problems ($C_{\max}$ Criterion)

Problem	Complexity	References	This book
$1\|p_j = b_j t\|C_{\max}$ (a)	$O(n)$	[216]	Theorem 6.1
$1\|p_{i,j} = b_{i,j} t, \theta_i, GT\|C_{\max}$	$O(n \log n)$	[44]	Theorem 6.4
$1\|p_{i,j} = b_{i,j} t, \theta_i = \delta_i t, GT\|C_{\max}$	$O(n)$	[300]	Theorem 6.5
$1\|p_j = b_j(A + Bt)\|C_{\max}$ (a)	$O(n \log n)$	[173, 304]	Theorem 6.20
$1\|p_{i,j} = b_{i,j}(A+Bt), \theta_i, GT\|C_{\max}$	$O(n \log n)$	[127]	Theorem 6.22
$1\|p_{i,j} \equiv (6.5),$ $\theta_i \equiv (6.9), GT\|C_{\max}$	$O(n)$	[284]	Theorem 6.23
$1\|p_j = a_j + b_j t\|C_{\max}$ (b)	$O(n \log n)$	[33, 128, 264, 277]	Theorem 6.24
$1\|p_j = a_j + b_j t\|C_{\max}$ (c)	$O(n \log n)$	[109]	Sect. 6.1.3, p. 70
$1\|p_j = a + b_j t,$ $b_j \in \{B_1, B_2\}, d_j\|C_{\max}$	$O(n \log n)$	[53]	Theorem 6.35
$1\|p_j = a_j + bt, d_j\|C_{\max}$	$O(n^5)$	[52]	Theorem 6.38
$1\|p_j = a_j + f(t), f(t) \nearrow \|C_{\max}$	$O(n \log n)$	[207]	Theorem 6.39
$1\|p_j = a_j + f(t), f(t) \searrow \|C_{\max}$ (d)	$O(n \log n)$	[207]	Theorem 6.41
$1\|p_j = b_j h(t)\|C_{\max}$ (e)	$O(n \log n)$	[173]	Theorem 6.43
$1\|p_j = a_j + \max\{0, b(t-d)\}\|C_{\max}$	$O(n \log n)$	[178]	Theorem 6.57
$1\|p_j = a_j + b_j \max\{t-t_0, 0\}\|C_{\max}$	$O(n \log n)$	[39]	Theorem 6.68
$1\|p_j = a + b_j \max\{t - t_0, 0\}\|C_{\max}$	$O(n \log n)$	[39]	Theorem 6.69 (a)
$1\|p_j = a_j + b \max\{t - t_0, 0\}\|C_{\max}$	$O(n \log n)$	[39]	Theorem 6.69 (b)
$1\|p_j = a_j +$ $b_j \max\{t - t_0, 0\}\|C_{\max}$ (f)	$O(n \log n)$	[39]	Theorem 6.69 (c)
$1\|p_j = a_j(1 - kt)\|C_{\max}$ (a,g)	$O(n)$	[289]	Theorem 6.82
$1\|p_j = a_j - b_j t, d_j = D\|C_{\max}$ (h)	$O(n \log n)$	[135]	Theorem 6.87 (c)
$1\|p_j = a - b_j t, d_j\|C_{\max}$ (i)	$O(n \log n)$	[53]	Theorem 6.92
$1\|p_j = a_j - bt, r_j\|C_{\max}$ (j)	$O(n^6 \log n)$	[49]	Theorem 6.96
$1\|p_j = a_j - b(t - y)\|C_{\max}$ (k)	$O(n \log n)$	[54]	Theorem 6.119 (a)
$1\|p_j = a - b_j(t - y)\|C_{\max}$ (k)	$O(n \log n)$	[54]	Theorem 6.119 (b)

(a) the value of $C_{\max}$ does not depend on schedule

(b) $S_1 \equiv t_0 = 0$

(c) $S_1 \equiv t_0 \geq 0$

(d) $|\frac{df(t)}{dt}| \leq 1$

(e) $h(t)$ is convex or concave, other assumptions see Theorem 6.43

(f) $a_j = b_j k$, where $k = const > 0$

(g) $k > 0, k(\sum_{j=1}^n a_j - a_{\min}) < 1$ for $1 \leq j \leq n$, where $a_{\min} := \min_{1 \leq j \leq n}\{a_j\}$

(h) $0 < b_j < 1, b_j D < a_j \leq D$ for $1 \leq j \leq n$

(i) $0 < b_j < 1, b_j d_j < a \leq d_j$ for $1 \leq j \leq n$

(j) $0 < b < 1, b_j(\sum_{i=1}^n a_i - a_j) < a_j$ for $1 \leq j \leq n$

(k) $y > 0, Y < \infty, 0 < b_j < 1$ for $1 \leq j \leq n$

Table 6.2: Intractable Single-Machine Time-Dependent Problems (C_{max} Criterion)

Problem	Class	References	This book
$1, h_{11}\|p_j = b_j t, nres\|C_{max}$	$\mathcal{NPC}$	[91, 155]	Theorem 6.11
$1, h_{1k}\|p_j = b_j t, nres\|C_{max}$	$\mathcal{SNPC}$	[91]	Theorem 6.15
$1, h_{11}\|p_j = b_j t, res\|C_{max}$	$\mathcal{NPC}$	[92]	Theorem 6.16
$1\|p_j = b_j t, r_j \in \{R_1, R_2\},$ $d_j \in \{D_1, D_2\}\|C_{max}$	$\mathcal{NPC}$	[91]	Theorem 6.18
$1\|p_j = b_j t, r_j, d_j\|C_{max}$	$\mathcal{SNPC}$	[91]	Theorem 6.19
$1\|p_j = a_j + b_j t, r_j, d_j\|C_{max}$	$\mathcal{SNPC}$	–	Theorem 6.33
$1\|p_j = a_j + b_j t, d_j\|C_{max}$	$\mathcal{SNPC}$	[52]	Theorem 6.34
$1\|p_j = 1 + b_j t, d_j\|-$	$\mathcal{SNPC}$	[53]	Theorem 6.36
$1\|p_j = 1 + b_j t, d_j \in \{D_1, D_2\}\|-$	$\mathcal{NPC}$	[53]	Theorem 6.37
$1\|p_j \in \{a_j, b_j : a_j \leq b_j\},$ $d_j = D\|C_{max}$	$\mathcal{NPC}$	[51, 217]	Theorem 6.50
$1\|p_j = a_j +$ $\max\{0, b_j(t - d_j)\}\|C_{max}$	$\mathcal{SNPC}$	[171]	Theorem 6.63 (a)
$1\|p_j = a_j + \max\{0, b_j(t - d_j)\},$ $b_j = B\|C_{max}$	$\mathcal{NPC}$	[171]	Theorem 6.63 (b)
$1\|p_j = a_j + \max\{0, b_j(t - d_j)\},$ $d_j = D\|C_{max}$	$\mathcal{NPC}$	[171]	Theorem 6.63 (c)
$1\|p_j = a_j + b_j \, \mathbf{1}_X\|C_{max}$ $^{(a,b)}$	$\mathcal{NPC}$	[39]	Theorem 6.66
$1\|p_j \equiv (6.39)\|C_{max}$ $^{(c)}$	$\mathcal{NPC}$	[177]	Theorem 6.72
$1\|p_j = a_j 2^{b_j(t-r_j)}, r_j\|C_{max}$	$\mathcal{SNPC}$	[146]	Theorem 6.79
$1\|p_j = a_j 2^{b_j(t-r_j)},$ $r_j \in \{0, R\}\|C_{max}$	$\mathcal{NPC}$	[146]	Theorem 6.80
$1\|p_j = a_j - b_j t, d_j\|C_{max}$ $^{(d)}$	$\mathcal{SNPC}$	[135]	Theorem 6.87 (a)
$1\|p_j = a_j - b_j t,$ $d_j \in \{D_1, D_2\}\|C_{max}$ $^{(d)}$	$\mathcal{NPC}$	[135]	Theorem 6.87 (b)
$1\|p_j = a_j - bt, d_j\|C_{max}$ $^{(e)}$	$\mathcal{SNPC}$	[50]	Theorem 6.88
$1\|p_j = 1 - b_j t, d_j\|C_{max}$ $^{(f)}$	$\mathcal{SNPC}$	[53]	Theorem 6.93
$1\|p_j = 1 - b_j t,$ $d_j \in \{D_1, D_2\}\|C_{max}$ $^{(f)}$	$\mathcal{NPC}$	[53]	Theorem 6.94
$1\|p_j = a_j - bt,$ $d_j \in \{D_1, D_2\}\|C_{max}$ $^{(e)}$	$\mathcal{NPC}$	[48]	Theorem 6.95
$1\|p_j = a_j + b_j t, r_j\|C_{max}$	$\mathcal{SNPC}$	[49]	Theorem 6.100
$1\|p_j = a_j + b_j t, r_j \in \{0, R\}\|C_{max}$	$\mathcal{SNPC}$	[49]	Theorem 6.101
$1\|p_j \in \{a_j, a_j - b_j\} :$ $0 \leq b_j \leq a_j\|C_{max}$	$\mathcal{NPC}$	[56]	Theorem 6.109
$1\|p_j = a_j - b_j(t - y)\|C_{max}$ $^{(g)}$	$\mathcal{NPC}$	[54]	Theorem 6.114

$^{(a)}$ $X := \{t : t - t_0 > 0\}$
$^{(b)}$ $\sum_{j=1}^{n} a_j > t_0$
$^{(c)}$ $d > 0, D = \infty$
$^{(d)}$ $0 < b_j < 1, b_j d_j < a_j \leq d_j$ for $1 \leq j \leq n$
$^{(e)}$ $0 < b < 1, b d_j < a_j \leq d_j$ for $1 \leq j \leq n$
$^{(f)}$ $0 < b_j < 1, b_j d_j < 1 \leq d_j$ for $1 \leq j \leq n$
$^{(g)}$ $y = 0, Y < \infty, 0 < b_j < 1$ for $1 \leq j \leq n$

Table 6.3: Tractable Single-Machine Time-Dependent Problems ($\sum C_j$ Criterion)

Problem	Complexity	References	This book
$1\|p_j = b_j t\|\sum C_j$	$O(n \log n)$	[216]	Theorem 6.120
$1\|p_{i,j} = b_{i,j} t, \theta_i = \delta_i, GT\|\sum C_j$	$O(n \log n)$	[300]	Theorem 6.122
$1\|p_j = b_j t, d_j\|\sum C_j$	$O(n \log n)$	[55]	Theorem 6.125
$1\|p_j = b_j(A + Bt)\|\sum C_j$	$O(n \log n)$	–	Theorem 6.127
$1\|p_j = 1 + b_j t,$	$O(n^{k+1})$	[106]	Theorem 6.139
$\quad b_j \in \{B_1, B_2, \ldots, B_k\}\|\sum C_j$			
$1\|p_j = a_j + bt, d_j\|\sum C_j$	$O(n^5)$	[48]	Theorem 6.145
$1\|p_j = a_j + f(t), f(t) \nearrow \|\sum C_j$	$O(n \log n)$	[89]	Theorem 6.146
$1\|p_j = a_j - bt\|\sum C_j$ $^{(a)}$	$O(n \log n)$	[226]	Property 6.155
$1\|p_j = a_j - ka_j t\|\sum C_j$ $^{(b)}$	$O(n \log n)$	[226]	Property 6.156
$1\|p_{i,j} = a_{i,j} - b_i t, \theta_i, GT\|\sum C_j$	$O(n \log n)$	[283]	Theorem 6.163

$^{(a)}$ $0 < b < 1$
$^{(b)}$ $k = const > 0$

Table 6.4: Intractable Single-Machine Time-Dependent Problems ($\sum C_j$ Criterion)

Problem	Complexity	References	This book
$1, h_{11}\|p_j = b_j t, nres\|\sum C_j$	$\mathcal{NPC}$	[155]	Theorem 6.123
$1\|p_j = a_j + b_j t, d_j = D\|\sum C_j$	$\mathcal{NPC}$	[55]	Theorem 6.141
$1\|p_j = a_j + b_j t, d_j\|\sum C_j$	$\mathcal{SNPC}$	[52]	Theorem 6.142
$1\|p_j \in \{a_j, b_j : a_j \leq b_j\},$	$\mathcal{NPC}$	[51]	Theorem 6.147
$\quad d_j = D\|\sum C_j$			
$1\|p_j = a_j - bt, d_j = D\|\sum C_j$ $^{(a)}$	$\mathcal{NPC}$	[55]	Theorem 6.161
$1\|p_j = a_j - b_j(t - y)\|\sum C_j$ $^{(b)}$	$\mathcal{NPC}$	[54]	Theorem 6.165
$1\|p_j = a_j - bt,$	$\mathcal{NPC}$	[48]	Theorem 6.168
$\quad d_j \in \{D_1, D_2\}\|\sum C_j$ $^{(a)}$			

$^{(a)}$ $0 < b < 1$
$^{(b)}$ $y = 0, Y < \infty, 0 < b_j < 1$ for $1 \leq j \leq n$

Table 6.5: Tractable Single-Machine Time-Dependent Problems ($L_{\max}$ Criterion)

Problem	Complexity	References	This book
$1\|p_j = b_j t\|L_{\max}$	$O(n \log n)$	[216]	Theorem 6.169
$1\|p_j = b_j(A + Bt)\|L_{\max}$	$O(n \log n)$	[173]	Theorem 6.170
$1\|p_j = a_j + bt\|L_{\max}$	$O(n^6 \log n)$	[52]	Theorem 6.179
$1\|p_j = b_j h(t)\|L_{\max}$ $^{(a)}$	$O(n \log n)$	[173]	Theorem 6.180

$^{(a)}$ $h(t)$ is convex or concave, other assumptions see Theorem 6.180

Table 6.6: Intractable Single-Machine Time-Dependent Problems ($L_{\max}$ Criterion)

Problem	Complexity	References	This book
$1\|p_j = a_j + b_jt\|L_{\max}$	$\mathcal{SNPC}$	[11, 172, 52]	Theorems 6.172– 6.174
$1\|p_j = a_j 2^{-b_jt}\|L_{\max}$	$\mathcal{SNPC}$	[146]	Theorem 6.181
$1\|p_j = a_j 2^{-b_jt}, d_j \in \{d, D\}\|L_{\max}$	$\mathcal{NPC}$	[146]	Theorem 6.182
$1\|p_j = a_j - bt,$ $d_j \in \{D_1, D_2\}\|L_{\max}$ $^{(a)}$	$\mathcal{NPC}$	[48]	Theorem 6.184

$^{(a)}$ $0 < b < 1$

Table 6.7: Tractable Single-Machine Time-Dependent Problems (Criteria other than $C_{\max}$, $\sum C_j$ and $L_{\max}$)

Problem	Complexity	References	This book
$1\|p_j = b_jt\|\sum w_j C_j$	$O(n \log n)$	[216]	Theorem 6.185
$1\|p_j = b_jt\|f_{\max}$	$O(n^2)$	–	Theorem 6.187
$1\|p_j = b_jt\|\sum L_j$	$O(n \log n)$	[216]	Theorem 6.194 (a)
$1\|p_j = b_jt\|T_{\max}$	$O(n \log n)$	[216]	Theorem 6.194 (b)
$1\|p_j = b_jt\|\sum U_j$	$O(n \log n)$	[216]	Theorem 6.196
$1\|p_j = b_j(A + Bt)\|\sum w_j C_j$	$O(n \log n)$	[173]	Theorem 6.197
$1\|p_j = b_j(A + Bt)\|\sum U_j$	$O(n \log n)$	[173]	Theorem 6.202
$1\|p_j = b_j(A + Bt)\|f_{\max}$	$O(n^2)$	[173]	Theorem 6.204
$1\|p_j = a_j + bt\|\sum w_j C_j$ $^{(a)}$	$O(n)$	[218]	Property 6.210
$1\|p_j = a_j + b_jt\|P_{\max}$ $^{(b)}$	$O(n \log n)$	[5]	Theorem 6.222
$1\|p_j = a_j + bt\|P_{\max}$ $^{(c)}$	$O(n \log n)$	[5]	Theorem 6.223
$1\|p_j = a_j + bt\|\sum(\alpha E_j + \beta T_j + \gamma d)$	$O(n \log n)$	[57]	Theorem 6.231
$1\|p_j = b_j h(t)\|\sum w_j C_j$ $^{(d)}$	$O(n \log n)$	[173]	Theorem 6.233
$1\|p_j = a_j + f(t)\|\sum C_j^k$ $^{(e)}$	$O(n \log n)$	[179]	Theorem 6.235
$1\|p_j \equiv (6.68), w_j \in \{W_1, W_2\}\|\sum w_j C_j$	$O(n \log n)$	[263]	Theorem 6.239
$1\|p_j \equiv (6.68), b_j \in \{B_1, B_2\}\|\sum w_j C_j$	$O(n \log n)$	[263]	Theorem 6.241
$1\|p_j \equiv (6.68) \wedge (6.69)\|\sum w_j C_j$ $^{(f)}$	$O(n \log n)$	[263]	Theorem 6.243
$1\|p_j = a_j(1 - kt)\|\sum U_j$ $^{(g)}$	$O(n \log n)$	[289]	Theorem 6.247
$1\|p_j = a_j(1 - kt)\|f_{\max}$ $^{(g)}$	$O(n^2)$	[289]	Theorem 6.248
$1\|p_j = a - bt\|\sum w_j C_j$ $^{(h)}$	$O(n \log n)$	[10]	Property 6.250
$1\|p_j = a_j - ka_jt\|\sum w_j C_j$ $^{(g)}$	$O(n \log n)$	[10]	Property 6.251
$1\|r_j = R, p_j = a_j - b_jt, d_j = D\|\sum U_j$ $^{(i)}$	$O(n^2)$	[42, 295]	Theorem 6.257
$1\|p_j = a_j - bt\|\sum g(E_j)$	$O(n \log n)$	[303]	Theorem 6.259
$1\|p_j = a_j - bt\|\sum(\alpha E_j + \beta T_j + \gamma d)$ $^{(h)}$	$O(n \log n)$	[59]	Theorem 6.261

$^{(a)}$ $w_j = \delta a_j$ for $1 \leq j \leq n$, $\delta = const > 0$, jobs are numbered in the $a_j \nearrow$ order
$^{(b)}$ $a_j > 0, b_j \geq 1$ for $1 \leq j \leq n$
$^{(c)}$ $b \geq 1, a_j > 0$ for $1 \leq j \leq n$
$^{(d)}$ $h(t)$ is convex or concave, other assumptions see Theorem 6.233
$^{(e)}$ for assumptions concerning the function $f(t)$ see Theorem 6.235
$^{(f)}$ $w_i \geq w_j \Rightarrow b_i \geq b_j$ for $1 \leq i, j \leq n$
$^{(g)}$ $k(\sum_{j=1}^n a_j - a_{\min}) < 1$, where $k = const > 0$ and $a_{\min} := \min_{1 \leq j \leq n}\{a_j\}$
$^{(h)}$ $0 \leq b < 1, b(\sum_{i=1}^n a_i - a_j) < a_j$ for $1 \leq j \leq n$
$^{(i)}$ $0 < b_j < 1, b_j D < a_j \leq D$ for $1 \leq j \leq n$

Table 6.8: Intractable Single-Machine Time-Dependent Problems (Criteria other than $C_{\max}$, $\sum C_j$ and $L_{\max}$)

Problem	Complexity	References	This book
$1\|p_j = a_j + b_j t\| \sum w_j C_j$	$\mathcal{NPC}$	[13]	Theorem 6.211
$1\|p_j = a_j 2^{-b_j t}\| \sum w_j C_j$	$\mathcal{NPC}$	[146]	Theorem 6.246
$1\|p_j = a_j - b_j t\| \sum w_j C_j$ [(a)]	$\mathcal{NPC}$	[10]	Theorem 6.249

[(a)] $0 \le b_j < 1, b_j(\sum_{i=1}^n a_i - a_j) < a_j$ for $1 \le j \le n$

Table 6.9: Polynomial Algorithms for Single-Machine Time-Dependent Problems ($C_{\max}$ Criterion)

Algorithm	Complexity	Problem	This book
A_1	$O(n \log n)$	$1\|p_{i,j} = b_{i,j} t, \theta_i, GT\|C_{\max}$	Sect. 6.1.1, p. 61
A_2	$O(n \log n)$	$1\|p_{i,j} = b_{i,j}(A + Bt), \theta_i, GT\|C_{\max}$	Sect. 6.1.2, p. 67
A_3	$O(n \log n)$	$1\|p_j = a_j + b_j t\|C_{\max}$	Sect. 6.1.3, p. 69
A_4	$O(n \log n)$	$1\|p_j = a_j + b_j t\|C_{\max}$	Sect. 6.1.3, p. 70
A_5	$O(n^2)$	$1\|p_j = a_j + b_j t\|C_{\max}$	Sect. 6.1.3, p. 70
A_{14}	$O(n \log n)$	$1\|p_j = a + b_j t, d_j, b_j \in \{B_1, B_2\}\|C_{\max}$	Sect. 6.1.7, p. 95
A_6	$O(n^5)$	$1\|p_j = a_j + bt, d_j\|C_{\max}$	Sect. 6.1.3, p. 74
A_7	$O(n \log n)$	$1\|p_j = a_j + f(t), f \nearrow \|C_{\max}$	Sect. 6.1.4, p. 76
A_8	$O(n \log n)$	$1\|p_j = a_j + f(t), f \searrow, \|\frac{df}{dt}\| \le 1\|C_{\max}$	Sect. 6.1.4, p. 77
A_9	$O(n \log n)$	$1\|p_j = b_j h(t)\|C_{\max}$	Sect. 6.1.4, p. 77
A_{10}	$O(n \log n)$	$1\|p_j = a_j + b_j \max\{t - t_0, 0\}\|C_{\max}$	Sect. 6.1.5, p. 86
A_{11}	$O(n \log n)$	$1\|p_j = a + b_j \max\{t - t_0, 0\}\|C_{\max}$	Sect. 6.1.5, p. 86
A_{12}	$O(n \log n)$	$1\|p_{i,j} = a_{i,j} - b_{i,j} t, \theta_i, GT\|C_{\max}$	Sect. 6.1.7, p. 91
A_{13}	$O(n \log n)$	$1\|p_j = a_j - b_j t, d_j = D\|C_{\max}$	Sect. 6.1.7, p. 93
A_{14}	$O(n \log n)$	$1\|p_j = a - b_j t, d_j, b_j \in \{B_1, B_2\}\|C_{\max}$	Sect. 6.1.7, p. 95
A_{15}	$O(n \log n)$	$1\|p_j = a_j - bt, r_j\|C_{\max}$	Sect. 6.1.7, p. 97

Table 6.10: Polynomial Algorithms for Single-Machine Time-Dependent Problems ($\sum C_j$ Criterion)

Algorithm	Complexity	Problem	This book
A_{16}	$O(n \log n)$	$1\|p_{i,j} = b_{i,j} t, \theta_i = \delta_i, GT\| \sum C_j$	Sect. 6.2.1, p. 104
A_{17}	$O(n \log n)$	$1\|p_j = b_j t, d_j\| \sum C_j$	Sect. 6.2.1, p. 105
A_{18}	$O(n \log n)$	$1\|p_j = b_j(A + Bt)\| \sum C_j$	Sect. 6.2.2, p. 106
A_{19}	$O(n \log n)$	$1 p_j = 1 + b_j t\| \sum C_j$	Sect. 6.2.3, p. 110
A_{20}	$O(n \log n)$	$1\|p_{i,j} = a_{i,j} - b_i t, \theta_i, GT\| \sum C_j$	Sect. 6.2.6, p. 116

Table 6.11: Polynomial Algorithms for Single-Machine Time-Dependent Problems ($L_{\max}$ Criterion)

Algorithm	Complexity	Problem	This book
A_{21}	$O(n \log n)$	$1\|p_j = b_j\|L_{\max}$	Sect. 6.3.1, p. 118
A_{22}	$O(n \log n)$	$1\|p_j = a_j + bt\|L_{\max}$	Sect. 6.3.3, p. 121
A_{23}	$O(n \log n)$	$1\|p_j = b_j h(t)\|L_{\max}$	Sect. 6.3.4, p. 121
A_{24}	$O(n \log n)$	$1\|p_j = b_j h(t)\|L_{\max}$	Sect. 6.3.4, p. 121

Table 6.12: Polynomial Algorithms for Single-Machine Time-Dependent Problems (Criteria other than $C_{\max}$, $\sum C_j$ and $L_{\max}$)

Algorithm	Complexity	Problem	This book
A_{25}	$O(n \log n)$	$1\|p_j = b_j t\|\sum w_j C_j$	Sect. 6.4.1, p. 123
A_{26}	$O(n \log n)$	$1\|p_j = b_j t\|\sum U_j$	Sect. 6.4.1, p. 125
A_{27}	$O(n \log n)$	$1\|p_j = a_j(A + Bt)\|\sum w_j C_j$	Sect. 6.4.2, p. 126
A_{28}	$O(n \log n)$	$1\|p_{i,j} \equiv (6.5),\ \theta_i \equiv (6.9), GT\|\sum w_j C_j$	Sect. 6.4.2, p. 126
A_{29}	$O(n^2)$	$1\|p_j = b_j(A + Bt)\|f_{\max}$	Sect. 6.4.2, p. 127
A_{30}	$O(n^2)$	$1\|p_j = a_j + b_j t,\ d_j = D\|\sum U_j$	Sect. 6.4.3, p. 132
A_{31}	$O(n \log n)$	$1\|p_j = a_j + bt\|\sum(\alpha E_j + \beta T_j + \gamma d)$	Sect. 6.4.3, p. 134
A_{32}	$O(\log n)$	$1\|p_j = a_j + bt\|\sum(\alpha E_j + \beta T_j + \gamma d)$	Sect. 6.4.3, p. 135
A_{33}	$O(n \log n)$	$1\|p_j = b_j h(t)\|\sum w_j C_j$	Sect. 6.4.4, p. 136
A_{34}	$O(n \log n)$	$1\|p_j = b_j h(t)\|\sum w_j C_j$	Sect. 6.4.4, p. 136
A_{35}	$O(n \log n)$	$1\|p_j \equiv (6.68),\ w_j \in \{W_1, W_2\}\|\sum w_j C_j$	Sect. 6.4.5, p. 138
A_{36}	$O(n \log n)$	$1\|p_j \equiv (6.68),\ b_j \in \{B_1, B_2\}\|\sum w_j C_j$	Sect. 6.4.5, p. 139
A_{37}	$O(n \log n)$	$1\|p_j \equiv (6.68)\|\sum w_j C_j$ [a]	Sect. 6.4.5, p. 140
A_{38}	$O(n \log n)$	$1\|p_j = a - bt\|\sum w_j C_j$	Sect. 6.4.7, p. 143
A_{39}	$O(n \log n)$	$1\|p_j = a_j - ka_j t\|\sum w_j C_j$	Sect. 6.4.7, p. 143
A_{40}	$O(n \log n)$	$1\|r_j = R, p_j = a_j - b_j t, d_j = D\|\sum U_j$	Sect. 6.4.7, p. 145
A_{41}	$O(n \log n)$	$1\|p_j = a_j - bt\|\sum(\alpha E_j + \beta T_j + \gamma d)$	Sect. 6.4.7, p. 146

[a] $w_i \geq w_j \Rightarrow b_i \geq w_j$ for $1 \leq i, j \leq n$

Table 6.13: Exact Algorithms for Single-Machine Time-Dependent Problems

Algorithm	Complexity	Problem	This book
E_1	$O(n^{k+1})$	$1\|p_j = 1 + b_j t,$ $b_j \in \{B_1, B_2, \ldots, B_k\}\|\sum C_j$	Sect. 6.2.3, p. 109
E_2	$O(n^k \log n)$	$1\|p_j \in \{a_j, b_j : a_j \leq b_j\},$ $d_j = D\|\sum C_j$	Sect. 6.2.5, p. 114
E_3	$O(n^k \log n)$	$1\|p_j \in \{a_j, b_j : a_j \leq b_j\},$ $d_j = D\|C_{\max}$	Sect. 6.2.3, p. 81
E_4	$O(n^k \log n)$	$1\|p_j \in \{a_j, b_j : a_j \leq b_j\},$ $d_j = D\|\sum w_j C_j$	Sect. 6.4.5, p. 141

7

Parallel-machine time-dependent scheduling

In the previous chapter, we discussed the complexity of single-machine time-dependent scheduling problems. In this chapter, we present the complexity results concerning time-dependent scheduling on parallel machines. In order to give the reader full insight into the subject, we give proofs or sketches of proofs of most discussed results. We also present pseudo-codes of algorithms for polynomially solvable problems.

Chapter 7 is composed of four sections. In Sect. 7.1, we present the results concerning parallel machines and minimization of the $C_{\max}$ criterion. In Sect. 7.2, we present the results concerning parallel machines and minimization of the $\sum C_j$ criterion. In Sect. 7.3, we present the results concerning parallel machines and minimization of other criteria than $C_{\max}$ and $\sum C_j$. The chapter is completed with Sect. 7.4 including the summary and tables.

7.1 Minimizing the maximum completion time

In this section, we consider parallel-machine time-dependent scheduling problems with the $C_{\max}$ criterion.

7.1.1 Proportional deterioration

Equal ready times and deadlines

The problem of multi-machine time-dependent scheduling with proportionally deteriorating jobs and the $C_{\max}$ criterion is computationally intractable.

Theorem 7.1. (Kononov [172], Mosheiov [219]) *The decision version of the problem $Pm|p_j = b_j t|C_{\max}$ is $\mathcal{NP}$-complete in the ordinary sense even if $m = 2$.*

Proof. Kononov [172] uses the following transformation from the SP problem (cf. Sect. 3.2): $n = p + 2$, $t_0 = 1$, $b_j = y_j - 1$ for $1 \leq j \leq p$, $b_{p+1} = \frac{2Y}{B} - 1$, $b_{p+2} = 2B - 1$, where $Y = \prod_{j=1}^{p} y_j$. The threshold value for the $C_{\max}$ criterion is $G = 2Y$.

Mosheiov [219] uses the following transformation from the EPP problem (cf. Sect. 3.2): $n = q$, $t_0 = 1$, $b_j = z_j - 1$ for $1 \leq j \leq q$. The threshold $G = \sqrt{\prod_{j=1}^{n} z_j}$.

To complete the proof, it is sufficient to show that the SP (ESP) problem has a solution if and only if for the above instance of the problem $P2|p_j = b_j t|C_{\max}$ there exists a schedule σ such that $C_{\max}(\sigma) \leq G$. (Example schedule for the first transformation is depicted in Fig. 7.1; see also Remark 6.12.)   □

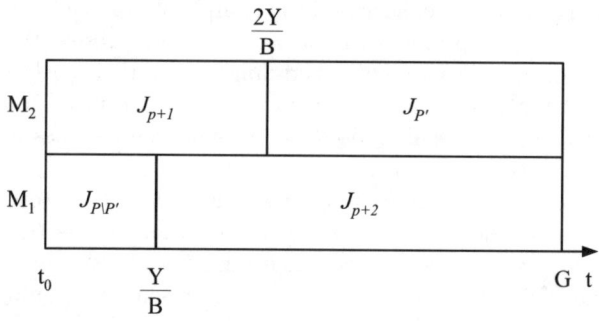

Fig. 7.1: Example schedule in the proof of Theorem 7.1

The following lower bound on the value of $C_{\max}$ for the problem with $m \geq 2$ machines is known.

Lemma 7.2. (Hsieh and Bricker [141], Mosheiov [219]) *The optimal value of the maximum completion time for the problem $Pm|p_j = b_j t|C_{\max}$ is not less than $LB := \sqrt[m]{\prod_{j=1}^{n}(1 + b_j)}$.*

Proof. Let $\mathcal{N}_{M_k}$, $1 \leq k \leq m$, denote the set of indices of jobs assigned to machine M_k in an optimal schedule. Then

$$C_{\max}^{\star} = \max_{1 \leq k \leq m} \left\{ \prod_{j \in \mathcal{N}_{M_k}} (1 + b_j) \right\}.$$

Since, by Lemma 1.1 (a), we have $C_{\max}^{\star} \geq \frac{1}{m} \sum_{k=1}^{m} \prod_{j \in \mathcal{N}_{M_k}} (1 + b_j)$ and, by Lemma 1.1 (b), we have $C_{\max}^{\star} \geq \sqrt[m]{\prod_{j=1}^{n}(1 + b_j)} = LB$, the result follows.

□

The following example shows that in the worst case the above bound LB can be arbitrarily large.

Example 7.3. (Mosheiov [219]) Let $n = 2$, $m = 2$, $t_0 = 1$, $b_1 = B > 0$, $b_2 = \epsilon > 0$. Then $C^\star_{\max} = 1 + B$, while $LB = \sqrt{(1+B)(1+\epsilon)} \to \sqrt{(1+B)}$, if $\epsilon \to 0$. Hence $\frac{C^\star_{\max}}{LB} \to \sqrt{(1+B)} \to \infty$, if $B \to \infty$ and $\epsilon \to 0$. ♦

It has been shown, however, that if the deterioration rates b_j, $1 \leq j \leq n$, are independently drawn from a distribution with a finite second moment and positive density at 0, then $\lim_{n \to \infty} \frac{C^\star_{\max}}{LB} = 1$ almost surely (see Mosheiov [219, Proposition 3]).

Remark 7.4. A sequence (x_n) converges *almost surely* to a limit x if the probability of the event $\lim_{n \to \infty} x_n = x$ is equal to 1.

If the number of machines is variable, then there holds the following result.

Theorem 7.5.   (Kononov [169])   *The decision version of the problem* $P|p_j = b_j t|C_{\max}$ *is $\mathcal{NP}$-complete in the strong sense.*

Proof. The transformation from the 4-P problem (cf. Sect. 3.2) is as follows: $n = 4p$, $t_0 = 1$ and $b_j = u_j - 1$ for $1 \leq j \leq 4p$. The threshold value is $G = D$.

To complete the proof, it is sufficient to show that the 4-P problem has a solution if and only if for the above instance of the problem $P|p_j = b_j t|C_{\max}$ there exists a schedule σ such that $C_{\max}(\sigma) \leq G$. □

In Chap. 9, we will consider a number of heuristics proposed for the problem $Pm|p_j = b_j t|C_{\max}$.

Lee and Wu [193] reformulated the problem $Pm|p_j = b_j t|C_{\max}$ as the problem $Pm, h_{m1}|p_j = b_j t|C_{\max}$ in which on each machine there is a known, single non-availablity period (cf. Remark 6.8). For the latter problem, they proposed the following two lower bounds on the optimal value of $C_{\max}$.

Lemma 7.6. (Lee and Wu [193]) *Let $W_{1,i}$ and $W_{2,i}$ denote the start time and the end time of the non-availability period on machine $M_i, 1 \leq i \leq m$, respectively. Let x_i be a binary variable such that $x_i := 1$ if*

$$W_{2,i} \leq \frac{t_0 \sqrt[m]{\prod_{j=1}^{n}(1 + b_j t)}}{\prod_{j=1}^{m-1}(1 + b_j)}$$

and $x_i := 0$ otherwise. Then the optimal value of the maximum completion time for the problem $Pm, h_{m1}|p_j = b_j t|C_{\max}$ is not less than

(a) $t_0 \left(\sqrt[m]{\prod_{j=1}^{n}(1 + b_j)} \right) \left(1 + \sqrt[m]{\prod_{i=1}^{m} \frac{x_i(W_{2,i} - W_{1,i})}{W_{1,i}(1 + b_i)}} \right)$, *if jobs are resumable,*

(b) $t_0 \left(\sqrt[m]{\prod_{j=1}^{n}(1 + b_j)} \right) \left(1 + \sqrt[m]{\prod_{i=1}^{m} \frac{x_i(W_{2,i} - W_{1,i})}{W_{1,i}}} \right)$, *if jobs are non-resumable.*

Proof. (a) By direct calculation, see [193, Proposition 1].
    (b) By direct calculation, see [193, Proposition 2]. ◇

7.1.2 Linear deterioration

Equal ready times and deadlines

Since proportional processing times are special cases of linear processing times, the results from Sect. 7.1.1 also hold for linear processing times.

Theorem 7.7.
(a) *The decision version of the problem $Pm|p_j = a_j + b_j t|C_{\max}$ is $\mathcal{NP}$-complete in the ordinary sense even if $m = 2$.*
(b) *The decision version of the problem $P|p_j = a_j + b_j t|C_{\max}$ is $\mathcal{NP}$-complete in the strong sense.*

Proof. (a) Since the special case of the problem $P2|p_j = a_j + b_j t|C_{\max}$, when $a_j = 0$ for $1 \le j \le n$, is $\mathcal{NP}$-complete in the ordinary sense by Theorem 7.1, the result follows.

(b) (a) Since the special case of the problem $P|p_j = a_j + b_j t|C_{\max}$, when $a_j = 0$ for $1 \le j \le n$, is $\mathcal{NP}$-complete in the strong sense by Theorem 7.5, the result follows.   □

In the case when $a_j = a > 0$ for $1 \le j \le n$, the following properties of the problem $Pm|p_j = a + b_j t| \sum C_j$ are known.

The first property is a multi-machine counterpart of Property 6.128.

Property 7.8. (Gawiejnowicz et al. [101]) Let $k_i = \arg\max\{b_j : j \in U\}$, where $i = 1, 2, \ldots, m$ and U is the set of indices of jobs not considered yet. Then k_i is the first job on machine M_i in the optimal schedule.

Proof. Let job J_{k_i}, $k_i \in U$, be scheduled as the first one on the machine M_i, $i \in \{1, 2, \ldots, m\}$. Then, the processing time and the completion time of this job are equal to $p_{k_i} = a + b_{k_i} \cdot 0 = a$ and $C_{k_i} = a$, respectively. Since C_{k_i} does not depend on b_{k_i}, it is easy to see that in an optimal schedule as the first job should be choosen such a job that its index $k_i = \arg\max\{b_j : j \in U\}$.   ∎

The second property does not have a single-machine counterpart.

Property 7.9. (Gawiejnowicz et al. [101]) If $n \ge 2m - 1$, then in any optimal schedule at least two jobs are scheduled on each machine.

Proof. Assume that there are given $n \ge 2m - 1$ jobs and that there exists an optimal schedule, σ^1, such that on some machine, M_l, is assigned only one job. Let M_k be a machine with the largest load in the schedule σ^1, j_k be the index of the job assigned to M_k as the last one and $S_{j_k} > a$ denote the starting time of this job. Then, $C_{j_k}(\sigma^1) = a + (1 + b_{j_k})S_{j_k}$ and the total completion time for the schedule σ^1 is $\sum C_j(\sigma^1) = T + a + (1 + b_{j_k})S_{j_k}$, where T denotes the total completion time for jobs other than J_{j_k}.

Construct now a new schedule, σ^2, by assigning the job with the index j_k on the machine M_l. Then, $C_{j_k}(\sigma^2) = a + (1 + b_{j_k})a$ and the total completion time for the schedule σ^2 is $\sum C_j(\sigma^2) = T + a + (1 + b_{j_k})a$. Since $\sum C_j(\sigma^2) -$

$\sum C_j(\sigma^1) = (1 + b_{j_k})(a - S_{j_k}) < 0$, the schedule σ^2 is better than σ^1. A contradiction.   ∎

The third property is a multi-machine counterpart of Property 6.130.

Property 7.10. (Gawiejnowicz et al. [101]) The total completion time for any sequence of indices of jobs assigned to a given machine and for the sequence in reversed order, starting from the second job, has the same value.

Proof. Consider any sequence of indices of jobs assigned to a given machine M_i, $i \in \{1, 2, \ldots, m\}$. The result follows, since starting from the second argument the $\sum C_j$ criterion is symmetric with respect to its arguments, $\sum_{j=0}^{n} \sum_{k=0}^{j} \prod_{l=k+1}^{j}(1 + b_k) = \sum_{j=0}^{n} \sum_{k=0}^{j} \prod_{l=k+1}^{j}(1 + b_{n-k+1})$.   ∎

Finally, the last result is a multi-machine counterpart of Theorem 6.133.

Theorem 7.11. (Gawiejnowicz et al. [101]) *The optimal schedule for the problem $Pm|p_j = a + b_j t| \sum C_j$ is composed of V-shaped subschedules.*

Proof. Assume that there exists such an optimal schedule that for some machine, M_k, the sequence of jobs assigned to the machine is not V-shaped. By rearranging the jobs in such a way that their sequence has a V-shape, we obtain a new schedule with decreased value of the criterion function since, by V-shape property for a single machine (cf. Theorem 6.133), we decreased the total completion time for the machine M_k. A contradiction.   ∎

7.1.3 Linear shortening

Distinct ready times and deadlines

Parallel-machine scheduling of jobs with decreasing step-linear processing times given by (6.54) is a computationally intractable problem.

Theorem 7.12. (Cheng et al. [54]) *The decision version of the problem $Pm|p_j = a_j - b(t - y), y = 0, Y = \infty|C_{\max}$ is $\mathcal{NP}$-complete in the ordinary sense even if $m = 2$.*

Proof. The main idea is to show that the two-machine problem with variable processing times, $P2|p_j = a_j - b(t - y), y = 0, Y = \infty|C_{\max}$, is equivalent to the problem with fixed procesing times, $P2||C_{\max}$, if b is sufficiently small. Let q_j, $1 \leq j \leq n$, and G denote job processing times and the threshold value of the $C_{\max}$ criterion in the decision version of the problem $P2||C_{\max}$.

Define job processing times and the value of the $C_{\max}$ criterion in the problem $P2|p_j = a_j - b(t - y), y = 0, Y = \infty|C_{\max}$ as follows: $a_j = q_j$ for $1 \leq j \leq n$, $b = 1 - (1 - \frac{1}{a_{\max}^2})^{\frac{1}{n}}$, where $a_{\max} := \max\{a_1, a_2, \ldots, a_n\}$.

Let $\mathcal{J}(M_k)$ and C_{M_k} denote the set of jobs assigned to machine M_k in an arbitrary schedule for the problem $P2||C_{\max}$ and the completion time of the last job from the set $\mathcal{J}(M_k)$, $1 \leq k \leq m$, respectively. Let us call the schedule σ.

Then

$$\sum_{J_j \in \mathcal{J}(M_k)} a_j - 1 < C_{M_k}(\sigma) < \sum_{J_j \in \mathcal{J}(M_k)} a_j$$

for $1 \leq k \leq m$. Hence, $C'_{\max}(\sigma) - 1 < C_{\max}(\sigma) < C'_{\max}(\sigma)$, where $C_{\max}(\sigma)$ and $C'_{\max}(\sigma)$ denote the maximum completion time for the schedule σ with fixed and variable job processing times, respectively.

Since all a_j are integers, we have $C_{\max}(\sigma) \leq G$ if and only if $C'_{\max}(\sigma) \leq G$. Hence, the result follows because the decision version of the problem $P2||C_{\max}$ is $\mathcal{NP}$-complete in the ordinary sense. $\qquad \square$

Theorem 7.13. (Cheng et al. [54]) *The decision version of the problem* $P|p_j = a_j - b(t - y), y = 0, Y = \infty|C_{\max}$ *is $\mathcal{NP}$-complete in the strong sense.*

Proof. Applying the reasoning from the proof of Theorem 7.12 to the problem $P||C_{\max}$, we obtain the result. $\qquad \square$

7.2 Minimizing the total completion time

In this section, we will consider the problems of parallel-machine time-dependent scheduling with the $\sum C_j$ criterion.

7.2.1 Proportional deterioration

Equal ready times and deadlines

Theorem 7.14. (Chen [43], Kononov [172]) *The decision version of the problem* $Pm|p_j = b_j t| \sum C_j$ *is $\mathcal{NP}$-complete in the ordinary sense even if $m = 2$.*

Proof. Chen [43] uses the following transformation from the SP problem (cf. Sect. 3.2): $n = p + 4$, $t_0 > 0$ arbitrary, $b_j = y_j - 1$ for $1 \leq j \leq p$, $b_{p+1} = \frac{Y^2}{B} - 1$, $b_{p+2} = YB - 1$, $b_{p+3} = b_{p+4} = Y^3 - 1$. The threshold $G = (2Y^5 + Y^4)t_0$, where $Y = \prod_{j=1}^{p} y_j$.

Kononov [172] uses the transformation from the same problem, SP, but his transformation is slightly different: $n = p + 4$, $t_0 = 1$, $b_j = y_j - 1$ for $1 \leq j \leq p$, $b_{p+1} = \frac{2Y}{B} - 1$, $b_{p+2} = 2B - 1$, $b_{p+3} = b_{p+4} = 6Y - 1$. The threshold $G = 24Y^2 + 8Y$, where Y is defined as previously.

To complete the proof, it is sufficient to show that the SP problem has a solution if and only if for the above instance of the problem $P2|p_j = b_j t| \sum C_j$ there exists a schedule σ such that $\sum C_j(\sigma) \leq G$. (An example schedule for the second transformation is depicted in Fig. 7.2; see also Remark 6.12.) $\qquad \square$

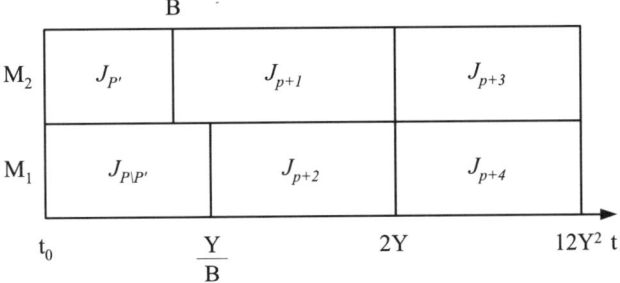

Fig. 7.2: Example schedule in the proof of Theorem 7.14

The problem with variable number of machines is hard to approximate.

Theorem 7.15. (Chen [43]) *If* $P \neq \mathcal{NP}$, *for the problem* $P|p_j = b_j t| \sum C_j$ *there does not exist a polynomial-time heuristic algorithm with a constant worst-case ratio.*

Proof. Assume that for the problem $P|p_j = b_j t| \sum C_j$ there exists an algorithm A such that its worst-case ratio $r \leq V$. Given any instance of the 3-P problem, construct an instance of the problem $P|p_j = b_j t| \sum C_j$ as follows: there are h machines, $n = 4h$ jobs, $t_0 = 1$, job deterioration rates $b_j = U^{c_j} - 1$ for $1 \leq j \leq 3h$, $b_{3h+i} = U^{hK+1} - 1$ for $1 \leq i \leq h$, where $U = (h+1)V$.

If we solved the instance by algorithm A with the relative error at most V, we would obtain a pseudopolynomial-time algorithm for the 3-P problem, which is strongly $\mathcal{NP}$-complete. A contradiction, since by Lemma 3.18 this is impossible if $P \neq \mathcal{NP}$.   □

Though from Theorems 7.14–7.15 it follows that $Pm|p_j = b_j t| \sum C_j$ is hard to solve, some properties of an optimal schedule for this problem are known.

Property 7.16. (Gawiejnowicz et al. [102]) In an optimal schedule for problem $Pm|p_j = b_j t| \sum C_j$, jobs assigned to a machine are arranged in the non-decreasing order of deterioration rates and scheduled without idle times.

Proof. Assume that σ is a schedule in which jobs assigned to a machine are not in a non-decreasing order of deterioration rates b_j, $1 \leq j \leq n$. By changing the order into a non-decreasing order, we obtain a schedule σ' such that $\sum C_j(\sigma') \leq \sum C_j(\sigma)$, since by Theorem 6.120 in an optimal schedule for a single machine, jobs have to be in non-decreasing order of b_j values. Changing, if necessary, the order of jobs on other machines, we obtain such an optimal schedule $\sigma^\star$ that on each machine jobs are arranged in non-decreasing order of b_j's. Since any idle time increases job completion times, the optimal schedule cannot contain idle times.   ■

Theorem 7.17. (Gawiejnowicz et al. [102]) *Let $\sigma^i = (b_1^i, b_2^i, \ldots, b_{n_i}^i)$ and $\bar{\sigma}^i = (b_{n_i}^i, b_{n_i-1}^i, \ldots, b_1^i)$ denote a subsequence of jobs assigned to machine M_i and the sequence reversed to σ^i, respectively, where b_j^i denotes deterioration rate of job J_j assigned to machine M_i, $1 \leq j \leq n_i$, $1 \leq i \leq m$ and $\sum_{i=1}^{m} n_i = n$. Then for every schedule $\sigma = (\sigma^1, \sigma^2, \ldots, \sigma^m)$ for the problem $Pm|p_j = b_j t| \sum C_j$ there exists a corresponding schedule $\bar{\sigma} = (\bar{\sigma}^1, \bar{\sigma}^2, \ldots, \bar{\sigma}^m)$ for the problem $Pm|p_j = 1 + b_j t| \sum C_{\max}^{(k)}$ and for every schedule $\bar{\sigma}$ for the problem $Pm|p_j = 1 + b_j t| \sum C_{\max}^{(k)}$ there exists a corresponding schedule σ for the problem $Pm|p_j = b_j t| \sum C_j$ such that $\sum C_j(\sigma) = \sum C_{\max}^{(k)}(\bar{\sigma}) - m$, provided that both these schedules start at time $t_0 = 1$.*

Proof. (Sufficiency) Assume that $t_0 = 1$ and let $\sigma = (\sigma^1, \sigma^2, \ldots, \sigma^m)$ be a schedule for the problem $Pm|p_j = b_j t| \sum C_j$, where $\sigma^i = (b_1^i, b_2^i, \ldots, b_{n_i}^i)$ for $1 \leq i \leq m$. Then, we have

$$\sum C_j(\sigma) = \sum_{k=1}^{m} \sum_{i=1}^{n_k} \prod_{j=1}^{i} (1 + b_j^k) = \sum_{k=1}^{m} \sum_{i=0}^{n_k} \prod_{j=1}^{i} (1 + b_j^k) - m =$$

$$\sum_{k=1}^{m} \sum_{i=0}^{n_k} \prod_{j=1}^{i} (1 + B_{n_k-j+1}^k) - m = \sum C_{\max}^{(k)}(\bar{\sigma}) - m,$$

where $B_{n_k-j+1}^k = b_j^k$ and $\bar{\sigma}^i = (b_{n_k}^i, b_{n_k-1}^i, \ldots, b_1^i) = (B_1^i, B_2^i, \ldots, B_{n_k}^i)$ for $1 \leq i \leq m$. The proof of necessity can be done in an analogous way. ∎

Remark 7.18. Problems which satisfy the conditions of Theorem 7.17 will be called *equivalent* problems. We will come back to this topic in Chap. 12.

In Chap. 9, we will consider a number of heuristic algorithms proposed for the problem $Pm|p_j = b_j t| \sum C_j$. Local search algorithms for the problem will be considered in Chap. 11.

7.2.2 Linear deterioration

Equal ready times and deadlines

Since proportional processing times are a special case of linear processing times, the results from Sect. 7.2.1 also hold for linear processing times.

Theorem 7.19. *The decision version of the problem $Pm|p_j = a_j + b_j t| \sum C_j$ is $\mathcal{NP}$-complete in the ordinary sense even if $m = 2$.*

Proof. Since the special case of the problem $Pm|p_j = a_j + b_j t| \sum C_j$, when $a_j = 0$ for $1 \leq j \leq n$ and $m = 2$, is $\mathcal{NP}$-complete in the ordinary sense by Theorem 7.14, the result follows. □

7.2.3 Simple non-linear deterioration

Equal ready times and deadlines

Results from Sects. 7.1.2–7.2.2 suggest that all problems concerning parallel machine scheduling subject to job deterioration are computationally intractable. There is known, however, a problem from this area which can be solved in polynomial time.

Theorem 7.20. (Gawiejnowicz [89]) *If $f(t)$ is an arbitrary increasing function such that $f(t) \geq 0$ for $t \geq 0$, then the problem $Pm|p_j = a_j + f(t)|\sum C_j$ is optimally solved in $O(n \log n)$ by scheduling jobs in the non-decreasing order of a_j values.*

Proof. Let σ be a schedule for the problem $Pm|p_j = a_j + f(t)|\sum C_j$ such that not all jobs are scheduled in the non-decreasing order of a_j values. Then there must exist a machine M_k, $1 \leq k \leq m$, such that some jobs assigned to the machine are scheduled in the non-increasing order of a_j values. By changing the order of the jobs into the non-decreasing one we obtain, by Theorem 6.146, a new schedule σ' such that $\sum C_j(\sigma') \leq \sum C_j(\sigma)$. Repeating, if necessary, the above change for other machines we obtain an optimal schedule $\sigma^\star$ in which all jobs are scheduled in the non-decreasing order of a_j values. ∎

By Theorem 7.20, the problem $Pm|p_j = a_j + f(t), f \nearrow |\sum C_j$ is solved by the algorithm $A_7 : (a_j|f) \mapsto (a_j \nearrow)$.

7.2.4 Linear shortening

Distinct ready times and deadlines

Some time-dependent scheduling problems with shortening job processing times (6.54) are polynomially solvable, if we deal with parallel uniform and parallel unrelated machines.

Theorem 7.21. (Cheng et âl. [54]) *The problem $Q|p_j = a_j - b(t - y)$, $y = 0, Y = \infty|\sum C_j$ is solvable in $O(n \log n)$ time by scheduling jobs in the non-increasing order of a_j values.*

Proof. If the machines are uniform, then $a_{l,j} = \frac{a_j}{s_l}$. In this case, the total completion time is a weighted sum of a_j values, with weights $b_{l,r} = \frac{1-(1-b)^r}{bs_l}$. No weight may be used more than once. Therefore, to minimize the value of $\sum C_j$ criterion, one should select the n smallest of all mn weights and match the selected weights with the largest a_j values. This can be done in $O(n \log n)$ time. □

By Theorem 7.21, the problem $Q|p_j = a_j - b(t - y)$, $y = 0, Y = \infty|\sum C_j$ is solved by the algorithm $A_7 : (a_j|b|y|Y) \mapsto (a_j \nearrow)$.

The next problem is solvable by an algorithm that solves a matching problem. Call the algorithm A_{42}.

Theorem 7.22. (Cheng et al. [54]) *The problem $R|p_{i,j} = a_{i,j} - b(t-y), y = 0,$ $Y = \infty| \sum C_j$ is solvable in $O(n^3)$ time by algorithm A_{42}.*

Proof. Introduce the variables $x_{(l,r),j}$ such that $x_{(l,r),j} = 1$ if job j is sequenced rth last on machine M_l, and $x_{(l,r),j} = 0$ otherwise. The problem under consideration is equivalent to the following weighted bipartite matching problem:

$$\text{minimize} \sum_{l,r} \sum_j x_{(l,r),j} a_{l,j} \frac{1 - (1-b)^r}{b}$$

subject to

$$\sum_{l,r} x_{(l,r),j} = 1, \quad j = 1, 2, \dots, n,$$

$$\sum_j x_{(l,r),j} \le 1, \quad l = 1, 2, \dots, m, r = 1, 2, \dots, n,$$

$$x_{(l,r),j} \in \{0,1\} \quad l = 1, 2, \dots, m, r = 1, 2, \dots, n,$$

where the summation is taken over all values of l and r or j. This matching problem can be solved in $O(n^3)$ by algorithm A_{42}. □

7.3 Other criteria

In this section, we consider the problems of parallel-machine time-dependent scheduling with the criteria other than $C_{\max}$ or $\sum C_j$.

7.3.1 Proportional deterioration

Equal ready times and deadlines

We start with a result concerning the total machine load criterion, $\sum C_{\max}^{(k)}$, introduced to time-dependent scheduling by Mosheiov, [219].

Lemma 7.23. (Mosheiov [219]) *The optimal total machine load for the problem $Pm|p_j = b_j t| \sum C_{\max}^{(k)}$ is not less than $m \times \sqrt[m]{\prod_{j=1}^n (1 + b_j)}$.*

Proof. Let $A_1, A_2, \dots, A_m$ and $C_{\max}^{(1)}, C_{\max}^{(2)}, \dots, C_{\max}^{(m)}$ denote the sets of jobs assigned to machines $M_1, M_2, \dots, M_m$ and the corresponding total loads, respectively. Then, by Lemma 1.1 (b), $\frac{1}{m} \sum_{i=1}^m C_{\max}^{(i)} = \frac{1}{m} \sum_{i=1}^m \prod_{j \in A_i} (1+b_j) \ge \sqrt[m]{\prod_{j=1}^n (1 + b_j)}$. □

Theorem 7.24. (Mosheiov [219], Gawiejnowicz et al. [102]) *The decision version of the problem $Pm|p_j = b_j t| \sum C_{\max}^{(k)}$ is $\mathcal{NP}$-complete in the ordinary sense even if $m = 2$.*

Proof. Mosheiov [219] gives the following idea of the proof. The problem $P2|p_j = b_j t| \sum C_{\max}^{(k)}$ is equivalent to finding $\min\{L_1 + L_2\}$ subject to $L_1 \times L_2 = A^2$ for some positive integer constant A, where L_1 and L_2 denote products of deterioration rates of the jobs assigned to machine M_1 and machine M_2, respectively. Since this is equivalent to the EPP problem (cf. Sect. 3.2), the result follows.

Gawiejnowicz et al. [102] proved the result using the notion of equivalent problems (cf. Theorem 7.17, see also Chap. 12).   □

Remark 7.25. Mosheiov [219] gives only a sketch of the proof of Theorem 7.24. The formal transformation may be the following. Given an instance of the EPP problem, define $n = q$, $b_j = z_j$ for $1 \leq j \leq q$ and the threshold $G = 2\sqrt{\prod_{j=1}^{q} z_j}$. Let $E^2 = \sqrt{\prod_{j=1}^{q} z_j}$. To complete the proof, it is sufficient to show that the EPP problem has a solution if and only if for the above instance of the problem $P2|p_j = b_j t| \sum C_{\max}^{(k)}$ there exists a schedule σ such that $\sum C_{\max}^{(k)}(\sigma) \leq 2E = G$.

7.3.2 Linear deterioration

Equal ready times and deadlines

The $\sum C_{\max}^{(k)}$ criterion. Since linear job processing times are generalization of proportional job processing times, there holds the following result.

Theorem 7.26. *The decision version of the problem $Pm|p_j = a_j + b_j t| \sum C_{\max}^{(k)}$ is $\mathcal{NP}$-complete in the ordinary sense even if $m = 2$.*

Proof. Since the special case of the $Pm|p_j = a_j + b_j t| \sum C_{\max}^{(k)}$, when $a_j = 0$ for $1 \leq j \leq n$ and $m = 2$, is $\mathcal{NP}$-complete in the ordinary sense by Theorem 7.24, the result follows.   □

For the problem $Pm|p_j = b_j t| \sum C_{\max}^{(k)}$, Mosheiov [219] proposed a heuristic algorithm. We will consider this algorithm in Chap. 9. Local search algorithms for the problem will be considered in Chap. 11.

Distinct ready times and deadlines

The $\sum(\alpha E_j + \beta T_j + \gamma d)$ criterion. Cheng et al. [58] considered the problem $Pm|p_j = a_j + bt| \sum(\alpha E_j + \beta T_j + \gamma d)$.

Theorem 7.27. (Cheng et al. [58]) *The decision version of the problem $Pm|p_j = a_j + bt| \sum(\alpha E_j + \beta T_j + \gamma d)$ is $\mathcal{NP}$-complete in the ordinary sense even if $m = 2$.*

Proof. Define $M := \frac{1}{2}\sum_{j=1}^{n} x_j$ and $e := \sum_{j=1}^{n} ja_j$. The reduction from the PP problem (cf. Sect. 3.2) is as follows: $n = k$ jobs, $a_j = x_j$ for $1 \le j \le n$, $b = (1 + \frac{1}{a_n^2})^{\frac{1}{2n^4}}$, $\alpha = 1$, $\beta = \frac{(2M+1)e+1}{b}$ and $\gamma = \frac{2e}{n}$. The threshold $G = (2M+1)e+1$.

To complete the proof, it is sufficient to show that the PP problem has a solution if and only if there exists a schedule σ for the above instance of the $P2|p_j = a_j + bt|\sum(\alpha E_j + \beta T_j + \gamma d)$ such that $\sum(\alpha E_j + \beta T_j + \gamma d) \le G$. $\square$

Cheng et al. [58] also proved that the problem under consideration can be solved in polynomial time, if $\gamma = 0$. Before we state the result, we will introduce new notation.

Let $\sigma = (\sigma_1, \sigma_2, \ldots, \sigma_m)$ be a schedule for the problem $Pm|p_j = a_j + bt|\sum(\alpha E_j + \beta T_j)$, n_i be the number of jobs scheduled on machine M_i and d_i be the optimal due-date for jobs scheduled on machine M_i, $1 \le i \le m$. (By Lemma 6.226 we have $d_i = C_{\sigma_i(K_i)}$, where $K_i = \lceil \frac{n_i\beta}{\alpha+\beta} \rceil$ and $C_{\sigma_i(K_i)}$ denotes the index of the job scheduled as the K_i-th in subschedule σ_i, $1 \le i \le m$.)

Define $m_{i,k_i} := b\sum_{j=k_i}^{K_i}(\alpha(j-1))B^{j-k_i} + b\sum_{j=K_i+1}^{n_i}\beta(n_i+1-j)B^{j-k_i}$ for $1 \le i \le m, 2 \le k_i \le K_i$ and $m_{i,k_i} := b\sum_{j=k_i}^{n_i}(\beta(n_i+1-k_i))B^{j-k_i}$ for $1 \le i \le m, K_i+1 \le k_i \le n_i$, where $B := 1+b$.

For $1 \le j \le n$ and $1 \le i \le m$ define $c_{j,(i,k_i)} := (\alpha(k-1) + m_{i,k_i+1})a_j$ if $1 \le k_i \le K_i$ and $c_{j,(i,k_i)} := (\beta(n_i+1-k_i) + m_{i,k_i+1})a_j$ if $K_i + 1 \le k_i \le n_i$.

Applying the notation, we have $d = \max\{C_{\sigma_i(K_i)} : 1 \le i \le m\}$ and $\sum(\alpha E_j + \beta T_j) \equiv \sum_{i=1}^{m}\sum_{k_i=1}^{n_i}c_{\sigma_i(k_i),(i,k_i)}$.

Let $A := \{(n_1, n_2, \ldots, n_m)\} \in \mathbb{Z}^m$ be the set of all m-elements sequences such that $1 \le n_i \le n - m$ and $\sum_{i=1}^{m} n_i = n$. For any $(n_1, n_2, \ldots, n_m) \in A$, $1 \le i \le m$, $1 \le j \le n$ and $1 \le k_i \le n_i$, let $x_{j,(i,k_i)}$ be the variable such that $x_{j,(i,k_i)} = 1$ if job J_j is scheduled as the k_i-th job on machine M_i and $x_{j,(i,k_i)} = 0$ otherwise. Then, the problem $Pm|p_j = a_j + bt|\sum(\alpha E_j + \beta T_j)$ is equivalent to the following weighted bipartite matching problem:

$$\text{Minimize} \quad \sum_{j}\sum_{(i,k_i)} c_{j,(i,k_i)}x_{j,(i,k_i)} \tag{7.1}$$

subject to

$$\sum_{(i,k_i)} x_{j,(i,k_i)} = 1 \quad \text{for} \quad j = 1, 2, \ldots, m; \tag{7.2}$$

$$\sum_{j} x_{j,(i,k_i)} = 1 \quad \text{for} \quad i = 1, 2, \ldots, m; j = 1, 2, \ldots, m; \tag{7.3}$$

$$x_{j,(i,k_i)} \in \{0,1\} \quad \text{for} \quad j = 1, 2, \ldots, n; i = 1, 2, \ldots, m; k_i = 1, 2, \ldots, n. \tag{7.4}$$

Now we can formulate an exact algorithm, proposed by Cheng et al. [58] for the problem $Pm|p_j = a_j + bt|\sum(\alpha E_j + \beta T_j)$. Let $g(n_1, n_2, \ldots, n_m) := \min\{\sum_{j}\sum_{(i,k_i)} c_{j,(i,k_i)}x_{j,(i,k_i)}\}$, $\sigma(n_1, n_2, \ldots, n_m)$ be the schedule corresponding to $(n_1, n_2, \ldots, n_m)$, and $d(n_1, n_2, \ldots, n_m) := \max\{C_{\sigma_i(K_i)} : 1 \le i \le m\}$.

(If $C_{\sigma_i(K_i)} < d(n_1, n_2, \ldots, n_m)$, then we change the starting time of machine M_i to $d(n_1, n_2, \ldots, n_m) - C_{\sigma_i(K_i)}$.)

Algorithm A_{43} for the problem $Pm|p_j = a_j + bt|\sum(\alpha E_j + \beta T_j)$ ([58])

Input: sequence $(a_1, a_2, \ldots, a_n)$, numbers b, α, β
Output: an optimal schedule $\sigma^\star$

▷ **Step 1:**

Construct set $A := \{(n_1, n_2, \ldots, n_m) \in \mathbb{Z}^m : 1 \le n_i \le n - m \wedge \sum_{i=1}^{m} n_i = n\}$;

▷ **Step 2:**

For all $(n_1, n_2, \ldots, n_m) \in A$ **do** solve the problem (7.1)–(7.4);

▷ **Step 3:**

Calculate $\min\{g(n_1, n_2, \ldots, n_m) : (n_1, n_2, \ldots, n_m) \in A\}$;

▷ **Step 4:**

$\sigma^\star \leftarrow$ the schedule corresponding to the minimum computed in Step 3;
return $\sigma^\star$.

Theorem 7.28. (Cheng et al. [58]) *The problem* $Pm|p_j = a_j + bt|\sum(\alpha E_j + \beta T_j)$ *is solvable in* $O(n^{m+1} \log n)$ *time by algorithm* A_{43}.

Proof. Algorithm A_{43} generates $O(n^m)$ all possible sequences $(n_1, n_2, \ldots, n_m)$ in which $1 \le n_i \le n-m$ and $\sum_{i=1}^{m} n_i = n$. Since each sequence $(n_1, n_2, \ldots, n_m)$ corresponds to a schedule for the problem $Pm|p_j = a_j + bt|\sum(\alpha E_j + \beta T_j)$ and since the solution of the problem (7.1)–(7.4) needs $O(n \log n)$ time for jobs arranged in the $a_j \nearrow$ order, the result follows. □

With this theorem, we end the review of parallel-machine time-dependent scheduling. In the next chapter, we will consider the results concerning time-dependent scheduling on dedicated machines.

7.4 Summary and tables

In this chapter, we reviewed parallel-machine time-dependent scheduling problems. We considered proportionally and linearly deteriorating or shortening job processing times. The criteria of schedule optimality include the most popular criteria such as the $C_{\max}$ or $\sum C_j$, as well as less popular criteria such as the $\sum C_{\max}^{(k)}$ or $\sum(\alpha E_j + \beta T_j + \gamma d)$.

Approximately one-third of the problems presented in the chapter are polynomially solvable. For these problems, we presented pseudo-codes of optimal algorithms. For the remaining problems that are ordinary or strongly $\mathcal{NP}$-complete, we presented $\mathcal{NP}$-completeness proofs or sketches of such proofs.

Below, we classify in the tabular form the parallel-machine time-dependent scheduling problems and polynomial algorithms considered in the chapter. As in Chap. 6, the results and algorithms are divided into groups with respect to the applied criterion.

Tables 7.1 and 7.2 present the problems concerning, respectively, tractable and intractable parallel-machine time-dependent scheduling problems with $C_{\max}$, $C_{\max}^{(k)}$, $\sum C_j$ and $\sum(\alpha E_j + \beta T_j + \gamma d)$ criteria.

Table 7.3 presents polynomial algorithms for the $\sum C_j$ criterion.

Table 7.4 presents polynomial algorithms for criteria other than $\sum C_j$.

Table 7.1: Tractable Parallel-Machine Time-Dependent Scheduling Problems ($\sum C_j$ Criterion)

Problem	Complexity	References	This book
$Pm\|p_j = a_j + f(t), f(t) \nearrow \| \sum C_j$	$O(n \log n)$	[89]	Theorem 7.20
$Q\|p_j = a_j - b(t - y)\| \sum C_j$ [a]	$O(n \log n)$	[54]	Theorem 7.21
$R\|p_{ij} = a_{ij} - b(t - y)\| \sum C_j$ [a]	$O(n^3)$	[54]	Theorem 7.22

[a] $y = 0, Y = \infty$

Table 7.2: Intractable Parallel-Machine Time-Dependent Scheduling Problems (Criteria $C_{\max}$, $\sum C_j$, $\sum C_{\max}^{(k)}$) and $\sum(\alpha E_j + \beta T_j + \gamma d)$

Problem	Complexity	References	This book
$P2\|p_j = b_j t\|C_{\max}$	$\mathcal{NPC}$	[172, 219]	Theorem 7.1
$P\|p_j = b_j t\|C_{\max}$	$\mathcal{SNPC}$	[169]	Theorem 7.5
$P2\|p_j = a_j + b_j t\|C_{\max}$	$\mathcal{NPC}$	—	Theorem 7.7 (a)
$P\|p_j = a_j + b_j t\|C_{\max}$	$\mathcal{SNPC}$	—	Theorem 7.7 (b)
$P2\|p_j = a_j - b(t - y)\|C_{\max}$ [a]	$\mathcal{NPC}$	[54]	Theorem 7.12
$P\|p_j = a_j - b(t - y)\|C_{\max}$ [a]	$\mathcal{SNPC}$	[54]	Theorem 7.13
$P2\|p_j = b_j t\| \sum C_j$	$\mathcal{NPC}$	[43, 172]	Theorem 7.14
$P2\|p_j = a_j + b_j t\| \sum C_j$	$\mathcal{NPC}$	—	Theorem 7.19
$P2\|p_j = b_j t\| \sum C_{\max}^{(k)}$	$\mathcal{NPC}$	[219, 102]	Theorem 7.24
$P2\|p_j = a_j + b_j t\| \sum C_{\max}^{(k)}$	$\mathcal{NPC}$	—	Theorem 7.26
$P2\|p_j = a_j + bt\| \sum(\alpha E_j + \beta T_j + \gamma d)$	$\mathcal{NPC}$	[58]	Theorem 7.27

[a] $y = 0, Y = \infty$

Table 7.3: Polynomial Algorithms for Parallel-Machine Time-Dependent Scheduling Problems ($\sum C_j$ Criterion)

Algorithm	Complexity	Problem	This book
A_7	$O(n \log n)$	$Pm\|p_j = a_j + f(t), f(t) \nearrow \| \sum C_j$	Sect. 7.2.3, p. 163
A_7	$O(n \log n)$	$Q\|p_j = a_j - b(t - y)\| \sum C_j$	Sect. 7.2.4, p. 163
A_{42}	$O(n^3)$	$R\|p_{ij} = a_{ij} - b(t - y)\| \sum C_j$	Sect. 7.2.4, p. 164

Table 7.4: Polynomial Algorithms for Parallel-Machine Time-Dependent Scheduling Problems (Criteria other than $\sum C_j$)

Algorithm	Complexity	Problem	This book
A_{43}	$O(n^{m+1} \log n)$	$Pm\|p_j = a_j + bt\| \sum (\alpha E_j + \beta T_j)$	Sect. 7.3.2, p. 167

8

Dedicated-machine time-dependent scheduling

This chapter completes the second part of the book, which is devoted to the complexity of time-dependent scheduling problems. In this chapter, we present the complexity results concerning time-dependent scheduling on dedicated machines. In order to give the reader full insight into the subject, we give proofs or sketches of proofs of most discussed results. We also present pseudo-codes of algorithms for polynomially solvable problems.

Chapter 8 is composed of five sections. In Sect. 8.1, we present the results concerning dedicated machines and minimization of the $C_{\max}$ criterion. In Sect. 8.2, we present the results concerning dedicated machines and minimization of the $\sum C_j$ criterion. In Sect. 8.3, we present the results concerning dedicated machines and minimization of the $L_{\max}$ criterion. In Sect. 8.4, we present the results concerning dedicated machines and minimization of other criteria than $C_{\max}$, $\sum C_j$ and $L_{\max}$. The chapter is completed with Sect. 8.5 including the summary and tables.

8.1 Minimizing the maximum completion time

In this section, we consider dedicated-machine time-dependent scheduling problems with the $C_{\max}$ criterion.

8.1.1 Proportional deterioration

Unlike parallel machine time-dependent scheduling problems with proportional job processing times (6.1), some dedicated machine time-dependent scheduling with these job processing times are solvable in polynomial time.

Flow shop problems

Lemma 8.1. (Mosheiov [220]) *There exists an optimal schedule for the problem $F2|p_{i,j} = b_{i,j}t|C_{\max}$, in which the job sequence is identical on both machines.*

Proof. The result is a special case of Lemma 8.15.   □

Before we state the next result, we define the notion of *isomorphic problems*, introduced by Kononov [169].

Let P be an optimization problem and I_P be an instance of P. Moreover, let $s = (s_1, s_2, \ldots, s_n)$ denote a feasible solution of the instance I_P and $f_P(s_1, s_2, \ldots, s_n)$ denote the value of criterion function f_P for this solution. Finally, let $\gamma : \mathbb{R}^+ \to \mathbb{R}^+$ be a strictly increasing function.

Definition 8.2. (Isomorphic problems)
A problem P_1 is said to be isomorphic *to problem P_2 (problems P_1 and P_2 are* isomorphic, *in short) with respect to function γ, if*
(a) for any instance I_{P_1} of problem P_1, there exists an instance I_{P_2} of problem P_2 such that function γ transforms any feasible solution $s = (s_1, s_2, \ldots, s_n)$ of instance I_{P_1} into a feasible solution $s' = (\gamma(s_1), \gamma(s_2), \ldots, \gamma(s_n))$ of instance I_{P_2}, and for any feasible solution $s' = (s'_1, s'_2, \ldots, s'_n)$ of instance I_{P_2} the solution $s = (\gamma^{-1}(s'_1), \gamma^{-1}(s'_2), \ldots, \gamma^{-1}(s'_n))$ is a feasible solution of instance I_{P_1};
(b) for any feasible solution s of an instance I_{P_1}, the criterion functions f_{P_1} and f_{P_2} satisfy the equality $f_{P_2}(\gamma(s_1), \gamma(s_2), \ldots, \gamma(s_n)) = \gamma(f_{P_1}(s_1, s_2, \ldots, s_n))$.

The basic tool used in proofs of the results concerning isomorphic problems is the following lemma about optimal solutions to such problems.

Lemma 8.3. (Kononov [169]) *Let problem P_2 be isomorphic to problem P_1 with respect to a function γ. Then if $s^\star = (s_1^\star, s_2^\star, \ldots, s_n^\star)$ is an optimal solution for an instance I_{P_1} of problem P_1, then $s^{\star\prime} = (\gamma(s_1^\star), \gamma(s_2^\star), \ldots, \gamma(s_n^\star))$ is an optimal solution for an instance I_{P_2} of problem P_2.*

Proof. Let $s^\star$ be an optimal solution to problem P_1 for an instance I_{P_1} and let P_1 be isomorphic to P_2. Then $s^{\star\prime}$, by condition (a) of Definition 8.2, is a feasible solution of problem P_2. Since $f_{P_2}(s^{\star\prime}) = \gamma(f_{P_1}(s^\star))$ and since γ is a strictly increasing function, $s^{\star\prime}$ is also optimal by condition (b) of Definition 8.2.   □

The next result shows that two-machine flow shop problems with fixed and proportional job processing times are isomorphic.

Theorem 8.4. (Kononov [169]) *The problem $F2|p_{i,j} = b_{i,j}t|C_{\max}$ is isomorphic to the problem $F2||C_{\max}$ with respect to function $\gamma = 2^x$.*

Proof. Let P_1 and P_2 denote the problem $F2||C_{\max}$ and $F2|p_{i,j} = b_{i,j}t|C_{\max}$, respectively. Let I_{P_1} be an arbitrary instance of the problem P_1.
Construct instance I_{P_2} of the problem P_2 as follows: $b_{i,j} = 2^{p_{i,j}} - 1$ for $1 \le i \le 2$ and $1 \le j \le n$.
Denote by $S_{i,j}(I_{P_1})$ and by $C_{i,j}(I_{P_1})$ ($S_{i,j}(I_{P_2})$ and $C_{i,j}(I_{P_2})$) the starting time and the completion time of operation $O_{i,j}$, $1 \le i \le 2$ and $1 \le j \le n$, in a schedule for instance I_{P_1} (I_{P_2}), respectively. Then, we have

$$C_{i,j}(I_{P_2}) = (1 + b_{i,j})S_{i,j}(I_{P_2}) = 2^{p_{i,j}}2^{S_{i,j}(I_{P_1})} = 2^{S_{i,j}(I_{P_1})+p_{i,j}} = 2^{C_{i,j}(I_{P_1})}.$$

Hence, condition (a) of Definition 8.2 is satisfied.

Equivalence of criterion functions follows from the equality

$$\max_{O_{i,j} \in \mathcal{J}} \left\{ 2^{C_{i,j}(I_{P_1})} \right\} = 2^{\max\limits_{O_{i,j} \in \mathcal{J}} \{C_{i,j}(I_{P_1})\}} .$$

Hence, condition (b) of Definition 8.2 is satisfied as well.   □

Now it is not difficult to prove that the problem $F2|p_{i,j} = b_{i,j}t|C_{\max}$ is polynomially solvable. The algorithm that solves the problem is an adapted version of Johnson's algorithm for the problem $F2||C_{\max}$ (Johnson [159]): instead of processing times $p_{i,j}$ we use deterioration rates $b_{i,j}$, where $1 \le i \le m$ and $1 \le j \le n$, and each addition (subtraction) we replace by a multiplication (division). The pseudo-code of the algorithm can be formulated as follows.

Algorithm A_{44} for the problem $F2|p_{i,j} = b_{i,j}t|C_{\max}$ ([169])

Input: sequence $((b_{1,1}, b_{2,1}), (b_{1,2}, b_{2,2}), \ldots, (b_{1,n}, b_{2,n}))$
Output: an optimal schedule $\sigma^\star$

▷ Step 1:
  $\mathcal{J}_1 \leftarrow \{J_j \in \mathcal{J} : b_{1,j} \le b_{2,j}\}$;
  $\mathcal{J}_2 \leftarrow \{J_j \in \mathcal{J} : b_{1,j} > b_{2,j}\}$;
▷ Step 2:
  Arrange jobs in $\mathcal{J}_1$ in the non-decreasing order of $b_{1,j}$ values;
  Call this sequence $\sigma^{(1)}$;
  Arrange jobs in $\mathcal{J}_2$ in the non-increasing order of $b_{2,j}$ values;
  Call this sequence $\sigma^{(2)}$;
▷ Step 3:
  $\sigma^\star \leftarrow (\sigma^{(1)}|\sigma^{(2)})$;
  return $\sigma^\star$.

Theorem 8.5. (Kononov [169], Mosheiov [220]) *The problem $F2|p_{i,j} = b_{i,j}t|C_{\max}$ is solvable in $O(n \log n)$ time by algorithm A_{44}.*

Proof. Kononov [169] proves the result applying Lemma 8.3 and Theorem 8.4; see [169, Sect. 5].

Mosheiov [220] analyses possible cases in which jobs are scheduled in another order than the one generated by algorithm A_{44} and proves that in each case, the respective schedule cannot be optimal (see [220, Theorem 1]).   ◇

The flow shop problem with proportional job processing times and $m \ge 3$ machines is computationally intractable.

Theorem 8.6. (Mosheiov [220]) *The decision version of the problem $F3|p_{i,j} = b_{i,j}t|C_{\max}$ is $\mathcal{NP}$-complete in the ordinary sense.*

Proof. The transformation from the EPP problem (cf. Sect. 3.2) is as follows. Let $n = q + 3$ jobs and $t_0 = \frac{1}{2}$. Job deterioration rates are defined in the following way: $b_{1,j} = 0$, $b_{2,j} = z_j - 1$, $b_{3,j} = 0$ for $1 \leq j \leq q$, $b_{1,q+1} = 0$, $b_{2,q+1} = 1$, $b_{3,q+1} = E + 1$, $b_{1,q+2} = 2E + 1$, $b_{2,q+2} = \frac{1}{E+1}$, $b_{3,q+2} = \frac{E^2 + 2E + 1}{E+2}$, $b_{1,q+3} = \frac{E^2 + 2E + 1}{E+1}$, $b_{2,q+3} = \frac{1}{E^2 + 3E + 2}$, $b_{3,q+3} = 0$. The threshold $G = E^2 + 3E + 3$.

To complete the proof, it is sufficient to show that the EPP problem has a solution if and only if for the above instance of the problem $F3|p_{i,j} = b_{i,j}t|C_{\max}$ there exists a schedule σ such that $C_{\max}(\sigma) \leq G$. $\square$

Theorem 8.7. (Kononov [169]) *The decision version of the problem $F3|p_{i,j} = b_{i,j}t|C_{\max}$ is $\mathcal{NP}$-complete in the strong sense.*

Proof. The transformation from the 4-P problem (cf. Sect. 3.2) is as follows. Let $n = 5p+1$ and $t_0 = 1$. Job deterioration rates are defined in the following way: $b_{1,j} = 0$, $b_{2,j} = u_j - 1$, $b_{3,j} = 0$ for $1 \leq j \leq 4p$, $b_{1,4p+1} = 0$, $b_{2,4p+1} = 0$, $b_{3,4p+1} = 2D - 1$, $b_{1,4p+2} = D - 1$, $b_{2,4p+2} = 1$, $b_{3,4p+2} = 2D - 1$, $b_{1,4p+k} = 2D - 1$, $b_{2,4p+k} = 1$, $b_{3,4p+k} = 2D - 1$ for $1 \leq k \leq p - 1$, $b_{1,5p} = 2D - 1$, $b_{2,5p} = 1$, $b_{3,5p} = D - 1$, $b_{1,5p+1} = 2D - 1$, $b_{2,5p+1} = 0$, $b_{3,5p+1} = 0$. The threshold $G = 2^{p-1}D^p$.

To complete the proof, it is sufficient to show that the 4-P problem has a solution if and only if for the above instance of the problem $F3|p_{i,j} = b_{i,j}t|C_{\max}$ there exists a schedule σ such that $C_{\max}(\sigma) \leq G$. $\square$

The problem $F3|p_{i,j} = b_{i,j}t|C_{\max}$ is hard to approximate, even if deterioration rates of the operations executed on machines M_1 and M_3 are equal.

Theorem 8.8. (Kononov and Gawiejnowicz [174]) *If $\mathcal{P} \neq \mathcal{NP}$, then for the problem $F3|p_{i,j} = b_{i,j}t, b_{i,1} = b_{i,3} = b|C_{\max}$ there does not exist a polynomial-time approximation algorithm with the worst case ratio bounded by a constant.*

Proof. Suppose that there exists a polynomial-time approximation algorithm A for the problem $F3|p_{i,j} = b_{i,j}t, b_{i,1} = b_{i,3} = b|C_{\max}$ such that the ratio $r < U = const$. We will show that this assumption leads to a contradiction.

Let Q be an instance of the 3-P problem (cf. Sect. 3.2). Construct instance Q_U of the $F3|p_{i,j} = b_{i,j}t, b_{i,1} = b_{i,3} = b|C_{\max}$ problem as follows: $t_0 = 1$, $n = 4h + 1$, $b_{1,j} = b_{3,j} = U^K - 1$ for $1 \leq j \leq 4h + 1$, $b_{2,j} = U^{c_j} - 1$ for $1 \leq j \leq 3h$, $b_{2,3h+k} = U^{3K} - 1$ for $1 \leq k \leq h + 1$.

If we applied algorithm A to the instance, we would obtain a pseudopolynomial algorithm for the strongly $\mathcal{NP}$-complete 3-P problem. A contradiction, since by Lemma 3.18 this is impossible if $\mathcal{P} \neq \mathcal{NP}$. $\square$

Remark 8.9. Wang and Xia [287] considered a number of flow shop problems with *dominating* machines (cf. Definitions 8.19–8.20). Since the results (see [287, Sect. 3]) are special cases of the results for job processing times given by (8.3), we do not present them here.

Open shop problems

We start the subsection with a lower bound on the optimal value of the maximum completion time in a two-machine open shop.

Lemma 8.10. (Mosheiov [220]) *The optimal value of the maximum completion time for the problem $O2|p_{i,j} = b_{i,j}t|C_{\max}$ is not less than*

$$\max\left\{\prod_{j=1}^{n}(1 + b_{1,j}), \prod_{j=1}^{n}(1 + b_{2,j}), \max_{1 \le j \le n}\{(1 + b_{1,j})(1 + b_{2,j})\}\right\}. \tag{8.1}$$

Proof. The first component of (8.1), $\prod_{j=1}^{n}(1 + b_{1,j})$, is equal to the machine load of machine M_1. The second component of (8.1), $\prod_{j=1}^{n}(1 + b_{2,j})$, is equal to the the machine load of machine M_2. The third component of (8.1), $\max_{1 \le j \le n}\{(1 + b_{1,j})(1 + b_{2,j})\}$, is equal to the total processing time of the largest jobs on both machines.

Since the maximum completion time is not less than each of the three components, the result follows. □

The two-machine open shop problem with proportional job processing times is polynomially solvable. The algorithm that solves the problem is an adapted version of Gonzales-Sahni's algorithm for the problem $O2||C_{\max}$ (Gonzales and Sahni [115]). Let $\sigma^{(\mathcal{J}_1 - p)}$ and $\sigma^{(\mathcal{J}_2 - q)}$ denote an arbitrary sequence of jobs from the set $\mathcal{J}_1 \setminus \{p\}$ and $\mathcal{J}_2 \setminus \{q\}$, respectively.

Algorithm A_{45} for the problem $O2|p_{i,j} = b_{i,j}t|C_{\max}$ ([169])

Input: sequence $((b_{1,1}, b_{2,1}), (b_{1,2}, b_{2,2}), \ldots, (b_{1,n}, b_{2,n}))$
Output: an optimal schedule

▷ **Step 1:**
 $\mathcal{J}_1 \leftarrow \{J_j \in \mathcal{J} : b_{1,j} \le b_{2,j}\}$;
 $\mathcal{J}_2 \leftarrow \mathcal{J} \setminus \mathcal{J}_1$;
▷ **Step 2:**
 In the set $\mathcal{J}_1$, find job J_p such that $b_{2,p} \ge \max_{J_j \in \mathcal{J}_1}\{b_{1,j}\}$;

 In the set $\mathcal{J}_2$, find job J_q such that $b_{1,q} \ge \max_{J_j \in \mathcal{J}_2}\{b_{2,j}\}$;

▷ **Step 3:**
 $\sigma^1 \leftarrow (\sigma^{(\mathcal{J}_1 - p)} \mid \sigma^{(\mathcal{J}_2 - q)} \mid q \mid p)$;
 $\sigma^2 \leftarrow (p \mid \sigma^{(\mathcal{J}_1 - p)} \mid \sigma^{(\mathcal{J}_2 - q)} \mid q)$;
 Schedule jobs on machine M_1 according to sequence σ^1
 ↪ and jobs on machine M_2 according to sequence σ^2
 ↪ in such a way that no two operations of the same job overlapp.

Theorem 8.11. (Kononov [169], Mosheiov [220]) *The problem $O2|p_{i,j} = b_{i,j}t| C_{\max}$ is solvable in $O(n)$ time by algorithm A_{45}.*

Proof. Kononov [169] proves the result using the notion of isomorphic problems (cf. Definition 8.2); the proof is similar to the proof of Theorem 8.5.   □

Mosheiov [220] analyses five possible types of schedules which can be generated by algorithm A_{45} and proved that in each case the respective schedule is optimal (see [220, Theorem 3]).   ◇

Theorem 8.12. (Kononov [169], Mosheiov [220]) *The decision version of the problem* $O3|p_{i,j} = b_{i,j}t|C_{\max}$ *is $\mathcal{NP}$-complete in the ordinary sense.*

Proof. Kononov [169] uses the following transformation from the SP problem (cf. Sect. 3.2): $n = p+3$, $t_0 = 1$, $b_{i,j} = y_j - 1$, $b_{i,p+1} = \frac{2Y}{B} - 1$, $b_{i,p+2} = 2B - 1$, $b_{i,p+3} = 2Y - 1$ for $1 \leq i \leq 3$ and $1 \leq j \leq p$, where $Y = \prod_{j=1}^{p} y_j$. The threshold $G = 8Y^3$.

Mosheiov [220] uses the following transformation from the EPP problem (cf. Sect. 3.2): $n = q+1$, $t_0 = 1$, $b_{1,j} = b_{2,j} = b_{3,j} = z_j - 1$ for $1 \leq j \leq q$, $b_{1,q+1} = b_{2,q+1} = b_{3,q+1} = E - 1$. The threshold $G = E^3$.

In order to complete the proof, it is sufficient to show that the SP (EPP) problem has a solution if and only if for the above instance of the problem $O3|p_{i,j} = b_{i,j}t|C_{\max}$ there exists a schedule σ such that $C_{\max}(\sigma) \leq G$.   □

Kononov and Gawiejnowicz proved that the problem $O3|p_{i,j} = b_{i,j}t|C_{\max}$ remains computationally intractable even if all deterioration rates on machine M_3 are equal.

Theorem 8.13. (Kononov and Gawiejnowicz [174]) *The decision version of the problem* $O3|p_{i,j} = b_{i,j}t, b_{i,3} = b|C_{\max}$ *is $\mathcal{NP}$-complete in the ordinary sense.*

Proof. The transformation from the SP problem is as follows. Let $\prod_{j=1}^{p} y_j = Y$, $\bar{Y} = 2Y$, $t_0 = 1$ and $n = p+4$. The job deterioration rates are as follows:
$b_{1,j} = 0, b_{2,j} = y_j^2 - 1, b_{3,j} = \bar{Y}^2 - 1$ for $1 \leq j \leq p$,
$b_{1,p+1} = 0, b_{2,p+1} = \frac{\bar{Y}^2}{Y^2} - 1, b_{3,p+1} = \bar{Y}^2 - 1$,
$b_{1,p+2} = 0, b_{2,p+2} = 4\bar{Y}^2 - 1, b_{3,p+2} = \bar{Y}^2 - 1$,
$b_{1,p+3} = \bar{Y}^{p+3} - 1, b_{2,p+3} = \bar{Y}^{p+3} - 1, b_{3,p+3} = \bar{Y}^2 - 1$,
$b_{1,p+4} = \bar{Y}^{p+5} - 1, b_{2,p+4} = \bar{Y}^{p+1} - 1, b_{3,p+4} = \bar{Y}^2 - 1$.
The threshold $G = \bar{Y}^{2p+8}$.

To complete the proof it is sufficient to show that the SP problem has a solution if and only if for the above instance of the problem $O3|p_{i,j} = b_{i,j}t, b_{i,3} = b|C_{\max}$ there exists a schedule σ such that $C_{\max}(\sigma) \leq G$.   □

Job shop problems

Kononov and Gawiejnowicz [174] stated the conjecture that the two-machine job shop with proportional processing times is a computationally intractable problem. The conjecture has been proved by Mosheiov.

Theorem 8.14. (Mosheiov [220]) *The decision version of the problem* $J2|p_{i,j} = b_{i,j}t|C_{\max}$ *is $\mathcal{NP}$-complete in the ordinary sense.*

Proof. The transformation from the EPP problem (cf. Sect. 3.2) has the following form. Let $n = q + 1$, $t_0 = 1$ and $b_{1,j} = z_j$ for $1 \leq j \leq q$. Job J_{q+1} consists of three operations: $O_{1,q+1}, O_{2,q+1}$ and $O_{3,q+1}$. The operations have to be done on machine M_2, machine M_1 and machine M_2, respectively. The deterioration rates for the job J_{q+1} are the following: $b_{1,q+1} = \frac{1}{E}$ and $b_{2,q+1} = E - 1$. The threshold value $G = (E + 1)E$.

To complete the proof, it is sufficient to show that the EPP problem has a solution if and only if for the above instance of the problem $J2|p_{i,j} = b_{i,j}t|C_{\max}$ there exists a schedule σ such that $C_{\max}(\sigma) \leq G$. □

8.1.2 Proportional-linear deterioration

Now we pass to the time-dependent scheduling problems on dedicated machines, in which the job processing times are *proportional-linear* ones, i.e.,

$$p_{i,j} = b_{i,j}(A + Bt) \tag{8.2}$$

for $A \geq 0, B \geq 0, b_{i,j} \geq 0$ and $1 \leq i \leq n_j, 1 \leq j \leq n$. This form of job deterioration has been introduced by Kononov [169].

Flow shop problems

We start this subsection with the following result.

Lemma 8.15. (Kononov and Gawiejnowicz [174]) *There exists an optimal schedule for the problem* $Fm|p_{i,j} = b_{i,j}(A + Bt)|C_{\max}$, *in which*
(a) *machines* M_1 *and* M_2 *perform jobs in the same order,*
(b) *machines* M_{m-1} *and* M_m *perform jobs in the same order.*

Proof. (a) Assume that in a schedule σ jobs executed on machine M_1 are processed according to sequence $\sigma^1 = (i_1, \ldots, i_n)$, while jobs executed on machine M_2 are processed according to sequence $\sigma^2 = (j_1 = i_1, j_2 = i_2, \ldots, j_p = i_p, j_{p+1} = i_{r+1}, \ldots, j_{p+q} = i_r, \ldots, j_n = i_n)$, $q > 1$, $r > p \geq 0$.

By Theorem 6.20, the following equations are satisfied for schedule σ :

$$S_{2,i_{r+1}}(\sigma) = \max\{\frac{A}{B}\prod_{k=1}^{r+1}(b_{2,i_k}B + 1) - \frac{A}{B}, C_{2,j_p}(\sigma)\},$$

$$S_{2,i_r}(\sigma) = \max\{\frac{A}{B}\prod_{k=1}^{r}(b_{2,i_k}B + 1) - \frac{A}{B}, C_{2,j_{p+q-1}}(\sigma)\} = C_{2,j_{p+q-1}}(\sigma) \geq$$

$$\geq C_{2,i_{r+1}}(\sigma) \geq C_{1,i_{r+1}}(\sigma) = \frac{A}{B}\prod_{k=1}^{r+1}(b_{2,i_k}B + 1) - \frac{A}{B}.$$

Consider schedule σ', differing from schedule σ only in that machine M_1 performs jobs according to sequence $\sigma^3 = (i_1, \ldots, i_{r-1}, i_{r+1}, i_r, i_{r+2}, \ldots, i_n)$.

For schedule σ', by Theorem 6.20, we obtain the following equalities:

$$S_{2,i_{r+1}}(\sigma') = \max\{\frac{A}{B}(b_{2,i_{r+1}}B+1)\prod_{k=1}^{r-1}(b_{2,i_k}B+1)-\frac{A}{B}, C_{2,j_p}(\sigma')\},$$

$$S_{2,i_r}(\sigma') = \max\{\frac{A}{B}\prod_{k=1}^{r}(b_{2,i_k}A+1)-\frac{A}{B}, C_{2,j_{p+q-1}}(\sigma')\}.$$

Since $C_{2,j_p}(\sigma) = C_{2,j_p}(\sigma')$, we have $S_{2,i_{r+1}}(\sigma) \geq S_{2,i_{r+1}}(\sigma')$. From this, it follows that $C_{2,j_{p+q-1}}(\sigma) \geq C_{2,j_{p+q-1}}(\sigma')$ and hence, $S_{2,i_r}(\sigma) \geq S_{2,i_r}(\sigma')$.

Repeating the above considerations no more than $O(n^2)$ times, we obtain a schedule $\bar{\sigma}$ in which machines M_1 and M_2 perform jobs in the same order, and such that inequality $C_{2,j}(\sigma) \geq C_{2,j}(\bar{\sigma})$ holds for all $J_j \in \mathcal{J}$. From this, it follows that $C_{\max}(\sigma) \geq C_{\max}(\bar{\sigma})$.

(b) Similar to the proof of (a). ∎

Lemma 8.16. (Kononov and Gawiejnowicz [174]) *If $m \in \{2,3\}$, then there exists an optimal schedule for the problem $Fm|p_{i,j} = b_{i,j}(A+Bt)|C_{\max}$, in which jobs are executed on all machines in the same order.*

Proof. Applying Lemma 8.15 for $m = 2, 3$, we obtain the result. □

Kononov generalized Theorem 8.5 to the case of proportional-linear job processing times given by (6.5).

Theorem 8.17. (Kononov [170]) *If there hold inequalities (6.6) and (6.7), then the problem $F2|p_{i,j} = b_{i,j}(A+Bt)|C_{\max}$ is solvable in $O(n\log n)$ time by algorithm A_{44}.*

Proof. Similar to the proof of Theorem 8.5, see [170, Theorem 33]. ◇

Remark 8.18. A special case of Theorem 8.17, without conditions (6.6) and (6.7) but with assumptions $A > 0, B > 0, b_{i,j} > 0$ for $1 \leq i \leq 2$ and $1 \leq j \leq n$, was given by Zhao et al. [304, Theorem 5].

Wang and Xia [288] considered job processing times

$$p_{i,j} = b_{i,j}(A+t), \tag{8.3}$$

where $1 \leq i \leq m$ and $1 \leq j \leq n$. (Notice that the job processing times (8.3) are special case of (8.2) for $B = 1$.) The authors, assuming that there hold some relations between the job processing times, defined the so-called *dominated* machines (see Definitions 8.19 and 8.20) and proved a number of results for multi-machine flow shop with machines of this type.

Definition 8.19. (Dominated machines)
The machine M_r is said to be dominated by machine M_k, $1 \leq k \neq r \leq m$, if and only if $\max\{b_{r,j} : 1 \leq j \leq n\} \leq \min\{b_{k,j} : 1 \leq j \leq n\}$.

If machine M_r is dominated by machine M_k, we will write $M_r \prec M_k$ or $M_k \succ M_r$.

Definition 8.20. (Series of dominating machines)
(a) *The machines* $M_1, M_2, \ldots, M_m$ *form an* increasing series of dominating machines (idm) *if and only if* $M_1 \prec M_2 \prec \ldots \prec M_m$;
(b) *The machines* $M_1, M_2, \ldots, M_m$ *form a* decreasing series of dominating machines (ddm) *if and only if* $M_1 \succ M_2 \succ \ldots \succ M_m$;
(c) *The machines* $M_1, M_2, \ldots, M_m$ *form an* increasing–decreasing series of dominating machines (idm-ddm) *if and only if* $M_1 \prec M_2 \prec \ldots \prec M_h$ *and* $M_h \succ M_{h+1} \succ \ldots \succ M_m$ *for some* $1 \le h \le m$;
(d) *The machines* $M_1, M_2, \ldots, M_m$ *form an* decreasing–increasing series of dominating machines (ddm-idm) *if and only if* $M_1 \succ M_2 \succ \ldots \succ M_h$ *and* $M_h \prec M_{h+1} \prec \ldots \prec M_m$ *for some* $1 \le h \le m$.

Wang and Xia [288] proposed the following algorithm for the case when a flow shop is of *idm-ddm, no-wait* and *idm-ddm*, or *no-idle* and *idm-ddm* type.

Algorithm A_{46} for the problem $Fm|p_{i,j} = b_{i,j}(A+t), \delta|C_{\max}$ ([288]),
where $\delta \in \{idm\text{-}ddm; no\text{-}wait, idm\text{-}ddm; no\text{-}idle, idm\text{-}ddm\}$

Input: sequences $(b_{1,j}, b_{2,j}, \ldots, b_{m,j})$ for $1 \le j \le n$, numbers A, h
Output: an optimal schedule $\sigma^\star$

▷ Step 1:
Find job J_u such that $\prod_{i=1}^{h-1}(1+b_{i,u}) = \min\left\{\prod_{i=1}^{h-1}(1+b_{i,j}) : 1 \le j \le n\right\}$;

Find job J_v such that $\prod_{i=h+1}^{m}(1+b_{i,v}) = \min\left\{\prod_{i=h+1}^{m}(1+b_{i,j}) : 1 \le j \le n\right\}$;

▷ Step 2:
if $(u = v)$ then Find job J_w such that
$$\hookrightarrow \prod_{i=1}^{h-1}(1+b_{i,w}) = \min\left\{\prod_{i=1}^{h-1}(1+b_{i,j}) : 1 \le j \ne u \le n\right\};$$
Find job J_z such that
$$\hookrightarrow \prod_{i=h+1}^{m}(1+b_{i,z}) = \min\left\{\prod_{i=h+1}^{m}(1+b_{i,j}) : 1 \le j \ne v \le n\right\};$$
if $\prod_{i=1}^{h-1}(1+b_{i,w}) \prod_{i=h+1}^{m}(1+b_{i,v}) \le \prod_{i=1}^{h-1}(1+b_{i,u}) \prod_{i=h+1}^{m}(1+b_{i,z})$
$$\hookrightarrow \text{then } u \leftarrow w$$
$$\text{else } v \leftarrow z;$$

▷ Step 3:
$\sigma^\star \leftarrow (u|(\mathcal{J} \setminus \{u,v\})|v)$;
return $\sigma^\star$.

Theorem 8.21. (Wang and Xia [288]) *The problems* $Fm|p_{i,j} = b_{i,j}(A + t)$, $\delta|C_{\max}$, *where* $\delta \in \{idm\text{-}ddm; no\text{-}wait, idm\text{-}ddm; no\text{-}idle, idm\text{-}ddm\}$, *are solvable in* $O(mn)$ *time by algorithm* A_{46}.

Proof. See [288, Theorem 1]. ◇

Remark 8.22. For other cases of the multi-machine flow shop with dominated machines, the authors proposed algorithms which are modifications of algorithm A_{46} (see [288, Theorems 2–4]). ◇

Open shop problems

Kononov [170] generalized Theorem 8.11 to the case of proportional-linear job processing times given by (6.5).

Theorem 8.23. (Kononov [170]) *If there hold inequalities* (6.6) *and* (6.7), *then the problem* $O2|p_{i,j} = b_{i,j}(A + Bt)|C_{\max}$ *is solvable in* $O(n)$ *time by algorithm* A_{45}.

Proof. Similar to the proof of Theorem 8.11, see [170, Theorem 33]. ◇

8.1.3 Linear deterioration

Flow shop problems

The two-machine flow shop problem with linear job processing times is computationally intractable.

Theorem 8.24. (Kononov and Gawiejnowicz [174]) *The decision version of the problem* $F2|p_{i,j} = a_{i,j} + b_{i,j}t|C_{\max}$ *is* $\mathcal{NP}$-*complete in the strong sense.*

Proof. The transformation from the 3-P problem (cf. Sect. 3.2) has the following form. Let $t_0 = 0$ and $n = 4h$. The job processing times are defined in the following way: $p_{1,1} = 0$, $p_{2,1} = 1 + nt$, $p_{1,j} = K + 1$, $p_{2,j} = \frac{1}{(j-1)(K+1)}t$ for $2 \leq j \leq h$, $p_{1,j} = 0$, $p_{2,j} = c_j$ for $h + 1 \leq j \leq 4h$. The threshold $G = hK + h$.

In order to complete the proof, it is sufficient to show that the 3-P problem has a solution if and only if for the above instance of the problem $F2|p_{i,j} = a_{i,j} + b_{i,j}t|C_{\max}$ there exists a schedule σ such that $C_{\max}(\sigma) \leq G$. □

Lee et al. [196] gave a lower bound on the optimal value of the total completion time for the problem $F2|p_{i,j} = a_{i,j} + bt|C_{\max}$. Let $B := 1 + b$.

Lemma 8.25. (Lee et al. [196]) *Let* k *be the number of already scheduled jobs in a subschedule* $\sigma^{(1)}$ *and let* $\mathcal{J}_{\sigma^{(1)}}$ *denote the set of jobs in* $\sigma^{(1)}$. *Then the maximum completion time* $C_{\max}(\sigma)$ *for the schedule* $\sigma = (\mathcal{J}_{\sigma^{(1)}}|\mathcal{J} \setminus \mathcal{J}_{\sigma^{(1)}})$ *for the problem* $F2|p_{i,j} = a_{i,j} + bt|C_{\max}$ *is not less than* $\max\{LB_1, LB_2\}$, *where*

$$LB_1 := B^{n-k+1}C_{1,[k]} + \sum_{i=1}^{n-k} B^{n-k-i+1}a_{1,(k+i)} + a_{2,(k+1)},$$

$$LB_2 := B^{n-k}C_{2,[k]} + \sum_{i=k+1}^{n} B^{n-i}a_{2,(i)},$$

$a_{1,(k+1)} \le a_{1,(k+2)} \le \cdots \le a_{1,(n)}$ and $a_{2,(k+1)} \le a_{2,(k+2)} \le \cdots \le a_{2,(n)}$ are non-decreasingly ordered basic processing times of unscheduled jobs on machine M_1 and M_2, respectively, and $a_{2,(k+1)} := \min\{a_{2,i} : i \in \mathcal{J} \setminus \mathcal{J}_{\sigma^{(1)}}\}$.

Proof. See [196, Sect. 3]. ◇

Lee et al. [196] also formulated a number of dominance properties (see [196, Properties 1–5]) and proposed a branch-and-bound algorithm for the problem (see [196, Sect. 4.1]). The branch-and-bound algorithm was tested on a number of instances with $8 \le n \le 32$ jobs (see [196, Sect. 5] for details).

The authors also proposed four heuristic algorithms for the problem. We will consider these heuristics in Chap. 9.

Cheng et al. [47] considered a few multi-machine flow shop problems with dominating machines (cf. Definition 8.20) and equal deterioration rates, $b_{i,j} = b$ for $1 \le i \le m$ and $1 \le j \le n$. Let $B := 1 + b$.

Lemma 8.26. (Cheng et al. [47]) *Let* $\sigma = ([1], [2], \ldots, [n])$ *be a schedule for the problem* $Fm|p_{i,j} = a_{i,j} + bt, \delta|C_{\max}$, *where* $\delta \in \{no\text{-}idle, idm; no\text{-}idle, ddm; no\text{-}idle, idm\text{-}ddm; no\text{-}idle, ddm\text{-}idm\}$. *Then*

(a) $C_{[j]} = \sum_{i=1}^{m} a_{i,[1]}B^{m+j-i+1} + \sum_{k=2}^{j} a_{m,[k]}B^{j-k}$, *if* $\delta \equiv no\text{-}idle, idm$;

(b) $C_{[j]} = \sum_{k=1}^{n-1} a_{1,[k]}B^{m+j-k+1} + \sum_{i=1}^{m} a_{i,[n]}B^{m-n+j-i} - \sum_{k=j+1}^{n} a_{m,[k]}B^{j-k}$, *if* $\delta \equiv no\text{-}idle, ddm$;

(c) $C_{[j]} = \sum_{i=1}^{h} a_{i,[1]}B^{m-i+j-1} + \sum_{k=2}^{n-1} a_{h,[k]}B^{m-h-k+j} + \sum_{i=h}^{m} a_{i,[n]}B^{m-n+j-i} - \sum_{k=j+1}^{n} a_{m,[k]}B^{j-k}$, *if* $\delta \equiv no - idle, idm - ddm$;

(d) $C_{[j]} = \sum_{k=1}^{n-1} a_{1,[k]}B^{m+j-k-1} + \sum_{i=1}^{h-1} a_{i,[n]}B^{m+j-n-i} - \sum_{k=2}^{n-1} a_{h,[k]}B^{m+j-h-k} + \sum_{i=h+1}^{m} a_{i,[1]}B^{m+j-i-1} + \sum_{k=2}^{j} a_{m,[k]}B^{j-k}$, *if* $\delta \equiv no - idle, ddm - idm$.

Proof. (a) Since machines $M_1, M_2, \ldots, M_m$ are of *no-idle, idm* type, in any feasible schedule σ, we have $C_{i,[j]} \le C_{i+1,[j]}$ for $1 \le i \le m$ and $1 \le j \le n$. Therefore, $C_{[j]} = \sum_{i=1}^{m} p_{i,[1]} + \sum_{k=2}^{j} p_{m,[k]}$.
(b)(c)(d) Similar to (a). □

Using Lemma 8.26, one can prove the next result.

Theorem 8.27. (Cheng et al. [47])

(a) *Let* $\sum_{i=1}^{m} a_{i,k} B^{m+n-i-1} = \min \left\{ \sum_{i=1}^{m} a_{i,j} B^{m+n-i-1} : 1 \leq j \leq n \right\}$ *for some* $1 \leq k \leq n$ *and* $\sigma^{(\mathcal{J}-k)}$ *be a subschedule obtained by scheduling jobs from the set* $\mathcal{J} \setminus \{J_k\}$ *in the non-decreasing order of* $a_{m,j}$ *values. Then the schedule* $(k|\sigma^{(\mathcal{J}-k)})$ *is optimal for the problem* $Fm|p_{i,j} = a_{i,j} + bt, no\text{-}idle, idm|C_{\max}$.

(b) *Let* $\sum_{i=1}^{m} a_{i,k} B^{m-i} = \min \left\{ \sum_{i=1}^{m} a_{i,j} B^{m-i} : 1 \leq j \leq n \right\}$ *for some* $1 \leq k \leq n$ *and* $\sigma^{(\mathcal{J}-k)}$ *be a subschedule obtained by scheduling jobs from the set* $\mathcal{J} \setminus \{J_k\}$ *in the non-decreasing order of* $a_{1,j}$ *values. Then the schedule* $(k|\sigma^{(\mathcal{J}-k)})$ *is optimal for the problem* $Fm|p_{i,j} = a_{i,j} + bt, no\text{-}idle, ddm|C_{\max}$.

Proof. (a) (b) The results are consequences of Lemma 8.26.   □

On the basis of Theorem 8.27, Cheng et al. [47] proposed for the problem $Fm|p_{i,j} = a_{i,j} + bt, no\text{-}idle, idm\text{-}ddm|C_{\max}$ the following algorithm.

Algorithm A_{47}
for the problem $Fm|p_{i,j} = a_{i,j} + bt, no\text{-}idle, idm\text{-}ddm|C_{\max}$ ([47])

Input: sequences $(a_{1,j}, a_{2,j}, \ldots, a_{m,j})$ for $1 \leq j \leq n$, number $B := 1 + b$
Output: an optimal schedule $\sigma^{\star}$

▷ **Step 1:**

Find job J_u such that

$\hookrightarrow \sum_{i=1}^{h} a_{i,u} B^{m+h-i-1} = \min \left\{ \sum_{i=1}^{h} a_{i,j} B^{m+h-i-1} : 1 \leq j \leq n \right\}$;

Find job J_v such that $\sum_{i=h}^{m} a_{i,v} B^{m-i} = \min \left\{ \sum_{i=h}^{m} a_{i,j} B^{m-i} : 1 \leq j \leq n \right\}$;

▷ **Step 2:**

if $(u = v)$ **then** Find job J_w such that

$\hookrightarrow \sum_{i=1}^{h} a_{i,w} B^{m+h-i-1} = \min \left\{ \sum_{i=1}^{h} a_{i,j} B^{m+h-i-1} : 1 \leq j \leq n \right\}$;

Find job J_z such that

$\hookrightarrow \sum_{i=h}^{m} a_{i,z} B^{m-i} = \min \left\{ \sum_{i=h}^{m} a_{i,j} B^{m-i} : 1 \leq j \neq v \leq n \right\}$;

if $\sum_{i=1}^{h} a_{i,u} B^{m+h-i-1} + \sum_{i=h}^{m} a_{i,z} B^{m-i} \leq \sum_{i=1}^{h} a_{i,w} B^{m+h-i-1} +$

$\hookrightarrow \sum_{i=h}^{m} a_{i,v} B^{m-i}$ **then** $v \leftarrow z$

else $u \leftarrow w$;

▷ **Step 3:**

$\sigma^{\star} \leftarrow (v|(\mathcal{J} \setminus \{u, v\})|u)$;

return $\sigma^{\star}$.

Theorem 8.28. (Cheng et al. [47]) *The problem* $Fm|p_{i,j} = a_{i,j} + bt, no\text{-}idle, idm\text{-}dd|C_{\max}$ *is solvable in* $O(n^3 \log n)$ *by algorithm* A_{47}.

Proof. The result is a consequence of Theorem 8.27 (a).   □

For the problem $Fm|p_{i,j} = a_{i,j} + bt, no - idle, ddm - idm|C_{\max}$ Cheng et al. [47] proposed the following algorithm.

Algorithm A_{48}
for the problem $Fm|p_{i,j} = a_{i,j} + bt, no\text{-}idle, ddm\text{-}idm|C_{\max}$ ([47])

Input: sequences $(a_{1,j}, a_{2,j}, \ldots, a_{m,j})$ for $1 \leq j \leq n$, number $B := 1 + b$
Output: an optimal schedule $\sigma^\star$

▷ Step 1:

Find job J_u such that $\sum\limits_{i=h+1}^{m} \left(a_{i,u} B^{1-i} + \frac{a_{1,u}}{m-h} \right) B^{m+n-2} =$

$\hookrightarrow \min \left\{ \sum\limits_{i=h+1}^{m} \left(a_{i,j} B^{1-i} + \frac{a_{1,j}}{m-h} \right) B^{m+n-2} : 1 \leq j \leq n \right\};$

Find job J_v such that

$\hookrightarrow \sum\limits_{i=1}^{h-1} \left(a_{i,v} B^{m-i} + \frac{a_{m,v}}{h-1} \right) = \min \left\{ \sum\limits_{i=1}^{h-1} \left(a_{i,j} B^{m-i} + \frac{a_{m,j}}{h-1} \right) : 1 \leq j \leq n \right\};$

▷ Step 2:

if $(u = v)$ then Find job J_w such that

$\hookrightarrow \sum\limits_{i=1}^{h} a_{i,w} B^{m+h-i-1} = \min \left\{ \sum\limits_{i=1}^{h} a_{i,j} B^{m+h-i-1} : 1 \leq j \leq n \right\};$

Find job J_z such that

$\hookrightarrow \sum\limits_{i=h}^{m} a_{i,z} B^{m-i} = \min \left\{ \sum\limits_{i=h}^{m} a_{i,j} B^{m-i} : 1 \leq j \neq v \leq n \right\};$

if $\sum\limits_{i=1}^{h} a_{i,u} B^{m+h-i-1} + \sum\limits_{i=h}^{m} a_{i,z} B^{m-i} \leq \sum\limits_{i=1}^{h} a_{i,w} B^{m+h-i-1} +$

$\hookrightarrow \sum\limits_{i=h}^{m} a_{i,v} B^{m-i}$ then $v \hookleftarrow z$

else $u \hookleftarrow w;$

▷ Step 3:

$\sigma^\star \hookleftarrow (v|(\mathcal{J} \setminus \{u, v\})|u);$

return $\sigma^\star$.

Theorem 8.29. (Cheng et al. [47]) *The problem $Fm|p_{i,j} = a_{i,j} + bt, no\text{-}idle,$ $ddm\text{-}idm|C_{\max}$ is solvable in $O(n^3 \log n)$ by algorithm A_{48}.*

Proof. The result is a consequence of Theorem 8.27 (b).   □

Open shop problems

The two-machine open shop problem with linear job processing times is computationally intractable.

Theorem 8.30. (Kononov and Gawiejnowicz [174]) *The decision version of the problem $O2|p_{i,j} = a_{i,j} + b_{i,j}t|C_{\max}$ is $\mathcal{NP}$-complete in the ordinary sense.*

Proof. The transformation from the PP problem (cf. Sect. 3.2) is as follows: $t_0 = 1$, $n = k + 1$, $a_{1,j} = a_{2,j} = x_j$ and $b_{1,j} = b_{2,j} = 0$ for $1 \leq j \leq k$, $a_{1,k+1} = a_{2,k+1} = 0$, $b_{1,k+1} = b_{2,k+1} = A$. The threshold $G = (A+1)^2 + A$.

In order to complete the proof, it is sufficient to show that the PP problem has a solution if and only if for the above instance of the problem $O2|p_{i,j} = a_{i,j} + b_{i,j}t|C_{\max}$ there exists a schedule σ such that $C_{\max}(\sigma) \leq G$ (see Fig. 8.1 and Remark 6.12). □

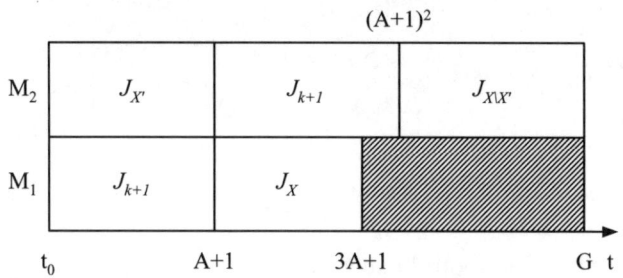

Fig. 8.1: Example schedule in the proof of Theorem 8.30

Job shop problems

As in the case of multi-machine flow shop and open shop problems, two-machine job shop problem with linear job processing times is already computationally intractable.

Theorem 8.31. *The decision version of the problem $J2|p_{i,j} = a_{i,j}+b_{i,j}t|C_{\max}$ is $\mathcal{NP}$-complete in the ordinary sense.*

Proof. The result is a corollary from Theorem 8.14. □

8.1.4 Simple non-linear deterioration

Flow shop problems

Melnikov and Shafransky [207] considered multi-machine flow shop with job processing times given by (6.15), $Fm|p_{i,j} = a_{i,j} + f(t)|C_{\max}$, where function $f(t)$ is differentiable, it satisfies (6.16) and

$$\frac{df(t)}{dt} \geq \lambda > 0 \quad \text{for} \quad t \geq 0. \tag{8.4}$$

For the above problem, the authors estimated the difference between the optimal maximum completion time $C_{max}^{\star}$ and the maximum completion time $C_{max}(\sigma^{\nearrow})$ for the schedule $\sigma^{\nearrow}$ obtained by scheduling jobs in non-decreasing order of $a_{1,j}$ values, $1 \leq j \leq n$.

Before we formulate the result, we state an auxiliary result.

Lemma 8.32. (Melnikov and Shafransky [207]) *For any differentiable function* $f(t)$, *satisfying conditions* (6.16) *and* (8.4), *there exists a finite number* N_0 *such that for all* $n \geq N_0$ *the inequality*

$$\sum_{j=1}^{n} a_{i,j} + f\left(\sum_{j=0}^{n-1} C_{1,j}\right) \geq \sum_{j=l-1}^{n-1} a_{i+1,j} + f\left(\sum_{j=0}^{l-2} C_{1,j}\right) \tag{8.5}$$

holds for any i *and* l, *where* $1 \leq i \leq m-1$, $2 \leq l \leq n$, $C_{1,0} := 0$ *and* $C_{1,j} = a_{1,j} + f(\sum_{k=0}^{j-1} C_{1,k})$ *for* $1 \leq j \leq n$.

Proof. See [207, Theorem 3].   ◇

Assuming that $f(t)$ is a differentiable function and it satisfies conditions (6.16) and (8.4), there holds the following result.

Theorem 8.33. (Melnikov and Shafransky [207]) *Let* $\sigma^{\nearrow}$ *denote the schedule for the problem* $Fm|p_{i,j} = a_{i,j} + f(t)|C_{max}$ *in which jobs are scheduled in non-decreasing order of* $a_{1,j}$ *values*, $1 \leq j \leq n$. *If the number of jobs* n *satisfies the inequality* (8.5), *then either* $\sigma^{\nearrow}$ *is an optimal schedule or the optimal schedule is one of* $k \leq n-1$ *schedules* π, *in which the last job satisfies the inequality*

$$\sum_{i=2}^{m} a_{i,\pi_n} < \sum_{i=2}^{m} a_{i,\sigma_n^{\nearrow}}$$

and the first $n-1$ *jobs are scheduled in non-decreasing order of* $a_{1,j}$ *values.*

Proof. By Lemma 8.32 and Theorem 6.39, the result follows; see [207, Theorem 4] for details.   ◇

If we pass to two-machine flow shop problem with job processing times given by (6.15), $F2|p_{i,j} = a_{i,j} + f(t)|C_{max}$, where $f(t)$ is defined as in Theorem 8.33, there holds the following result.

Theorem 8.34. (Melnikov and Shafransky [207]) *Let* $\sigma^{\nearrow}$ *denote the schedule for the problem* $F2|p_{i,j} = a_{i,j} + f(t)|C_{max}$ *in which jobs are scheduled in non-decreasing order of* $a_{1,j}$ *values*, $1 \leq j \leq n$. *Then there holds the inequality*

$$C_{max}^{\star} - C_{max}(\sigma^{\nearrow}) \leq a_{2,\sigma_n^{\nearrow}} - \min_{1 \leq j \leq n}\{a_{2,j}\}.$$

Proof. See [207, Theorem 5].   ◇

8.1.5 Proportional-linear shortening

Flow shop problems

Wang and Xia [289] considered multi-machine flow shop with job processing times given by (6.44). Let $b_{\min} := \min_{i,j}\{b_{i,j}\}$.

Theorem 8.35. (Wang and Xia [289]) *The problem* $F2|p_{i,j} = b_{i,j}(1 - kt)$, $k > 0$, $k\left(\sum_{i=1}^{m}\sum_{j=1}^{n} b_{i,j} - b_{\min}\right) < 1|C_{\max}$ *is solvable in* $O(n\log n)$ *time by algorithm* A_{44}.

Proof. Similar to the proof of Theorem 8.5.   □

Assuming that $b_{i,j} = b_j$ for all $1 \leq i \leq m$, the authors obtained the following result.

Theorem 8.36. (Wang and Xia [289]) *For the problem* $F2|p_{i,j} = b_{i,j}(1-kt)$, $b_{i,j} = b_j, k > 0$, $k\left(n\sum_{i=1}^{m} b_i - b_{\min}\right) < 1|C_{\max}$ *the maximum completion time does not depend on the schedule of jobs.*

Proof. The result follows from Theorem 6.1.   □

Remark 8.37. Wang [280] considered multi-machine flow shop problems with job processing times in the form of (6.44) and dominated machines (cf. Definitions 8.19–8.20). Since all the problems are solved by algorithms which are similar to algorithms A_{47} and A_{48} (see [280, Theorems 1–4] for details), we do not present the algorithms here.

8.2 Minimizing the total completion time

In this section, we consider dedicated-machine time-dependent scheduling problems with the $\sum C_j$ criterion.

8.2.1 Proportional deterioration

Flow shop problems

Wang et al. [286] considered the problem $F2|p_{i,j} = b_{i,j}t|\sum C_j$, where deterioration rates $b_{i,j} \in (0,1)$ for $1 \leq i \leq 2$ and $1 \leq j \leq n$.

Lemma 8.38. (Wang et al. [286]) *There exists an optimal schedule for the problem* $F2|p_{i,j} = b_{i,j}t, 0 < b_{i,j} < 1|\sum C_j$ *in which the job sequence is identical on both machines.*

Proof. Similar to the proof of Lemma 8.15 for $m = 2$.   □

The next two results concern the cases when some deterioration rates in the problem $F2|p_{i,j} = b_{i,j}t|\sum C_j$ are equal.

Theorem 8.39. (Wang et al. [286]) *If $b_{2,j} = b$ for $1 \leq j \leq n$, then the problem $F2|p_{i,j} = b_{i,j}t, 0 < b_{i,j} < 1|\sum C_j$ is solvable in $O(n \log n)$ time by scheduling jobs in the non-decreasing order of $b_{1,j}$ values.*

Proof. Consider schedule σ in which jobs are scheduled in the non-decreasing order of $b_{1,j}$ values and an arbitrary other schedule τ. By showing that $C_j(\sigma) \leq C_{[j]}(\tau)$ for $1 \leq j \leq n$, the result follows (see [286, Theorem 1]).   □

By Theorem 8.39, if $b_{2,j} = b$ for $1 \leq j \leq n$, then the problem $F2|p_{i,j} = b_{i,j}t, 0 < b_{i,j} < 1|\sum C_j$ is solved by the algorithm $A_7 : (b_{1,j}|b_{2,j}) \mapsto (b_{1,j} \nearrow)$.

Theorem 8.40. (Wang et al. [286]) *If $b_{1,j} = b_{2,j}$ for $1 \leq j \leq n$, then the problem $F2|p_{i,j} = b_{i,j}t, 0 < b_{i,j} < 1|\sum C_j$ is solvable in $O(n \log n)$ time by scheduling jobs in the non-decreasing order of $b_{1,j}$ values.*

Proof. By pairwise job interchange argument.   □

By Theorem 8.40, if $b_{1,j} = b_{2,j}$ for $1 \leq j \leq n$, then the problem $F2|p_{i,j} = b_{i,j}t, 0 < b_{i,j} < 1|\sum C_j$ is solved by the algorithm $A_5 : (b_{1,j}|b_{2,j}) \mapsto (b_{1,j} \nearrow)$.

The next two results concern the cases when flow shop machines are dominating (cf. Definitions 8.19–8.20).

Lemma 8.41. (Wang et al. [286]) *If in an instance of the problem $F2|p_{i,j} = b_{i,j}t, \ 0 < b_{i,j} < 1, M_1 \prec M_2|\sum C_j$ the first scheduled job is fixed, then in an optimal schedule the remaining jobs are in the non-decreasing order of $b_{2,j}$ values.*

Proof. By direct calculation and Lemma 6.120, the result follows.   □

By Lemma 8.41 Wang et al. [286] constructed the following algorithm.

Algorithm A_{49}
for the problem $F2|p_{i,j} = b_{i,j}t, 0 < b_{i,j} < 1, M_1 \prec M_2|\sum C_j$ ([286])

Input: sequences $(b_{1,j}, b_{2,j})$ for $1 \leq j \leq n$
Output: an optimal schedule $\sigma^\star$

▷ Step 1:
  for $j \leftarrow 1$ **to** n **do**
    Schedule first job J_j;
    Schedule remaining jobs in the non-decreasing order of $b_{2,j}$ values;
    Call the obtained schedule σ^j;
▷ Step 2:
  $\sigma^\star \leftarrow$ the best schedule from the schedules $\sigma^1, \sigma^2, \ldots, \sigma^n$;
▷ Step 3:
  return $\sigma^\star$.

Theorem 8.42. (Wang et al. [286]) *The problem $F2|p_{i,j} = b_{i,j}t, 0 < b_{i,j} < 1$, $M_1 \prec M_2| \sum C_j$ is solvable in $O(n \log n)$ time by algorithm A_{49}.*

Proof. By Lemma 8.41 we can construct n distinct schedules for the problem, with a fixed first job in each of them. Since each such a schedule can be constructed in $O(n \log n)$ time, the result follows. □

Theorem 8.43. (Wang et al. [286]) *The problem $F2|p_{i,j} = b_{i,j}t, 0 < b_{i,j} < 1$, $M_1 \succ M_2| \sum C_j$ is solvable in $O(n \log n)$ time by scheduling jobs in the non-decreasing order of $\frac{b_{1,j}}{(1+b_{1,j})(1+b_{2,j})}$ ratios.*

Proof. By pairwise job interchange argument. □

By Theorem 8.43, the problem $F2|p_{i,j} = b_{i,j}t, 0 < b_{i,j} < 1, M_1 \succ M_2| \sum C_j$ is solved by the algorithm $A_{50} : (b_{1,j}|b_{2,j}) \mapsto (\frac{b_{1,j}}{(1+b_{1,j})(1+b_{2,j})} \nearrow)$.

Wang et al. [286] formulated also dominance properties (see [286, Propositions 1–8; Theorems 5–6]) and proposed a branch-and-bound algorithm for the problem $F2|p_{i,j} = b_{i,j}t, 0 < b_{i,j} < 1| \sum C_j$ (see [286, Sect. 5.3]). The algorithm is based on the properties and the following lower bound.

Lemma 8.44. (Wang et al. [286]) *Let k be the number of already scheduled jobs in a subschedule $\sigma^{(1)}$ and let $\mathcal{J}_{\sigma^{(1)}}$ denote the set of jobs in $\sigma^{(1)}$. Then the total completion time $\sum C_j(\sigma)$ for the schedule $\sigma = (\mathcal{J}_{\sigma^{(1)}}|\mathcal{J} \setminus \mathcal{J}_{\sigma^{(1)}})$ for the problem $F2|p_{i,j} = b_{i,j}t, 0 < b_{i,j} < 1| \sum C_j$ is not less than $\max\{LB_1, LB_2\}$, where*

$$LB_1 := \sum_{j=1}^{k} C_{[j]}(\sigma^{(1)}) + C_{[k]}(\sigma^{(1)}) \sum_{j=k+1}^{n} \prod_{i=k+1}^{j} (1 + b_{2,(i)}),$$

$$LB_2 := \sum_{j=1}^{k} C_{[j]}(\sigma^{(1)}) + t_0 \left(1 + \min_{k+1 \leq j \leq n} \{b_{2,j}\} \right) \prod_{j=1}^{k} (1 + b_{1,[j]}) \times$$
$$\times \left(\sum_{i=k+1}^{n} \prod_{j=k+1}^{i} (1 + b_{1,(j)}) \right),$$

$b_{1,(k+1)} \leq b_{1,(k+2)} \leq \ldots \leq b_{1,(n)}$ and $b_{2,(k+1)} \leq b_{2,(k+2)} \leq \ldots \leq b_{2,(n)}$ are non-decreasingly ordered basic processing times of unscheduled jobs on machine M_1 and machine M_2, respectively.

Proof. See [286, Sect. 5.2]. ◇

The branch-and-bound algorithm proposed by Wang et al. was tested on 100 instances with $n \leq 14$ jobs (see [286, Sect. 6] for details).

The same authors also proposed a heuristic for the problem $F2|p_{i,j} = b_{i,j}t$, $0 < b_{i,j} < 1| \sum C_j$. We will consider this algorithm in Chap. 9.

Shiau et al. [257] gave a lower bound on the optimal value of the total completion time for the general problem $F2|p_{i,j} = b_{i,j}t| \sum C_j$.

Lemma 8.45. (Shiau et al. [257]) *Let k be the number of already scheduled jobs in a subschedule $\sigma^{(1)}$ and let $\mathcal{J}_{\sigma^{(1)}}$ denote the set of jobs in $\sigma^{(1)}$. Let*
$$B := \prod_{i=1}^{k} (1 + b_{1,[i]}). \text{ Then the total completion time } \sum C_j(\sigma) \text{ for the schedule}$$
$\sigma = (\mathcal{J}_{\sigma^{(1)}} | \mathcal{J} \setminus \mathcal{J}_{\sigma^{(1)}})$ *for the problem $F2|p_{i,j} = b_{i,j}t| \sum C_j$ is not less than* $\max\{LB_1, LB_2, LB_3\}$, *where*

$$LB_1 := \sum_{j=1}^{k} C_{[j]}(\sigma^{(1)}) + C_{[k]}(\sigma^{(1)}) \sum_{j=k+1}^{n} \prod_{i=k+1}^{j} (1 + b_{2,(i)}),$$

$$LB_2 := \sum_{j=1}^{k} C_{[j]}(\sigma^{(1)}) + B \times \sum_{i=k+1}^{n} (1 + b_{2,(n+k+1-i)}) \prod_{j=k+1}^{i} (1 + b_{1,(j)}),$$

$$LB_3 := \sum_{j=1}^{k} C_{[j]}(\sigma^{(1)}) + (n-k)B \times \sqrt[n-k]{\prod_{i=k+1}^{n} (1 + b_{2,i}) \prod_{i=k+1}^{n} (1 + b_{1,(i)})^{n+1-i}},$$

$b_{1,(k+1)} \le b_{1,(k+2)} \le \cdots \le b_{1,(n)}$ *and* $b_{2,(k+1)} \le b_{2,(k+2)} \le \cdots \le b_{2,(n)}$ *are non-decreasingly ordered basic processing times of unscheduled jobs on machine M_1 and machine M_2, respectively.*

Proof. See [257, Sect. 4].   ◇

Remark 8.46. The lower bound LB_1 from Lemma 8.45 is a generalization of the lower bound LB_1 from Lemma 8.44.

Shiau et al. [257] also formulated a number of dominance properties (see [257, Propositions 1–11]) and proposed a branch-and-bound algorithm for the problem $F2|p_{i,j} = b_{i,j}t| \sum C_j$ (see [257, Sect. 5.1]). The branch-and-bound algorithm was tested on 450 instances with $6 \le n \le 14$ jobs (see [257, Sect. 7] for details).

The authors also proposed three heuristics and a simulated annealing algorithm for the problem. We will consider these heuristics and this algorithm in Chaps. 9 and 11, respectively.

8.2.2 Linear deterioration

Flow shop problems

The two-machine flow shop problem with linear job processing times and the total completion time criterion is computationally intractable.

Theorem 8.47. *The decision version of the problem $F2|p_{i,j} = a_{i,j} + b_{i,j}t| \sum C_j$ is $\mathcal{NP}$-complete in the strong sense.*

Proof. Since the special case of the problem is the problem $F2|| \sum C_j$, which is $\mathcal{NP}$-complete in the strong sense (Garey et al. [86]), the result follows. ∎

Theorem 8.48. (Wu and Lee [299]) *The decision version of the problem $F2|p_{i,j} = a_{i,j} + bt| \sum C_j$ is $\mathcal{NP}$-complete in the strong sense.*

Proof. Similar to the proof of Theorem 8.47.   □

Wu and Lee proposed for the problem $F2|p_{i,j} = a_{i,j} + bt| \sum C_j$ a branch-and-bound algorithm, based on a number of dominance properties and a lower bound. First, we briefly describe some of these properties. Let $\Theta := C_{2,[n]} - (b+1)C_{1,[n]}$, where $b > 0$ is the common job deterioration rate.

Property 8.49. (Wu and Lee [299]) *If jobs J_i and J_j are scheduled consecutively, $a_{1,i} \geq \Theta$, $\max\{a_{1,i}, a_{2,i}\} < a_{1,j}$ and $a_{2,j} \leq \min\{a_{1,i}, a_{2,i}\}$, then there exists an optimal schedule for the problem $F2|p_{i,j} = a_{i,j} + bt| \sum C_j$ in which job J_i is the immediate predecessor of job J_j.*

Proof. By pairwise job interchange argument; see [299, Appendix A].   ◇

Property 8.50. (Wu and Lee [299]) *If jobs J_i and J_j are scheduled consecutively, $a_{1,i} \geq \Theta$, $a_{1,j} \geq a_{1,i}$ and*

$$\min\left\{ a_{1,j}, a_{2,j}, \frac{b+1}{b+2}a_{1,j} + \frac{1}{b+2}a_{2,j}, \frac{ba_{1,j}}{b+1} + \frac{ba_{2,j} + a_{2,i}}{(b+1)^2} \right\} > a_{1,i},$$

then there exists an optimal schedule for the problem $F2|p_{i,j} = a_{i,j} + bt| \sum C_j$ in which job J_i is the immediate predecessor of job J_j.

Proof. The result is given without proof; see [299, Property 2].   ◇

Other properties given by Wu and Lee without proof (see [299, Properties 3–12]) are similar to Property 8.50.

The authors proposed also the following lower bound for the problem.

Lemma 8.51. (Wu and Lee [299]) *Let k be the number of already scheduled jobs in a subschedule $\sigma^{(1)}$, $J_{\sigma^{(1)}}$ denote the set of jobs in $\sigma^{(1)}$ and $r := n - k$. Then the total completion time $\sum C_j(\sigma)$ for the schedule $\sigma = (J_{\sigma^{(1)}}|(\mathcal{J} \setminus J_{\sigma^{(1)}}))$ for the problem $F2|p_{i,j} = a_{i,j} + b_{i,j}t| \sum C_j$ is not less than $\max\{LB_1, LB_2\}$, where*

$$LB_1 := \sum_{j=1}^{k} C_{2,[j]}(\sigma) + C_{1,[k]}(\sigma^{(1)}) \sum_{j=1}^{r} (1+b)^{j+1} + \sum_{j=1}^{r} \sum_{i=1}^{r-j+1} a_{1,(k+j)}(1+b)^j +$$
$$+ \sum_{j=1}^{r} a_{2,k+j},$$

$$LB_2 := \sum_{j=1}^{k} C_{2,[j]}(\sigma) + C_{2,[k]}(\sigma) \sum_{j=1}^{r} (1+b)^j + \sum_{j=1}^{r} \sum_{i=1}^{r-j} a_{2,(k+j)}(1+b)^{i-1},$$

$a_{1,(k+1)} \leq a_{1,(k+2)} \leq \cdots \leq a_{1,(k+r)}$ *and* $a_{2,(k+1)} \leq a_{2,(k+2)} \leq \cdots \leq a_{2,(k+r)}$ *are non-decreasingly ordered basic processing times of unscheduled jobs on machine M_1 and M_2, respectively, and $\sum_{j=1}^{r} a_{2,k+j}$, is the sum of the basic processing times of the unscheduled jobs on machine M_2.*

Proof. See [299, Sect. 4].   ◇

The branch-and-bound algorithm, based on the mentioned properties and Lemma 8.51, was tested on instances with $n \leq 27$ jobs (see [299, Sect. 6]).

Cheng et al. [47] considered two multi-machine flow shop problems with dominating machines (cf. Definition 8.20) and equal deterioration rates, $b_{i,j} = b$ for $1 \leq i \leq m$ and $1 \leq j \leq n$. Let $B := 1 + b$.

Theorem 8.52. (Cheng et al. [47])
(a) Let $\sum_{i=1}^{m} a_{i,k} B^{m-i} = \min \left\{ \sum_{i=1}^{m} a_{i,j} B^{m-i} : 1 \leq j \leq n \right\}$ for some $1 \leq k \leq n$ and let $\sigma^{(\mathcal{J}-k)}$ be a subschedule obtained by scheduling jobs from the set $\mathcal{J} \setminus \{J_k\}$ in the non-decreasing order of $a_{m,j}$ values. Then the schedule $(k|\sigma^{(\mathcal{J}-k)})$ is optimal for the problem $Fm|p_{i,j} = a_{i,j} + bt, no\text{-}idle, idm| \sum C_j$.

(b) Let $N := \frac{B^n - 1}{B - 1}$ and $m - n + 1 := r$. Let $N \sum_{i=1}^{m-1} a_{i,k} B^{r-i} + a_{m,k} = \min \left\{ N \sum_{i=1}^{m-1} a_{i,j} B^{r-i} + a_{m,k} : 1 \leq j \leq n \right\}$ for some $1 \leq k \leq n$ and let $\sigma^{(\mathcal{J}-k)}$ be a subschedule obtained by scheduling jobs from the set $\mathcal{J} \setminus \{J_k\}$ in the non-decreasing order of $(a_{1,j} B^m - a_{m,j})(B^n - 1) + a_{m,j} B^{n+1}$ values. Then the schedule $(k|\sigma^{(\mathcal{J}-k)})$ is optimal for the problem $Fm|p_{i,j} = a_{i,j} + bt, no\text{-}idle, ddm| \sum C_j$.

Proof. (a) See [47, Theorem 10].
(b) See [47, Theorem 11].   ◇

On the basis of Theorem 8.52, for the problem $Fm|p_{i,j} = a_{i,j} + bt, no\text{-}idle, idm\text{-}ddm| \sum C_j$, Cheng et al. [47] proposed two algorithms (see [47, Sect. 4]) that are modifications of algorithms A_{47} nd A_{48} (cf. Sect. 8.1.3). We will call the algorithms A_{51} and A_{52}.

Theorem 8.53. (Cheng et al. [47])
(a) *The problem* $Fm|p_{i,j} = a_{i,j} + bt, no\text{-}idle, idm| \sum C_j$ *is solvable in* $O(n^3 \log n)$ *by algorithm* A_{51}.
(b) *The problem* $Fm|p_{i,j} = a_{i,j} + bt, no\text{-}idle, ddm| \sum C_j$ *is solvable in* $O(n^3 \log n)$ *by algorithm* A_{52}.

Proof. (a) The result is a consequence of Theorem 8.52 (a).
(b) The result is a consequence of Theorem 8.52 (b).   □

8.2.3 Proportional-linear shortening

Flow shop problems

Wang and Xia [289] considered multi-machine flow shop with job processing times given by (6.44). Let $b_{\min} := \min_{i,j} \{b_{i,j}\}$.

Theorem 8.54. (Wang and Xia [289]) *The problem $Fm|p_{i,j} = b_{i,j}(1 - kt)$,*
$b_{i,j} = b_j, k > 0, k \left(\sum_{j=1}^{n} b_j - b_{\min} \right) < 1 | \sum C_j$ *is solvable in $O(n \log n)$ time by scheduling jobs in the non-decreasing order of b_j values.*

Proof. The result follows from Theorem 6.120.   □

By Theorem 8.54, the problem $Fm|p_{i,j} = b_{i,j}(1 - kt)$, $b_{i,j} = b_j, k > 0$, $k \left(\sum_{j=1}^{n} b_j - b_{\min} \right) < 1 | \sum C_j$ is solved by the algorithm $A_7 : (b_{1,j}|b_{2,j}) \mapsto (b_j \nearrow)$.

8.3 Minimizing the maximum lateness

In this section, we consider dedicated-machine time-dependent scheduling problems with the $L_{\max}$ criterion.

8.3.1 Proportional deterioration

Flow shop problems

Theorem 8.55. (Kononov [169]) *The decision version of the problem $F2|p_{i,j} = b_{i,j}t|L_{\max}$ is $\mathcal{NP}$-complete in the strong sense.*

Proof. The transformation is from the 4-P problem (cf. Sect. 3.2): $n = 5p - 1$ and $t_0 = 1, b_{1,j} = 0, b_{2,j} = u_j - 1, d_j = 2^{p-1}D^p$ for $1 \le j \le 4p, b_{1,4p+1} = D - 1$, $b_{2,4p+1} = 1, d_{4p+1} = 2D, b_{1,4p+k} = 2D - 1, b_{2,4p+k} = 1, d_{4p+k} = 2^k D^k$ for $2 \le k \le m - 1$. The threshold $G = 0$.

To complete the proof, it is sufficient to show that the 4-P problem has a solution if and only if for the above instance of the problem $F2|p_{i,j} = b_{i,j}t|L_{\max}$ there exists a schedule σ such that $L_{\max}(\sigma) \le G$.   □

Remark 8.56. Wang and Xia [287] considered a number of flow shop problems with dominating machines (cf. Definitions 8.19–8.20) and the $L_{\max}$ criterion. Since the results (see [287, Sect. 5]) are special cases of the results for job processing times given by (8.3), we do not present them here.

Open shop problems

Theorem 8.57. (Kononov [169]) *The decision version of the problem $O2|p_{i,j} = b_{i,j}t|L_{\max}$ is $\mathcal{NP}$-complete in the ordinary sense.*

Proof. The transformation from the SP problem (cf. Sect. 3.2) is as follows: $n = p + 3$ and $t_0 = 1$, $b_{i,j} = y_j - 1$, $b_{i,p+1} = \frac{2Y}{B} - 1$, $b_{i,p+2} = 2B - 1$, $b_{i,p+3} = 2Y - 1$ for $1 \le i \le 2$, $d_j = 8Y^3$ for $1 \le j \le p + 2$ and $d_{p+3} = 4Y^2$, where $Y = \prod_{j=1}^{p} y_j$. The threshold $G = 0$.

To complete the proof, it is sufficient to show that the SP problem has a solution if and only if for the above instance of the problem $O2|p_{i,j} = b_{i,j}t|L_{\max}$ there exists a schedule σ such that $L_{\max}(\sigma) \le G$. $\qquad\square$

8.3.2 Proportional-linear deterioration

Flow shop problems

Wang and Xia [288] considered job processing times given by (8.3), assuming that flow shop machines are dominating (cf. Definitions 8.19–8.20).

Lemma 8.58. (Wang and Xia [288]) *If in an instance of the problem $Fm|p_{i,j} = b_{i,j}(A + t), \delta|L_{\max}$, where $\delta \in \{idm; no\text{-}wait, idm; no\text{-}idle, idm\}$, the first scheduled job is fixed, then in an optimal schedule the remaining jobs are in the EDD order.*

Proof. By a contradiction; see [288, Theorem 10]. $\qquad\diamond$

By Lemma 8.58, Wang and Xia [288] proposed an algorithm for the problem $Fm|p_{i,j} = b_{i,j}(A + t), \delta|L_{\max}$, where $\delta \in \{idm; no\text{-}wait, idm; no\text{-}idle, idm\}$. The algorithm is an adaptation of algorithm A_{47} in which in Step 1 all jobs except the first one are scheduled in the EDD order instead of in the non-decreasing order of $b_{2,j}$ values. Call the new algorithm A_{53}.

Theorem 8.59. (Wang and Xia [288]) *The problems $Fm|p_{i,j} = b_{i,j}(A + t), \delta|L_{\max}$, where $\delta \in \{idm; no\text{-}wait, idm; no\text{-}idle, idm\}$, are solvable in $O(n \log n)$ time by algorithm A_{53}.*

Proof. By Lemma 8.58, in order to find an optimal schedule for the problems $Fm|p_{i,j} = b_{i,j}(A + t), \delta|L_{\max}$, where $\delta \in \{idm; no\text{-}wait, idm; no\text{-}idle, idm\}$, it is sufficient to construct n distinct schedules by inserting at the first position in the j-th schedule the j-th job and scheduling the remaining jobs in the EDD order. Since each such a schedule can be obtained in $O(n \log n)$ time, the result follows. $\qquad\square$

Remark 8.60. For other cases of the multi-machine flow shop with dominated machines and the $L_{\max}$ criterion, the authors applied a similar approach; see [288, Theorems 11–12]. $\qquad\diamond$

8.3.3 Linear deterioration

Flow shop problems

Theorem 8.61. *The decision version of the problem $F2|p_{i,j} = a_{i,j} + b_{i,j}t|L_{\max}$ is $\mathcal{NP}$-complete in the ordinary sense.*

Proof. The result is a corollary from Theorem 8.55. $\qquad\blacksquare$

8.3.4 Proportional-linear shortening

Flow shop problems

Wang and Xia considered two-machine flow shop with job processing times given by (6.44).

Theorem 8.62. (Wang and Xia [289]) *The problem* $F2|p_{i,j} = b_{i,j}(1 - kt)$, $b_{i,j} = b_j, k > 0, k \left(\sum_{j=1}^{n} b_j - b_{\min} \right) < 1|L_{\max}$ *is solvable in* $O(n \log n)$ *time by scheduling jobs in the non-decreasing order of* d_j *values.*

Proof. The result follows from Theorem 6.169.   □

By Theorem 8.62, the problem $F2|p_{i,j} = b_{i,j}(1 - kt)$, $b_{i,j} = b_j, k > 0$, $k \left(\sum_{j=1}^{n} b_j - b_{\min} \right) < 1|L_{\max}$ is solved by the algorithm $A_{17} : (b_{1,j}|b_{2,j}|d_j) \mapsto (d_j \nearrow)$.

Remark 8.63. Wang [280] considered multi-machine flow shop problems with job processing times in the form of (6.44) and dominated machines (cf. Definitions 8.19–8.20). Since all the problems are solved by algorithms which are similar to algorithm A_{53} (see [280, Theorems 10–12] for details), we do not present the algorithms here.

8.4 Other criteria

In this section, we consider the problems of time-dependent scheduling on dedicated machines with criteria other than $C_{\max}$, $\sum C_j$ or $L_{\max}$.

8.4.1 Proportional deterioration

Flow shop problems

Wang and Xia [287] considered a number of flow shop problems with dominating machines (cf. Definitions 8.19–8.20) and the $\sum w_j C_j$ criterion. Since the results (see [287, Sect. 4]) are special cases of the results for job processing times given by (8.3), we do not present them here.

8.4.2 Proportional-linear deterioration

Flow shop problems

Wang and Xia [288] considered multi-machine flow shop with dominating machines (cf. Definitions 8.19–8.20) and job processing times given by (8.3). The authors prove a number of results for the $\sum w_j C_j$ criterion.

Lemma 8.64. (Wang and Xia [288]) *If in an instance of the problem $Fm|p_{i,j} = b_{i,j}(A+t), \delta| \sum w_j C_j$, where $\delta \in \{idm; no\text{-}wait, idm; no\text{-}idle, idm\}$, the first scheduled job is fixed, then in an optimal schedule the remaining jobs are scheduled in the non-decreasing order of $\frac{b_{m,j}}{w_j(1+b_{m,j})}$ ratios.*

Proof. The result follows from the formula for $\sum w_j C_j$ and Theorem 6.185. $\square$

By Lemma 8.64, Wang and Xia [288] proposed an algorithm for the problem $Fm|p_{i,j} = b_{i,j}(A + t), \delta| \sum w_j C_j$, where $\delta \in \{idm; no\text{-}wait, idm; no\text{-}idle, idm\}$. The algorithm is an adaptation of algorithm A_{47} in which in Step 1 all jobs except the first one are scheduled in the non-decreasing order of $\frac{b_{m,j}}{w_j(1+b_{m,j})}$ ratios instead of in the non-decreasing order of $b_{2,j}$ values. Call the new algorithm A_{54}.

Theorem 8.65. (Wang and Xia [288]) *The problems $Fm|p_{i,j} = b_{i,j}(A + t), \delta| \sum w_j C_j$, where $\delta \in \{idm; no\text{-}wait, idm; no\text{-}idle, idm\}$, are solvable in $O(n \log n)$ time by algorithm A_{54}.*

Proof. By Lemma 8.64, in order to find an optimal schedule for the problems $Fm|p_{i,j} = b_{i,j}(A+t), \delta| \sum w_j C_j$, where $\delta \in \{idm; no\text{-}wait, idm; no\text{-}idle, idm\}$, it is sufficient to construct n distinct schedules by inserting at the first position in the j-th schedule the j-th job. Since each such a schedule can be obtained in $O(n \log n)$ time, the result follows. $\square$

Remark 8.66. For other cases of the multi-machine flow shop with dominated machines and the $\sum w_j C_j$ criterion, the authors applied similar approach; see [288, Theorems 6–9]. $\diamond$

8.4.3 Proportional-linear shortening

Flow shop problems

Wang and Xia [289] considered multi-machine flow shop with job processing times given by (6.44). For this problem modified algorithm A_{29} (see [289, Modified Algorithm 1]) can be used appropriately. Let $b_{\min} := \min_{i,j}\{b_{i,j}\}$.

Theorem 8.67. (Wang and Xia [289]) *The problem $Fm|p_{i,j} = b_{i,j}(1 - kt)$, $k > 0$, $k\left(\sum\limits_{j=1}^{n} b_j - b_{\min}\right) < 1, b_{i,j} = b_j|f_{\max}$ is solvable in $O(n^2)$ time by the modified algorithm A_{29}.*

Proof. The result follows from Theorem 13.35. $\square$

For the problem with the $\sum U_j$ criterion, the authors used appropriately modified algorithm A_{26} (see [289, Algorithm 2]).

Theorem 8.68. (Wang and Xia [289]) *The problem* $Fm|p_{i,j} = b_{i,j}(1 - kt)$, $k > 0$, $k\left(\sum_{j=1}^{n} b_j - b_{\min}\right) < 1, b_{i,j} = b_j| \sum U_j$ *is solvable in* $O(n \log n)$ *time by the modified algorithm* A_{26}.

Proof. The result follows from Theorem 6.196.   □

Remark 8.69. Wang [280] considered a number of multi-machine flow shop problems with job processing times (6.44) and dominated machines (cf. Definitions 8.19–8.20). Since all the problems are solved by algorithms that are similar to algorithm A_{54} (see [280, Theorems 5–9] for details), we do not present the algorithms here.

With this remark, we end the review of dedicated-machine time-dependent scheduling. This chapter also ends the second part of the book, which is devoted to the complexity of time-dependent scheduling problems. In the next part of the book, we will consider different classes of algorithms for computationally intractable time-dependent scheduling problems.

8.5 Summary and tables

In this chapter, we reviewed dedicated-machine time-dependent scheduling problems. We considered different deteriorating and shortening job processing times. The criteria of schedule optimality include the most popular criteria such as the $C_{\max}$, $\sum C_j$ or $L_{\max}$, as well as less popular criteria such as the $\sum w_j C_j$, $f_{\max}$ or $\sum U_j$.

Approximately two-thirds of the problems presented in the chapter are polynomially solvable. For these problems, we presented pseudo-codes of optimal algorithms. For the remaining problems that are ordinary or strongly $\mathcal{NP}$-complete, we presented $\mathcal{NP}$-completeness proofs or sketches of such proofs.

Below, we classify in the tabular form the dedicated-machine time-dependent scheduling problems and algorithms considered in the chapter. As in Chaps. 6 and 7, the results are divided into groups with respect to the applied criterion.

Tables 8.1 and 8.2 present, respectively, tractable and intractable dedicated-machine time-dependent scheduling problems with the $C_{\max}$ criterion.

Table 8.11 presents polynomial algorithms for dedicated-machine time-dependent scheduling problems with criteria other than $C_{\max}$, $\sum C_j$ and $L_{\max}$.

Tables 8.3 and 8.4 present, respectively, tractable and intractable dedicated-machine time-dependent scheduling problems with the $\sum C_j$ criterion.

Tables 8.5 and 8.6 present, respectively, tractable and intractable dedicated-machine time-dependent scheduling problems with the $L_{\max}$ criterion.

Table 8.7 presents tractable dedicated-machine time-dependent scheduling problems with other criteria than $C_{\max}$, $\sum C_j$ or $L_{\max}$.

Tables 8.8, 8.9 and 8.10 present, respectively, polynomial algorithms for dedicated-machine time-dependent scheduling problems with the $C_{\max}$, $\sum C_j$ and $L_{\max}$ criteria.

Table 8.11 presents polynomial algorithms for dedicated-machine time-dependent scheduling problems with criteria other than $C_{\max}$, $\sum C_j$ and $L_{\max}$.

Table 8.1: Tractable Dedicated-Machine Time-Dependent Scheduling Problems ($C_{\max}$ Criterion)

Problem	Complexity	References	This book
$F2\|p_{i,j} = b_{i,j}t\|C_{\max}$	$O(n\log n)$	[169], [220]	Theorem 8.5
$O2\|p_{i,j} = b_{i,j}t\|C_{\max}$	$O(n)$	[169], [220]	Theorem 8.11
$F2\|p_{i,j} = b_{i,j}(a+bt)\|C_{\max}$	$O(n\log n)$	[170]	Theorem 8.17
$Fm\|p_{i,j} = b_{i,j}(a+t),$	$O(mn)$	[288]	Theorem 8.21
$idm\text{-}ddm\|C_{\max}$			
$Fm\|p_{i,j} = b_{i,j}(a+t),$	$O(mn)$	[288]	Theorem 8.21
$no\text{-}wait, idm\text{-}ddm\|C_{\max}$			
$Fm\|p_{i,j} = b_{i,j}(a+t),$	$O(mn)$	[288]	Theorem 8.21
$no\text{-}idle, idm\text{-}ddm\|C_{\max}$			
$O2\|p_{i,j} = b_{i,j}(a+bt)\|C_{\max}$	$O(n\log n)$	[170]	Theorem 8.23
$Fm\|p_{i,j} = a_{i,j} + bt, no\text{-}idle, idm\text{-}dd\|C_{\max}$	$O(n^3\log n)$	[47]	Theorem 8.28
$Fm\|p_{i,j} = a_{i,j} + bt, no\text{-}idle, ddm\text{-}idm\|C_{\max}$	$O(n^3\log n)$	[47]	Theorem 8.29
$F2\|p_{i,j} = b_{i,j}(1-kt)\|C_{\max}$ [a]	$O(n\log n)$	[289]	Theorem 8.35
$F2\|p_{i,j} = b_{i,j}(1-kt)\|C_{\max}$ [b,c]	$O(n)$	[289]	Theorem 8.36

[a] $k > 0, k(\sum_{i=1}^{m}\sum_{j=1}^{n} b_{i,j} - b_{\min}) < 1$, where $b_{\min} := \min\{b_{i,j}\}$
[b] $b_{i,j} = b_j, k > 0, k(n\sum_{i=1}^{m} b_i - b_{\min}) < 1$, where $b_{\min} := \min\{b_{i,j}\}$
[c] the value of $C_{\max}$ does not depend on schedule

Table 8.2: Intractable Dedicated-Machine Time-Dependent Scheduling Problems ($C_{\max}$ Criterion)

Problem	Complexity	References	This book
$F3\|p_{i,j} = b_{i,j}t\|C_{\max}$	$\mathcal{SNPC}$	[169], [220]	Theorems 8.6–8.7
$O3\|p_{i,j} = b_{i,j}t\|C_{\max}$	$\mathcal{NPC}$	[169], [220]	Theorem 8.12
$O3\|p_{i,j} = b_{i,j}t, b_{i,3} = b\|C_{\max}$	$\mathcal{NPC}$	[174]	Theorem 8.13
$J2\|p_{i,j} = b_{i,j}t\|C_{\max}$	$\mathcal{NPC}$	[220]	Theorem 8.14
$F2\|p_{i,j} = a_{i,j} + b_{i,j}t\|C_{\max}$	$\mathcal{SNPC}$	[174]	Theorem 8.24
$O2\|p_{i,j} = a_{i,j} + b_{i,j}t\|C_{\max}$	$\mathcal{NPC}$	[174]	Theorem 8.30
$J2\|p_{i,j} = a_{i,j} + b_{i,j}t\|C_{\max}$	$\mathcal{NPC}$	–	Theorem 8.31

Table 8.3: Tractable Dedicated-Machine Time-Dependent Scheduling Problems ($\sum C_j$ Criterion)

Problem	Complexity	References	This book
$F2\|p_{i,j} = b_{i,j}t, b_{2,j} = b\| \sum C_j$ [a]	$O(n \log n)$	[286]	Theorem 8.39
$F2\|p_{i,j} = b_{i,j}t, b_{1,j} = b_{2,j}\| \sum C_j$ [a]	$O(n \log n)$	[286]	Theorem 8.40
$F2\|p_{i,j} = b_{i,j}t, M_1 \ll M_2\| \sum C_j$ [a]	$O(n \log n)$	[286]	Theorem 8.42
$F2\|p_{i,j} = b_{i,j}t, M_1 \gg M_2\| \sum C_j$ [a]	$O(n \log n)$	[286]	Theorem 8.43
$Fm\|p_{i,j} = a_{i,j} + bt, no\text{-}idle, idm\| \sum C_j$	$O(n^3 \log n)$	[47]	Theorem 8.53 (a)
$Fm\|p_{i,j} = a_{i,j} + bt, no\text{-}idle, ddm\| \sum C_j$	$O(n^3 \log n)$	[47]	Theorem 8.53 (b)
$Fm\|p_{i,j} = b_{i,j}(1 - kt)\| \sum C_j$ [b]	$O(n \log n)$	[289]	Theorem 8.54

[a] $0 < b_{i,j} < 1$
[b] $k > 0, k(\sum_{j=1}^{n} b_j - b_{\min}) < 1, b_{i,j} = b_j$, where $b_{\min} := \min\{b_{i,j}\}$

Table 8.4: Intractable Dedicated-Machine Time-Dependent Scheduling Problems ($\sum C_j$ Criterion)

Problem	Complexity	References	This book
$F2\|p_{i,j} = a_{i,j} + b_{i,j}t\| \sum C_j$	$\mathcal{SNPC}$	–	Theorem 8.47
$F2\|p_{i,j} = a_{i,j} + bt\| \sum C_j$	$\mathcal{SNPC}$	[299]	Theorem 8.48

Table 8.5: Tractable Dedicated-Machine Time-Dependent Scheduling Problems ($L_{\max}$ Criterion)

Problem	Complexity	References	This book
$Fm\|p_{i,j} = b_{i,j}(A + t), idm\|L_{\max}$	$O(n \log n)$	[288]	Theorem 8.59
$Fm\|p_{i,j} = b_{i,j}(A + t), no\text{-}wait, idm\|L_{\max}$	$O(n \log n)$	[288]	Theorem 8.59
$Fm\|p_{i,j} = b_{i,j}(A + t), no\text{-}idle, idm\|L_{\max}$	$O(n \log n)$	[288]	Theorem 8.59
$F2\|p_{i,j} = b_{i,j}(1 - kt)\|L_{\max}$ [a]	$O(n \log n)$	[289]	Theorem 8.62

[a] $k > 0, k(\sum_{j=1}^{n} b_j - b_{\min}) < 1, b_{i,j} = b_j$, where $b_{\min} := \min\{b_{i,j}\}$

Table 8.6: Intractable Dedicated-Machine Time-Dependent Scheduling Problems ($L_{\max}$ Criterion)

Problem	Complexity	References	This book
$F2\|p_{i,j} = b_{i,j}t\|L_{\max}$	$\mathcal{SNPC}$	[169]	Theorem 8.55
$O2\|p_{i,j} = b_{i,j}t\|L_{\max}$	$\mathcal{NPC}$	[169]	Theorem 8.57
$F2\|p_{i,j} = a_{i,j} + b_{i,j}t\|L_{\max}$	$\mathcal{SNPC}$	–	Theorem 8.61

Table 8.7: Tractable Dedicated-Machine Time-Dependent Scheduling Problems (Criteria other than $C_{\max}$, $\sum C_j$ and $L_{\max}$)

Problem	Complexity	References	This book
$Fm\|p_{i,j} = b_{i,j}(A + t),$ $idm\|\sum w_j C_j$	$O(n\log n)$	[288]	Theorem 8.65
$Fm\|p_{i,j} = b_{i,j}(A + t),$ $no\text{-}wait, idm\|\sum w_j C_j$	$O(n\log n)$	[288]	Theorem 8.65
$Fm\|p_{i,j} = b_{i,j}(A + t),$ $no\text{-}idle, idm\|\sum w_j C_j$	$O(n\log n)$	[288]	Theorem 8.65
$Fm\|p_{i,j} = b_{i,j}(1 - kt)\|f_{\max}$ [a]	$O(n^2)$	[289]	Theorem 8.67
$Fm\|p_{i,j} = b_{i,j}(1 - kt)\|\sum U_j$ [a]	$O(n\log n)$	[289]	Theorem 8.68

[a] $k > 0, k(\sum_{j=1}^{n} b_j - b_{\min}) < 1, b_{i,j} = b_j$, where $b_{\min} := \min\{b_{i,j}\}$

Table 8.8: Polynomial Algorithms for Dedicated-Machine Time-Dependent Scheduling Problems ($C_{\max}$ Criterion)

Algorithm	Complexity	Problem	This book
A_{44}	$O(n\log n)$	$F2\|p_{i,j} = b_{i,j}t\|C_{\max}$	Sect. 8.1.1, p. 173
A_{44}	$O(n\log n)$	$F2\|p_{i,j} = b_{i,j}(A + Bt)\|C_{\max}$	Sect. 8.1.2, p. 178
A_{44}	$O(n\log n)$	$Fm\|p_{i,j} = b_{i,j}(1 - kt)\|C_{\max}$	Sect. 8.1.5, p. 186
A_{45}	$O(n)$	$O2\|p_{i,j} = b_{i,j}t\|C_{\max}$	Sect. 8.1.1, p. 175
A_{45}	$O(n)$	$O2\|p_{i,j} = b_{i,j}(A + Bt)\|C_{\max}$	Sect. 8.1.2, p. 180
A_{46}	$O(mn)$	$Fm\|p_{i,j} = b_{i,j}(A + t), \delta\|C_{\max}$ [a]	Sect. 8.1.2, p. 179
A_{48}	$O(n\log n)$	$Fm\|p_{i,j} = a_{i,j} + bt,$ $no\text{-}idle, ddm\text{-}idm\|C_{\max}$	Sect. 8.1.3, p. 183

[a] $\delta \in \{idm\text{-}ddm; no\text{-}wait, idm\text{-}ddm; no\text{-}idle, idm\text{-}ddm\}$

Table 8.9: Polynomial Algorithms for Dedicated-Machine Time-Dependent Scheduling Problems ($\sum C_j$ Criterion)

Algorithm	Complexity	Problem	This book
A_7	$O(n\log n)$	$F2\|p_{i,j} = b_{i,j}t\|\sum C_j$	Sect. 8.2.1, p. 187
A_7	$O(n\log n)$	$Fm\|p_{i,j} = b_{i,j}(1 - kt)\|\sum C_j$	Sect. 8.2.3, p. 192
A_{47}	$O(n\log n)$	$F2\|p_{i,j} = b_{i,j}t, M_1 \prec M_2\|\sum C_j$ [a]	Sect. 8.2.1, p. 182
A_{49}	$O(n\log n)$	$F2\|p_{i,j} = b_{i,j}t, M_1 \prec M_2\|\sum C_j$	Sect. 8.2.1, p. 187
A_{50}	$O(n\log n)$	$F2\|p_{i,j} = b_{i,j}t, M_1 \succ M_2\|\sum C_j$	Sect. 8.2.1, p. 188

[a] $0 < b_{i,j} < 1$

Table 8.10: Polynomial Algorithms for Dedicated-Machine Time-Dependent Scheduling Problems (L_{max} Criterion)

Algorithm	Complexity	Problem	This book
A_{17}	$O(n \log n)$	$F2\|p_{i,j} = b_{i,j}(1 - kt)\|L_{max}$	Sect. 8.3.4, p. 194
A_{51}	$O(n^3 \log n)$	$Fm\|p_{i,j} = a_{i,j} + bt,$ $no\text{-}idle, idm\| \sum C_j$	Sect. 8.3.2, p. 191
A_{52}	$O(n^3 \log n)$	$Fm\|p_{i,j} = a_{i,j} + bt,$ $no\text{-}idle, ddm\| \sum C_j$	Sect. 8.3.2, p. 191
A_{53}	$O(n \log n)$	$Fm\|p_{i,j} = b_{i,j}(A + t), \delta\|L_{max}$ [a]	Sect. 8.3.2, p. 193

[a] $\delta \in \{idm; no\text{-}wait, idm; no\text{-}idle, idm\}$

Table 8.11: Polynomial Algorithms for Single-Machine Time-Dependent Problems (Criteria other than C_{max}, $\sum C_j$ and L_{max})

Algorithm	Complexity	Problem	This book
A_{26} [a]	$O(n \log n)$	$Fm\|p_{i,j} = b_{i,j}(1 - kt), b_{i,j} = b_j\| \sum U_j$	Sect. 8.4.3, p. 196
A_{29}	$O(n^2)$	$Fm\|p_{i,j} = b_{i,j}(1 - kt), b_{i,j} = b_j\|f_{max}$	Sect. 8.4.3, p. 195
A_{54}	$O(n \log n)$	$Fm\|p_{i,j} = b_{i,j}(A + t), \delta\| \sum w_j C_j$ [b]	Sect. 8.4.2, p. 195

[a] modified version
[b] $\delta \in \{idm; no\text{-}wait, idm; no\text{-}idle, idm\}$

Part III

ALGORITHMS

9

Approximation and heuristic algorithms

Computationally intractable time-dependent scheduling problems may be solved by different near-optimal algorithms. Therefore, the third part of the book is devoted to the main classes of such algorithms.

This part is composed of three chapters. In Chap. 9, we present approximation and heuristic algorithms, which construct the final schedule in a step-by-step manner. Greedy algorithms based on signatures of sequences of job deterioration rates are discussed in Chap. 10. Local search algorithms are presented in Chap. 11.

Chapter 9 is composed of five sections. In Sect. 9.1, we present approximation and heuristic algorithms for the $C_{\max}$ criterion. In Sects. 9.2 and 9.3, we present the algorithms for the $\sum C_j$ and $L_{\max}$ criteria, respectively. In Sect. 9.4, we discuss heuristic algorithms for criteria other than $C_{\max}$, $\sum C_j$ and $L_{\max}$. Concluding remarks and tables are given in Sect. 9.5.

9.1 Minimizing the maximum completion time

In this section, we present approximation and heuristic algorithms for the $C_{\max}$ criterion.

9.1.1 Proportional deterioration

Equal ready times and no deadlines

Single-machine problems. Recall that the problem of minimizing the $C_{\max}$ criterion for a single machine and proportionally deteriorating jobs is solvable in $O(n)$ time (cf. Theorem 6.1), while the problem with a single period of the machine non-availability is computationally intractable (cf. Theorem 6.11). Therefore, unless $\mathcal{P} = \mathcal{NP}$, only approximation or heuristic polynomial-time algorithms can be found for the latter problem (cf. Remark 3.7).

There exist various approaches to constructing polynomial-time heuristics for an intractable time-dependent scheduling problem. For example, we can adapt algorithms proposed for scheduling problems with fixed job processing times. We will show on examples that this approach fails in the case of time-dependent scheduling if job deterioration rates are unbounded.

A scheduling algorithm may be an offline (cf. Definition 2.17 (a)), an online (cf. Definition 2.17 (b)) or a semi-online algorithm (cf. Remark 2.18). The same classification will be applied to time-dependent scheduling algorithms. In the previous two parts of this book, we considered only offline time-dependent scheduling algorithms. In this chapter, we also present online and semi-online time-dependent scheduling algorithms.

As the first algorithm in the chapter, we consider an online algorithm, which we will call H_1. In this algorithm, as long as there are jobs to be scheduled, the first available job is assigned to the first available machine. (Notice that H_1 is an adaptation of the algorithm proposed by Graham [122] for scheduling jobs with fixed processing times.)

Ji et al. [155] applied algorithm H_1 to the computationally intractable problem $1, h_{11}|p_j = b_j t, nres|C_{\max}$ (cf. Theorem 6.11) and established the competitive ratio (cf. Definition 2.19) for the algorithm. (The symbols h_{11} and $nres$ are explained in Remark 6.10).

Theorem 9.1. (Ji et al. [155]) *Let* $W_{1,1} \geq t_0 \equiv S_{[1]}$ *denote the start time of the non-availability period in the problem* $1, h_{11}|p_j = b_j t, nres|C_{\max}$. *Then algorithm* H_1 *is* $\frac{W_{1,1}}{t_0}$*-competitive for the problem.*

Proof. Let $N_{\mathcal{J}_1}$ ($N_{\mathcal{J}_2}$) denote the set of indices of jobs scheduled before (after) the non-availability period in an optimal schedule. Let $C_{\max}^\star$ and $C_{\max}^{H_1}$ denote the length of the optimal schedule and the schedule constructed by H_1 algorithm, respectively.

If $N_{\mathcal{J}_2} = \emptyset$, then by (6.2) we have $C_{\max}^\star = C_{\max}^{H_1} = t_0 \prod_{j=1}^{n}(1 + b_j)$ and $\frac{C_{\max}^{H_1}}{C_{\max}^\star} = 1 \leq \frac{W_{1,1}}{t_0}$.

Assume that $N_{\mathcal{J}_2} \neq \emptyset$. Then

$$t_0 \prod_{j \in N_{\mathcal{J}_1}} (1 + b_j) \leq W_{1,1} \tag{9.1}$$

and

$$C_{\max}^\star = W_{1,2} \prod_{j \in N_{\mathcal{J}_2}} (1 + b_j), \tag{9.2}$$

where $W_{1,2} > W_{1,1}$ denotes the end time of the non-availability period.

From (9.2) it follows that

$$W_{1,2} \prod_{j=1}^{n}(1 + b_j) = C_{\max}^\star \prod_{j \in N_{\mathcal{J}_1}} (1 + b_j). \tag{9.3}$$

Hence, by (9.2) and (9.3), we have

$$C_{\max}^{H_1} \leq W_{1,2} \prod_{j=1}^{n} (1 + b_j). \tag{9.4}$$

From (9.4), by (9.1) and (9.3), it follows that $C_{\max}^{H_1} \leq \frac{W_{1,1}}{t_0} C_{\max}^{\star}$. ∎

Remark 9.2. Analysis of online algorithms is a new topic in time-dependent scheduling. In the classic scheduling (cf. Sect. 5.1), however, the matter has been studied since the mid-1960s and it has an extensive literature; see the references given in Sect. 2.3 for details.

For the problem $1, h_{11}|p_j = b_j t, nres|C_{\max}$, Ji et al. [155] also proposed an offline algorithm, H_2. The algorithm is an adaptation of another algorithm for scheduling jobs with fixed processing times, LPT, proposed by Graham [123]. In algorithm LPT, jobs first are arranged in a list in the non-increasing order of the processing times of the jobs. Next, as long as there are unscheduled jobs, the first available job from the list is assigned to the first available machine. In algorithm H_2, job processing times are replaced by job deterioration rates. The pseudo-code of the algorithm can be formulated as follows.

Algorithm H_2 for the problem $1, h_{11}|p_j = b_j t, nres|C_{\max}$ ([155])

Input: sequence $(b_1, b_2, \ldots, b_n)$
Output: a suboptimal schedule

▷ **Step 1:**
   Arrange jobs in the non-increasing order of b_j values;
▷ **Step 2:**
   Apply algorithm H_1 to the list of jobs obtained in Step 1.

Theorem 9.3. (Ji et al. [155]) *If $b_{\min} := \min_{1 \leq j \leq n}\{b_j\}$, then algorithm H_2 is an approximation algorithm for the problem $1, h_{11}|p_j = b_j t, nres|C_{\max}$ with the worst-case ratio $R_{H_2}^a = 1 + b_{\min}$ if $1 + b_{\min} \leq \frac{W_{1,1}}{t_0}$ and $R_{H_2}^a = 1$ otherwise.*

Proof. See [155, Lemmata 4–5, Theorem 4]. ◇

Parallel-machine problems. Algorithm H_1 has also been applied to the computationally intractable problem $Pm|p_j = bjt|C_{\max}$ (cf. Theorem 7.1). We will show now that if job deterioration rates in the problem are unbounded, algorithm H_1 can produce schedules which are arbitrarily bad.

Example 9.4. (Gawiejnowicz [89]) Consider the following instance I of the problem $P2|p_j = b_j t|C_{\max}$. Let $p_1 = p_2 = Kt, p_3 = K^2 t$ for some constant $K > 0$. Let both machines start at time $t_0 > 0$. Then $\frac{C_{\max}^{H_1}(I)}{C_{\max}^{\star}(I)} = \frac{K^2+1}{K+1} \to \infty$ as $K \to \infty$. ◆

Cheng and Sun [46] considered the performance of algorithm H_1 for the problem $Pm|p_j = b_j t|C_{\max}$ in the case when $m = 2$ and $b_j \in (0, 1\rangle$ for all j.

Theorem 9.5. (Cheng and Sun [46]) *If $b_j \in (0, 1\rangle$ for $1 \leq j \leq n$, then algorithm H_1 is $\sqrt{2}$-competitive for the problem $P2|p_j = b_j t|C_{\max}$.*

Proof. Let σ^{H_1} and σ^* denote the schedule constructed by algorithm H_1 and an optimal schedule for the considered problem, respectively. Let $N_{\mathcal{J}_1}$ and $N_{\mathcal{J}_2}$, where $N_{\mathcal{J}_1} \cap N_{\mathcal{J}_2} = \emptyset$ and $N_{\mathcal{J}_1} \cup N_{\mathcal{J}_2} = N_{\mathcal{J}} := \{1, 2, \ldots, n\}$, denote the set of indices of jobs assigned to machine M_1 and M_2, respectively. Then,

$$C_{\max}(\sigma^*) = \max \left\{ \prod_{j \in N_{\mathcal{J}_1}} (1 + b_j), \prod_{j \in N_{\mathcal{J}_2}} (1 + b_j) \right\} \geq \sqrt{\prod_{j=1}^{n}(1 + b_j)}. \quad (9.5)$$

Let J_k be the job that determines the maximum completion time in σ^*. Then,

$$S_k \leq \sqrt{\prod_{j=1, j \neq k}^{n} (1 + b_j)}. \quad (9.6)$$

Hence, by (9.5) and (9.6), we have

$$\frac{C_{\max}(\sigma^{H_1})}{C_{\max}(\sigma^*)} = \frac{S_k(1 + b_k)}{C_{\max}(\sigma^*)} \leq \frac{\sqrt{\prod_{j=1, j \neq k}^{n} (1 + b_j)(1 + b_k)}}{\sqrt{\prod_{j=1}^{n} (1 + b_j)}} \leq \sqrt{1 + b_k} \leq \sqrt{2}. \quad \blacksquare$$

Remark 9.6. Cheng and Sun stated without proof a similar result for an arbitrary m : if $b_j \in (0, 1\rangle$ for $1 \leq j \leq n$, then algorithm H_1 is $2^{\frac{m-1}{m}}$-competitive for the problem $Pm|p_j = b_j t|C_{\max}$; see [46, Theorem 2].   ◇

Remark 9.7. In the case when only the largest deterioration rate, $b_{\max} := \max_{1 \leq j \leq n} \{b_j\}$, is known in advance, algorithm H_1 is $(1 + b_{\max})^{1 - \frac{1}{m}}$-competitive for the problem $Pm|p_j = b_j t, b_j \leq b_{\max}|C_{\max}$; see [46, Theorem 6].   ◇

Cheng and Sun also proposed a semi-online algorithm (cf. Remark 2.18) for the problem $P2|p_j = b_j t|C_{\max}$ in the case when only the largest possible deterioration rate, $b_{\max}$, is known in advance.

Algorithm H_3 for the problem $P2|p_j = b_j t, b_j \leq b_{\max}|C_{\max}$ ([46])

Input: sequence $(b_1, b_2, \ldots, b_n)$; numbers $t_0, b_{\max}$
Output: a suboptimal schedule

▷ Step 1:
$\quad L \leftarrow t_0;$
$\quad b \leftarrow$ deterioration rate of the current job;

▷ **Step 2:**
  while $((b \neq b_{max}) \wedge ((1 + b)L > (1 + b_{max})^2))$ **do**
     Assign the current job to machine M_1;
     $L \leftarrow (1 + b)L$;
     $b \leftarrow$ deterioration rate of the current job;
▷ **Step 3:**
  Assign the current job to machine M_2;
▷ **Step 4:**
  Apply algorithm H_1 to all subsequent jobs.

Theorem 9.8. (Cheng and Sun [46]) *If $b_j \in (0, 1)$ for $1 \leq j \leq n$ and $b_{max} := \max_{1 \leq j \leq n}\{b_j\}$, then algorithm H_3 is $\sqrt{1 + b_{max}}$-competitive for the problem $P2|p_j = b_j t, b_j \leq b_{max}|C_{max}$.*

Proof. See [46, Theorem 5]. ◇

For the problem $Pm|p_j = b_j t|C_{max}$, Mosheiov [219] proposed to use algorithm H_2. Unfortunately, if deterioration rates are unbounded, in this case also, the algorithm can produce arbitrarily bad schedules.

Example 9.9. (Mosheiov [219]) Consider the following instance I of the problem $P2|p_j = b_j t|C_{max}$: $n = 5$, $b_1 = b_2 = K^{\frac{1}{2}} - 1$, $b_3 = b_4 = b_5 = K^{\frac{1}{3}} - 1$ for some constant $K > 1$. Then $R^a_{H_2}(I) = \frac{C^{H_2}_{max}(I)}{C^*_{max}(I)} = \frac{K^{\frac{7}{6}}}{K} = K^{\frac{1}{6}} \to \infty$ as $K \to \infty$. ◆

Examples 9.4 and 9.9 show that well-known scheduling algorithms for jobs with fixed processing times are a risky choice for problems with unbounded deterioration. The situation changes if we bound job deterioration rates.

Lemma 9.10. (Hsieh and Bricker [141]) *If I is an arbitrary instance of the problem $Pm|p_j = b_j t|C_{max}$, then*

$$C^{H_2}_{max}(I \setminus \{J_n\}) \leq \left(\prod_{j=1}^{n-1}(1 + b_j)\right)^{\frac{1}{m}}.$$

Proof. See [141, Proposition 1 (b)]. ◇

Theorem 9.11. (Hsieh and Bricker [141]) *Let I be an arbitrary instance of the problem $Pm|p_j = b_j t|C_{max}$. If $b_j \in (0, 1)$ for $1 \leq j \leq n$ and job J_n is assigned to the machine whose maximum completion time determines the overall maximum completion time, then*

$$R^a_{H_2}(I) \leq (1 + b_n)^{1 - \frac{1}{m}}.$$

Proof. Let I be an arbitrary instance of the problem $Pm|p_j = b_j t|C_{\max}$. First, by Lemma 7.2 we have $C^*_{\max}(I) \geq \left(\prod_{j=1}^{n}(1 + b_j)\right)^{\frac{1}{m}}$. Second, there holds the equality $C^{H_2}_{\max}(I) = C^{H_2}_{\max}(I \setminus \{J_n\})(1 + b_n)$. Finally, by Lemma 9.10, $C^{H_2}_{\max}(I \setminus \{J_n\}) \leq \left(\prod_{j=1}^{n-1}(1 + b_j)\right)^{\frac{1}{m}}$. Therefore, $R^a_{H_2}(I)$ is not greater than

$$\frac{C^{H_2}_{\max}(I \setminus \{J_n\})(1 + b_n)}{\left(\prod_{j=1}^{n}(1 + b_j)\right)^{\frac{1}{m}}} \leq \frac{\left(\prod_{j=1}^{n-1}(1 + b_j)\right)^{\frac{1}{m}}(1 + b_n)}{\left(\prod_{j=1}^{n}(1 + b_j)\right)^{\frac{1}{m}}} = (1 + b_n)^{1 - \frac{1}{m}}.$$

■

Theorem 9.12. (Hsieh and Bricker [141]) *Let I be an arbitrary instance of the problem $Pm|p_j = b_j t|C_{\max}$. If $b_j \in (0, 1)$ for $1 \leq j \leq n$, and if job J_k, $1 < k < n$, is assigned to the machine whose maximum completion time determines the overall maximum completion time, then*

$$R^a_{H_2}(I) \leq (1 + b_k)^{1 - \frac{1}{m}}(1 + b_n)^{-\frac{n-k}{m}}.$$

Proof. Let I be an arbitrary instance of the problem $Pm|p_j = b_j t|C_{\max}$. First, there holds the equality $C^{H_2}_{\max}(I) = C^{H_2}_{\max}(I \setminus \{J_k\})(1 + b_k)$. Second, by Lemma 9.10, we have $C^{H_2}_{\max}(I \setminus \{J_n\}) \leq \left(\prod_{j=1}^{k-1}(1 + b_j)\right)^{\frac{1}{m}}$. Hence,

$$R^a_{H_2}(I) \leq \frac{C^{H_2}_{\max}(I \setminus \{J_k\})(1 + b_k)}{\left(\prod_{j=1}^{n}(1 + b_j)\right)^{\frac{1}{m}}} \leq \frac{\left(\prod_{j=1}^{k-1}(1 + b_j)\right)^{\frac{1}{m}}(1 + b_k)}{\left(\prod_{j=1}^{n}(1 + b_j)\right)^{\frac{1}{m}}} =$$

$$(1 + b_k)^{1 - \frac{1}{m}}\left(\prod_{k+1}^{n}(1 + b_j)\right)^{-\frac{1}{m}} \leq (1 + b_k)^{1 - \frac{1}{m}}(1 + b_n)^{-\frac{n-k}{m}}.$$

■

Remark 9.13. In the formulation of Theorem 9.12, one can assume $1 < k \leq n$ instead of $1 < k < n$. The new formulation covers formulations of both Theorem 9.11 and Theorem 9.12.

Remark 9.14. From Theorem 9.12 it follows that if the values of b_j are uniformly distributed in the $(0, 1)$ interval, then $\lim_{n \to \infty} \frac{C_{\max}(H_2)}{C^*_{\max}} = 1$, i.e., algorithm H_2 is asymptotically optimal.

Hsieh and Bricker [141] conducted a computational experiment in which instances with up to $n = 500$ jobs were solved by algorithm H_2. The results of the experiment confirmed the observation from Remark 9.14.

Fully polynomial-time approximation schemata

Single-machine problems. Ji et al. [155] proposed an FPTAS for the problem $1, h_{11}|p_j = b_j t, nres|C_{\max}$. The main idea of the scheme is to construct from a given instance of the problem an instance of the KP problem (cf. Sect. 3.2) and apply a standard FPTAS for the KP problem. The proposed FPTAS runs in $O(n^2 \epsilon^{-1})$ time; see [155, Theorem 7] for details.   ◇

Parallel-machine problems. Ren and Kang [243] proposed an FPTAS (cf. Definition 2.16) for the problem $Pm|p_j = b_j t|C_{\max}$. The scheme, based on the same idea as the scheme proposed by Kovalyov and Kubiak [176], is a modification of the FPTAS proposed by Kang et al. [160] for the problem $Pm|p_j = a_j + b_j t|C_{\max}$ (cf. Sect. 9.1.2).

In the modified FPTAS, the variables x_j, $1 \le j \le n$, and the set X are defined as those from Sect. 9.1.2. By Theorem 6.1 we can assume that all jobs, available starting from time $t_0 > 0$, have been indexed in an arbitrary way. Definitions of the functions F_j^i and $Q(x)$ are as follows: $F_0^i(x) := t_0$ for $1 \le i \le m$, $F_j^i(x) := F_{j-1}^i(x) + b_j F_{j-1}^i(x)$ for $i = x_j$, $F_j^i(x) := F_{j-1}^i(x)$ for $i \ne x_j$ and $Q(x) := \max\{F_n^i(x) : 1 \le i \le m\}$. The remaining parts of this FPTAS are organized similarly to those in the FPTAS from Sect. 9.1.2.

For $L := \log \max\{n, \frac{1}{\epsilon}, 1 + b_{\max}, S_1\}$, where $b_{\max} := \max_{1 \le j \le n}\{b_j\}$, the modified scheme runs in $O(n^{2m+1} L^{m+1} \epsilon^{-m})$ time. We refer the reader to [243, Sect. 3] for more details on this FPTAS.   ◇

Remark 9.15. Ren and Kang [243] also proposed an FPTAS for the problem $P2|p_j = b_j t|C_{\max}$, running in $O(n^5 L^3 \epsilon^{-2})$ time; see [243, Sect. 2].   ◇

9.1.2 Linear deterioration

Equal ready times and no deadlines

Parallel-machine problems. Hsieh and Bricker [141] proposed three heuristic algorithms for the problem $Pm|p_j = a_j + b_j t|C_{\max}$, where $b_j \in (0, 1)$ for $1 \le j \le n$. All of these algorithms are adaptations of algorithm H_2.

The first heuristic algorithm exploits Theorem 6.24.

Algorithm H_4 for the problem $Pm|p_j = a_j + b_j t|C_{\max}$ ([141])

Input: sequences $(a_1, a_2, \ldots, a_n)$, $(b_1, b_2, \ldots, b_n)$
Output: a suboptimal schedule

▷ Step 1:
   Arrange jobs in the non-increasing order of $\frac{a_j}{b_j}$ ratios;
▷ Step 2:
   for $i \leftarrow 1$ to n do
      Assign job $J_{[i]}$ to the machine with the smallest completion time.

The second heuristic algorithm uses slightly another rule of jobs ordering.

Algorithm H_5 for the problem $Pm|p_j = a_j + b_j t|C_{\max}$ ([141])

Input: sequences $(a_1, a_2, \ldots, a_n)$, $(b_1, b_2, \ldots, b_n)$
Output: a suboptimal schedule

▷ **Step 1:**
   Arrange jobs in the non-increasing order of $\frac{1-b_j}{a_j}$ ratios;
▷ **Step 2:**
   for $i \leftarrow 1$ **to** n **do**
      Assign job $J_{[i]}$ to the machine with the smallest machine load.

The third heuristic algorithm contains a random step and is based on the following observation. The optimal schedule for the problem $Pm|p_j = a_j + b_j t|C_{\max}$ is often very similar to the schedules generated by algorithm H_5. Since H_5 assigns jobs with smaller values of the ratio $\frac{a_j}{b_j}$ before those with larger ratios, it seems reasonable to choose the jobs to be assigned with a probability inversely proportional to the ratio $\frac{a_j}{b_j}$.

Algorithm H_6 for the problem $Pm|p_j = a_j + b_j t|C_{\max}$ ([141])

Input: sequences $(a_1, a_2, \ldots, a_n)$, $(b_1, b_2, \ldots, b_n)$
Output: a suboptimal schedule

▷ **Step 1:**
   $N_{\mathcal{J}} \leftarrow \{1, 2, \ldots, n\}$;
   $Z_0 \leftarrow 0$;
   $Z_{n+1} \leftarrow 1$;
   $j \leftarrow 1$;
▷ **Step 2:**
   while $(j \leq n)$ **do**
      $tmp \leftarrow \sum\limits_{k \in N_{\mathcal{J}}} \frac{b_k}{a_k}$;
      for all $i \in N_{\mathcal{J}}$ **do**
      $z_i \leftarrow \frac{b_i}{a_i * tmp}$;
      $Z_i \leftarrow \sum\limits_{\substack{j \in N_{\mathcal{J}} \\ j \leq i}} z_j$;
      Generate a random number $r \in (0, 1)$;
      if $((Z_{i-1} < r \leq Z_i) \wedge (i \in N_{\mathcal{J}}))$ **then** $T_j \leftarrow i$
      else if $((Z_{i-1} < r \leq Z_i) \wedge (i = n+1))$ **then** $T_j \leftarrow n+1$;
      $N_{\mathcal{J}} \leftarrow N_{\mathcal{J}} \setminus \{i\}$;
      $j \leftarrow j + 1$
▷ **Step 3:**
   Rearrange jobs according to table T.

Computational experiments conducted for $n = 10$ and $n = 15$ jobs show that for these values of n, algorithms H_4, H_5 and H_6 generate schedules that are close to optimal ones; see [141, Sect. 5].

Dedicated-machine problems. Lee et al. [196] proposed four heuristics for the problem $F2|p_{i,j} = a_{i,j} + bt|C_{max}$. The first of them is algorithm A_{39}. The next two algorithms are in the form of $H_7 : (a_{1,j}, a_{2,j}, b) \mapsto (a_{1,j}) \nearrow$ and $H_8 : (a_{1,j}, a_{2,j}, b) \mapsto (a_{2,j}) \nearrow$, respectively. All these heuristic algorithms run in $O(n \log n)$. The fourth heuristic algorithm, H_9, is composed of two phases. In the first phase, one of the above three heuristics is applied to obtain an initial schedule. In the second phase, this schedule is improved by all possible exchanges of two jobs. This algorithm runs in $O(n^2)$ time.

The results of computational experiments suggest that the best schedules are generated by heuristic H_9, if the initial schedule was generated by heuristic H_8; we refer the reader to [196, Sect. 4–5] for more details.

Fully polynomial-time approximation schemata

Parallel-machine problems. Kang and Ng [161] proposed an FPTAS (cf. Definition 2.16) for the problem $Pm|p_j = a_j + b_j t|C_{max}$. The scheme, based on the same idea as the scheme proposed by Kovalyov and Kubiak [176], is a modification of the FPTAS proposed by Kang et al. [160] for the problem $Pm|p_j = a_j - b_j t|C_{max}$ (cf. Sect. 9.1.4).

In the modified FPTAS, the variables x_j, $1 \leq j \leq n$, and the set X are defined as those from Sect. 9.1.4. By Theorem 6.24 we can assume that all jobs have been indexed in such a way that $\frac{a_1}{b_1} \leq \frac{a_2}{b_2} \leq \ldots \leq \frac{a_n}{b_n}$. Definitions of the functions F_j^i and $Q(x)$ are as follows: $F_0^i(x) := 0$ for $1 \leq i \leq m$, $F_j^i(x) := F_{j-1}^i(x) + a_j + b_j F_{j-1}^i(x)$ for $i = x_j$, $F_j^i(x) := F_{j-1}^i(x)$ for $i \neq x_j$ and $Q(x) := \max\{F_n^i(x) : 1 \leq i \leq m\}$. The remaining parts of this FPTAS are organized similarly to those in the FPTAS presented in Sect. 9.1.4.

For $L := \log \max\{n, \frac{1}{\epsilon}, a_{max}, 1 + b_{max}\}$, where $a_{max} := \max_{1 \leq j \leq n}\{a_j\}$ and $b_{max} := \max_{1 \leq j \leq n}\{b_j\}$, the modified scheme runs in $O(n^{2m+1} L^{m+1} \epsilon^{-m})$ time. We refer the reader to [161, Sect. 2] for more details on this FPTAS.   ◇

Distinct ready times and deadlines

Lee et al. [194] proposed two heuristics for the problem $1|p_j = a_j + bt, r_j|C_{max}$. Both these algorithms run in $O(n^2)$ time and are composed of two phases.

The first of these algorithms constructs a schedule by selecting at each stage the job with the smallest completion time. The initial schedule constructed in the first phase, is iteratively improved in the second phase.

For a given $1 < k < n$, let $\tau(J_{[k]} \to J_{[l]})$ denote a schedule τ in which job $J_{[k]}$ has been moved immediately before job $J_{[l]}$, while all other jobs are scheduled as in τ. The pseudo-code of the first algorithm is as follows.

Algorithm H_{10} for the problem $1|p_j = a_j + bt, r_j|C_{\max}$ ([194])

Input: sequences $(a_1, a_2, \ldots, a_n)$, $(r_1, r_2, \ldots, r_n)$, number b
Output: a suboptimal schedule σ

▷ Step 1:
  $k \leftarrow 1$;
  $C_{[0]} \leftarrow 0$;
  $r_{\max} \leftarrow \max\limits_{1 \le j \le n} \{r_j\}$;
  $N_{\mathcal{J}} \leftarrow \{1, 2, \ldots, n\}$;
  $\sigma \leftarrow (\phi)$;
▷ Step 2: Phase I
  while $(k < n)$ **do**
      if $C_{[k-1]} < r_{\max}$ **then** Choose $i \in N_{\mathcal{J}}$ with the smallest value of
                      $\hookrightarrow (1 + b) \max\{r_i, C_{[k-1]}\} + a_i$
      else Choose $i \in N_{\mathcal{J}}$ with the smallest a_i;
      $\sigma_k \leftarrow i$;
      $C_{[k]} \leftarrow (1 + b) \max\{r_i, C_{[k-1]}\} + a_i$;
      $N_{\mathcal{J}} \leftarrow N_{\mathcal{J}} \setminus \{i\}$;
      $k \leftarrow k + 1$;
  $\sigma = (\sigma_1, \sigma_2, \ldots, \sigma_{n-1}, N_{\mathcal{J}})$;   ▷ initial schedule σ
▷ Step 3: Phase II
  $\tau \leftarrow \sigma$;
  $l \leftarrow 1$;
  $k \leftarrow l + 1$;
  while $(k < n)$ **do**
      $\sigma \leftarrow \tau(J_{[k]} \rightarrow J_{[l]})$;
      if $C_{\max}(\sigma) < C_{\max}(\tau)$ **then** $\tau \leftarrow \sigma$;
      if $k < n$ **then** $k \leftarrow k + 1$
      else $l \leftarrow l + 1$;
▷ Step 4:
  return σ.

The second heuristic also generates the final schedule in two phases. The algorithm, however, in a slightly different way constructs the schedule generated in the first phase. The pseudo-code of the algorithm is as follows.

Algorithm H_{11} for the problem $1|p_j = a_j + bt, r_j|C_{\max}$ ([194])

Input: sequences $(a_1, a_2, \ldots, a_n)$, $(r_1, r_2, \ldots, r_n)$, number b
Output: a suboptimal schedule σ

▷ Step 1:
  $k \leftarrow 1$;
  $C_{[0]} \leftarrow 0$;

$r_{\max} \leftarrow \max\limits_{1 \leq j \leq n} \{r_j\}$;

$N_{\mathcal{J}} \leftarrow \{1, 2, \ldots, n\}$;

$\sigma \leftarrow (\phi)$;

▷ Step 2: Phase I

   while $(k < n)$ **do**

      if $C_{[k-1]} < r_{\max}$ **then** $A \leftarrow \{i \in N_{\mathcal{J}} : r_i < C_{[k-1]}\}$;

      if $(A \neq \emptyset)$ **then** Choose $i \in A$ with the smallest a_i;

$$C_{[k]} \leftarrow (1 + b)C_{[k-1]} + a_i$$

      else Choose $i \in N_{\mathcal{J}}$ with the smallest r_i;

        $C_{[k]} \leftarrow (1 + b)r_i + a_i$;

      $\sigma_k \leftarrow i$;

      $N_{\mathcal{J}} \leftarrow N_{\mathcal{J}} \setminus \{i\}$;

      $k \leftarrow k + 1$;

   $\sigma = (\sigma_1, \sigma_2, \ldots, \sigma_{n-1}, N_{\mathcal{J}})$;   ▷ initial schedule σ

▷ Step 3: Phase II

   $\tau \leftarrow \sigma$;

   $l \leftarrow 1$;

   $k \leftarrow l + 1$;

   while $(k < n)$ **do**

      $\sigma \leftarrow \tau(J_{[k]} \rightarrow J_{[l]})$;

      if $C_{\max}(\sigma) < C_{\max}(\tau)$ **then** $\tau \leftarrow \sigma$;

      if $k < n$ **then** $k \leftarrow k + 1$

      else $l \leftarrow l + 1$;

▷ Step 4:

   return σ.

The performance of algorithms H_{10} and H_{11} was tested on instances with $12 \leq n \leq 28$ jobs. The results of the computational experiment suggest that on average the algorithms produce schedules of satisfactory quality; see [194, Sect. 5] for details.

9.1.3 General non-linear deterioration

Equal ready times and no deadlines

Single-machine problems. Gupta and Gupta [128] proposed two heuristics for a single machine time-dependent scheduling problem with job processing times in the form of (6.23), defined by a polynomial of $m \geq 2$ degree. Both heuristics are based on the calculation of a set of ratios.

In the first algorithm, the ratios are independent of time. Since the processing times of jobs are polynomials of $m \geq 2$ degree, the algorithm first calculates m different ratios in a loop. Next, it generates m different schedules by arranging jobs in the non-increasing order of the j-th ratio values, $1 \leq j \leq m$. The pseudo-code of the algorithm can be formulated as follows.

Algorithm H_{12} for the problem $1|p_j = a_j + b_j t + \ldots + n_j t^m|C_{\max}$ ([128])

Input: sequences $(a_j, b_j, \ldots, n_j), 1 \le j \le n$
Output: a suboptimal schedule

▷ Step 1:
$N_{\mathcal{J}} \leftarrow \{1, 2, \ldots, n\}$;
▷ Step 2:
for each $j \in N_{\mathcal{J}}$ do
$q_j(a) \leftarrow a_j$;
$q_j(b) \leftarrow \frac{a_j}{b_j}$;

$\vdots \quad \vdots \quad \vdots$

$q_j(m) \leftarrow \frac{a_j}{m_j}$;
▷ Step 3:
Generate m schedules by arranging jobs in the non-increasing order
$\hookrightarrow$ of the $q_j(x)$ values for $a \le x \le m$;
▷ Step 4:
Select the schedule with the minimal completion time.

In the second algorithm, the calculated parameters $q_j(x)$ are functions of time. We formulate this algorithm for the case of *quadratic* job processing times, $p_j = a_j + b_j t + c_j t^2$ for $1 \le j \le n$.

Algorithm H_{13} for the problem $1|p_j = a_j + b_j t + c_j t^2|C_{\max}$ ([128])

Input: sequences $(a_j, b_j, c_j), 1 \le j \le n$
Output: a suboptimal schedule σ

▷ Step 1:
$\sigma \leftarrow (\phi)$;
$N_{\mathcal{J}} \leftarrow \{1, 2, \ldots, n\}$;
$T \leftarrow 0$;
▷ Step 2:
while $(N_{\mathcal{J}} \ne \emptyset)$ do
for each $j \in N_{\mathcal{J}}$ do
$q_j(a) \leftarrow \frac{a_j}{b_j + c_j T}$;
$q_j(b) \leftarrow \frac{a_j + b_j T + c_j T^2}{b_j + 2c_j T}$;
$q_j(c) \leftarrow a_j + b_j T + c_j T^2$;
Find $k \in N_{\mathcal{J}}$ such that $q_k(x) = \min\{q_j(x) : j \in N_{\mathcal{J}} \wedge x \in \{a, b, c\}\}$;
$\sigma \leftarrow (\sigma|k)$;
$N_{\mathcal{J}} \leftarrow N_{\mathcal{J}} \setminus \{k\}$;
$T \leftarrow T + C_k$;
▷ Step 3:
return σ.

The performance of algorithms H_{12} and H_{13} was tested on small instances with $4 \leq n \leq 7$ jobs. Since the heuristics were found to be unpredictable in producing schedules close to the optimal one, a technique for improvement of the final schedule has also been proposed; see [128, Tables 6–7].

The problem $1|p_j = f_j(t)|C_{\max}$ with general non-linear processing times was also considered by Alidaee [4]. On the basis of Theorem 6.47, Alidaee proposed the following algorithm.

Algorithm H_{14} for the problem $1|p_j = f_j(t)|C_{\max}$ ([4])

Input: sequence $(f_1, f_2, \ldots, f_n)$
Output: a suboptimal schedule σ

▷ Step 1:
$T \leftarrow 0$;
$N_{\mathcal{J}} \leftarrow \{1, 2, \ldots, n\}$;
$K \leftarrow n$;
$\sigma \leftarrow (\phi)$;
▷ Step 2:
while $(N_{\mathcal{J}} \neq \emptyset)$ **do**
$\quad z \leftarrow T + \frac{1}{2} K \sum\limits_{k \in N_{\mathcal{J}}} f_k(T)$;

$\quad$ Choose the smallest $i \in N_{\mathcal{J}}$ such that $\dfrac{f_i'(z)}{f_i(T)} = \max\limits_{j \in N_{\mathcal{J}}} \left\{ \dfrac{f_j'(z)}{f_j(T)} \right\}$;

$\quad \sigma \leftarrow (\sigma|i)$;
$\quad T \leftarrow T + f_i(T)$;
$\quad N_{\mathcal{J}} \leftarrow N_{\mathcal{J}} \setminus \{i\}$;
$\quad K \leftarrow K - 1$;
▷ Step 3:
return σ.

Algorithm H_{14} has been tested on two sets of instances. In the first set of instances, the job processing times were quadratic, i.e. $p_j = a_j + b_j t + c_j t^2$ for $1 \leq j \leq n$, where $5 \leq n \leq 8$ (see [4, Table 1]). In the second set, the job processing times were exponential, $p_j = e^{a_j t}$ for $1 \leq j \leq n$, where $5 \leq n \leq 9$. Twenty instances were generated for each value of n.

For quadratic processing times, algorithm H_{14} gave better results than algorithm H_{13}. However, the results became worse for a polynomial of degree higher than two, and for the case when the coefficients $a_j, b_j, \ldots, n_j$ were considerably different in size for $1 \leq j \leq n$.

Another algorithm for the problem with exponential job processing times, $p_j = e^{a_j t}$ for $1 \leq j \leq n$, has been proposed by Hsieh [140]. Before we formulate the algorithm, we will state an auxiliary result.

Lemma 9.16. (Hsieh [140]) *Let function $h(t)$ be in the form of*

$$h(t) \equiv f_{j+1}(t) + f_j(t + f_{j+1}(t)) - f_j(t) - f_{j+1}(t + f_j(t)),$$

where $f_j(t) := e^{a_j t}$. Then the equation $h(t) = 0$
(a) *has exactly one solution in* $(0, \infty)$ *for* $a_i \in (0, a_j)$, *where* $a_j \in (0, 1)$;
(b) *has no solution for* $t \in [0, 1]$.

Proof. (a) See [140, Lemma 2].
    (b) See [140, Lemma 4].   ◇

On the basis of Lemma 9.16 and some other results (see [140, Sect. 3]), Hsieh proposed the following algorithm.

$$\textbf{Algorithm } H_{15} \textbf{ for the problem } 1|p_j = e^{a_j t}|C_{\max} \text{ ([140])}$$

Input: sequence $(a_1, a_2, \ldots, a_n)$
Output: a suboptimal schedule σ

▷ Step 1:
  Arrange jobs in the non-decreasing order of a_j values;
  $N_{\mathcal{J}} \leftarrow \{1, 2, \ldots, n\}$;
▷ Step 2:
  while $(N_{\mathcal{J}} \neq \emptyset)$ **do**
    for $i \leftarrow 1$ **to** n **do**
      for $j \leftarrow 1$ **to** n **do**
        $T_{a_{[i]}, a_{[j]}} \leftarrow$ solution of the equation $h(t) = 0$; ▷ cf. Lemma 9.16
    $\sigma \leftarrow (n, n-1)$;
    $N_{\mathcal{J}} \leftarrow N_{\mathcal{J}} \setminus \{n, n-1\}$;
    $T \leftarrow 1 + e^{a_{n-1}}$;
    if $(T < T_{a_{[i]}, a_{[j]}}$ for all $i, j \in N_{\mathcal{J}})$ **then** $k = \arg \max_{i \in N_{\mathcal{J}}} \{a_{[i]}\}$
    else $k \leftarrow p$, where $(p, q) = \arg \min_{(i,j)} \{T - T_{a_{[i]}, a_{[j]}} | T > T_{a_{[i]}, a_{[j]}}\}$;
    $T \leftarrow T + f_k(T)$;
    $N_{\mathcal{J}} \leftarrow N_{\mathcal{J}} \setminus \{i\}$;
▷ Step 3:
  return σ.

For exponential job processing times given by (6.40), Janiak and Kovalyov [146] proposed three heuristics. Applying the notation introduced in Remark 6.29, the heuristics can be denoted as $H_{16} : (a_j|b_j|r_j|d_j) \mapsto (r_j \nearrow)$, $H_{17} : (a_j|b_j|r_j|d_j) \mapsto (r_j \nearrow |b_j \searrow)$ and $H_{18} : (a_j|b_j|r_j|d_j) \mapsto (r_j \nearrow |a_j \searrow)$. According to the authors (see [146, Sect. 4]), algorithm H_{16} gives, on average, the best results.

Distinct ready times and deadlines

Single-machine problems. For step deteriorating processing times given by (6.27), Mosheiov [217] proposed the following heuristic algorithm.

Algorithm H_{19} for the problem $1|p_j \equiv (6.27)|C_{\max}$ ([217])

Input: sequences $(a_1, a_2, \ldots, a_n)$, $(b_1, b_2, \ldots, b_n)$, $(d_1, d_2, \ldots, d_n)$
Output: a suboptimal schedule σ

▷ Step 1:
 $L \leftarrow \emptyset$;
 $E \leftarrow \{1, 2, \ldots, n\}$;
 Arrange jobs in the non-increasing order of d_j values;
 Schedule jobs in the obtained order;

▷ Step 2:
 while (there exist $j \in E$ such that $S_j > d_j$) **do**
  $k \leftarrow \min\{j \in E : S_j > d_j\}$;
  $l \leftarrow \arg\min\{\frac{b_j - a_j}{a_j} : j \in E, j \le k, a_j \ge S_k - d_k\}$;
  $L \leftarrow L \cup \{l\}$;
  $E \leftarrow E \setminus \{l\}$;

▷ Step 3:
 $\sigma \leftarrow (E|L)$;
 return σ.

Algorithm H_{19} is an adaptation of Moore-Hodgson's algorithm for the problem $1||\sum U_j$ (cf. [213]). Though in the worst case algorithm H_{19} may produce arbitrarily bad schedules (see [217, Example 1]), the results of the computational experiment reported in [217, Sect. 2.3] suggest that on average the performance of the algorithm is quite satisfactory.

For multi-step deteriorating processing times given by (6.34), Mosheiov proposed the following heuristic, which is an adaptation of algorithm H_{19}. For simplicity of presentation, we formulate the algorithm for the case of two-step deterioration.

Algorithm H_{20} for the problem $1|p_j \equiv (6.34)|C_{\max}$ ([217])

Input: sequences $(a_1^1, a_2^1, \ldots, a_n^1)$, $(a_1^2, a_2^2, \ldots, a_n^2)$,
     $(d_1^1, d_2^1, \ldots, d_n^1)$, $(d_1^2, d_2^2, \ldots, d_n^2)$,
Output: a suboptimal schedule σ

▷ Step 1:
 $L^1 \leftarrow \emptyset$;
 $E^1 \leftarrow \{1, 2, \ldots, n\}$;
 $L^2 \leftarrow \emptyset$;
 $E^2 \leftarrow \{1, 2, \ldots, n\}$;
 Arrange jobs in the non-increasing order of d_j^1 values;
 $T \leftarrow 0$; ▷ **the current time**

▷ Step 2:

  while (there exist $j \in E^1$ such that $S_j > d_j^1$) **do**

    $l_1 \leftarrow \min\{j \in E^1 : S_j > d_j^1\}$;

    $l_2 \leftarrow \arg \min\{\frac{a_j^2 - a_j^1}{a_j^1} : j \in E^1, j \leq k, a_j^1 \geq S_{l_1} - d_{l_1}^1\}$;

    $L^1 \leftarrow L^1 \cup \{l_1\}$;

    $E^1 \leftarrow E^1 \setminus \{l_1\}$;

    $T \leftarrow T + C_{l_1}$;

▷ Step 3:

  Arrange jobs in L^1 in the non-decreasing order of a_j^2 values;

  Call the obtained sequence τ;

  Starting at time T, schedule jobs in L^1 according to τ;

  while (there exist $j \in E^2$ such that $S_j > d_j^2$) **do**

    $l_1 \leftarrow \min\{j \in E^2 : S_j > d_j^2\}$;

    $l_2 \leftarrow \arg \min\{\frac{a_j^3 - a_j^2}{a_j^2} : j \in E^2, j \leq l_1, a_j^2 \geq S_{l_1} - d_{l_1}^2\}$;

    $L^2 \leftarrow L^2 \cup \{l_1\}$;

    $E^2 \leftarrow E^2 \setminus \{l_2\}$;

▷ Step 4:

  $\sigma \leftarrow (E^1|(L^1 \setminus L^2)|L^2)$;

  return σ.

The reported results of computational experiments (see [217, Sect. 4.2]) suggest that on average the schedules generated by algorithm H_{20} are about 10% worse than the optimal ones.

For job processing times given by (6.37), Kunnathur and Gupta [178] proposed five heuristic algorithms. All the algorithms construct the final schedule iteratively, but they differently select the job to be added to the subschedule constructed so far.

The first heuristic algorithm proposed by the authors iteratively chooses the job with the minimal value of the function L, where $L(\sigma|k) := C_{\max}(\sigma|k) + \sum_{j \in N_{\mathcal{J}}}(a_j + \max\{0, b_j(C_{\max}(\sigma|j) - d_j)\})$ for $\sigma \in \hat{\mathbb{S}}_n$ and $k \in \{1, 2, \ldots, n\}$.

Algorithm H_{21} for the problem $1|p_j \equiv (6.37)|C_{\max}$ ([178])

Input: sequences $(a_j, b_j, d_j), 1 \leq j \leq n$

Output: a suboptimal schedule σ

▷ Step 1:

  $\sigma \leftarrow (\phi)$;

  $N_{\mathcal{J}} \leftarrow \{1, 2, \ldots, n\}$;

  $T \leftarrow 0$;

▷ Step 2:

  while ($N_{\mathcal{J}} \neq \emptyset$) **do**

    Find $j \in N_{\mathcal{J}}$ such that $L(\sigma|j) = \min\{L(\sigma|k) : k \in N_{\mathcal{J}}\}$;

    ▷ Break ties by using the smallest d_k, then the largest b_k

    ▷ ↪ and then the smallest a_k

$\sigma \leftarrow (\sigma|j)$;
$N_{\mathcal{J}} \leftarrow N_{\mathcal{J}} \setminus \{j\}$;
$T \leftarrow T + p_j$;
▷ Step 3:
return σ.

The second heuristic algorithm iteratively chooses the job with the minimal value of the ratio $\frac{a_j}{b_j}$.

Algorithm H_{22} for the problem $1|p_j \equiv (6.37)|C_{\max}$ ([178])

Input: sequences $(a_j, b_j, d_j), 1 \le j \le n$
Output: a suboptimal schedule σ

▷ Step 1:
$\sigma \leftarrow (\phi)$;
$N_{\mathcal{J}} \leftarrow \{1, 2, \ldots, n\}$;
$T \leftarrow 0$;
▷ Step 2:
while $(N_{\mathcal{J}} \ne \emptyset)$ do
 repeat
  $L \leftarrow \{k \in N_{\mathcal{J}} : d_k < T\}$;
  $F \leftarrow N_{\mathcal{J}} \setminus L$;
  Find $j \in F$ such that $\frac{a_j}{b_j} = \min\limits_{k \in F}\{\frac{a_k}{b_k}\}$;
  ▷ Break ties by selecting the job with the smallest d_k
  $\sigma \leftarrow (\sigma|j)$;
  $N_{\mathcal{J}} \leftarrow N_{\mathcal{J}} \setminus \{j\}$;
  $T \leftarrow T + p_j$;
 until $(F = \emptyset)$
 Arrange L in the non-decreasing order of $\frac{a_j - d_j b_j}{b_j}$ ratios;
 $\sigma \leftarrow (\sigma|L)$;
 $T \leftarrow T + \sum\limits_{i \in L} p_i$;
▷ Step 3:
return σ.

The remaining three heuristics proposed in [178] differ from algorithm H_{22} only in Step 2 (see [178, Sect. 4] for details). The reported results of computational experiments (see [178, Sect. 6]) suggest that algorithm H_{21} is the best of all the five algorithms.

Parallel-machine problems. Mosheiov [217] extended algorithm H_{19} to parallel-machine settings. Let $\mathcal{J}_{M_i}$ and C_{M_i} denote the set of all jobs assigned to machine M_i and the completion time of the last job assigned to machine M_i, $1 \le i \le m$, respectively. The pseudo-code of the algorithm can be formulated as follows.

Algorithm H_{23} for the problem $Pm|p_j \equiv (6.27)|C_{\max}$ ([217])

Input: sequences $(a_1, a_2, \ldots, a_n)$, $(b_1, b_2, \ldots, b_n)$, $(d_1, d_2, \ldots, d_n)$
Output: a suboptimal schedule σ

▷ Step 1:
  Arrange jobs in the non-decreasing order of a_j values;
  $L \leftarrow \emptyset$;
  $E \leftarrow \{1, 2, \ldots, n\}$;
  $j \leftarrow 0$;
  for $i \leftarrow 1$ **to** m **do**
    $C_{M_i} \leftarrow 0$;

▷ Step 2:
  while $(j \leq n)$ **do**
    $j \leftarrow j + 1$;
    $k \leftarrow \arg \min\{C_{M_i} : 1 \leq i \leq m\}$;
    if $(C_{M_k} \leq d_j)$ **then** $C_{M_k} \leftarrow C_{M_k} + a_j$
    else
        $l \leftarrow \arg \min\{\frac{b_i - a_i}{a_i} : i \leq j, J_i \in \mathcal{J}_{M_k}, a_i \geq C_{M_k} - d_j\}$;
        $L \leftarrow L \cup \{l\}$;
        $E \leftarrow E \setminus \{l\}$;
        $C_{M_k} \leftarrow C_{M_k} - a_i$;

▷ Step 3:
  Arrange jobs in L in the non-increasing order of b_j values;
  $j \leftarrow 0$;
  repeat
      $j \leftarrow j + 1$;
      $k \leftarrow \arg \min\{C_{M_k} : 1 \leq i \leq m\}$;
      $C_{M_k} \leftarrow C_{M_k} + a_j$;
  until $(j > r)$;
  $\sigma \leftarrow (\mathcal{J}_{M_1}|\mathcal{J}_{M_2}|\ldots|\mathcal{J}_{M_m})$;
  return σ.

Algorithm H_{23} runs in $O(n(n + m)) \approx O(n^2)$ time, since usually $n \gg m$.
In the worst case, algorithm H_{23}, similarly to H_{19}, may produce arbitrarily bad schedules (see [217, Example 2]). Computational experiments for $m = 2$ machines suggest, however, that average behaviour of H_{23} is quite satisfactory (see [217, Sect. 3.3] for details).

Fully polynomial-time approximation schemata

Single-machine problems. Cai et al. [39] proposed a fully polynomial-time approximation scheme (an FPTAS, see Definition 2.16) for the problem $1|p_j = a + b_j \max\{t - d_0, 0\}| C_{\max}$, where $d_0 > 0$ is the time after which job processing times start to deteriorate. The scheme is based on the observation that finding an optimal schedule is equivalent to finding the set of jobs that are completed before d_0 and the first job, J_{d_0}, that is completed after d_0. (By Theorem 6.68, all jobs which start after time $t = d_0$ should be arranged in the non-decreasing order of $\frac{a_j}{b_j}$ ratios.) Iteratively constructing, for each possible choice of J_{d_0}, a polynomial number of sets of schedules which differ by the factor $1 + \frac{\epsilon}{2n}$, and choosing from each such set the schedule with the minimal value of a certain function, we obtain an approximate schedule differing from the optimal one by the factor $1 + \epsilon$.

For a given $\epsilon > 0$ and for n jobs, the FPTAS runs in $O(n^6 \epsilon^{-2})$ time. We refer the reader to [39, Sect. 5] for more details on this FPTAS.   ◇

Another FPTAS, for the single-machine problem with job processing times given by (6.39) and the maximum completion time criterion, has been developed by Kovalyov and Kubiak [176]. The main idea of the scheme is as follows.

Let x_j, for $1 \le j \le n$, be a 0-1 variable such that $x_j := 0$ if job J_j is early, and $x_j := 1$ if job J_j is tardy or suspended (cf. Sect. 6.1.5). Let X be the set of all 0-1 vectors $x = [x_1, x_2, \ldots, x_{n-1}]$. Define functions F_j, G_j and P_j as follows: $F_0(x) := s$, $G_0(x) := s - d$, $P_0(x) := 0$; $F_j(x) := F_{j-1}(x) + x_j(a_j + b_j G_{j-1}(x))$, $G_j(x) := \min\{F_j(x), D\} - d$ and $P_j(x) := \sum_{i=1}^j a_i(1 - x_i)$, where $x \in X$, $1 \le j \le n-1$, and s is the starting time of the earliest tardy job.

The FPTAS iteratively constructs a sequence of sets $Y_1, Y_2, \ldots, Y_{n-1}$, where $Y_j \subseteq X_j := \{x \in X : x_i = 0, j + 1 \le i \le n - 1\}$ for $1 \le j \le n - 1$. In each iteration of the algorithm, set Y_j is partitioned into subsets in such a way that for any two vectors from the same subset the values of functions F_j and G_j are close enough. Next, from each such subset, only the solution with the minimal value of function P_j is chosen and used in the next iteration, while all remaning solutions are discarded. The final solution is the vector $x^\circ \in Y_{n-1}$ such that $F_{n-1}(x^\circ) = \min\{F_{n-1}(x) : x \in Y_{n-1}\}$.

For $L := \log \max\{n, D, \frac{1}{\epsilon}, a_{\max}, b_{\max}\}$, where $a_{\max} := \max_{1 \le j \le n}\{a_j\}$ and $b_{\max} := \max_{1 \le j \le n}\{b_j\}$, the FPTAS runs in $O(n^5 L^4 \epsilon^{-2})$. We refer the reader to [176, Sects. 2–3] for more details.   ◇

Remark 9.17. Woeginger [296, Sect. 8.4] proved the existence of an FPTAS for the single-machine problem with job processing times given by (6.39) and the $C_{\max}$ criterion applying dynamic programming approach and the notion of *cc-benevolent problem* (cf. [296, Sect. 7]).

9.1.4 Linear shortening

Equal ready times and deadlines

Single-machine problems. For the problem $1|p_j = a_j - b_j(t - y), y > 0$, $0 < b_j < 1, Y < \infty|C_{\max}$, Cheng et al. [54] proposed three heuristic algorithms. The time complexity of all these algorithms is $O(n \log n)$.

The first algorithm is based on Property 6.116 and is equivalent to the algorithm $A_8 : (a_j|b_j|y|Y) \mapsto (\frac{a_j}{b_j} \searrow)$.

In the second algorithm, the jobs starting by time $t = y$ are sequenced in the non-decreasing order of $\frac{a_j}{b_j}$ ratios and Property 6.116 is applied only to jobs that start after time y.

Algorithm H_{24} for the problem
$$1|p_j = a_j - b_j(t - y), y > 0, 0 < b_j < 1, Y < \infty|C_{\max} \text{ ([54])}$$

Input: sequences $(a_1, a_2, \ldots, a_n), (b_1, b_2, \ldots, b_n)$, numbers y, Y
Output: a suboptimal schedule σ

▷ Step 1:
Arrange jobs in the non-decreasing order of $\frac{a_j}{b_j}$ ratios;
$i \leftarrow 1$;
$j \leftarrow n$;
$\sigma \leftarrow (\phi)$;
$T \leftarrow 0$;
▷ Step 2:
while $(T < y)$ **do**
$\quad \sigma \leftarrow (\sigma|[i])$;
$\quad T \leftarrow T + a_{[i]}$;
$\quad i \leftarrow i + 1$;
▷ Step 3:
while $(i < j)$ **do**
$\quad \sigma \leftarrow (\sigma|[j])$;
$\quad$ **if** $(y \leq T < Y)$ **then** $T \leftarrow T + a_{[j]} - b_{[j]}(T - y)$;
$\quad$ **if** $(T \geq Y)$ **then** $T \leftarrow T + a_{[j]} - b_{[j]}(Y - y)$;
$\quad j \leftarrow j - 1$;
▷ Step 4:
return σ.

The third algorithm can be formulated as follows.

Algorithm H_{25} for the problem
$$1|p_j = a_j - b_j(t - y), y > 0, 0 < b_j < 1, Y < \infty|C_{\max} \text{ ([54])}$$

Input: sequences $(a_1, a_2, \ldots, a_n), (b_1, b_2, \ldots, b_n)$, numbers y, Y
Output: a suboptimal schedule σ

▷ Step 1:
Arrange jobs in the non-decreasing order of $\frac{a_j}{b_j}$ ratios;
$\sigma^{(1)} \leftarrow (\phi)$;
$\sigma^{(2)} \leftarrow (\phi)$;
$N_{\mathcal{J}} \leftarrow \{1, 2, \ldots, n\}$;

▷ Step 2:
   while $(N_{\mathcal{J}} \neq \emptyset)$ **do**
      Choose $k \in N_{\mathcal{J}}$ such that $a_k = \max\limits_{j \in N_{\mathcal{J}}}\{a_j\}$;
      $\sigma^{(1)} \leftarrow (\sigma^{(1)}|k)$;
      $N_{\mathcal{J}} \leftarrow N_{\mathcal{J}} \setminus \{k\}$;
      Choose $l \in N_{\mathcal{J}}$ such that $b_l = \max\limits_{j \in N_{\mathcal{J}}}\{b_j\}$;
      $\sigma^{(2)} \leftarrow (l|\sigma^{(2)})$;
      $N_{\mathcal{J}} \leftarrow N_{\mathcal{J}} \setminus \{l\}$;
▷ Step 3:
   $\sigma \leftarrow (\sigma^{(1)}|\sigma^{(2)})$;
   return σ.

These three algorithms were tested on instances with $n = 10, 50, 100, 500$ and $1,000$ jobs. The results of computational experiments suggest (see [54, Sect. 2.3]) that the algorithm $A_8 : (a_j|b_j|y|Y) \mapsto (\frac{a_j}{b_j} \searrow)$ is the best one.

Fully polynomial-time approximation schemata

Parallel-machine problems. Kang et al. [160] proposed an FPTAS for the problem $Pm|p_j = a_j - b_j t|C_{\max}$, where deterioration rates satisfy inequality (6.48). The scheme is based on the same idea as the scheme proposed by Kovalyov and Kubiak [176].

The main idea is as follows. By Theorem 6.87 (c), we can assume that all jobs have been indexed in such a way that $\frac{a_1}{b_1} \geq \frac{a_2}{b_2} \geq \ldots \geq \frac{a_n}{b_n}$. Let x_j, for $1 \leq j \leq n$, be a variable such that $x_j := k$ if job J_j is executed on machine M_k, where $k \in \{1, 2, \ldots, m\}$. Let X be the set of all vectors $x = [x_1, x_2, \ldots, x_n]$ such that $x_j := k$, $1 \leq j \leq n$, $1 \leq k \leq m$. Define functions F_j and Q as follows: $F_0^i(x) := 0$ for $1 \leq i \leq m$; $F_j^k(x) := F_{j-1}^k(x) + a_j - b_j F_{j-1}^k(x)$ for $x_j = k$; $F_j^i := F_{j-1}^i(x)$ for $x_j \neq k$ and $Q(x) := \max\{F_n^j(x) : 1 \leq j \leq m\}$.

Starting from the set $Y_0 := \{(0, 0, \ldots, 0)\}$, the FPTAS iteratively constructs a sequence of sets $Y_1, Y_2, \ldots, Y_n$, where Y_j, $1 \leq j \leq n$, is obtained from Y_{j-1} by adding k, $k = 1, 2, \ldots, m$, in the j-th position of each vector in Y_{j-1} and by applying to all obtained vectors the functions $F_j^k(x)$ and $F_j^i(x)$. Next, Y_j is partitioned into subsets in such a way that any two solutions in the same subset are close enough. From each such subset, only the solution with the minimal value of a certain function is chosen as the subset's representative for the next iteration (all remaning solutions are discarded). The final solution is the vector $x^\circ \in Y_n$ such that $Q(x^\circ) = \min\{Q(x) : x \in Y_n\}$.

For $L := \log \max\{n, \frac{1}{\epsilon}, a_{\max}\}$, where $a_{\max} := \max_{1 \leq j \leq n}\{a_j\}$, the scheme runs in $O(n^{m+1}L^{m+1}\epsilon^{-m})$ time. We refer the reader to [160, Sect. 3] for more details on this FPTAS.   ◇

Remark 9.18. Kang et al. [160] also proposed an FPTAS for the problem $P2|p_j = a_j - b_j t|C_{\max}$, running in $O(n^3 L^3 \epsilon^{-2})$ time; see [160, Sect. 2].   ◇

9.1.5 General non-linear shortening

Equal ready times and deadlines

Single-machine problems. Now, we come back to the single-machine problem $1|p_j \in \{a_j, a_j - b_j : 0 \le b_j \le a_j\}|C_{\max}$. Cheng et al. [56] proposed for the problem the following online algorithm (cf. Definition 2.17): schedule every new job J_{j+1} after the jobs $J_1, J_2, \ldots, J_j$ without any idle time, $1 \le j \le n-1$. (Note that the algorithm is an adaptation of the H_1 algorithm discussed in Sect. 9.1.1.) We will call the algorithm H_{26}.

Theorem 9.19. (Cheng et al. [56]) *Algorithm H_{26} is 2-competitive for the problem* $1|p_j \in \{a_j, a_j - b_j\} : 0 \le b_j \le a_j\}|C_{\max}$.

Proof. We start with the following three observations. First, note that if $D > \sum_{j=1}^{n} a_j$ the problem is trivial: we can schedule all jobs in an arbitrary order before time D. Hence, without loss of generality, we can assume that $D \le \sum_{j=1}^{n} a_j$. Second, from the assumption that $D \le \sum_{j=1}^{n} a_j$ it follows that there exists a unique index k such that $\sum_{j=1}^{k-1} a_j < D \le \sum_{j=1}^{k} a_j$. Third, algorithm H_{26} schedules the jobs with indices from set N_1 (N_2) before (after) time D, where $N_1 := \{1, 2, \ldots, k-1\}$ and $N_2 := \{k+1, k+2, \ldots, n\}$.

Consider now job J_k. There are two possible cases: J_k is executed either in the interval $\langle D, D + a_k - b_k \rangle$ or in the interval $\langle \sum_{j \in N_1} a_j, \sum_{j \in N_1} + a_k \rangle$.

In the first case, the maximum online completion time is equal to $D + \sum_{j \in N_2 \cup \{k\}} (a_j - b_j) \le D + \sum_{j=1}^{n} (a_j - b_j)$. In the second case, the maximum online completion time is equal to $\sum_{j \in N_1} a_j + a_k + \sum_{j \in N_2} (a_j - b_j) \le D - b_k + a_k + \sum_{j \in N_2} (a_j - b_j) \le D + \sum_{j=1}^{n} (a_j - b_j)$.

Since the maximum offline completion time is not less than D and not less than $\sum_{j=1}^{n} (a_j - b_j)$, the result follows. $\qquad \square$

Remark 9.20. A c-competitive algorithm is introduced in Definition 2.19.

Ji et al. [156] proposed offline version of the online algorithm H_{26}. We will call the modified algorithm H_{27}.

Algorithm H_{27}
for the problem $1|p_j \in \{a_j, a_j - b_j : 0 \le b_j \le a_j\}|C_{\max}$ ([56])

Input: sequences $(a_1, a_2, \ldots, a_n), (b_1, b_2, \ldots, b_n)$, number D
Output: a suboptimal schedule σ

▷ Step 1:
  Arrange jobs in the non-decreasing order of $\frac{b_j}{a_j}$ ratios;
  Find $k := \min\{j : \sum_{i=1}^{j} a_i > D\}$;
▷ Step 2:
  Schedule jobs $J_1, J_2, \ldots, J_{k-1}$ before time D;

▷ **Step 3:**
  If $(C_{k-1} > D - b_k)$ **then** schedule jobs $J_k, J_{k+1}, \ldots, J_n$
      $\hookrightarrow$ starting from time D
  else schedule jobs $J_k, J_{k+1}, \ldots, J_n$ starting from time C_{k-1};
▷ **Step 4:**
  $\sigma \leftarrow (1, 2, \ldots, k, k+1, \ldots, n)$;
  return σ.

Theorem 9.21. (Ji et al. [156]) *For an arbitrary instance I of the problem* $1|p_j \in \{a_j, a_j - b_j : 0 \leq b_j \leq a_j\}|C_{\max}$ *there holds the inequality* $R^a_{H_{27}}(I) \leq \frac{5}{4}$.

Proof. See [156, Theorem 2].   ◇

Fully polynomial-time approximation schemata

Single-machine problems. Cheng et al. [56] proved that for the problem $1|p_j \in \{a_j, a_j - b_j : 0 \leq b_j \leq a_j\}|C_{\max}$ there exists an FPTAS. The proof is based on Lemma 6.110 and exploits the fact that for the KP problem there exists an FPTAS (see, e.g., Kellerer et al. [163]); we refer the reader to [56, Theorem 3] for more details.   ◇

Parallel-machine problems. Ji and Cheng [153] proposed an FPTAS for the problem $Pm|p_j \equiv (9.7)|C_{\max}$ in which for a given $D > 0$, we have

$$p_j = a_j - b_j \min\{t, D\} \tag{9.7}$$

and $0 < b_j \leq \frac{a_j}{2D}$ for $1 \leq j \leq n$.

The scheme for the problem $Pm|p_j = a_j - b_j \min\{t, D\}|C_{\max}$ is based on the same idea as the scheme proposed by Kovalyov and Kubiak [176].

The main idea is as follows. By Theorem 6.87 (c), we can assume that all jobs have been indexed in such a way that $\frac{a_1}{b_1} \geq \frac{a_2}{b_2} \geq \ldots \geq \frac{a_n}{b_n}$. Let x_j, for $1 \leq j \leq 2m$, be a variable such that $x_j := 2k - 1$ ($x_j := 2k$) if job J_j is processed on machine M_k, $k \in \{1, 2, \ldots, m\}$, and its starting time is less (no less) than D. Let X be the set of all vectors $x = [x_1, x_2, \ldots, x_n]$ with $x_j = k$, $1 \leq j \leq n$, $1 \leq k \leq m$. Define functions F^i_j, G^i_j and Q as follows: $F^i_0(x) := 0$ for $1 \leq i \leq m$, $G^i_0(x) := 0$ for $1 \leq i \leq m$, $F^k_j(x) := F^k_{j-1}(x) + a_j - b_j G^k_{j-1}(x)$ if $x_j = 2k - 1$, $F^k_j(x) := F^k_{j-1}(x) + a_j - b_j D$ if $x_j = 2k$, $F^i_j(x) := F^i_{j-1}(x)$ if $x_j = 2k - 1$ or $x_j = 2k$, $i \neq k$, $G^k_j(x) := \min\{G^k_{j-1}(x) + a_j - b_j G^k_{j-1}(x), D\}$ if $x_j = 2k - 1$, $G^i_j(x) := G^i_{j-1}(x)$ if $x_j = 2k - 1$ or $x_j = 2k$, $i \neq k$, and $Q(x) := \max\{F^i_n(x) : 1 \leq i \leq m\}$.

The remaining parts of the FPTAS are organized similarly to those from other FPTASes based on the scheme proposed by Kovalyov and Kubiak [176] (see Sects. 9.1.1, 9.1.2 and 9.1.4 for details).

For $L := \log \max\{n, \frac{1}{\epsilon}, a_{\max}\}$, where $a_{\max} := \max_{1 \leq j \leq n}\{a_j\}$, the modified scheme runs in $O(n^{2m+1} L^{2m+1} \epsilon^{-2m})$ time. We refer the reader to [153, Sect. 3] for more details on this FPTAS.   ◇

Remark 9.22. Ji and Cheng [153] proposed also an FPTAS for the problem $1|p_j = a_j - b_j \min\{t, D\}|C_{\max}$ with $0 < b_j \leq \frac{a_j}{2D}$ for $1 \leq j \leq n$.

For $L := \log \max\{n, \frac{1}{\epsilon}, a_{\max}\}$, where $a_{\max} := \max_{1 \leq j \leq n}\{a_j\}$, the scheme runs in $O(n^3 L^3 \epsilon^{-2})$ time. We refer the reader to [153, Sect. 2] for more details on this FPTAS.   ◇

9.2 Minimizing the total completion time

In this section, we present approximation and heuristic algorithms for the $\sum C_j$ criterion.

9.2.1 Proportional deterioration

Equal ready times and no deadlines

Parallel-machine problems. Since the problem $1|p_j = b_j t|\sum C_j$ is solvable in $O(n \log n)$ time by the algorithm $A_{11} : (b_j) \mapsto (b_j \nearrow)$ (cf. Theorem 6.120), a natural question is how the algorithm will perform for $m \geq 2$ parallel identical machines. (Since by Theorem 7.14 the parallel-machine problem is computationally intractable already for $m = 2$ machines, it is clear that the adapted algorithm can be only a heuristic.) Algorithm A_{11} has been applied by Mosheiov [219] to the two-machine problem, $P2|p_j = b_j t|\sum C_j$. (The generalization of the algorithm for $m > 2$ machines is straightforward.)

Algorithm H_{28} for the problem $P2|p_j = b_j t|\sum C_j$ ([219])

Input: sequence $(b_1, b_2, \ldots, b_n)$
Output: a suboptimal schedule σ

▷ Step 1:
Arrange jobs in the non-increasing order of b_j values;
$t_{M_1} \leftarrow t_0$;
$t_{M_2} \leftarrow t_0$;
$\sigma^{(1)} \leftarrow (\phi)$;
$\sigma^{(2)} \leftarrow (\phi)$;
$N_J \leftarrow \{1, 2, \ldots, n\}$;

▷ Step 2:
while $(N_J \neq \emptyset)$ **do**
    $k \leftarrow$ the index of the job with the smallest b_j value;
    if $(t_{M_1} \leq t_{M_2})$ **then** $\sigma^{(1)} \leftarrow (\sigma^{(1)}|k)$;
                $t_{M_1} \leftarrow (1 + b_k)t_{M_1}$
    else $\sigma^{(2)} \leftarrow (\sigma^{(2)}|k)$;
        $t_{M_2} \leftarrow (1 + b_k)t_{M_2}$;
    $N_J \leftarrow N_J \setminus \{k\}$;

▷ **Step 3:**
  $\sigma \leftarrow (\sigma^{(1)} | \sigma^{(2)});$
  return σ.

The following example shows that in the case of unbounded deterioration rates algorithm H_{28} can produce arbitrarily bad schedules.

Example 9.23. (Chen [43]) Consider the following instance I of the problem $P2|p_j = b_j t| \sum C_j : m = 2, n = 3, S_1 = S_2 = 1, p_1 = p_2 = (B-1)t,$
$p_3 = (B^2 - 1)t$, where $B > 0$ is a constant. Then $R^a_{H_{28}}(I) = \frac{\sum C_j^{H_{27}}(I)}{\sum C_j^\star(I)} = \frac{B^3 + 2B}{2B^2 + B} \to \infty$ as $B \to \infty$.   ♦

Chen showed that even for the two-machine case, the absolute worst-case ratio of algorithm H_{28} is unbounded.

Theorem 9.24. (Chen [43]) *For an arbitrary instance I of the problem $P2|p_j = b_j t| \sum C_j$ there holds the inequality*

$$R^a_{H_{28}}(I) \leq \max \left\{ \frac{1+b_n}{1+b_1}, \frac{2}{n-1} + \frac{(1+b_1)(1+b_n)}{1+b_2} \right\}.$$

Proof. First, note that without loss of generality we can assume that jobs have been rearranged so that $b_1 \leq b_2 \leq \ldots \leq b_n$.

Second, if $n = 2k$ for some $k \in \mathbb{N}$, the schedule generated by algorithm H_{28} is in the form of $(1, 3, \ldots, 2k-1)$ for machine M_1 and $(2, 4, \ldots, 2k)$ for machine M_2. The total completion time of the schedule is equal to

$$\sum_{j=1}^{k} \left(\prod_{i=1}^{j} (1 + b_{2i-1}) \right) + \sum_{j=1}^{k} \left(\prod_{i=1}^{j} (1 + b_{2i}) \right) \leq 2 \sum_{j=1}^{k} \left(\prod_{i=1}^{j} (1 + b_{2i}) \right). \quad (9.8)$$

Similarly, if $n = 2k + 1$ for some $k \in \mathbb{N}$, the schedule generated by algorithm H_{28} is in the form of $(1, 3, \ldots, 2k-1, 2k+1)$ for machine M_1 and $(2, 4, \ldots, 2k)$ for machine M_2. The total completion time of the schedule is equal to

$$\sum_{j=0}^{k} \left(\prod_{i=0}^{j} (1 + b_{2i+1}) \right) + \sum_{j=1}^{k} \left(\prod_{i=1}^{j} (1 + b_{2i}) \right) \leq 2 \sum_{j=0}^{k} \left(\prod_{i=0}^{j} (1 + b_{2i}) \right). \quad (9.9)$$

Third, by direct calculations (see [43, Lemma 4.2] for details) and by Lemma 1.1 (b), we have that if $n = 2k$ for some $k \in \mathbb{N}$, the minimal total completion time is not less than

$$2 \sum_{j=1}^{k} \left(\prod_{i=1}^{j} (1 + b_{2i-1}) \right). \quad (9.10)$$

Similarly, if $n = 2k + 1$ for some $k \in \mathbb{N}$, then the minimal total completion time is not less than

$$2\sum_{j=1}^{k}(\prod_{i=1}^{j}(1+b_{2i})). \tag{9.11}$$

By calculating the ratio of the right sides of (9.8) and (9.10) and the ratio of the right sides of (9.9) and (9.11), the result follows.   □

Theorem 9.24 concerns only algorithm H_{28}, so we may hope to find for the considered problem a better algorithm, with a bounded absolute the worst-case ratio. This, however, is impossible, since the problem $Pm|p_j = b_jt|\sum C_j$ with an arbitrary number of machines is difficult to approximate.

Theorem 9.25. (Chen [43]) *There is no polynomial-time approximation algorithm with a constant worst-case bound for the problem $Pm|p_j = b_jt|\sum C_j$ with an arbitrary number of machines, unless $\mathcal{P} = \mathcal{NP}$.*

Proof. Assuming that there is a polynomial-time approximation algorithm for the problem $Pm|p_j = b_jt|\sum C_j$ with an arbitrary number of machines, we would be able to solve the strongly $\mathcal{NP}$-complete 3-P problem (cf. Sect. 3.2) in pseudopolynomial time. A contradiction, since by Lemma 3.18 a strongly $\mathcal{NP}$-complete problem cannot be solved by a pseudopolynomial-time algorithm, unless $\mathcal{P} = \mathcal{NP}$.   □

Jeng and Lin [150] observed that a modification of algorithm H_2 leads to a new heuristic algorithm for the problem $Pm|p_j = b_jt|\sum C_j$. We will call the modified algorithm H_{29}. The pseudo-code of the new algorithm is as follows.

Algorithm H_{29} for the problem $Pm|p_j = b_jt|\sum C_j$ ([150])

Input: sequence $(b_1, b_2, \ldots, b_n)$
Output: a suboptimal schedule

▷ Step 1:
Arrange jobs in the non-increasing order of b_j values;
▷ Step 2:
for $i \leftarrow 1$ to n do
Assign job $J_{[i]}$ to the machine with the smallest machine load;
▷ Step 3:
Reverse the job sequence on each machine.

Since in Step 3 algorithm, H_{29} reverses the sequence of jobs on each machine, in the final schedule the jobs are arranged in the non-decreasing order of deterioration rates. The time complexity of the algorithm is $O(n \log n)$.

Dedicated-machine problems. Wang et al. [286] proposed the following heuristic algorithm for the problem $F2|p_{i,j} = b_{i,j}t, 0 < b_{i,j} < 1|\sum C_j$.

Algorithm H_{30} for the problem $F2|p_{i,j} = b_{i,j}t, 0 < b_{i,j} < 1|\sum C_j$ ([286])

Input: sequences $(b_{1,1}, b_{1,2}, \ldots, b_{1,n})$, $(b_{2,1}, b_{2,2}, \ldots, b_{2,n})$
Output: a suboptimal schedule σ

▷ Step 1: the beginning of Phase I
$k \leftarrow 1$;
$N_{\mathcal{J}} \leftarrow \{1, 2, \ldots, n\}$;
$\sigma \leftarrow (\phi)$;
▷ Step 2:
Find J_i such that $(1 + b_{1,i})(1 + b_{2,i}) = \min\{(1 + b_{1,k})(1 + b_{2,k}) : k \in N_{\mathcal{J}}\}$;
$\sigma_k \leftarrow i$; ▷ Schedule job J_i in position k
$A_k \leftarrow t_0(1 + b_{1,i})$;
$C_{[k]} \leftarrow t_0(1 + b_{1,i})(1 + b_{2,i})$;
$N_{\mathcal{J}} \leftarrow N_{\mathcal{J}} \setminus \{i\}$;
▷ Step 3:
while $(N_{\mathcal{J}} \neq \emptyset)$ do
$\quad E \leftarrow \{J_l \in \mathcal{J} : A_k(1 + b_{1,l}) \leq C_{[k]}\}$;
$\quad$ if $(|E| \geq 1)$ then
$\quad\quad$ Find $J_i \in E$ such that $b_{1,i} = \min\{b_{1,k} : J_k \in E\}$
$\quad$ else
$\quad\quad$ Find $J_i \in \mathcal{J}$ such that $b_{1,i} = \min\{b_{1,k} : k \in N_{\mathcal{J}}\}$;
$\quad \sigma_{k+1} \leftarrow i$;
$\quad C_{[k+1]} \leftarrow \max\{A_k(1 + b_{1,i}), C_{[k]}\}(1 + b_{2,i})$;
$\quad A_{k+1} \leftarrow A_k(1 + b_{1,i})$;
$\quad N_{\mathcal{J}} \leftarrow N_{\mathcal{J}} \setminus \{i\}$;
$\quad k \leftarrow k + 1$;
▷ Step 4: the beginning of Phase II
repeat
$\quad k \leftarrow 1$;
$\quad i \leftarrow k + 1$;
$\quad$ while $(k \leq n)$ do
$\quad\quad \sigma' \leftarrow$ the schedule obtained from σ by moving $J_{[i]}$
$\quad\quad\quad \hookrightarrow$ forward to position k;
$\quad\quad$ if $(\sum C_j(\sigma') < \sum C_j(\sigma))$ then $\sigma \leftarrow \sigma'$;
$\quad\quad i \leftarrow i + 1$;
$\quad\quad k \leftarrow k + 1$;
until $(k \geq n)$.
▷ Step 5:
return σ.

Algorithm H_{30} was tested on 100 instances with $n \leq 14$ jobs. In the experiments (see [286, Sect. 6] for details), the mean and the maximum error of generated schedules did not exceed 6.9% and 21.4%, respectively.

Remark 9.26. Shiau et al. [257] proposed three heuristic algorithms for the problem $F2|p_{i,j} = b_{i,j}t| \sum C_j$. However, since the schedules generated by the algorithms are worse compared to the schedules generated by the simulated annealing algorithm H_{60} (cf. Sect. 11.3) proposed by the same authors for the same problem, we do not present them here. We refer the reader to [257, Sects. 5–6] for more details on these heuristics.   ◇

Fully polynomial-time approximation schemata

Parallel-machine problems. Woeginger [296, Sect. 6.4] proved the existence of an FPTAS for the problem $P2|p_j = b_jt| \sum C_j$, applying dynamic programming approach and the notion of *ex-benevolent problem* (cf. [296, Sect. 5]). The main idea of the proof is to show that the proposed dynamic programming algorithm satisfies some structural conditions.

The dynamic programming algorithm goes through n phases. In the k-th phase, given input vector $X_k = [b_k]$, it produces set $\mathcal{S}_k$ of states, where $1 \leq k \leq n$. Any state $s \in \mathcal{S}_k$ is a vector, $s := [s_1, s_2, s_3]$, where s_1 is the total completion time on machine M_1, s_2 is the total completion time on machine M_2 and s_3 is the criterion value of the current schedule. Elements of set $\mathcal{S}_k$ are constructed from the elements of set $\mathcal{S}_{k-1}$ using two functions, F_1 and F_2, where $F_1(b_k, s_1, s_2, s_3) := [s_1(1+b_k), s_2, s_3+s_1(1+b_k)]$ and $F_2(b_k, s_1, s_2, s_3) := [s_1, s_2(1+b_k), s_3+s_2(1+b_k)]$. The initial state $\mathcal{S}_0 := \{[1,1,1]\}$ and the final schedule is the one which gives $\min\{G(s) : s \in \mathcal{S}_n\}$, where $G(s_1, s_2, s_3) := s_3$. We refer the reader to [296, Sect. 6.4] for more details on this FPTAS.   ◇

Remark 9.27. The above approach can also be carried over to the problems $Pm|p_j = b_jt| \sum C_j$, $Qm|p_j = b_jt| \sum C_j$, $Pm|p_j = b_jt| \sum w_jC_j$ and $Qm|p_j = b_jt| \sum w_jC_j$, (cf. [296, Sect. 6.4]).   ◇

Ji and Cheng [154] proposed for the problem $Pm|p_j = b_jt| \sum C_j$ an FPTAS, based on the idea from the scheme proposed by Kovalyov and Kubiak [176].

The main idea is as follows. By Theorem 6.120, we can assume that all jobs, available starting from time $t_0 > 0$, are indexed in such a way that $b_1 \leq b_2 \leq \ldots \leq b_n$. The variables $x_j, 1 \leq j \leq n$, and the set X are defined as in Sect. 9.1.4. Define functions F_j^i and G_j as follows: $F_0^i(x) := t_0$ for $1 \leq i \leq m$, $G_0(x) := 0$, $F_j^k(x) := F_{j-1}^k(x) + b_j F_{j-1}^k(x)$ for $x_j = k$, $F_j^k(x) := F_{j-1}^k(x)$ for $x_j = k$ and $i \neq k$, $G_j(x) := G_{j-1} + \sum_{i=1}^{m} F_j^i(x)$, where $x \in X$.

The remaining parts of the FPTAS are organized similarly to those from other FPTASes based on the scheme proposed by Kovalyov and Kubiak [176] (see Sects. 9.1.1, 9.1.2, 9.1.4 and 9.1.5 for details).

For $L := \log \max\{n, \frac{1}{\epsilon}, 1 + b_{max}, S_0\}$, where $b_{max} := \max_{1 \leq j \leq n}\{b_j\}$, the scheme runs in $O(n^{2m+3}L^{m+2}\epsilon^{-(m+1)})$ time. We refer the reader to [154, Sect. 2] for more details on this FPTAS.   ◇

9.2.2 Linear deterioration

Equal ready times and deadlines

Single-machine problems. As mentioned in Sect. 6.2, the complexity of the problem $1|p_j = a_j + b_j t| \sum C_j$ is still unknown, even if $a_j = 1$ for $1 \leq j \leq n$. For the problem $1|p_j = 1 + b_j t| \sum C_j$, a few heuristic algorithms are known. Two of them have been proposed by Mosheiov [215].

The first of these heuristics, acting on the non-increasingly ordered sequence of deterioration rates $b_1, b_2, \ldots, b_n$, adds the job corresponding to a given b_j either to the left or to the right branch of the constructed V-shaped sequence.

Algorithm H_{31} for the problem $1|p_j = 1 + b_j t| \sum C_j$ ([215])

Input: sequence $(b_1, b_2, \ldots, b_n)$
Output: a suboptimal V-shaped schedule σ

▷ Step 1:
  Arrange jobs in the non-increasing order of b_j values;
▷ Step 2:
  $l \leftarrow (\phi)$;
  $r \leftarrow (\phi)$;
  $i \leftarrow 1$;
  while $(i \leq n)$ **do**
    if (i is odd) **then** $l \leftarrow (l|b_{[i]})$
    else $r \leftarrow (b_{[i]}|r)$;
    $i \leftarrow i + 1$;
▷ Step 3:
  $\sigma \leftarrow (l|r)$;
  return σ.

The next algorithm is as follows. Step 1 is the same as in H_{31}. In Step 2, an element is joined either to the left or to the right branch of the constructed V-shaped sequence if the sum of deterioration rates of the left (right) branch is lower than the sum of deterioration rates of the right (left) branch.

Algorithm H_{32} for the problem $1|p_j = 1 + b_j t| \sum C_j$ ([215])

Input: sequence $(b_1, b_2, \ldots, b_n)$
Output: a suboptimal V-shaped schedule σ

▷ Step 1:
  Arrange jobs in the non-increasing order of b_j values;
▷ Step 2:
  $l \leftarrow (\phi)$;
  $sum_l \leftarrow 0$;
  $r \leftarrow (\phi)$;
  $sum_r \leftarrow 0$;

$i \leftarrow 1$;

while $(i \leq n)$ **do**

  if $(sum_l \leq sum_r)$ **then** $l \leftarrow (l|b_{[i]})$;

        $sum_l \leftarrow sum_l + b_{[i]}$

  else $r \leftarrow (b_{[i]}|r)$;

    $sum_r \leftarrow sum_r + b_{[i]}$;

  $i \leftarrow i + 1$;

▷ **Step 3:**

  $\sigma \leftarrow (l|r)$;

 return σ.

Both H_{31} and H_{32} run in $O(n \log n)$ time and are easy to implement. Furthermore, it has been shown (see [215, Sect. 4]) that they are asymptotically optimal, under the assumption that all deterioration rates are independent, identically distributed random variables. The reported results of computational experiments (see [215, Sect. 5]) suggest that the heuristics are, on average, quite efficient, with much better behaviour of H_{32} than of H_{31}.

Remark 9.28. In Chap. 10, we will present two other heuristics for the problem, H_{51} and H_{52}, based on the so-called *signatures* of the deterioration rates sequence (cf. Gawiejnowicz et al. [95]). It will be shown that for certain classes of instances algorithms H_{51} and H_{52} give better results than algorithm H_{32}.

Remark 9.29. Sharma [256] proposed a heuristic for the single machine problem of minimizing the variance of the completion times of jobs with fixed processing times. According to the author, this heuristic can also be applied to the problem $1|p_j = 1 + b_j t| \sum C_j$ (see [256, Sects. 4–5] for details).

Dedicated-machine problems. For the problem $F2|p_{i,j} = a_{i,j} + bt| \sum C_j$ Wu and Lee [299] proposed six heuristics. The first three of them are $O(n \log n)$ algorithms: $H_{33} : (a_{1,j}|a_{2,j}|b) \mapsto (a_{j,1} \nearrow)$, $H_{34} : (a_{1,j}|a_{2,j}|b) \mapsto (a_{j,2} \nearrow)$ and $H_{35} : (a_{1,j}|a_{2,j}|b) \mapsto ((a_{j,1} + a_{j,2}) \nearrow)$. The remaining three $O(n^2)$ algorithms are modified versions of the above algorithms. The modification consists in adding a procedure PI which uses pairwise job interchange to improve a given schedule. The modified algorithms have been evaluated by an experiment in which 4440 instances with $n \leq 27$ jobs have been tested (see [299, Sect. 6] for details). The experiment has shown that the heuristic H_{33} combined with the procedure PI is the best one.

Fully polynomial-time approximation schemata

Single-machine problems. Gawiejnowicz et al. [100], applying the approach proposed by Woeginger [296], proved that for the problem $1|p_j = 1 + b_j t| \sum C_j$ there exists an FPTAS. The authors have shown that the problem is cc-benevolent (cf. [296, Sect. 7]) and hence an FPTAS for this problem can be constructed using a dynamic programming approach. We refer the reader to [100, Sect. 7] for more details on this FPTAS.

                       ◇

9.2.3 Linear shortening

Equal ready times and deadlines

Single-machine problems. For the problem $1|p_j = a_j - b_j(t - y), y > 0,$ $0 < b_j < 1, Y < \infty| \sum C_j$ Cheng et al. [54] proposed four heuristic algorithms.

The first algorithm is equivalent to the algorithm $A_5 : (a_j|b_j|y|Y) \mapsto (a_j \nearrow)$. The second algorithm is based on Properties 6.166–6.167. The algorithm first arranges jobs in the non-decreasing order of a_j values. Next, it iteratively constructs the final schedule. The pseudo-code of this algorithm is as follows.

Algorithm H_{36} for the problem
$$1|p_j = a_j - b_j(t - y), y > 0, 0 < b_j < 1, Y < \infty| \sum C_j \ ([54])$$

Input: sequences $(a_1, a_2, \ldots, a_n), (b_1, b_2, \ldots, b_n)$, numbers y, Y
Output: a suboptimal schedule

▷ Step 1:
　Arrange jobs in the non-decreasing order of a_j values;
　$C \leftarrow 0$;
　$k \leftarrow 1$;
　$N_{\mathcal{J}} \leftarrow \{1, 2, \ldots, n\}$;
▷ Step 2:
　while $(C \leq y)$ **do**
　　　Schedule job $J_{[k]}$;
　　　$C \leftarrow C + p_{[k]}(C)$;
　　　$N_{\mathcal{J}} \leftarrow N_{\mathcal{J}} \setminus \{[k]\}$;
　　　$k \leftarrow k + 1$;
▷ Step 3:
　while $(C \leq Y)$ **do**
　　　for all $j \in N_{\mathcal{J}}$ **do**
　　　　$r_j \leftarrow \frac{C - a_j}{1 - b_j}$;
　　　　Choose job $J_{[k]}$ such that $r_{[k]} = \min\{r_j : j \in N_{\mathcal{J}}\}$;
　　　　Schedule job $J_{[k]}$;
　　　　$C \leftarrow C + p_{[k]}(C)$;
　　　　$N_{\mathcal{J}} \leftarrow N_{\mathcal{J}} \setminus \{[k]\}$;
　　　　$k \leftarrow k + 1$;
▷ Step 4:
　Schedule remaining jobs in the non-decreasing order of $a_j - b_j(Y - y)$ values.

The third algorithm is a modification of algorithm H_{36}. The two algorithms differ mainly in Step 2 and Step 3 in which are selected and arranged jobs satisfying some conditions.

The pseudo-code of the third algorithm can be formulated as follows.

Algorithm H_{37} for the problem
$$1|p_j = a_j - b_j(t - y), y > 0, 0 < b_j < 1, Y < \infty| \sum C_j \ ([54])$$

Input: sequences $(a_1, a_2, \ldots, a_n), (b_1, b_2, \ldots, b_n)$, numbers y, Y
Output: a suboptimal schedule σ

▷ **Step 1:**
$\quad \sigma^{(1)} \leftarrow (\phi);$
$\quad \sigma^{(2)} \leftarrow (\phi);$
$\quad N_{\mathcal{J}} \leftarrow \{1, 2, \ldots, n\};$

▷ **Step 2:**
$\quad$ **while** $(N_{\mathcal{J}} \neq \emptyset)$ **do**
$\quad\quad$ Choose job J_k such that $a_k = \min_{j \in N_{\mathcal{J}}} \{a_j\};$
$\quad\quad \sigma^{(1)} \leftarrow (\sigma^{(1)}|k);$
$\quad\quad N_{\mathcal{J}} \leftarrow N_{\mathcal{J}} \setminus \{k\};$
$\quad\quad$ Choose job J_l such that $b_l = \max_{j \in N_{\mathcal{J}}} \{a_j - b_j(Y - y)\};$
$\quad\quad \sigma^{(2)} \leftarrow (l|\sigma^{(2)});$
$\quad\quad N_{\mathcal{J}} \leftarrow N_{\mathcal{J}} \setminus \{l\};$

▷ **Step 3:**
$\quad$ Arrange $\sigma^{(1)}$ in the non-decreasing order of a_j values;
$\quad$ Arrange $\sigma^{(2)}$ in the non-decreasing order of $a_j - b_j(Y - y)$ values;

▷ **Step 4:**
$\quad \sigma \leftarrow (\sigma^{(1)}|\sigma^{(2)});$
$\quad$ **return** $\sigma.$

The fourth algorithm proposed by Cheng et al. [54] for the problem $1|p_j = a_j - b_j(t - y), y > 0, 0 < b_j < 1, \ Y < \infty| \sum C_j$ is algorithm H_{24}. All the mentioned algorithms run in $O(n \log n)$ time, except algorithm H_{36}, which runs in $O(n^2)$ time.

These four algorithms were tested on instances with $n = 10, 50, 100, 500$ and 1000 jobs. Computational experiments have shown (see [54, Sect. 2.3]) that the algorithm equivalent to $A_5 : (a_j|b_j|y|Y) \mapsto (a_j \nearrow)$ is the best one.

9.3 Minimizing the maximum lateness

In this section, we present heuristic algorithms for time-dependent scheduling problems with the $L_{\max}$ criterion.

9.3.1 Linear deterioration

Distinct ready times and deadlines

Single-machine problems. Bachman and Janiak [11] proposed two heuristics for the problem $1|p_j = a_j + b_j t|L_{\max}$. The first one is equivalent to the algorithm $A_{15} : (a_j|b_j|d_j) \mapsto (d_j \nearrow)$, which schedules jobs in the EDD order. The second one, given below, first arranges jobs in the non-decreasing order of $\frac{a_j}{b_j}$ ratios and then iteratively improves the solution.

Algorithm H_{38} for the problem $1|p_j = a_j + b_j t|L_{\max}$ ([11])

Input: sequences $(a_1, a_2, \ldots, a_n)$, $(b_1, b_2, \ldots, b_n)$, $(d_1, d_2, \ldots, d_n)$
Output: a suboptimal schedule σ

▷ Step 1:
  Arrange jobs in the non-decreasing order of $\frac{a_j}{b_j}$ ratios;
  Call the obtained schedule σ;
  $\sigma' \leftarrow \sigma$;

▷ Step 2:
  while $(L_{\max}(\sigma) \leq L_{\max}(\sigma'))$ **do**
      Find in σ the position k such that $k = \arg\max\{C_j - d_j\}$;
      while $(TRUE)$ **do**
          Denote the subset of jobs before job J_k by T_k;
          Find in $\{J_j \in T_k : 0 \leq j \leq k-1\}$ job J_i such that $d_i > d_k$;
          if $(i = 0)$ **then return** σ;
          Construct schedule σ' by inserting job J_i after job J_k;
          if $(L_{\max}(\sigma') < L_{\max}(\sigma))$ **then** $\sigma \leftarrow \sigma'$
          else if $(i > 0)$ **then** $k \leftarrow i$.

▷ Step 3:
  return σ.

The results of the computational experiment conducted for instances with $n = 10$ and $n = 50$ jobs (see [11, Sect. 4]) have shown that algorithm H_{38} in most cases is better than A_{15}.

For the same problem, $1|p_j = a_j + b_j t|L_{\max}$, Hsu and Lin [142, Sect. 4] proposed two heuristic algorithms. The algorithms combine algorithm H_{38} with the so-called *hill-climbing* procedure. However, the reported results of a computational experiment (see [142, Sect. 5]) do not allow to formulate a clear conclusion about the performance of these algorithms.

9.3.2 General non-linear deterioration

Distinct ready times and deadlines

Single-machine problems. Janiak and Kovalyov [146] proposed three following heuristics for exponential job processing times given by (6.40): $A_{15} : (a_j|b_j|r_j|d_j) \mapsto (d_j \nearrow)$, $H_{39} : (a_j|b_j|r_j|d_j) \mapsto (\frac{a_j}{b_j+d_j} \nearrow)$ and $H_{40} : (a_j|b_j|r_j|d_j) \mapsto (b_j \nearrow |d_j \searrow)$. According to the authors (see [146, Sect. 4]), algorithms A_{15} and H_{40} give the best results.

9.4 Other criteria

In this section, we present heuristics for time-dependent scheduling problems with criteria other than C_{max}, $\sum C_j$ or L_{max}.

9.4.1 Proportional deterioration

Equal ready times and deadlines

Single-machine problems. For the problem $1|p_j = b_j t| \sum (C_i - C_j)$ Oron [229] proposed two heuristic algorithms.

Both these heuristics construct the final schedule in two steps. In the first step, jobs are arranged in the non-increasing order of deterioration rates. In the second step, the jobs are assigned to predetermined positions in schedule. The indices of the positions change in dependence of the number of jobs n is even or odd (cf. Properties 6.188–6.192, Theorem 6.193). The running time of both the algorithms is $O(n \log n)$.

The pseudo-code of the first of these heuristic algorithms is as follows.

Algorithm H_{41} for the problem $1|p_j = b_j t| \sum (C_i - C_j)$ ([229])

Input: sequence $(b_1, b_2, \ldots, b_n)$
Output: a suboptimal schedule σ

▷ Step 1:
  Arrange jobs in the non-increasing order of b_j values;
  $\sigma \leftarrow (\phi)$;

▷ Step 2:
  if n is even **then**
  for $i \leftarrow 1$ **to** $\frac{n}{2} + 1$ **do**
      Schedule job $J_{\frac{n}{2}+2-i}$ in the i-th position in σ;
  for $i \leftarrow \frac{n}{2} + 2$ **to** n **do**
      Schedule job J_i in the i-th position in σ;
  ▷ the final schedule is $(\frac{n}{2} + 1, \frac{n}{2}, \ldots, 3, 2, 1, \frac{n}{2} + 2, \frac{n}{2} + 3, \ldots, n)$
  else   ▷ n is odd
  for $i \leftarrow 1$ **to** $\frac{n+3}{2}$ **do**
      Schedule job $J_{\frac{n+5}{2}-i}$ in the i-th position in σ;
  for $i \leftarrow \frac{n+5}{2}$ **to** n **step** 2 **do**
      Schedule job J_i in the i-th position in σ;
  ▷ the final schedule is $(\frac{n+3}{2}, \frac{n+1}{2}, \ldots, 3, 2, 1, \frac{n+5}{2}, \frac{n+7}{2}, \ldots, n)$
▷ Step 3:
  return σ.

The second of these heuristic algorithms can be formulated as follows.

Algorithm H_{42} for the problem $1|p_j = b_j t| \sum (C_i - C_j)$ ([229])

Input: sequence $(b_1, b_2, \ldots, b_n)$
Output: a suboptimal schedule σ

▷ Step 1:
  Arrange jobs in the non-increasing order of b_j values;
  $\sigma \leftarrow (\phi)$;

▷ Step 2:
  if n is even **then**
  for $i \leftarrow 1$ **to** $\frac{n}{2}$ **do**
      Schedule job $J_{\frac{n}{2}+2-i}$ in the i-th position in σ;
      Schedule job J_{2i-1} in the $\frac{n}{2} + i$-th position in σ;
      ▷ the final schedule is $(n, n-2, \ldots, 4, 2, 1, 3, \ldots, n-3, n)$
  else   ▷ n is odd
  for $i \leftarrow 1$ **to** $\frac{n+1}{2}$ **do**
      Schedule job J_{n+2-2i} in the i-th position in σ;
  for $i \leftarrow 1$ **to** $\frac{n-1}{2}$ **do**
      Schedule job J_{2i} in the $\frac{n+1}{2} + i$-th position in σ;
      ▷ the final schedule is $(n, n-2, \ldots, 5, 3, 1, 2, 4, \ldots, n-3, n-1)$
▷ Step 3:
  return σ.

The performance of algorithms H_{41} and H_{42}, compared to the lower bound from Property 6.192, was tested on instances with $n = 20, 50$ or $n = 100$ jobs. For each value of n, 1000 random instances were generated, with deterioration rates b_j from random uniform distribution $U(0.05, 1)$. The results of these computational experiments (see [229, Sect. 4]) suggest that the algorithm H_{41}, on average, is more effective than H_{42}.

Parallel-machine problems.   For the problem $Pm|p_j = b_j t| \sum C_{\max}^{(k)}$, Mosheiov [219] proposed to apply heuristic H_2. Moreover, he proved that if $n \rightarrow \infty$, then the absolute worst-case ratio of the heuristic for the $\sum C_{\max}^{(k)}$ criterion is bounded and is asymptotically close to 1.

9.4.2 Linear deterioration

Equal ready times and deadlines

Single-machine problems.   Alidaee and Landram [5] proposed an $O(n^2)$ algorithm for the problem $1|p_j = g_j(t)|P_{\max}$. The algorithm is based on the algorithm for the problem $1|prec|f_{\max}$ (Lawler [182]). Let $C_{\max}(\mathcal{J})$ denote the maximum completion time of the last job in set $\mathcal{J}$.

Algorithm H_{43} for the problem $1|p_j = g_j(t)|P_{\max}$ ([5])

Input: sequence $(g_1, g_2, \ldots, g_n)$
Output: a suboptimal schedule σ

▷ Step 1:
$\quad N_{\mathcal{J}} \leftarrow \{1, 2, \ldots, n\};$
$\quad \sigma \leftarrow (\phi);$

▷ Step 2:
$\quad$ **for** $i \leftarrow n$ **downto** 1 **do**
$\quad\quad$ **for** all $j \in N_{\mathcal{J}}$ **do**
$\quad\quad\quad$ Schedule job J_j in the i-th position in σ;
$\quad\quad\quad T \leftarrow C_{\max}(N_{\mathcal{J}} \setminus \{j\});$
$\quad\quad\quad p_j \leftarrow g_j(T);$
$\quad\quad k \leftarrow \arg \min_{j \in N_{\mathcal{J}}} \{p_j\};$
$\quad\quad \sigma \leftarrow (k|\sigma);$
$\quad\quad N_{\mathcal{J}} \leftarrow N_{\mathcal{J}} \setminus \{k\};$

▷ Step 3:
$\quad$ **return** σ.

To evaluate the quality of schedules generated by algorithm H_{43}, Alidaee and Landram conducted a computational experiment in which $g_j(t) = a_j + b_j t$. For each n, where $6 \leq n \leq 9$, 50 random instances were generated, with $a_j \in (0, 1)$ and $b_j \in (1, 10)$. Since the performance was rather poor (see [5, Sect. 3]), H_{43} is not recommended for arbitrary a_j's and b_j's. The algorithm, however, is optimal when either $a_j > 0 \wedge b_j \geq 1$ or $b_j = b$ for $1 \leq j \leq n$ (see Sect. 6.4 for details).

For the problem $1|p_j = a_j + bt| \sum w_j C_j$, Mosheiov [218] proposed a heuristic algorithm, based on the following observation. Applying pairwise job interchange argument, one can show that for a given starting time t, jobs in an optimal schedule will be executed in the increasing order of $\frac{a_j}{w_j} + \frac{b}{w_j} t$ values. Therefore, for small values of t the order $\frac{a_j}{w_j} \nearrow$ seems to be more attractive, while for large t values the order $w_j \searrow$ is better.

Algorithm H_{44} for the problem $1|p_j = a_j + bt| \sum w_j C_j$ ([218])

Input: sequences $(a_1, a_2, \ldots, a_n)$, $(w_1, w_2, \ldots, w_n)$, number b
Output: a suboptimal schedule σ

▷ Step 1:
$\quad$ Schedule jobs in the increasing order of $\frac{a_j}{w_j}$ ratios;
$\quad$ Call the obtained schedule σ^1;
$\quad G_1 \leftarrow \sum w_j C_j(\sigma^1);$

▷ Step 2:
  Schedule jobs in the decreasing order of w_j values;
  Call the obtained schedule σ^2;
  $G_2 \leftarrow \sum w_j C_j(\sigma^2)$;
▷ Step 3:
  if $(G_1 < G_2)$ **then** $\sigma \leftarrow \sigma^1$
  else $\sigma \leftarrow \sigma^2$;
  return σ.

The results of a computational experiment (see [218, Sect. 'The case of general weights']) suggest that algorithm H_{44} is, on average, quite good.

Distinct ready times and deadlines

Parallel-machine problems. Cheng et al. [58] proposed a heuristic algorithm for the problem $Pm|p_j = a_j + bt| \sum(\alpha E_j + \beta T_j + \gamma d)$. The heuristic exploits the idea of algorithm A_{41} for the problem $Pm|p_j = a_j + bt| \sum(\alpha E_j + \beta T_j)$ and has been tested on a few instances with $5 \le n \le 8$. We refer the reader to [58, Table 1] for more details.

9.4.3 General non-linear deterioration

Distinct ready times and deadlines

Single-machine problems. Janiak and Kovalyov [146] proposed three heuristic algorithms for processing times given by (6.40) and the criterion $\sum w_j C_j$: $H_{45} : (a_j|b_j|r_j|w_j) \mapsto (\frac{p_j}{w_j} \nearrow)$, $H_{46} : (a_j|b_j|r_j|w_j) \mapsto (w_i \searrow |b_j \nearrow)$ and $H_{47} : (a_j|b_j|r_j|w_j) \mapsto (\frac{a_i}{w_i} \nearrow |b_j \searrow)$. In the opinion of the authors (see [146, Sect. 4]), algorithm H_{45} gives, on average, the best results.

Sundararaghavan and Kunnathur proposed an algorithm for job processing times given by (6.68) and the criterion $\sum w_j C_j$. Applying the notation introduced in Sect. 6.4.5, the algorithm can be formulated as follows.

Algorithm H_{48} for the problem $1|p_j \equiv (6.68)| \sum w_j C_j$ ([263])

Input: sequences $(b_1, b_2, \ldots, b_n)$, $(w_1, w_2, \ldots, w_n)$, numbers a, D
Output: a suboptimal schedule

▷ Step 1:
  Arrange jobs in the non-increasing order of w_j values;
  Call the obtained sequence σ;
▷ Step 2:
  $k \leftarrow \lfloor \frac{D}{a} \rfloor + 1$;
  Schedule k jobs according to the order given by σ;
  $E \leftarrow \{J_k \in \mathcal{J} : C_k \le D\}$;

$L \leftarrow \mathcal{J} \setminus E$;

Arrange jobs in L in the non-decreasing order of $\frac{a+b_j}{w_j}$ ratios;

▷ **Step 3:**

   repeat

      if $(J_{[i]} \in E \wedge J_{[j]} \in L \wedge \Delta(J_{[i]} \leftrightarrow J_{[j]}) > 0)$ **then** exchange $J_{[i]}$ with $J_{[j]}$;

      until (no more exchange $J_{[i]} \leftrightarrow J_{[j]}$ exists for $J_{[i]} \in E$ and $J_{[j]} \in L$).

Sundararaghavan and Kunnatur [263] conjectured that algorithm H_{48} is optimal. Cheng and Ding presented the following counter-example which refutes this conjecture.

Example 9.30. (Cheng and Ding [51]) Given a sufficiently large number $Y > 0$, define $Z = 2Y^4$. Let $n = 4$, $a = 1$, $D = 1$, $w_1 = 2Y + 3 + \frac{1}{3Y}$, $w_2 = 2Y$, $w_3 = Y + 1$, $w_4 = Y$, $b_1 = YZ$, $b_2 = (Y+1)Z$, $b_3 = 2YZ$, $b_4 = (2Y+3+\frac{2}{3Y})Z$.

Since algorithm H_{48} generates schedule $\sigma = (1, 3, 2, 4)$, and for schedule $\sigma' = (2, 4, 1, 3)$ we have $\sum w_j C_j(\sigma') < \sum w_j C_j(\sigma)$, the algorithm cannot be optimal (see [51, Sect. 4]). ◆

9.4.4 Linear shortening

Equal ready times and deadlines

Single-machine problems. For the problem $1 | p_j = a_j - b_j t$, $0 \leq b_j < 1$, $b_i(\sum_{j=1}^n a_j - a_i) < a_i | \sum w_j C_j$, Bachman et al. [10] have proposed two heuristics. In the first of the two, which is based on Property 6.251, jobs are scheduled by algorithm $H_{49} : (a_j | b_j | w_j) \mapsto (\frac{a_j}{w_j(1-b_j)} \nearrow)$. In the second one, H_{50}, which is based on Properties 6.253-6.255, an even-odd V-shaped schedule is constructed. The results of a computational experiment (see [10, Sect. 'Heuristic algorithms']) suggest that H_{50} outperforms H_{49}.

With this remark, we end the review of heuristic and approximation algorithms for computationally intractable time-dependent scheduling problems. In Chaps. 10 and 11, we will consider heuristic algorithms based on signatures of deterioration rates and local search time-dependent scheduling algorithms, respectively.

9.5 Concluding remarks

In this chapter, we considered approximation and heuristic algorithms for time-dependent scheduling problems.

The algorithms presented in the chapter have a few common features. First, the algorithms have low running times, e.g., $O(n \log n)$ or $O(n^2)$. Second, they construct the final schedule either step by step or using a list of jobs ordered in some way. Finally, though the heuristics can, on average, produce schedules of acceptable quality, there are instances of scheduling problems for which the worst-case behaviour of the algorithms is especially bad.

Below, we classify, in the tabular form, the considered heuristic algorithms for time-dependent scheduling problems. The problems are divided into groups with respect to the applied optimality criterion.

Tables 9.1 and 9.2 present the heuristics concerning single- and parallel-machine time-dependent scheduling problems with the $C_{\max}$ criterion.

Table 9.3 presents the heuristics concerning dedicated-machine time-dependent scheduling problems with the $C_{\max}$ criterion.

Tables 9.4, 9.5 and 9.6 present the heuristics concerning, respectively, single-, parallel- and dedicated-machine time-dependent scheduling problems with the $\sum C_j$ criterion.

Tables 9.7 and 9.8 present the heuristics concerning, respectively, single-machine time-dependent scheduling problems with the $L_{\max}$ criterion and with criteria other than $C_{\max}$, $\sum C_j$ and $L_{\max}$.

Finally, Tables 9.9 and 9.10 present fully polynomial-time approximation schemata for single-machine and parallel-machine problems, respectively.

Table 9.1: Heuristic Algorithms for Single-Machine Problems ($C_{\max}$ Criterion)

Heuristic	Complexity	Problem	References	This book
H_1	$O(n)$	$1, h_{11}\|p_j = b_j t, nres\|C_{\max}$	[155]	Sect. 9.1.1, p. 204
H_2	$O(n \log n)$	$1, h_{11}\|p_j = b_j t, nres\|C_{\max}$	[155]	Sect. 9.1.1, p. 205
H_{10}	$O(n^2)$	$1\|p_j = a_j + bt, r_j\|C_{\max}$	[194]	Sect. 9.1.2, p. 212
H_{11}	$O(n^2)$	$1\|p_j = a_j + bt, r_j\|C_{\max}$	[194]	Sect. 9.1.2, p. 212
H_{12}	$O(n \log n)$	$1\|p_j \equiv (6.24)\|C_{\max}$	[128]	Sect. 9.1.3, p. 214
H_{13}	$O(n^2)$	$1\|p_j = a_j + b_j t + c_j t^2\|C_{\max}$	[128]	Sect. 9.1.3, p. 214
H_{14}	$O(n^2)$	$1\|p_j = f_j(t)\|C_{\max}$	[4]	Sect. 9.1.3, p. 215
H_{15}	$O(n^3)$	$1\|p_j = e^{a_j t}\|C_{\max}$	[140]	Sect. 9.1.3, p. 216
H_{16}	$O(n \log n)$	$1\|p_j = a_j 2^{b_j (t-r_j)}\|C_{\max}$	[146]	Sect. 9.1.3, p. 216
H_{17}	$O(n \log n)$	$1\|p_j = a_j 2^{b_j (t-r_j)}\|C_{\max}$	[146]	Sect. 9.1.3, p. 216
H_{18}	$O(n \log n)$	$1\|p_j = a_j 2^{b_j (t-r_j)}\|C_{\max}$	[146]	Sect. 9.1.3, p. 216
H_{19}	$O(n \log n)$	$1\|p_j \equiv (6.27)\|C_{\max}$	[217]	Sect. 9.1.3, p. 217
H_{20}	$O(n \log n)$	$1\|p_j \equiv (6.34)\|C_{\max}$	[217]	Sect. 9.1.3, p. 217
H_{21}	$O(n^2)$	$1\|p_j \equiv (6.37\|C_{\max}$	[178]	Sect. 9.1.3, p. 218
H_{22}	$O(n^2)$	$1\|p_j \equiv (6.37)\|C_{\max}$	[178]	Sect. 9.1.3, p. 219
H_{24}	$O(n \log n)$	$1\|p_j = a_j - b_j(t - y)\|C_{\max}$ [a]	[54]	Sect. 9.1.4, p. 222
H_{25}	$O(n \log n)$	$1\|p_j = a_j - b_j(t - y)\|C_{\max}$ [a]	[54]	Sect. 9.1.4, p. 222
H_{26}	$O(n)$	$1\|p_j \in \{a_j, a_j - b_j :$ $0 \le b_j \le a_j\}\|C_{\max}$	[56]	Sect. 9.1.5, p. 224
H_{27}	$O(n)$	$1\|p_j \in \{a_j, a_j - b_j\}\|C_{\max}$ [b]	[56]	Sect. 9.1.5, p. 224

[a] $0 < b_j < 1$ for $1 \le i \le n$, $y > 0, Y < \infty$
[b] $0 \le b_j \le a_j$

Table 9.2: Heuristic Algorithms for Parallel-Machine Problems (C_{max} Criterion)

Heuristic	Complexity	Problem	References	This book
H_1	$O(n)$	$Pm\|p_j = b_j t\|C_{max}$	[46]	Sect. 9.1.1, p. 204
H_2	$O(n \log n)$	$Pm\|p_j = b_j t\|C_{max}$ [a]	[141],[219]	Sect. 9.1.1, p. 205
H_3	$O(n)$	$Pm\|p_j = b_j t\|C_{max}$	[46]	Sect. 9.1.1, p. 206
H_4	$O(n \log n)$	$Pm\|p_j = a_j + b_j t\|C_{max}$	[141]	Sect. 9.1.2, p. 209
H_5	$O(n \log n)$	$Pm\|p_j = a_j + b_j t\|C_{max}$	[141]	Sect. 9.1.2, p. 210
H_6	$O(n^2)$	$Pm\|p_j = a_j + b_j t\|C_{max}$	[141]	Sect. 9.1.2, p. 210
H_{23}	$O(n(n+m))$	$Pm\|p_j \equiv (6.27)\|C_{max}$	[217]	Sect. 9.1.3, p. 220

[a] $0 < b_j < 1$ for $1 \leq i \leq n$

Table 9.3: Heuristic Algorithms for Dedicated-Machine Problems (C_{max} Criterion)

Heuristic	Complexity	Problem	References	This book
H_7	$O(n \log n)$	$F2\|p_{i,j} = a_{i,j} + bt\|C_{max}$	[196]	Sect. 9.1.2, p. 211
H_8	$O(n \log n)$	$F2\|p_{i,j} = a_{i,j} + bt\|C_{max}$	[196]	Sect. 9.1.2, p. 211
H_9	$O(n^2)$	$F2\|p_{i,j} = a_{i,j} + bt\|C_{max}$	[196]	Sect. 9.1.2, p. 211

[a] $0 < b_j < 1$ for $1 \leq i \leq n$

Table 9.4: Heuristic Algorithms for Single-Machine Problems ($\sum C_j$ Criterion)

Heuristic	Complexity	Problem	References	This book
H_{31}	$O(n \log n)$	$1\|p_j = 1 + b_j t\|\sum C_j$	[215]	Sect. 9.2.2, p. 231
H_{32}	$O(n \log n)$	$1\|p_j = 1 + b_j t\|\sum C_j$	[215]	Sect. 9.2.2, p. 231
H_{36}	$O(n \log n)$	$1\|p_j = a_j - b_j(t - y)\|\sum C_j$ [a]	[54]	Sect. 9.2.3, p. 233
H_{37}	$O(n \log n)$	$1\|p_j = a_j - b_j(t - y)\|\sum C_j$ [a]	[54]	Sect. 9.2.3, p. 234
H_{51}	$O(n \log n)$	$1\|p_j = 1 + b_j t\|\sum C_j$	[95]	Sect. 9.2.2, p. 232
H_{52}	$O(n \log n)$	$1\|p_j = 1 + b_j t\|\sum C_j$	[95]	Sect. 9.2.2, p. 232

[a] $y > 0$, $0 < b_j < 1$ for $1 \leq j \leq n$, $Y < \infty$

Table 9.5: Heuristic Algorithms for Parallel-Machine Problems ($\sum C_j$ Criterion)

Heuristic	Complexity	Problem	References	This book
H_{28}	$O(n \log n)$	$P2\|p_j = b_j t\|\sum C_j$	[219]	Sect. 9.2.1, p. 226
H_{29}	$O(n \log n)$	$Pm\|p_j = b_j t\|\sum C_j$	[150]	Sect. 9.2.1, p. 228

Table 9.6: Heuristic Algorithms for Dedicated-Machine Problems (Criterion $\sum C_j$)

Heuristic	Complexity	Problem	References	This book
H_{30}	$O(n^2)$	$F2\|p_{i,j} = b_{i,j} t\|\sum C_j$	[286]	Sect. 9.2.1, p. 229
H_{33}	$O(n \log n)$	$F2\|p_{i,j} = a_{i,j} + bt\|\sum C_j$	[299]	Sect. 9.2.2, p. 232
H_{34}	$O(n \log n)$	$F2\|p_{i,j} = a_{i,j} + bt\|\sum C_j$	[299]	Sect. 9.2.2, p. 232
H_{35}	$O(n \log n)$	$F2\|p_{i,j} = a_{i,j} + bt\|\sum C_j$	[299]	Sect. 9.2.2, p. 232

Table 9.7: Heuristic Algorithms for Single-Machine Problems (L_{max} Criterion)

Heuristic	Complexity	Problem	References	This book
H_{38}	$O(n^2)$	$1\|p_j = a_j + b_j t\|L_{max}$	[11]	Sect. 9.3.1, p. 235
H_{39}	$O(n \log n)$	$1\|p_j = a_j 2^{b_j(t-r_j)}\|L_{max}$	[146]	Sect. 9.3.2, p. 235
H_{40}	$O(n \log n)$	$1\|p_j = a_j 2^{b_j(t-r_j)}\|L_{max}$	[146]	Sect. 9.3.2, p. 235

Table 9.8: Heuristic Algorithms for Single-Machine Problems (Criteria other than C_{max}, $\sum C_j$ and L_{max})

Heuristic	Complexity	Problem	References	This book
H_{41}	$O(n \log n)$	$1\|p_j = b_j(t)\|\sum(C_i - C_j)$	[229]	Sect. 9.4.1, p. 236
H_{42}	$O(n \log n)$	$1\|p_j = b_j + bt\|\sum(C_i - C_j)$	[229]	Sect. 9.4.1, p. 237
H_{43}	$O(n^2)$	$1\|p_j = g_j(t)\|P_{max}$	[5]	Sect. 9.4.2, p. 238
H_{44}	$O(n \log n)$	$1\|p_j = a_j + bt\|\sum w_j C_j$	[218]	Sect. 9.4.2, p. 238
H_{45}	$O(n \log n)$	$1\|p_j = a_j 2^{b_j(t-r_j)}\|\sum w_j C_j$	[146]	Sect. 9.4.3, p. 239
H_{46}	$O(n \log n)$	$1\|p_j = a_j 2^{b_j(t-r_j)}\|\sum w_j C_j$	[146]	Sect. 9.4.3, p. 239
H_{47}	$O(n \log n)$	$1\|p_j = a_j 2^{b_j(t-r_j)}\|\sum w_j C_j$	[146]	Sect. 9.4.3, p. 239
H_{48}	$O(n^2)$	$1\|p_j \equiv (6.68)\|\sum w_j C_j$	[263]	Sect. 9.4.3, p. 239
H_{49}	$O(n \log n)$	$1\|p_j = a_j - b_j t\|\sum w_j C_j^{(a)}$	[10]	Sect. 9.4.4, p. 240
H_{50}	$O(n \log n)$	$1\|p_j = a_j - b_j t\|\sum w_j C_j^{(a)}$	[10]	Sect. 9.4.4, p. 240

$^{(a)}$ $0 \leq b_j < 1, b_i(\sum_{j=1}^{n} a_j - a_i) < a_i$ for $1 \leq i \leq n$

Table 9.9: Fully Polynomial-Time Approximation Schemata (Single Machine Problems)

Problem	Running Time	References	This book
$1, h_{11}\|p_j = b_j t, nres\|C_{max}$	$O(n^2 \epsilon^{-1})$	[155]	Sect. 9.1.1, p. 209
$1\|p_j = a + b_j \max\{t - d_0, 0\}\|C_{max}$	$O(n^6 \epsilon^{-2})$	[39]	Sect. 9.1.3, p. 221
$1\|p_j \equiv (6.39)\|C_{max}$	$O(n^5 L^4 \epsilon^{-2})^{(a)}$	[176],[296]	Sect. 9.1.3, p. 221
$1\|p_j = a_j - b_j \min\{t, D\}\|C_{max}$ $^{(b)}$	$O(n^3 L^3 \epsilon^{-2})^{(c)}$	[153]	Sect. 9.1.5, p. 226

$^{(a)}$ $L := \log \max\{n, D, \frac{1}{\epsilon}, a_{max}, b_{max}\}$

$^{(b)}$ $0 < b_j \leq \frac{a_j}{2D}$ for $1 \leq j \leq n$,

$^{(c)}$ $L := \log \max\{n, \frac{1}{\epsilon}, 1 + b_{max}, S_1\}$, where $a_{max} := \max_{1 \leq j \leq n}\{a_j\}$ and $b_{max} := \max_{1 \leq j \leq n}\{b_j\}$

Table 9.10: Fully Polynomial-Time Approximation Schemata (Parallel Machine Problems)

Problem	Running Time	References	This book
$Pm\|p_j = b_j t\|C_{\max}$	$O(n^{2m+1} L^{m+1} \epsilon^{-m})^{(a)}$	[243]	Sect. 9.1.1, p. 209
$Pm\|p_j = a_j + b_j t\|C_{\max}$	$O(n^{2m+1} L^{m+1} \epsilon^{-m})^{(b)}$	[161]	Sect. 9.1.2, p. 211
$Pm\|p_j = a_j - b_j t\|C_{\max}$	$O(n^{m+1} L^{m+1} \epsilon^{-m})^{(c)}$	[160]	Sect. 9.1.4, p. 223
$Pm\|p_j \equiv (9.7)\|C_{\max}$	$O(n^{2m+1} L^{2m+1} \epsilon^{-2m})^{(c)}$	[153]	Sect. 9.1.5, p. 225
$Pm\|p_j = b_j\| \sum C_j$	$O(n^{2m+3} L^{m+2} \epsilon^{-(m+1)})^{(a)}$	[154], [296]	Sect. 9.2.1, p. 230

(a) $L := \log \max\{n, \frac{1}{\epsilon}, 1 + b_{\max}, S_1\}$

(b) $L := \log \max\{n, \frac{1}{\epsilon}, a_{\max}, 1 + b_{\max}\}$

(c) $L := \log \max\{n, \frac{1}{\epsilon}, a_{\max}\}$, where $a_{\max} := \max_{1 \leq j \leq n}\{a_j\}$ and $b_{\max} := \max_{1 \leq j \leq n}\{b_j\}$

10

Greedy algorithms based on signatures

Heuristic algorithms for intractable time-dependent scheduling problems may be constructed in many various ways. In this chapter, we present two greedy heuristic time-dependent scheduling algorithms which exploit certain properties of the so-called 'signatures' of job deterioration rates.

Chapter 10 is composed of five sections. In Sect. 10.1, we formulate the problem that is the subject of this chapter and introduce the signatures. In Sect. 10.2, we present basic properties of signatures. In Sect. 10.3, we formulate the first greedy algorithm based on the properties of signatures. In Sect. 10.4, we introduce the so-called *regular* sequences and formulate the second greedy algorithm based on signatures. We also give arguments for the conjecture that the greedy algorithms find optimal schedules for regular sequences. Concluding remarks are given in Sect. 10.5.

10.1 Preliminaries

In this section, we give the formulation of the problem under consideration and define the notion of signatures of job deterioration rates.

10.1.1 Problem formulation

We consider the following version of the problem $1|p_j = 1 + b_j| \sum C_j$. A set $\mathcal{J}$ of $n + 1$ deteriorating jobs, $J_0, J_1, J_2, \ldots, J_n$, is to be processed on a single machine, which is available from time $t_0 = 0$. The job processing times are in the form of $p_j = 1 + b_j t$, where $b_j > 0$ for $0 \leq j \leq n$. The criterion of schedule optimality is the total completion time, $\sum C_j = \sum_{j=0}^{n} C_j$, where $C_0 := 1$ and $C_{j-1} + p_j(C_{j-1}) = 1 + (1 + b_j)C_{j-1}$ for $1 \leq j \leq n$. For simplicity of further presentation, let $\beta_j = 1 + b_j$ for $0 \leq j \leq n$ and $\hat{\beta} := (\beta_0, \beta_1, \beta_2, \ldots, \beta_n)$.

Recall that the time complexity of the problem is still unknown (cf. Sect. 6.2.3), though there exists a hypothesis that it is at least $\mathcal{NP}$-complete in the ordinary sense (Cheng et al. [55, Sect. 3]). Therefore, in accordance with a recommendation by Garey and Johnson [85, Chap. 4], the consideration of special cases of the problem may lead to finding polynomially solvable cases of the problem and delineating the border between its easy and hard cases. Hence, through the chapter we assume that job deterioration rates are of a special form, e.g., they are consecutive natural numbers.

10.1.2 Definition of signatures

Lemma 10.1. (Gawiejnowicz et al. [95]) *Let $C(\hat{\beta}) = [C_0, C_1, C_2, \ldots, C_n]$ be the vector of job completion times in the form of (6.11) for a given sequence $\hat{\beta} = (\beta_0, \beta_1, \beta_2, \ldots, \beta_n)$. Then*

(a) $\|C(\hat{\beta})\|_1 := \sum\limits_{j=0}^{n} C_j(\hat{\beta}) = \sum\limits_{j=0}^{n}(1 + \sum\limits_{i=1}^{j} \prod\limits_{k=i}^{j} \beta_k) = \sum\limits_{j=1}^{n} \sum\limits_{i=1}^{j} \prod\limits_{k=i}^{j} \beta_k + (n+1),$

(b) $\|C(\hat{\beta})\|_\infty := \max\limits_{0 \le j \le n} \{C_j(\hat{\beta})\} = 1 + \sum\limits_{i=1}^{n} \prod\limits_{k=i}^{n} \beta_k.$

Proof. (a) The result follows from equality (6.11) and Definition 1.18 of the norm l_p for $p = 1$.

(b) The result follows from equality (6.11), Definition 1.18 of the norm l_p for $p = \infty$ and the fact that $\max_{0 \le j \le n}\{C_j(\hat{\beta})\} = C_n(\hat{\beta})$. $\qquad\square$

Remark 10.2. By Lemma 10.1, $\|C(\hat{\beta})\|_1 \equiv \sum C_j(\hat{\beta})$ and $\|C(\hat{\beta})\|_\infty \equiv C_{\max}(\hat{\beta})$. Hence, minimizing the norm l_1 (l_∞) is equivalent to minimizing the criterion $\sum C_j$ ($C_{\max}$); see Gawiejnowicz et al. [105] for details.

Remark 10.3. Notice that since $S_{[0]} = t_0 = 0$, the coefficient $\beta_{[0]}$ has no influence on the value of $C_j(\hat{\beta})$ for $0 \le j \le n$. Moreover, $C_j(\hat{\beta})$ depends on β_i in a monotone non-decreasing way for each $1 \le i \le j$. Therefore, given any permutation of sequence $\hat{\beta}$, the best strategy for minimizing $\sum C_j(\hat{\beta})$ is to set as $\beta_{[0]}$ the maximal element in this sequence (cf. Theorem 6.128). In other words, if we start at $t_0 = 0$, the subject of our interest is the remaining n-element subsequence $\beta = (\beta_1, \beta_2, \ldots, \beta_n)$ of $\hat{\beta}$, with $\beta_{[0]}$ maximal. Hence, from now on, we assume that $\beta_{[0]} = \max_{0 \le j \le n}\{\beta_j\}$ and consider mainly sequence β.

Definition 10.4. (Gawiejnowicz et al. [95]) *Let $C(\beta) = [C_1, C_2, \ldots, C_n]$ be the vector of job completion times (6.11) for a given sequence $\beta = (\beta_1, \beta_2, \ldots, \beta_n)$. Functions $F(\beta)$ and $M(\beta)$ are defined as follows:*

$$F(\beta) := \sum_{j=1}^{n}\sum_{i=1}^{j}\prod_{k=i}^{j} \beta_k$$

and

$$M(\beta) := 1 + \sum_{i=1}^{n}\prod_{k=i}^{n} \beta_k.$$

Remark 10.5. We refer to the minimizing of functions $F(\beta)$ and $M(\beta)$ as to the *F-problem* and *M-problem*, respectively. Since, by Lemma 10.1, $\|C(\hat{\beta})\|_1 = F(\beta) + (n+1)$ and $\|C(\hat{\beta})\|_\infty = M(\beta)$, the M-problem and F-problem are closely related to the problems of minimization of the C_{max} criterion (cf. Sect. 6.1.3) and $\sum C_j$ criterion (cf. Sect. 6.2.3), respectively.

Now, we define the basic notion in the chapter (cf. Gawiejnowicz et al. [95]).

Definition 10.6. (Signatures of deterioration rate sequence β)
For a given sequence $\beta = (\beta_1, \beta_2, \ldots, \beta_n)$, *signatures* $S^-(\beta)$ *and* $S^+(\beta)$ *of sequence* β *are defined as follows:*

$$S^-(\beta) := M(\bar{\beta}) - M(\beta) = \sum_{i=1}^{n} \prod_{j=1}^{i} \beta_j - \sum_{i=1}^{n} \prod_{j=i}^{n} \beta_j \qquad (10.1)$$

and

$$S^+(\beta) := M(\bar{\beta}) + M(\beta), \qquad (10.2)$$

where $\bar{\beta} := (\beta_n, \beta_{n-1}, \ldots, \beta_1)$ *is the reverse permutation of elements of* β.

Since the signatures (10.1) and (10.2) are essential in further considerations, we now prove some of their properties.

10.2 Basic properties of signatures

Let us introduce the following notation. Given a sequence $\beta = (\beta_1, \ldots, \beta_n)$ and any two numbers $\alpha > 1$ and $\gamma > 1$, let $(\alpha|\beta|\gamma)$ and $(\gamma|\beta|\alpha)$ denote concatenations of α, β and γ in the indicated orders, respectively. Let $\mathcal{B} := \prod_{j=1}^{n} \beta_j$.

We start with a lemma which shows how to calculate the values of function $F(\cdot)$ for sequence β extended with the elements α and γ if we know the values of $F(\beta)$, $M(\beta)$ and $M(\bar{\beta})$.

Lemma 10.7. (Gawiejnowicz et al. [95]) *For a given sequence* β *and any numbers* $\alpha > 1$ *and* $\gamma > 1$, *there hold the following equalities:*

$$F(\alpha|\beta|\gamma) = F(\beta) + \alpha \, M(\bar{\beta}) + \gamma \, M(\beta) + \alpha \, \mathcal{B} \, \gamma \qquad (10.3)$$

and

$$F(\gamma|\beta|\alpha) = F(\beta) + \gamma \, M(\bar{\beta}) + \alpha \, M(\beta) + \alpha \, \mathcal{B} \, \gamma. \qquad (10.4)$$

Proof. Let $\beta = (\beta_1, \ldots, \beta_n)$, $\beta_0 = \alpha > 1$, $\beta_{n+1} = \gamma > 1$. Then $F(\beta_0|\beta|\beta_{n+1}) = \sum_{j=0}^{n+1} \sum_{i=0}^{j} \beta_i \beta_{i+1} \cdots \beta_j = F(\beta) + \sum_{j=0}^{n} \beta_0 \beta_1 \cdots \beta_j + \sum_{i=0}^{n+1} \beta_i \beta_{i+1} \cdots \beta_{n+1} = F(\beta) + \beta_0 (1 + \sum_{j=1}^{n} \beta_1 \beta_2 \cdots \beta_j) + \beta_{n+1} (1 + \sum_{i=1}^{n} \beta_i \beta_{i+1} \cdots \beta_n) + \beta_0 \beta_1 \cdots \beta_{n+1}$, and equality (10.3) follows. To prove (10.4), it is sufficient to exchange α and γ in (10.3) and to note that the last term remains unchanged. ∎

By Lemma 10.7, we obtain general formulae concerning the difference and the sum of values of $F(\cdot)$ for sequences $(\alpha|\beta|\gamma)$ and $(\gamma|\beta|\alpha)$.

Lemma 10.8. (Gawiejnowicz et al. [95]) *For a given sequence β and any numbers $\alpha > 1$ and $\gamma > 1$, there hold the following equalities:*

$$F(\alpha|\beta|\gamma) - F(\gamma|\beta|\alpha) = (\alpha - \gamma)S^-(\beta) \tag{10.5}$$

and

$$F(\alpha|\beta|\gamma) + F(\gamma|\beta|\alpha) = (\alpha + \gamma)S^+(\beta) + 2(F(\beta) + \alpha\,\mathcal{B}\,\gamma). \tag{10.6}$$

Proof. Let $\beta = (\beta_1, \ldots, \beta_n)$, $\alpha > 1$ and $\gamma > 1$ be given. Then by subtracting the left and the right sides of equalities (10.3) and (10.4), respectively, and by applying Definition 10.6, equation (10.5) follows.

Similarly, by adding the left and the right sides of equalities (10.3) and (10.4), respectively, and by applying Definition 10.6, we obtain equation (10.6). ∎

Lemma 10.8 shows the relation which holds between a signature and a change of the value of function $F(\cdot)$, if the first and the last element in sequence β have been mutually exchanged.

From identities (10.5) and (10.6), we can obtain another pair of equalities, expressed in terms of signatures $S^-(\cdot)$ and $S^+(\cdot)$.

Lemma 10.9. (Gawiejnowicz et al. [95]) *For a given sequence β and any numbers $\alpha > 1$ and $\gamma > 1$, there hold the following equalities:*

$$F(\alpha|\beta|\gamma) = F(\beta) + \frac{1}{2}((\alpha + \gamma)S^+(\beta) + (\alpha - \gamma)S^-(\beta)) + \alpha\,\mathcal{B}\,\gamma \tag{10.7}$$

and

$$F(\gamma|\beta|\alpha) = F(\beta) + \frac{1}{2}((\alpha + \gamma)S^+(\beta) - (\alpha - \gamma)S^-(\beta)) + \alpha\,\mathcal{B}\,\gamma. \tag{10.8}$$

Proof. Indeed, by adding the left and the right sides of equalities (10.5) and (10.6), respectively, we obtain equality (10.7).

Similarly, by subtracting the left and the right sides of equalities (10.6) and (10.5), respectively, we obtain equality (10.8). ∎

The next result shows how to concatenate new elements α and γ with a given sequence β in order to decrease the value of function $F(\cdot)$.

Theorem 10.10. (Gawiejnowicz et al. [95]) *Let there be given a sequence β related to the F-problem and the numbers $\alpha > 1$ and $\gamma > 1$. Then there holds the following equivalence:*

$$F(\alpha|\beta|\gamma) \leq F(\gamma|\beta|\alpha) \text{ iff } (\alpha - \gamma)S^-(\beta) \leq 0. \tag{10.9}$$

Moreover, there holds a similar equivalence, if in equivalence (10.9) the symbol '$\leq$' is replaced with '$\geq$'.

Proof. The result follows from identity (10.5) in Lemma 10.8.   ∎

From Theorem 10.10, it follows that in order to decrease the value of $F(\cdot|\beta|\cdot)$ we should choose $(\alpha|\beta|\gamma)$ instead of $(\gamma|\beta|\alpha)$ when $(\alpha - \gamma)S^-(\beta) \leq 0$, and $(\gamma|\beta|\alpha)$ instead of $(\alpha|\beta|\gamma)$ in the opposite case. Therefore, the behaviour of function $F(\cdot)$ for such concatenations is determined by the sign of the signature $S^-(\beta)$ of the original sequence β.

In the next theorem, we give a greedy strategy for solving the F-problem. This strategy is based on the behaviour of the signature $S^-(\cdot)$ only.

Theorem 10.11. (Gawiejnowicz et al. [95]) *Let $\beta = (\beta_1, \ldots, \beta_n)$ be a non-decreasingly ordered sequence for the F-problem, let $u = (u_1, \ldots, u_{k-1})$ be a V-sequence constructed from the first $k - 1$ elements of β, let $\alpha = \beta_k > 1$ and $\gamma = \beta_{k+1} > 1$, where $1 < k < n$, and let $\alpha \leq \gamma$. Then there holds the following implication:*

$$\text{if } S^-(u) \geq 0, \text{ then } F(\alpha|u|\gamma) \leq F(\gamma|u|\alpha). \tag{10.10}$$

Moreover, there holds a similar implication, if in implication (10.10) the symbol '$\geq$' is replaced by '$\leq$' and the symbol '$\leq$' is by replaced '$\geq$'.

Proof. Assume that the sign of the signature $S^-(u)$ is known. Then it is sufficient to note that by equivalence (10.9) the sign of the difference $F(\alpha|u|\gamma) - F(\gamma|u|\alpha)$ is determined by the sign of the difference $\alpha - \gamma$.   ∎

Theorem 10.11 indicates which one of the two sequences, $(\alpha|u|\gamma)$ or $(\gamma|u|\alpha)$, should be chosen if the sign of the signature $S^-(u)$ is known.

The next result shows a relation between signatures of sequences $(\alpha|\beta|\gamma)$ and $(\gamma|\beta|\alpha)$ and the values of function $M(\cdot)$ for sequences β and $\bar{\beta}$.

Theorem 10.12. (Gawiejnowicz et al. [95]) *For a given sequence β and any numbers $\alpha > 1$ and $\gamma > 1$, there hold the following equalities:*

$$S^-(\alpha|\beta|\gamma) = \alpha\, M(\bar{\beta}) - \gamma\, M(\beta) \tag{10.11}$$

and

$$S^-(\gamma|\beta|\alpha) = \gamma\, M(\bar{\beta}) - \alpha\, M(\beta). \tag{10.12}$$

Proof. Let $\beta = (\beta_1, \ldots, \beta_n)$, $\beta_0 = \alpha > 1$ and $\beta_{n+1} = \gamma > 1$. Then we have $S^-(\alpha|\beta|\gamma) = \sum_{i=0}^{n+1} \beta_0\beta_1 \cdots \beta_i - \sum_{i=0}^{n+1} \beta_i \cdots \beta_n\beta_{n+1} = \beta_0(1+\sum_{i=1}^{n} \beta_1 \cdots \beta_i) - \beta_{n+1}(1+\sum_{i=1}^{n} \beta_i \cdots \beta_n)$. Since the expressions in the brackets are nothing else than $M(\bar{\beta})$ and $M(\beta)$, respectively, identity (10.11) follows.

Similarly, by exchanging α and γ in (10.11), we obtain (10.12).   ∎

From Theorem 10.12, there follow identities which determine the behaviour of subsequently calculated signatures $S^-(\cdot)$.

Theorem 10.13. (Gawiejnowicz et al. [95]) *For a given sequence β and any numbers $\alpha > 1$ and $\gamma > 1$, there hold the following identities:*

$$S^-(\alpha|\beta|\gamma) + S^-(\gamma|\beta|\alpha) = (\alpha + \gamma)S^-(\beta), \qquad (10.13)$$

$$S^-(\alpha|\beta|\gamma) - S^-(\gamma|\beta|\alpha) = (\alpha - \gamma)S^+(\beta) \qquad (10.14)$$

and

$$\begin{aligned} S^-(\alpha|\beta|\gamma)^2 - S^-(\gamma|\beta|\alpha)^2 &= (\alpha^2 - \gamma^2)\left(M(\bar{\beta})^2 - M(\beta)^2\right) \\ &= (\alpha^2 - \gamma^2)\, S^-(\beta)\, S^+(\beta). \qquad (10.15) \end{aligned}$$

Proof. Indeed, by adding the left and right sides of equalities (10.11) and (10.12), respectively, we obtain identity (10.13).

Similarly, by subtracting the left and right sides of equalities (10.11) and (10.12), respectively, we obtain identity (10.14).

Multiplying the left and the right sides of identities (10.13) and (10.14), respectively, we obtain identity (10.15). ∎

Remark 10.14. The analysis of these identities shows that in general, we cannot determine uniquely the sign of signatures $S^-(\alpha|\beta|\gamma)$ and $S^-(\gamma|\beta|\alpha)$ in terms of the sign of the signature $S^-(\beta)$ only even if we know that $\alpha \leq \gamma$ (or $\alpha \geq \gamma$). Indeed, if we know the sign of $F(\alpha|\beta|\gamma) - F(\gamma|\beta|\alpha)$ or, equivalently, the sign of $(\alpha - \gamma)S^-(\beta)$, then from identities (10.13), (10.14) and (10.15), it follows that for the consecutive signatures we only know the sign of $|S^-(\alpha|\beta|\gamma)| - |S^-(\gamma|\beta|\alpha)|$.

Finally, by Theorem 10.13, we can prove one more pair of identities.

Theorem 10.15. (Gawiejnowicz et al. [95]) *For a given sequence β and any numbers $\alpha > 1$ and $\gamma > 1$, there hold the following identities:*

$$S^-(\alpha|\beta|\gamma) = \frac{1}{2}((\alpha + \gamma)S^-(\beta) + (\alpha - \gamma)S^+(\beta)) \qquad (10.16)$$

and

$$S^-(\gamma|\beta|\alpha) = \frac{1}{2}((\alpha + \gamma)S^-(\beta) - (\alpha - \gamma)S^+(\beta)). \qquad (10.17)$$

Proof. Indeed, by adding the left and right sides of identities (10.13) and (10.14), respectively, we obtain identity (10.16).

Similarly, by subtracting the left and right sides of identities (10.14) and (10.13), respectively, we obtain identity (10.17). ∎

Remark 10.16. Considering sequence $\bar{\beta}$ instead of β, we can formulate and prove counterparts of Lemmata 10.7–10.9 and Theorems 10.10–10.13 and 10.15. We omit the formulations of these results, since they do not introduce new insights into the problem. Note only that there holds the equality $S^-(\beta) + S^-(\bar{\beta}) = 0$, i.e., the signatures $S^-(\beta)$ and $S^-(\bar{\beta})$ have opposite signs.

10.3 A greedy algorithm

In this section, we introduce the first greedy heuristic algorithm for the problem $1|p_j = 1+b_jt| \sum C_j$. The algorithm is based on the properties of signatures presented in Sect. 10.2.

Let u denote a V-shaped sequence composed of the first $k \geq 1$ elements of sequence β, which have been ordered non-decreasingly. Let $\alpha = \beta_{k+1} > 1$ and $\gamma = \beta_{k+2} > 1$ be two consecutive elements of β, where $\alpha \leq \gamma$. Then there are two ways of extending sequence u: by concatenating α at the beginning of the left branch and γ at the end of the right branch of the constructed sequence, or conversely. On the basis of the results from Sect. 10.2, we can formulate the following algorithm.

Algorithm H_{51} for the problem $1|p_j = 1 + b_jt| \sum C_j$ ([94])

Input: sequence $\hat{\beta} = (\beta_0, \beta_1, \ldots, \beta_n)$
Output: a suboptimal sequence u

▷ **Step 1:**
   Arrange sequence $\hat{\beta}$ in the non-decreasing order;
▷ **Step 2:**
   if (n is odd) **then**
      $u \leftarrow (\beta_{[1]})$;
      for $i \leftarrow 2$ **to** $n - 1$ **step** 2 **do**
         if $(S^-(u) \leq 0)$ **then** $u \leftarrow (\beta_{[i+1]}|u|\beta_{[i]})$
         else $u \leftarrow (\beta_{[i]}|u|\beta_{[i+1]})$
   else   ▷ n **is even**
      $u \leftarrow (\beta_{[1]}, \beta_{[2]})$;
      for $i \leftarrow 3$ **to** $n - 1$ **step** 2 **do**
         if $(S^-(u) \leq 0)$ **then** $u \leftarrow (\beta_{[i+1]}|u|\beta_{[i]})$
         else $u \leftarrow (\beta_{[i]}|u|\beta_{[i+1]})$;
▷ **Step 3:**
   $u \leftarrow (\beta_0|u)$;
   return u.

The greedy algorithm, starting from an initial sequence, iteratively constructs a new sequence, concatenating the previous sequence with new elements according to the sign of the signature $S^-(u)$ of this sequence.

Notice that since Step 1 runs in $O(n \log n)$ time, Step 2 is a loop running in $O(n)$ time and Step 3 runs in a constant time, the total running time of algorithm H_{51} is $O(n \log n)$.

We illustrate the performance of algorithm H_{51} with two examples (cf.[94]).

Example 10.17. Let $\beta = (2, 3, 4, 6, 8, 16, 21)$. The optimal V-shaped sequence is $\beta^* = (21, 8, 6, 3, 2, 4, 16)$, with $\sum C_j(\beta^*) = 23226$. Algorithm H_{51} generates the sequence $u_{H_{51}} = (21, 8, 6, 2, 3, 4, 16)$, with $\sum C_j(u_{H_{51}}) = 23240$.

Other algorithms, e.g., algorithms H_{31} and H_{32} (cf. Sect. 9.2.2) give worse results: $u_{H_{31}} = (21, 8, 4, 2, 3, 6, 16)$, $u_{H_{32}} = (21, 6, 3, 2, 4, 8, 16)$,   with $\sum C_j(u_{H_{31}}) = 23418$ and $\sum C_j(u_{H_{32}}) = 24890$, respectively.   ♦

Thus, in general, algorithm H_{51} is not optimal. The following example shows, however, that this algorithm can be optimal for sequences of consecutive natural numbers.

Example 10.18. Let $\beta = (2, 3, 4, 5, 6, 7, 8)$. Algorithm H_{51} generates the optimal V-sequence $\beta^\star = (8, 6, 5, 2, 3, 4, 7)$ with $\sum C_j(\beta^\star) = 7386$.

The sequences generated by algorithms H_{31} and H_{32} are the following: $\beta_{H_{31}} = (8, 6, 4, 2, 3, 5, 7)$ and $\beta_{H_{32}} = (8, 5, 4, 2, 3, 6, 7)$, with $\sum C_j(\beta_{H_{31}}) = 7403$ and $\sum C_j(\beta_{H_{32}}) = 7638$, respectively.   ♦

In order to evaluate the quality of schedules generated by algorithm H_{51}, a number of computational experiments have been conducted.

Table 10.1: Experiment results for consecutive integer deterioration rates

n	$OPT(I)$	$R_{H_{51}}^r(I)$	$R_{H_{31}}^r(I)$	$R_{H_{32}}^r(I)$
2	8	⋆	⋆	⋆
3	21	⋆	⋆	0.142857142857
4	65	⋆	0.015384615385	0.138461538462
5	250	⋆	0.008000000000	0.084000000000
6	1,232	⋆	0.008928571429	0.060876623377
7	7,559	⋆	0.003571901045	0.053049345151
8	55,689	⋆	0.002621702670	0.033884609169
9	475,330	⋆	0.000995098142	0.020871815370
10	4,584,532	⋆	0.000558835667	0.014906428835
11	49,111,539	⋆	0.000244423210	0.011506155407
12	577,378,569	⋆	0.000142137247	0.009070282967
13	7,382,862,790	⋆	0.000080251254	0.007401067385
14	101,953,106,744	⋆	0.000052563705	0.006210868342
15	1,511,668,564,323	⋆	0.000035847160	0.005304460215
16	23,947,091,701,857	⋆	0.000025936659	0.004588979235
17	403,593,335,602,130	⋆	0.000019321905	0.004013033262
18	7,209,716,105,574,116	⋆	0.000014779355	0.003541270022
19	136,066,770,200,782,755	⋆	0.000011522779	0.003149229584
20	2,705,070,075,537,727,250	⋆	0.000009131461	0.002819574105

In the first experiment, the schedules generated by algorithm H_{51} have been compared with schedules obtained by algorithms H_{31} and H_{32}. In this experiment, job deterioration coefficients were consecutive natural numbers,

$\beta_j = j + 1$ for $0 \leq j \leq n$. The results of the experiment are summarized in Table 10.1 (cf. [95]). The star ('$\star$') denotes that for a particular value of n, instance I and algorithm A the ratio $R_A^r(I)$ (cf. Remark 2.15) is equal to 0.

The aim of the second computational experiment was to find optimal solutions to the problem $1|p_j = 1 + b_j t| \sum C_j$, where $b_j = j + 1$ for $1 \leq j \leq 20$. The results of the experiment are given in Table 10.2 (the case $n = 2m$) and Table 10.3 (the case $n = 2m - 1$), where $1 \leq m \leq 10$.

Table 10.2: Solutions of the problem $1|p_j = 1 + (1 + j)t| \sum C_j$ for $n \leq 20, n$ even

m	$n = 2m$
1	$(1, 2)$
2	$(4, 1, 2, 3)$
3	$(5, 4, 1, 2, 3, 6)$
4	$(8, 5, 4, 1, 2, 3, 6, 7)$
5	$(9, 8, 5, 4, 1, 2, 3, 6, 7, 10)$
6	$(12, 9, 8, 5, 4, 1, 2, 3, 6, 7, 10, 11)$
7	$(13, 12, 9, 8, 5, 4, 1, 2, 3, 6, 7, 10, 11, 14)$
8	$(16, 13, 12, 9, 8, 5, 4, 1, 2, 3, 6, 7, 10, 11, 14, 15)$
9	$(17, 16, 13, 12, 9, 8, 5, 4, 1, 2, 3, 6, 7, 10, 11, 14, 15, 18)$
10	$(20, 17, 16, 13, 12, 9, 8, 5, 4, 1, 2, 3, 6, 7, 10, 11, 14, 15, 18, 19)$

Table 10.3: Solutions of the problem $1|p_j = 1 + (1 + j)t| \sum C_j$ for $n \leq 20, n$ odd

m	$n = 2m - 1$
1	(1)
2	$(3, 1, 2)$
3	$(4, 3, 1, 2, 5)$
4	$(7, 4, 3, 1, 2, 5, 6)$
5	$(8, 7, 4, 3, 1, 2, 5, 6, 9)$
6	$(11, 8, 7, 4, 3, 1, 2, 5, 6, 9, 10)$
7	$(12, 11, 8, 7, 4, 3, 1, 2, 5, 6, 9, 10, 13)$
8	$(15, 12, 11, 8, 7, 4, 3, 1, 2, 5, 6, 9, 10, 13, 14)$
9	$(16, 15, 12, 11, 8, 7, 4, 3, 1, 2, 5, 6, 9, 10, 13, 14, 17)$
10	$(19, 16, 15, 12, 11, 8, 7, 4, 3, 1, 2, 5, 6, 9, 10, 13, 14, 17, 18)$

Remark 10.19. In sequences presented in Tables 10.2 and 10.3 are omitted, by Remark 10.3, the indices that correspond to the value of $\beta_{[0]}$.

Remark 10.20. The solutions given in Tables 10.2–10.3 have been found by an exact algorithm. Since by Theorem 6.133 any optimal sequence for the problem $1|p_j = 1 + b_j t| \sum C_j$ has to be V-shaped, the algorithm for a given n constructed all possible V-shaped sequences and selected the optimal one.

The results of these experiments suggest that for certain types of β sequences, which will be called *regular*, algorithm H_{51} constructs optimal schedules. A regular sequence is, e.g., the sequence composed of consecutive natural numbers or elements of an arithmetic (a geometric) progression. (Two latter sequences will be called *arithmetic* and *geometric* sequences, respectively.) Moreover, it seems that for regular sequences it is possible to construct an optimal schedule knowing only the form of sequence β, *without* calculation of the signature $S^-(u)$. The arguments supporting the conjectures are given in the next section.

10.4 Signatures of regular sequences

In this section, we present some results which strongly support the conjecture that algorithm H_{51} is optimal for regular sequences of job deterioration rates. We start with the sequence of consecutive natural numbers.

10.4.1 Sequences of consecutive natural numbers

Let us define the following two sequences:

$$\beta = (r_m + (-1)^m, \ldots, r_2 + 1, r_1 - 1, r_1, r_2, \ldots, r_m) \text{ for } n = 2m, \quad (10.18)$$

$$\beta = (s_{m-1} + 2, \ldots, s_2 + 2, s_1 + 2, s_1, s_2, \ldots, s_m) \text{ for } n = 2m - 1, \quad (10.19)$$

where

$$r_k = 2k - \tfrac{1}{2}((-1)^k + 3) + 1, \ k = 1, 2, \ldots, m \text{ for } n = 2m, \quad (10.20)$$

$$s_k = 2k - \tfrac{1}{2}((-1)^k + 3), \ k = 1, 2, \ldots, m \text{ for } n = 2m - 1. \quad (10.21)$$

We will refer to sequences r_k and s_k, and to the related sequence β, as to the *even* and *odd* sequence, respectively.

Remark 10.21. Since sequences given in Tables 10.2 and 10.3 correspond to sequences (10.20) and (10.21) for $1 \leq m \leq 10$, respectively, formulae (10.18) and (10.19) can be considered as generalizations of these sequences for an arbitrary $m \geq 1$.

Now, we prove a formula that can be derived from Definition (10.1) of the signature $S^-(\beta)$. For simplicity of notation, if sequence β is fixed, we will write S_n^- instead of $S^-(\beta)$:

$$S_n^- = \sum_{i=1}^{m} \beta_1 \cdots \beta_i - \sum_{i=1}^{n-m} \beta_{n-i+1} \cdots \beta_n + \sum_{i=1}^{n-m} \beta_1 \cdots \beta_{m+i} - \sum_{i=1}^{m} \beta_i \cdots \beta_n, \quad (10.22)$$

where $1 \leq m \leq n$.

From formula (10.22), we can obtain the following representation of the signature for $n = 2m$ and $n = 2m - 1$, respectively.

Lemma 10.22. (Gawiejnowicz et al. [95]) *Let $\beta = (\beta_1, \ldots, \beta_n)$. If $n = 2m$, then*

$$S_{2m}^- = \sum_{i=1}^{m} \eta_i(m) \left(\prod_{j=1}^{m-i+1} \beta_j - \prod_{j=m+i}^{2m} \beta_j \right), \quad (10.23)$$

where $\eta_1(m) = 1$ and $\eta_i(m) = 1 + \prod_{j=m-i+2}^{m+i-1} \beta_j$ for $i = 2, 3, \ldots, m$.

If $n = 2m - 1$, then

$$S_{2m-1}^- = \sum_{i=1}^{m-1} \omega_i(m) \left(\prod_{j=1}^{m-i} \beta_j - \prod_{j=m+i}^{2m-1} \beta_j \right), \quad (10.24)$$

where $\omega_i(m) = 1 + \prod_{j=m-i+1}^{m+i-1} \beta_j$ for $i = 1, 2, \ldots, m - 1$.

Proof. Let $n = 2m$. Then

$$S_{2m}^- = \sum_{i=1}^{m} (\beta_1 \cdots \beta_i - \beta_{2m-i+1} \cdots \beta_{2m}) + \sum_{i=1}^{m} \beta_1 \cdots \beta_{m+i} - \sum_{i=1}^{m} \beta_i \cdots \beta_{2m}.$$

Reducing the last term in the second sum with the first one in the third sum we have $S_{2m}^- = \sum_{i=1}^{m} (\beta_1 \cdots \beta_i - \beta_{2m-i+1} \cdots \beta_{2m}) + \sum_{i=1}^{m-1} \beta_{i+1} \cdots \beta_{2m-i} \times (\beta_1 \cdots \beta_i - \beta_{2m-i+1} \cdots \beta_{2m})$. Next, by joining both sums and by changing the index of summation according to $i := m - i + 1$, we have $S_{2m}^- = \sum_{i=2}^{m} (1 + \beta_{m-i+2} \cdots \beta_{m+i-1}) \times (\beta_1 \cdots \beta_{m-i+1} - \beta_{m+i} \cdots \beta_{2m}) + \beta_1 \cdots \beta_m - \beta_{m+1} \cdots \beta_{2m}$. Hence, taking into account definitions of the coefficients η_i, formula (10.23) follows.

To prove formula (10.24), we proceed in the same way. Let $n = 2m - 1$. Then $S_{2m-1}^- = \sum_{i=1}^{m} \beta_1 \cdots \beta_i - \sum_{i=1}^{m-1} \beta_{m+i} \cdots \beta_{2m-1} + \sum_{i=1}^{m-1} \beta_1 \cdots \beta_{m+i} - \sum_{i=1}^{m} \beta_i \cdots \beta_{2m-1}$. Changing the index of the summation in the first sum according to $i := m - i$, in the third sum according to $i := i - 1$ and in the last sum according to $i := m - i + 1$, we obtain $S_{2m-1}^- = \sum_{i=0}^{m-1} \beta_1 \cdots \beta_{m-i} - \sum_{i=1}^{m-1} \beta_{m+i} \cdots \beta_{2m-1} + \sum_{i=2}^{m-1} (\beta_1 \cdots \beta_{m-i}) \times (\beta_{m-i+1} \cdots \beta_{m+i-1}) - \sum_{i=1}^{m-1} (\beta_{m-i+1} \cdots \beta_{m+i-1}) \times (\beta_{m+i} \cdots \beta_{2m-1})$. By moving in the first sum the term with the index $i = 0$ to the third one under the index $i = 1$ and applying definitions of coefficients ω_i, we obtain formula (10.24). ∎

Lemma 10.23. (Gawiejnowicz et al. [95]) *Let* $n = 2m$, *and let* β *be an even sequence. Then for each integer* $m \geq 1$, *there holds the following equality:*

$$S_{2m}^- = \sum_{i=1}^m \eta_i \left(\prod_{j=i}^m (r_j + (-1)^j) - \prod_{j=i}^m r_j \right), \qquad (10.25)$$

where

$$\eta_1 = 1 \text{ and } \eta_i = 1 + \prod_{j=1}^{i-1} r_j \prod_{j=1}^{i-1} (r_j + (-1)^j) \qquad (10.26)$$

for $i = 2, 3, \ldots, m$.

Proof. Applying Lemma 10.22 to sequence β given by formulae (10.18) and (10.20), we obtain formula (10.25) for the signature S_{2m}^-. ∎

Now, on the basis of (10.25), we can state the following result.

Theorem 10.24. (Gawiejnowicz et al. [95]) *Let* $n = 2m$, *and let* β *be an even sequence. Then for the signatures* S_{2m+2}^- *and* S_{2m}^-, *there holds the following formula:*

$$S_{2m+2}^- = r_{m+1} S_{2m}^- + (-1)^{m+1} \sum_{i=1}^{m+1} \eta_i \prod_{j=i}^m (r_j + (-1)^j), \qquad (10.27)$$

where the signature S_{2m}^- *and coefficients* η_i *are defined by formulae (10.25) and (10.26), respectively. Moreover, there holds the following identity:*

$$S_{2m+2}^- = R_m \left((-1)^{m+1} + \Theta_m \right), \qquad (10.28)$$

where $\Theta_m = \frac{S_{2m}^-}{R_m} (r_{m+1} + (-1)^{m+1})$ *and* $R_m = \sum_{i=1}^{m+1} \eta_i \prod_{j=i}^m r_j$.

Proof. Applying Lemma 10.23, we obtain

$$S_{2m+2}^- = \sum_{i=1}^{m+1} \eta_i \left((r_i + (-1)^i) \cdots (r_{m+1} + (-1)^{m+1}) - r_i \cdots r_{m+1} \right)$$

$$= \sum_{i=1}^m \eta_i \left((r_i + (-1)^i) \cdots (r_{m+1} + (-1)^{m+1}) - r_i \cdots r_{m+1} \right)$$

$$+ \eta_{m+1} \left((r_{m+1} + (-1)^{m+1}) - r_{m+1} \right)$$

$$= r_{m+1} S_{2m}^- + (-1)^{m+1} \sum_{i=1}^m \eta_i \left((r_i + (-1)^i) \cdots (r_m + (-1)^m) \right)$$

$$+ \eta_{m+1} \left((r_{m+1} + (-1)^{m+1}) - r_{m+1} \right)$$

$$= r_{m+1} S_{2m}^- + (-1)^{m+1} \sum_{i=1}^{m+1} \eta_i \left((r_i + (-1)^i) \cdots (r_m + (-1)^m) \right),$$

and formula (10.27) follows in view of the definition of coefficients η_i. Formula (10.28) follows from the assumed notation and formula (10.27).   ∎

Now we consider the case of an *odd* sequence. Applying Lemma 10.22 to sequence β given by formulae (10.19) and (10.21), we obtain the following formula for the signature S_{2m-1}^-.

Lemma 10.25. (Gawiejnowicz et al. [95]) *Let β be an odd sequence and let $n = 2m - 1$. Then for every integer $m \geq 1$ there holds the following equality:*

$$S_{2m-1}^- = \sum_{i=1}^{m-1} \omega_i \left(\prod_{j=i}^{m-1} (s_j + 2) - \prod_{j=i+1}^{m} s_j \right) \tag{10.29}$$

where $\omega_i = 1 + \prod_{j=1}^{i} s_j \prod_{j=1}^{i-1} (s_j + 2)$ for $i = 1, 2, \ldots, m - 1$.

Proof. Let $n = 2m - 1$, and let β be an odd sequence. Then we have $\beta_{m-i} = s_i + 2$ for $i = 1, 2, \ldots, m - 1$ and $\beta_{m+i-1} = s_i$ for $i = 1, 2, \ldots, m$. Substituting these values in formula (10.24) and noticing that $\omega_i = 1 + \beta_{m-i+1} \cdots \beta_{m+i-1} = 1 + s_1 \cdots s_i (s_1 + 2) \cdots (s_{i-1} + 2)$, formula (10.29) follows.   ∎

On the basis of formula (10.29), we can state the following result, concerning the behaviour of the signature S_n^- for $n = 2m + 1$.

Theorem 10.26. (Gawiejnowicz et al. [95]) *Let $n = 2m + 1$ and let β be an odd sequence. Then for the signatures S_{2m+1}^- and S_{2m-1}^-, there holds the following formula:*

$$S_{2m+1}^- = (s_m + 2) S_{2m-1}^- + (-1)^{m+1} \sum_{i=1}^{m} \omega_i \prod_{j=i+1}^{m} s_j, \tag{10.30}$$

where $\omega_i = 1 + \prod_{j=1}^{i} s_j \prod_{j=1}^{i-1} (s_j + 2)$ for $i = 1, 2, \ldots, m$. Moreover, for S_{2m+1}^- there holds the following identity:

$$S_{2m+1}^- = Q_m \left((-1)^{m+1} + \Gamma_m \right), \tag{10.31}$$

where $\Gamma_m = \frac{S_{2m-1}^-}{Q_m}(s_m + 2)$ and $Q_m = \sum_{i=1}^{m} \omega_i \prod_{j=i+1}^{m} s_j$.

Proof. By Lemma 10.25, for $\omega_i = 1 + (s_1 \cdots s_i)(s_1 + 2) \cdots (s_{i-1} + 2)$, we obtain

$$S_{2m+1}^- = \sum_{i=1}^{m} \omega_i \left((s_i + 2) \cdots (s_m + 2) - s_{i+1} \cdots s_{m+1} \right)$$

$$= \sum_{i=1}^{m-1} \omega_i \left((s_i + 2) \cdots (s_m + 2) - q_i + q_i - s_{i+1} \cdots s_{m+1} \right)$$

$$+ \omega_m \left((s_m + 2) - s_{m+1} \right),$$

where $q_i \equiv (s_{i+1} \cdots s_m)(s_m + 2)$. Hence, by Lemma 10.25, we have

$$S_{2m+1}^- = (s_m + 2) \sum_{i=1}^{m-1} \omega_i \left((s_i + 2) \cdots (s_{m-1} + 2) - s_{i+1} \cdots s_m \right)$$

$$+ \sum_{i=1}^{m-1} \omega_i (s_{i+1} \cdots s_m) \left((s_m + 2) - s_{m+1} \right)$$

$$+ \omega_m \left((s_m + 2) - s_{m+1} \right).$$

Collecting the last terms, applying identity (10.25) and using the equality $(s_m + 2) - s_{m+1} = (-1)^{m+1}$, we obtain formula (10.30).

Formula (10.31) is an immediate consequence of formula (10.30) and the assumed notation. ∎

We will now prove that for an arbitrary m, the sign of signatures S_{2m}^- and S_{2m-1}^- varies periodically. Knowing the behaviour of the signatures, we are able to simplify algorithm H_{51}, since we will not have to calculate the signatures in Step 2 of the algorithm.

Theorem 10.27. (Gawiejnowicz et al. [95]) *Let there be given V-sequences (10.18) and (10.19) of sequence $\beta := (1, 2, \ldots, n)$. Then for each integer $m \geq 1$ the sign of the signatures S_{2m}^- and S_{2m-1}^- for these sequences varies periodically according to the formulae $\mathrm{sign}(S_{2m}^-) = (-1)^m$ and $\mathrm{sign}(S_{2m-1}^-) = (-1)^m$, respectively.*

Before we prove Theorem 10.27, we will prove some technical lemmata.

Lemma 10.28. (Gawiejnowicz et al. [95]) *For every integer $m \geq 1$, there hold the following recurrence relations:*

$$\Theta_1 = -\frac{4}{5}, \Theta_{m+1} = D_m(\Theta_m - (-1)^m) \text{ for } n = 2m \tag{10.32}$$

and

$$\Gamma_1 = 0, \Gamma_2 = \frac{8}{11}, \Gamma_{m+1} = E_m(\Gamma_m - (-1)^m) \text{ for } n = 2m - 1, \tag{10.33}$$

where

$$D_m = (r_{m+1} + 2) \frac{R_m}{R_{m+1}} \tag{10.34}$$

and

$$E_m = (s_{m+1} + 2) \frac{Q_m}{Q_{m+1}}. \tag{10.35}$$

Proof. Recurrence relations (10.32) and (10.33) follow from Theorem 10.24 and Theorem 10.26, respectively. In the case of formula (10.34), we apply the equality $r_{m+2} + (-1)^{m+2} = r_{m+1} + 2$. In both formulae, (10.34) and (10.35), it is sufficient to apply definitions of Θ_{m+1} and Γ_{m+1}, and the recurrence formulae (10.25) and (10.29) for S_{2m}^- and S_{2m-1}^-, respectively. Clearly, the definitions of R_m (cf. Theorem 10.24) and Q_m (cf. Theorem 10.26) must be also applied. ∎

Remark 10.29. Note that we have $|\Theta_1| < 1$, $\Gamma_1 = 0$ and $|\Gamma_2| < 1$. Moreover, $\Theta_1 < 0$ and $\Gamma_2 > 0$.

The next two lemmata are needed in proofs of inequalities $0 < D_m < 1$ and $0 < E_m < 1$.

Lemma 10.30. (Gawiejnowicz et al. [95]) *For every integer $m \geq 1$ there holds $R_m < \frac{1}{2}\eta_{m+2}$.*

Proof. We will proceed by induction. The case $m = 1$ is immediate, since $R_1 = \eta_1 r_1 + \eta_2 = r_1 + (1 + r_1(r_1 - 1))$ and $\eta_3 = 1 + r_1 r_2(r_1 + 2)$, where $r_1 = 2$ and $r_2 = 3$.

Now, assume that $R_{m-1} < \frac{1}{2}\eta_{m+1}$. Hence

$$R_m = r_m R_{m-1} + \eta_{m+1} < \frac{1}{2}(r_m + 2)\eta_{m+1}.$$

Thus, it is sufficient to prove that $(r_m + 2)\eta_{m+1} < \eta_{m+2}$. To prove this, note first that $(r_m + 2)\eta_{m+1} = (r_m + 2) + \frac{1}{r_{m+1}}(\eta_{m+2} - 1)$. Consequently, we have to prove that $(r_m + 1) + (1 - \frac{1}{r_{m+1}}) < (1 - \frac{1}{r_{m+1}})\eta_{m+2}$, or that

$$r_m + 1 < (1 - \frac{1}{r_{m+1}})(\eta_{m+2} - 1) = \frac{r_{m+1} - 1}{r_{m+1}}(r_1 \cdots r_m r_{m+1})(r_1 + 2)\cdots(r_m + 2).$$

Since $r_{m+1} = r_m + 2 + (-1)^m$, it is sufficient to check the latter inequality in the expression

$$\frac{r_m + 1}{r_m + 1 + (-1)^m} \leq 1 + \frac{1}{r_m} < (r_1 \cdots r_m)(r_1 + 2)\cdots(r_m + 2).$$

Finally, since $r_m \nearrow$, it is sufficient to check the case $m = 1$, which is evident. ∎

Lemma 10.31. (Gawiejnowicz et al. [95]) *For every integer $m \geq 1$ there holds $Q_m < \frac{1}{2}\omega_{m+1}$.*

Proof. First, we prove the inequality

$$(s_m + 2)\omega_m < \omega_{m+1}. \tag{10.36}$$

We have

$$(s_m + 2)\omega_m = (s_m + 2)(1 + (s_1 \cdots s_m)(s_1 + 2) \cdots (s_{m-1} + 2))$$
$$= (s_m + 2) + \frac{1}{s_{m+1}}((1 + (s_1 \cdots s_{m+1})(s_1 + 2) \cdots (s_m + 2)) - 1)$$
$$= (s_m + 2) + \frac{1}{s_{m+1}}(\omega_{m+1} - 1).$$

It is sufficient to check that $(s_m + 2) + \frac{1}{s_{m+1}}(\omega_{m+1} - 1) < \omega_{m+1}$, or that

$$\frac{s_m + 1}{s_m + (1 + (-1)^m)} \leq 1 + \frac{1}{s_m} < (s_1 \cdots s_m)(s_1 + 2) \cdots (s_m + 2). \quad (10.37)$$

To obtain these inequalities, we have applied the equality $s_{m+1} - 1 = s_m + (1 + (-1)^m)$. Since $s_i \geq 1$, inequalities (10.37) are obviously satisfied, which completes the proof of inequality (10.36).

To prove the lemma, we will proceed by induction. Let $m = 1$. Since $2Q_1 = 2\omega_1 = 2(1 + s_1)$ and $\omega_2 = 1 + (s_1 s_2)(s_1 + 2)$, we obtain $2Q_1 < \omega_2$, since $s_1 = 1$ and $s_2 = 2$. Now, assume that $Q_{m-1} < \frac{1}{2}\omega_m$ holds. Since

$$Q_m = \sum_{i=1}^{m} \omega_i \, s_{i+1} \cdots s_m = s_m Q_{m-1} + \omega_m,$$

we obtain $Q_m < \frac{1}{2}(s_m + 2)\omega_m$. Now, applying inequality (10.36), we obtain

$$Q_m < \frac{1}{2}(s_m + 2)\omega_m < \frac{1}{2}\omega_{m+1}.$$

∎

Lemma 10.32. (Gawiejnowicz et al. [95]) *For every integer $m \geq 1$, there hold the following inequalities*:

$$0 < D_m < 1 \quad \text{and} \quad 0 < E_m < 1.$$

Proof. It is easy to see that $D_m > 0$ and $E_m > 0$. From the definition of D_m, the inequality $D_m < 1$ is equivalent to $R_m < \frac{1}{2}\eta_{m+2}$, which is satisfied in view of Lemma 10.30. To prove that $E_m < 1$, we apply Lemma 10.31. Indeed, in view of the definition of E_m, the inequality $E_m < 1$ is equivalent to the inequality $Q_m < \frac{1}{2}\omega_{m+1}$ from Lemma 10.31. ∎

Lemma 10.33. (Gawiejnowicz et al. [95]) *For every integer $m \geq 1$ there holds the inequality $|\Theta_m| < 1$ and the equality $\mathrm{sign}(\Theta_m) = (-1)^m$.*

Proof. Taking $m = 2k$ or $m = 2k - 1$, for every integer $k \geq 1$ we obtain, by Lemma 10.28, respectively:

$$\Theta_{2k+1} = D_{2k}(\Theta_{2k} - (-1)^{2k}) = D_{2k}(\Theta_{2k} - 1)$$

and

$$\Theta_{2k} = D_{2k-1}(\Theta_{2k-1} - (-1)^{2k-1}) = D_{2k-1}(\Theta_{2k-1} + 1).$$

After substituting the value $2k - 1$ for odd indices we have

$$\Theta_{2k+1} = D_{2k}(D_{2k-1}(\Theta_{2k-1} + 1) - 1).$$

To prove that $|\Theta_{2k-1}| < 1$ and $\Theta_{2k-1} < 0$ for every integer $k \geq 1$, we will proceed by induction.

For $k = 1$ we have $|\Theta_1| < 1$ and $\Theta_1 < 0$, since $\Theta_1 = -\frac{4}{5}$ by definition. Assume that $|\Theta_{2k-1}| < 1$ and $\Theta_{2k-1} < 0$. We will prove that $|\Theta_{2k+1}| < 1$ and $\Theta_{2k+1} < 0$. By induction assumption, $0 < D_{2k-1}(\Theta_{2k-1}+1) < 1$ and consequently $0 < 1 - D_{2k-1}(\Theta_{2k-1}+1) < 1$. Hence, $|\Theta_{2k+1}| < 1$ and $\Theta_{2k+1} < 0$.

Now, consider the case of even indices $2k$. Applying the odd case, we obtain

$$\Theta_{2k} = D_{2k-1}(\Theta_{2k-1} + 1) > 0$$

with $D_{2k-1}(\Theta_{2k-1} + 1) < 1$. Consequently, $|\Theta_{2k}| < 1$ with $Q_{2k} > 0$.   ■

Lemma 10.34. (Gawiejnowicz et al. [95]) *For every integer $m > 1$ there holds the inequality $|\Gamma_m| < 1$ and the equality $\mathrm{sign}(\Gamma_m) = (-1)^m$.*

Proof. Taking $m = 2k$ or $m = 2k - 1$, for every integer $k \geq 1$, we obtain, by Lemma 10.28, respectively:

$$\Gamma_{2k+1} = E_{2k}(\Gamma_{2k} - (-1)^{2k}) = E_{2k}(\Gamma_{2k} - 1)$$

and

$$\Gamma_{2k} = E_{2k-1}(\Gamma_{2k-1} - (-1)^{2k-1}) = E_{2k-1}(\Gamma_{2k-1} + 1).$$

After substituting the value $2k - 1$ for odd indices we have

$$\Gamma_{2k+1} = E_{2k}(E_{2k-1}(\Gamma_{2k-1} + 1) - 1).$$

To prove that $|\Gamma_{2k-1}| < 1$ and $\Gamma_{2k-1} < 0$ for every integer $k \geq 2$, we proceed by induction.

Note that for $k = 1$, we have $\Gamma_1 = 0$. For $k = 2$ there holds $|\Gamma_3| < 1$ and $\Gamma_3 < 0$ since $\Gamma_3 = E_2(\Gamma_2 - 1) = E_2(\frac{8}{11} - 1)$ and $0 < E_m < 1$.

Now, let $|\Gamma_{2k-1}| < 1$ and $\Gamma_{2k-1} < 0$ for an arbitrary $k > 2$. Then

$$-1 < E_{2k-1}(\Gamma_{2k-1} + 1) - 1 < 0,$$

since $0 < E_{2k-1}(\Gamma_{2k-1}+1) < 1$. Finally, we obtain $|\Gamma_{2k+1}| < 1$ and $\Gamma_{2k+1} < 0$, which finishes the induction step. This result implies that

$$\Gamma_{2k} = E_{2k-1}(\Gamma_{2k-1} + 1) > 0 \quad \text{and} \quad \Gamma_{2k} < 1$$

for each integer $k \geq 2$. Moreover, $\Gamma_2 = \frac{8}{11}$, i.e., $\Gamma_2 > 0$ and $\Gamma_2 < 1$.   ■

Lemmata 10.33 and 10.34 allow us to prove Theorem 10.27.

Proof of Theorem 10.27. In view of the formula $S_{2m+2}^- = R_m((-1)^{m+1} + \Theta_m)$ for an arbitrary integer $m \geq 2$, from the fact that $\text{sign}(S_2^-) = 1$ and from Lemma 10.33, we have that $\text{sign}(S_{2m}^-) = (-1)^m$ for an arbitrary integer $m \geq 1$.

Similarly, in view of the formula $S_{2m+1}^- = Q_m((-1)^{m+1} + \Gamma_m)$ for an arbitrary integer $m \geq 1$, from the fact that $\text{sign}(S_1^-) = -1$ and from Lemma 10.34, we have that $\text{sign}(S_{2m-1}^-) = (-1)^m$ for an arbitrary integer $m \geq 1$. ∎

The results of this section lead us to the following.

Conjecture 10.35. Algorithm H_{51} is optimal for the $1|p_j = 1 + b_j t| \sum C_j$ problem in the case when $b_j = j + 1$ for $j = 0, 1, 2, \ldots, n$.

Remark 10.36. If Conjecture 10.35 is true, then in Step 2 of algorithm H_{51} it is not necessary to check the sign of the signature $S^-(u)$, since the sign varies periodically. The simplified version of algorithm H_{51} will be called H_{52}.

Algorithm H_{52} for the problem $1|p_j = 1 + b_j t| \sum C_j$ ([94])

Input: sequence $\hat{\beta} = (\beta_0, \beta_1, \ldots, \beta_n)$
Output: a suboptimal sequence u

▷ Step 1:
   Arrange sequence $\hat{\beta}$ in the non-decreasing order;
▷ Step 2:
  if (n is odd) **then**
     $u \leftarrow (\beta_{[1]})$;
     $sgn \leftarrow (-1)$; ▷ the sign of signature of u
     for $i \leftarrow 2$ **to** $n - 1$ **step** 2 **do**
        $sgn \leftarrow sgn \times (-1)$;
        if ($sgn < 0$) **then** $u \leftarrow (\beta_{[i+1]}|u|\beta_{[i]})$
        else $u \leftarrow (\beta_{[i]}|u|\beta_{[i+1]})$
  else ▷ n is even
     $u \leftarrow (\beta_{[1]}, \beta_{[2]})$;
     $sgn \leftarrow 1$;
     for $i \leftarrow 3$ **to** $n - 1$ **step** 2 **do**
     $sgn \leftarrow sgn \times (-1)$;
        if ($sgn < 0$) **then** $u \leftarrow (\beta_{[i+1]}|u|\beta_{[i]})$
        else $u \leftarrow (\beta_{[i]}|u|\beta_{[i+1]})$;
▷ Step 3:
   $u \leftarrow (\beta_0|u)$;
   return u.

In the remaining part of the section, the above results will be extended to cover arithmetic and geometric sequences (cf. Gawiejnowicz et al. [99]).

10.4.2 Arithmetic sequences

We start the subsection with two examples that illustrate the behaviour of algorithms H_{51} and H_{52} for arithmetic sequences.

Let α_A and ρ_A denote, respectively, the *first term* and the *common difference* in arithmetic sequence $\beta_j = \alpha_A + j\rho_A$, where $0 \leq j \leq n$.

Example 10.37. Let $\beta = (1.5, 2.0, \ldots, 9.0)$ be an arithmetic sequence in which $n = 15$, $\alpha_A = 1.5$ and $\rho_A = 0.5$. Then the optimal V-sequence is

$$\beta^\star = (9.0, 8.5, 7.0, 6.5, 5.0, 4.5, 3.0, 2.5, 1.5, 2.0, 3.5, 4.0, 5.5, 6.0, 7.5, 8.0),$$

with $\sum C_j(\beta^\star) = 7071220899.8750$.   ◆

Example 10.38. Let $\beta = (1.5, 1.8, \ldots, 6.3)$ be an arithmetic sequence in which $n = 16$, $\alpha_A = 1.5$ and $\rho_A = 0.3$, Then the optimal V-sequence is

$$\beta^\star = (6.3, 6.0, 5.1, 4.8, 3.9, 3.6, 2.7, 2.4, 1.5, 1.8, 2.1, 3.0, 3.3, 4.2, 4.5, 5.4, 5.7),$$

with $\sum C_j(\beta^\star) = 642302077.7853$.   ◆

Since, in both cases, algorithms H_{51} and H_{52} generate the optimal schedules, the examples suggest that in the case of arithmetic sequences the algorithms behave similarly to the case of consecutive natural numbers.

Let us now introduce V-sequences of arithmetic sequences by the formulae

$$\beta = (u_m + (-1)^m \alpha_A, \ldots, u_2 + \alpha_A, u_1 - \alpha_A, u_1, u_2, \ldots, u_m) \qquad (10.38)$$

$$\beta = (v_{m-1} + 2\alpha_A, \ldots, v_2 + 2\alpha_A, v_1 + 2\alpha_A, v_1, v_2, \ldots, v_m) \qquad (10.39)$$

for $n = 2m$ and $n = 2m - 1$, respectively, where sequences $u_k = \alpha_A r_k + \rho_A$, $v_k = \alpha_A s_k + \rho_A$ are such that $u_k \geq 1$ and $v_k \geq 1$ for $1 \leq k \leq m$, and sequences (r_k) and (s_k) are defined by (10.20) and (10.21), respectively.

In this case, there holds the following counterpart of Theorem 10.27.

Theorem 10.39. (Gawiejnowicz et al. [99]) *Let $\rho_A \geq 0$ and $\alpha_A + \rho_A \geq 1$. Then the sign of signatures S^- for the arithmetic sequences (10.38) with $\alpha_A \geq 0.11$ and the sequences (10.39) with $\alpha_A \geq 0.50$ varies according to formulae* $\mathrm{sign}(S_{2m}^-) = (-1)^m$ *and* $\mathrm{sign}(S_{2m-1}^-) = (-1)^m$, *respectively, where $m \geq 1$.*

Proof. Let $n = 2m$, $\alpha_A \geq 0.11$, $\rho_A \geq 0$ and $\alpha_A + \rho_A \geq 1$. Then there holds the recurrence relation $\Theta_{m+1} = D_m(\Theta_m + (-1)^{m+1}\alpha_A)$ with

$$\Theta_1 = -\frac{\alpha_A(4\alpha_A + \rho_A)}{(2\alpha_A + \rho_A)(\alpha_A + \rho_A + 1) + 1},$$

where $D_m = (u_{m+1} + 2\alpha_A)\frac{R_m}{R_{m+1}}$.

Similarly, if $n = 2m - 1$, $\alpha_A \geq 0.50$, $\rho_A \geq 0$ and $\alpha_A + \rho_A \geq 1$, then there holds the recurrence relation $\Gamma_{m+1} = F_m(\Gamma_m + (-1)^{m+1}\alpha_A)$, with $\Gamma_1 = 0$ and

$$\Gamma_2 = \frac{\alpha_A(\alpha_A + \rho_A + 1)(4\alpha_A + \rho_A)}{(2\alpha_A + \rho_A)(1 + (\alpha_A + \rho_A)(3\alpha_A + \rho_A + 1)) + 1},$$

where $F_m = (v_{m+1} + 2\alpha_A)\frac{Q_m}{Q_{m+1}}$.

To end the proof, it is sufficient to show that $0 < D_m < 1$ and $\Theta_m < \alpha_A$, if $n = 2m$ and that $0 < F_m < 1$ and $\Gamma_n < \alpha_A$, if $n = 2m - 1$. $\qquad\square$

10.4.3 Geometric sequences

We start the subsection with two examples that illustrate the behaviour of algorithms H_{51} and H_{52} for geometric sequences.

Let ρ_G denote the *ratio* in geometric sequence $\beta_j = \rho_G^j$, where $1 \leq j \leq n$.

Example 10.40. Let $\beta = (3, 9, \ldots, 19683)$ be a geometric sequence for $n = 8$ and $\rho_G = 3$. Then, the optimal V-sequence is

$$\beta^\star = (19683, 6561, 243, 81, 3, 9, 27, 729, 2187),$$

with $\sum C_j(\beta^\star) = 150186346871598597$.   ◆

Example 10.41. Let $\beta = (2, 4, \ldots, 2048)$ be a geometric sequence for $n = 10$ and $\rho_G = 2$. Then the optimal V-sequence is

$$\beta^\star = (2048, 256, 128, 16, 8, 2, 4, 32, 64, 512, 1024)$$

with $\sum C_j(\beta^\star) = 36134983945485585$.   ◆

As previously, algorithms H_{51} and H_{52} generate the optimal schedules.

We now define two sequences which are counterparts of sequences (10.38) and (10.39) for geometric sequences. We will distinguish the case $n = 2m$ and the case $n = 2m - 1$. Let for some $\rho_G > 1$

$$\beta = (\rho_G^{r_m + (-1)^m}, \ldots, \rho_G^{r_2 + 1}, \rho_G^{r_1 - 1}, \rho_G^{r_1}, \rho_G^{r_2}, \ldots, \rho_G^{r_m}) \tag{10.40}$$

and

$$\beta = (\rho_G^{s_{m-1} + 2}, \ldots, \rho_G^{s_2 + 2}, \rho_G^{s_1 + 2}, \rho_G^{s_1}, \rho_G^{s_2}, \ldots, \rho_G^{s_m}) \tag{10.41}$$

for $n = 2m$ and $n = 2m - 1$, respectively.

In this case, there holds the following counterpart of Theorem 10.39. (We omit a technical proof.)

Theorem 10.42. (Gawiejnowicz et al. [99]) *The sign of signatures* $S^-(\beta)$ *of the geometric sequences* (10.40) *and* (10.41) *varies according to formulae* $\text{sign}(S^-_{2m}) = (-1)^m$ *and* $\text{sign}(S^-_{2m-1}) = (-1)^m$, *respectively, where* $m \geq 1$.

A computational experiment has been conducted in order to evaluate the quality of schedules generated by algorithms H_{51} and H_{52} for arithmetic and geometric sequences (see Gawiejnowicz et al. [99] for details). In the experiment, random instances of arithmetic and geometric sequences were generated. Fifty instances were generated for each value of n, where $n = 10, 15, 20$. Algorithms H_{51} and H_{52} found an optimal schedule for all 150 instances.

10.4.4 Arbitrary sequences

From Example 10.17, we know that algorithm H_{51} (and hence H_{52}) is not optimal for arbitrary β sequences. A computational experiment has been conducted in order to evaluate the quality of schedules generated by algorithms H_{51} and H_{52} for arbitrary sequences (see Gawiejnowicz et al. [99] for details).

In the experiment, random instances of β sequence were generated. Fifty instances were generated for each value of $n = 10, 15, 20$. The average ratio $R_{H_{51}}(I)$, calculated for 50 instances, was equal to 6654×10^{-8}, 5428×10^{-8} and 1695×10^{-8} for $n = 10$, $n = 15$ and $n = 20$, respectively. The average ratio $R^r_{H_{52}}(I)$ was equal to 26988×10^{-8}, 12927×10^{-8} and 2698×10^{-8} for $n = 10$, $n = 15$ and $n = 20$, respectively.

Hence, the performance of algorithms H_{51} and H_{52} for random sequences of average size is quite satisfactory.

With this remark, we end the presentation of heuristic algorithms based on signatures of sequences of job deterioration rates. In Chap. 11, we will consider local search algorithms for time-dependent scheduling problems.

10.5 Concluding remarks

In this chapter, we considered two $O(n \log n)$ greedy heuristic algorithms for the problem $1|p_j = 1 + b_j t| \sum C_j$. Both these algorithms, H_{51} and H_{52}, are based on signatures of sequences of job deterioration rates.

We have shown that algorithm H_{51} generates V-shaped sequences (10.18) and (10.19), which are optimal for the problem with $b_j = j + 1$ for $n \leq 20$ (see Tables 10.2–10.3). We also proved (cf. Theorem 10.27) that signatures of these sequences vary periodically. (In the latter case, we simplified algorithm H_{51} and formulated algorithm H_{52}.) Finally, we formulated the conjecture that algorithm H_{51} is optimal when job deterioration rates constitute the sequence of subsequent natural numbers, an arithmetic or a geometric sequence.

Experiments have shown that the proposed algorithms perform very well for random sequences with $n \leq 20$. In general, algorithm H_{51} is better than algorithm H_{52}. The former algorithm is recommended for arbitrary sequences, while the latter one is recommended for regular sequences.

The formal proof of optimality of both these algorithms still remains an open problem.

11

Local search algorithms

$\mathbf{S}$ uboptimal schedules for intractable time-dependent scheduling problems may be found by using various algorithms. In this chapter, which completes the third part of the book, we consider local search algorithms.

Chapter 11 is composed of four sections. In Sect. 11.1, we recall basic definitions concerning local search algorithms. In Sect. 11.2, we briefly review basic types of local search algorithms. In Sect. 11.3, we discuss local search algorithms for time-dependent scheduling problems. Conclusions and one table are given in Sect. 11.4.

11.1 Preliminaries

In the section, we introduce basic definitions and present general concepts related to local search algorithms.

11.1.1 Basic definitions

An optimization problem P is specified by a collection of instances of the problem. An instance is defined by the implicit specification of a pair $(\mathfrak{F}, f)$, where the *solution space* $\mathfrak{F}$ is the set of all feasible solutions and $f : \mathfrak{F} \rightarrow \mathbb{R}$ is a criterion function. (Without loss of generality, cf. Sect. 2.1, we restrict the further discussion to minimization problems.) A solution $s^\star \in \mathfrak{F}$ is *optimal* (a *global minimum*), if $f(s^\star) \leq f(s)$ for all $s \in \mathfrak{F}$. The set $\mathfrak{F}_{opt} := \{s \in \mathfrak{F} : f(s) = f^\star\}$ is called *the set of all optimal solutions* (*global minima*). The problem P is solved, if a solution $s \in \mathfrak{F}_{opt}$ has been found.

For a given optimization problem P, a *neighbourhood function* $\mathcal{N} : \mathfrak{F} \rightarrow 2^{\mathfrak{F}}$ may be defined. For each solution $s \in \mathfrak{F}$, the function specifies a set $\mathcal{N}(s) \subseteq \mathfrak{F}$ of *neighbours* of s. The set $\mathcal{N}(s)$ is called the *neighbourhood* of solution s. A solution $s^\circ \in \mathfrak{F}$ is called a *local minimum with respect to* $\mathcal{N}$, if $f(s^\circ) \leq f(s)$ for all $s \in \mathcal{N}(s^\circ)$. A neighbourhood function $\mathcal{N}$ is called *exact*, if every local minimum with respect to $\mathcal{N}$ is also a global minimum.

11.1.2 General concepts in local search

The term *local search* refers to a general approach applied for finding suboptimal solutions to intractable optimization problems. The main idea is to start from an initial solution, $s_0 \in \mathfrak{F}$, construct its neighbourhood $\mathcal{N}(s_0)$ and look for better solutions there. Basically, it is assumed that the neighborhood includes only feasible and complete solutions (cf. Remark 2.8) which are 'close', in the problem-specific sense, to the solution s_0. The pseudo-code of a general local search algorithm is as follows.

Algorithm *GeneralLocalSearch*

Input: initial solution s_0, neighbourhood function $\mathcal{N}$, criterion f
Output: a locally optimal solution s_{act}

▷ Step 1:
$\quad s_{act} \leftarrow s_0$;
$\quad Initialization$;
▷ Step 2:
$\quad$ **repeat**
$\quad\quad$ Generate $\mathcal{N}(s_{act})$;
$\quad\quad$ **for all** $s \in \mathcal{N}(s_{act})$ **do**
$\quad\quad\quad$ **if** $f(s) < f(s_{act})$ **then** $s_{act} \leftarrow s$;
$\quad\quad$ **until** $(stop\_condition)$;
▷ Step 3:
$\quad$ **return** s_{act}.

The pseudo-code given above is a generic template of any local search algorithm. Therefore, some remarks are necessary.

Remark 11.1. The *Initialization* procedure includes preliminary operations such as initialization of counters, setting *control parameters* used during the construction of the neighbourhood $\mathcal{N}(s_{act})$, etc.

Remark 11.2. The neighbourhood function $\mathcal{N}$ is problem-specific and can be defined in various ways.

Remark 11.3. The form of a *stop_condition* depends on the applied variant of the local search algorithm.

Remark 11.4. Some steps in the template (e.g. the *Initialization* procedure) may be dropped and some (e.g., $stop\_condition \equiv FALSE$) may be trivial.

The basic assumptions of the *GeneralLocalSearch* algorithm may be modified in various ways. The most often encountered modifications are as follows.

First, the search must not necessarily be conducted in the set $\mathfrak{F}$ of all feasible solutions. In some variants of local search, the solutions are searched in the set $E(\mathfrak{F})$, which is an image of the set $\mathfrak{F}$ under some mapping E.

Second, the neighbourhood $\mathcal{N}(s_{act})$ can be composed not only of feasible solutions, if the set $E(\mathfrak{F})$ is considered instead of $\mathfrak{F}$.

Finally, not only complete solutions can be elements of the neighbourhood $\mathcal{N}(s_{act})$. In general, partial solutions may also belong to the neighbourhood.

11.1.3 Applicability of local search algorithms

Local search algorithms have the following main advantages. First, due to its generality, the *GeneralLocalSearch* template can be applied to various optimization problems, unlike the constructive heuristics, which use problem-specific properties and therefore, usually, are not versatile.

Second, since local search algorithms search only the set $\mathcal{N}(s_{act})$, they are capable of solving problems of larger sizes.

Third, local search algorithms, in most cases, produce solutions of acceptable quality, even if no special attention has been paid to choosing appropriate values of the control parameters of the algorithms.

Local search algorithms also have some disadvantages. First, the minimal exact neighbourhood may be exponential with respect to the size of the input of a given optimization problem.

Second, exponential time may be needed to find a local minimum.

Third, the solutions obtained by a local search algorithm may deviate arbitrarily far from the elements of the set $\mathfrak{F}_{opt}$.

Despite the above disadvantages, local search algorithms have been applied successfully to many intractable optimization problems; see the references given in Sect. 2.3 for details.

11.2 Selected types of local search algorithms

There exist a great number of types and variants of local search algorithms. In this section, we shortly describe only those which have been applied to time-dependent scheduling problems, i.e., iterative improvement, steepest descent search, simulated annealing, genetic and evolutionary algorithms.

11.2.1 Iterative improvement algorithms

The simplest local search algorithm is the *iterative improvement* algorithm. In this case, starting from an initial solution $s_0 \in \mathfrak{F}$, we iteratively search for a neighbour s_{loc} of the current solution s_{act}, which has the best value of the criterion function f. The neighbourhood function $\mathcal{N}$ defines the way in which the neighbour s_{loc} is generated. The set $\mathcal{N}(s_{act})$ is composed only of one solution. The pseudo-code of the algorithm is as follows.

Algorithm *IterativeImprovement*

Input: solution s_0, neighbourhood function $\mathcal{N}$, criterion f
Output: a locally optimal solution s_{act}

▷ Step 1:

    $s_{act} \leftarrow s_0$;

▷ Step 2:

   repeat

       Generate a neighbour s_{loc} of s_{act}; ▷ $|\mathcal{N}(s_{act})| = 1$

       if $f(s_{loc}) < f(s_{act})$ **then** $s_{act} \leftarrow s_{loc}$;

   until (*stop_condition*);

▷ Step 3:

   return s_{act}.

Since the solution generated by the iterative improvement algorithm is the first-encountered local minimum, the quality of the minimum may be arbitrarily bad. Therefore, some modifications of the iterative improvement have been proposed. One of them leads to the so-called *steepest descent search*.

11.2.2 Steepest descent search algorithms

Unlike the iterative improvement, in the *steepest descent search* all possible neighbours of the current solution s_{act} are generated. The best neighbour is accepted as the final solution. This improves the quality of the final solution at the cost of increasing the time complexity of the algorithm. The pseudo-code of the algorithm is as follows.

Algorithm *SteepestDescentSearch*

Input: solution s_0, neighbourhood function $\mathcal{N}$, criterion f
Output: a locally optimal solution s_{act}

▷ Step 1:

    $s_{act} \leftarrow s_0$;

▷ Step 2:

   repeat

       Generate $\mathcal{N}(s_{act})$; ▷ $|\mathcal{N}(s_{act})| \gg 1$

       for all $s \in \mathcal{N}(s_{act})$ **do**

          if $f(s) < f(s_{act})$ **then** $s_{act} \leftarrow s$;

   until (*stop_condition*);

▷ Step 3:

   return s_{act}.

The iterative improvement and the steepest descent search are simple local search algorithms, which, in many cases, produce solutions that are of poor quality. The main reason of this fact is the so-called *trap of local optimum*: after finding a solution that is locally optimal, the local search heuristics are not able to find better solutions, since they cannot move out from the neighbourhood of a locally optimal solution. Therefore, more powerful local search algorithms, called *metaheuristics*, have been proposed in the literature.

Metaheuristics apply more sophisticated strategies of constructing the neighbourhood $\mathcal{N}(s_{act})$ than the ones used in iterative improvement or steepest descent search algorithms. Moreover, instead of a single solution s_{act} they use *populations* of solutions. One of metaheuristics is the *simulated annealing*.

11.2.3 Simulated annealing algorithms

The *simulated annealing* (SA, in short) is based on a procedure that imitates the annealing (slow cooling) of a solid after it has been heated to its melting point. The SA algorithm is a non-deterministic algorithm, since the current solution s_{act} is selected from the neighbourhood $\mathcal{N}(s_{act})$, and accepted, in a random way. (So far the only non-deterministic algorithm considered by us was algorithm H_6 presented in Chap. 9.) The behaviour of the SA algorithm is determined by a number of parameters such as the *initial temperature*, the *cooling rate* and the function mirroring the decrease in temperature during the annealing. The pseudo-code of the metaheuristic SA is as follows.

Algorithm *SimulatedAnnealing*

Input: initial solution s_0, neighbourhood function $\mathcal{N}$, criterion f,
      acceptance probability function $p(i)$
Output: a locally optimal solution s_{best}

▷ Step 1:
   $s_{act} \leftarrow s_0$;
   $s_{best} \leftarrow s_0$;
   $f_{best} \leftarrow f(s_{best})$;
   $i \leftarrow 1$; ▷ procedure *Initialization*
▷ Step 2:
   repeat
      Generate $\mathcal{N}(s_{act})$;
      Choose at random $s_{loc} \in \mathcal{N}(s_{act})$;
      if $f(s_{loc}) \leq f(s_{act})$ **then** $s_{act} \leftarrow s_{loc}$;
      if $f(s_{loc}) < f_{best}$ **then** $f_{best} \leftarrow f(s_{loc})$;
             $s_{best} \leftarrow s_{loc}$
      else choose at random $p \in \langle 0, 1 \rangle$;
      if $p \leq p(i)$ **then** $s_{act} \leftarrow s_{loc}$; ▷ accept s_{loc} with probability p
      $i \leftarrow i + 1$;
   until (*stop_condition*);
▷ Step 3:
   return s_{best}.

The *acceptance probability function* $p(i)$ is usually in the form of

$$p(i) := \exp\left(-\frac{1}{T(i)}\Delta f_i\right),$$

where $\Delta f_i := f(s_{loc}) - f(s_{act})$ and $T(i)$ is a non-increasing function of time. The function $T(i)$, called the *function of temperature* or a *cooling scheme*, is usually defined as follows. Starting from T_0, the temperature is constant for L consecutive steps, and next it is decreased according to the formula $T(iL) \equiv T_i := c^i T_0$, where $0 < c < 1$ is a fixed constant factor. The parameters T_0, c and L are called the *initial temperature*, *cooling rate* and *length of plateau*, respectively. We refer the reader to the literature given in Sect. 2.3 for more details on the SA metaheuristic.

11.2.4 Genetic and evolutionary algorithms

Another metaheuristic is the *genetic algorithm* (GA, in short), based on some mechanisms (selection, crossover, mutation) known from the nature. The GA, in turn, evolved to the *evolutionary algorithm* (EA, in short), which is a combination of the GA and *genetic programming*, *evolutionary strategy* and *evolutionary programming* (see, e.g., Calégari et al. [40] for more details).

The pseudo-code of the simplest EA, called the *SimpleEvolutionaryAlgorithm* (SEA), can be formulated as follows.

Algorithm *SimpleEvolutionaryAlgorithm*

Input: procedures *Initialization*, *Evaluation*,
          operators *Preselection*, *CrossOver_and_Mutation*, *Postselection*
Output: a locally optimal solution

▷ Step 1:
  $i \leftarrow 0$;
  Initialization($\mathbf{P}_0$); ▷ base population
  Evaluation($\mathbf{P}_0$);
▷ Step 2:
  repeat
      $\mathbf{T}_i \leftarrow$ *Preselection*($\mathbf{P}_i$); ▷ temporary population
      $\mathbf{O}_i \leftarrow$ *CrossOver_and_Mutation*($\mathbf{T}_i$);
      Evaluation($\mathbf{O}_i$);
      $\mathbf{P}_{i+1} \leftarrow$ *Postselection*($\mathbf{P}_i, \mathbf{O}_i$); ▷ offspring population
      $t \leftarrow i + 1$;
  until (*stop_condition*).

The SEA works on the *base*, *offspring* and *temporary* populations of *individuals*, and produces *generations* of solutions, indexed by the variable i.

In the *initialization* step of the SEA, the base population $\mathbf{P}_0$ is created in a random way. Next, the *evaluation* of $\mathbf{P}_0$ is performed, i.e., for each individual from $\mathbf{P}_0$ a *fitness function* is calculated.

In the main step of the SEA, the offspring population $\mathbf{P}_i$, $i > 0$, is created. First, in the process of *preselection*, the temporary population $\mathbf{T}_i$ is built from the best individuals of $\mathbf{P}_i$. Next, individuals from $\mathbf{T}_i$ are *crossed* and *mutated*, which leads through the *postselection* to the next offspring population $\mathbf{P}_{i+1}$. The process is continued until a certain *stop_condition* is met.

Evolutionary algorithms have many variants, whose description is beyond the scope of the book; see the literature given in Sect. 2.3 for more details.

11.3 Local search time-dependent scheduling algorithms

In this section, we present a few local search algorithms that have been applied to time-dependent scheduling problems.

11.3.1 Steepest descent search algorithms

The $C_{\max}$ criterion. For the parallel-machine problem $Pm|p_j = b_j t|C_{\max}$, Hindi and Mhlanga [134] proposed a steepest descent search algorithm. Since in any schedule for this problem, by Theorem 6.1, the order of jobs assigned to a machine is immaterial, every subschedule for the problem can be represented by a subset of indices of jobs assigned to the machine. Hence, the neighbourhood $\mathcal{N}(s_{act})$ is defined as the set of all partitions of the set $N_{\mathcal{J}} := \{1, 2, \ldots, n\}$ into m parts. Every new partition is obtained from another partition by a single move. A single *move* is either a transfer of one job from one subset of a partition to another subset, or a mutual exchange of two jobs belonging to two different subsets of a partition. Let σ^0 and σ_{act} denote the initial schedule and the current best schedule, respectively.

Algorithm H_{53} for the problem $Pm|p_j = b_j t|C_{\max}$ ([134])

Input: initial schedule σ^0, neighbourhood function $\mathcal{N}$
Output: a locally optimal schedule σ_{act}
▷ Step 1:
    $\sigma_{act} \leftarrow \sigma^0$;
▷ Step 2:
    repeat
        $T \leftarrow C_{\max}(\sigma_{act})$;
        Find machine $\mathcal{M} \in \{M_1, M_2, \ldots, M_m\}$ such that $C_{\max}(\mathcal{M}) = T$;
        $\mathcal{N}(\sigma_{act}) \leftarrow$ set of schedules obtained by all possible moves of jobs
        $\hookrightarrow$ assigned to machine $\mathcal{M}$;
        for all $\sigma \in \mathcal{N}(\sigma_{act})$ **do**
            if $C_{\max}(\sigma) < C_{\max}(\sigma_{act})$ **then** $\sigma_{act} \leftarrow \sigma$;
    until (no improvement of σ is possible);
▷ Step 3:
    return σ_{act}.

Algorithm H_{53} has been tested on a set of 320 instances in which deterioration rates were randomly generated values from the $(0, 1)$ interval. The reported results of the experiment (see [134, Sect. 5]) suggest that the schedules generated by the algorithm are, on average, quite satisfactory.

The $\sum C_j$ criterion. For the problem $Pm|p_j = b_j t| \sum C_j$, Gawiejnowicz et al. [102] proposed a steepest descent search algorithm. The algorithm checks iteratively if the necessary condition for the optimality of a schedule σ_{act} is satisfied, i.e., if the inequality $\sum C_j(\sigma_{act}) - \sum C_j(\tau) \le 0$ holds for any schedule $\tau \in \mathcal{N}(\sigma_{act})$. Let J_{b_j} denote the job corresponding to deterioration rate b_j. The pseudo-code of the algorithm is as follows.

Algorithm H_{54} for the problem $Pm|p_j = b_j t| \sum C_j$ ([102])

Input: sequence $(b_1, b_2, \ldots, b_n)$, neighbourhood function $\mathcal{N}$
Output: a locally optimal schedule σ_{act}

▷ **Step 1: Construction of the initial schedule σ^0**
Arrange all jobs in the non-increasing order of b_j values;
Assign $m - 1$ jobs with greatest b_j values to machines $M_2, M_3, \ldots, M_m$;
Assign the remaining $n - m$ jobs to machine M_1;
▷ **Step 2: Construction of the set $\mathcal{N}(s^0)$**
$\sigma_{act} \leftarrow \sigma^0$;
repeat
    $\sigma_{last} \leftarrow \sigma_{act}$;
    $\mathcal{N}(\sigma_{act}) \leftarrow \emptyset$;
    for jobs assigned to machine M_1 in σ_{act} **do**
        Choose a job J_{b_j};
        for $\mathcal{M} \in \{M_2, M_3, \ldots, M_m\}$ **do**
            Construct schedule σ' by moving job J_{b_j} to machine $\mathcal{M}$;
            $\mathcal{N}(\sigma_{act}) \leftarrow \mathcal{N}(\sigma_{act}) \cup \sigma'$;
▷ **Step 3: Selection of the best schedule in $\mathcal{N}(\sigma_{act})$**
        Choose $\tau \in \mathcal{N}(\sigma_{act})$ such that
        $\hookrightarrow \tau = \arg\max\{\sum C_j(\sigma_{last}) - \sum C_j(\sigma') : \sigma' \in \mathcal{N}(\sigma_{act})\}$;
        if $(\sum C_j(\sigma_{last}) - \sum C_j(\tau) > 0)$ **then** $\sigma_{act} \leftarrow \tau$;
    until (no improvement of σ_{act} is possible);
▷ **Step 4:**
    return σ_{act}.

The time complexity of algorithm H_{54} depends on the number of iterations of **repeat-until** loop, which is $O(nm)$, and the cost of checking the condition $\sum C_j(\sigma_{last}) - \sum C_j(\tau) > 0$. (Notice that the construction of σ' in Step 2 can be done very efficiently, since the jobs assigned to each machine should be in the non-increasing order of b_j values, cf. Theorem 6.120.) Since this latter cost is $O(n)$, the algorithm runs in $O(n^2 m) \equiv O(n^2)$ time for fixed m.

The $\sum w_j C_j$ criterion. Wu et al. [301] proposed three heuristic algorithms for the problem $1|p_j = a_j + b_j t| \sum w_j C_j$. All the algorithms generate an initial schedule and try to find a better one by iterative improvement of the best schedule constructed so far.

The first algorithm tries to improve the schedule in which jobs are in the non-decreasing order of $\frac{a_j}{w_j}$ ratios. The pseudo-code of the algorithm can be formulated as follows.

Algorithm H_{55} for the problem $1|p_j = a_j + b_j t| \sum w_j C_j$ ([301])

Input: sequences $(a_1, a_2, \ldots, a_n)$, $(b_1, b_2, \ldots, b_n)$, $(w_1, w_2, \ldots, w_n)$
Output: a suboptimal schedule σ

▷ Step 1:
  Arrange jobs in the non-decreasing order of $\frac{a_j}{w_j}$ ratios;
  Call the obtained sequence σ^0;

▷ Step 2:
  Make pairwise interchanges in σ^0 until no improvement can be made;
  Call the final schedule σ;

▷ Step 3:
  return σ.

The second algorithm tries to improve the schedule in which jobs are in the non-decreasing order of $\frac{b_j}{w_j}$ ratios. The pseudo-code of this algorithm can be formulated as follows.

Algorithm H_{56} for the problem $1|p_j = a_j + b_j t| \sum w_j C_j$ ([301])

Input: sequences $(a_1, a_2, \ldots, a_n)$, $(b_1, b_2, \ldots, b_n)$, $(w_1, w_2, \ldots, w_n)$
Output: a suboptimal schedule σ

▷ Step 1:
  Arrange jobs in the non-decreasing order of $\frac{b_j}{w_j}$ ratios;
  Call the obtained sequence σ^0;

▷ Step 2:
  Make pairwise interchanges in σ^0 until no improvement can be made;
  Call the final schedule σ;

▷ Step 3:
  return σ.

Finally, the third algorithm tries to improve the schedule in which jobs are in the non-decreasing order of $\frac{a_j + (1 + b_j) S_j}{w_j}$ ratios. The pseudo-code of this algorithm is as follows.

Algorithm H_{57} for the problem $1|p_j = a_j + b_j t| \sum w_j C_j$ ([301])

Input: sequences $(a_1, a_2, \ldots, a_n)$, $(b_1, b_2, \ldots, b_n)$, $(w_1, w_2, \ldots, w_n)$,
number t_0
Output: a suboptimal schedule σ

▷ Step 1:
$N_J \leftarrow \{1, 2, \ldots, n\}$;
$\sigma \leftarrow (\phi)$;

▷ Step 2:
for $k \leftarrow 1$ **to** n **do**
Choose job J_i for which $\min\{\frac{a_j + (1 + b_j)t_{k-1}}{w_j} : j \in N_J\}$ is achieved;
$\sigma_k \leftarrow i$;
$N \leftarrow N \setminus \{i\}$;
$t_k \leftarrow a_i + (1 + b_i)t_{k-1}$;
Call the final schedule σ^0;

▷ Step 3:
Make pairwise interchanges in σ^0 until no improvement can be made;
Call the final schedule σ;

▷ Step 4:
return σ.

Algorithms H_{55}–H_{57} have been tested on a set of 600 instances in which basic processing times, deterioration rates and job weights were randomly generated. The reported results of the experiment (see [301, Sect. 5]) suggest that the best schedules, on average, are generated by the algorithm H_{55}.

11.3.2 Iterative improvement algorithms

The $\sum C_j$ criterion. For the problem $Pm|p_j = 1 + b_j t| \sum C_j$ Gawiejnowicz et al. [102] proposed an iterative improvement algorithm, based on the following idea. If an initial schedule for the problem is known, it can be improved by successively moving jobs between machines in order to find such an assignment of jobs that gives the smaller total completion time than the initial one.

The initial schedule can be constructed by an arbitrary heuristic algorithm for the problem $Pm|p_j = 1 + b_j t| \sum C_j$. The authors used algorithm H_{54}. Moreover, they assumed that *stop_condition* in the algorithm takes into account not only the increase of the citerion function but also limits the number of performed iterations of the algorithm.

Let $\sigma(J_i \leftrightarrow J_k)$ and $ind(\sigma, J_i)$ denote schedule σ in which jobs J_i and J_k have been mutually replaced and the index of a machine to which job J_i has been assigned in schedule σ, respectively. The pseudo-code of the algorithm can be formulated as follows.

Algorithm H_{58} for the problem $Pm\|p_j = 1 + b_j t\| \sum C_j$ ([102])

Input: sequence $(b_1, b_2, \ldots, b_n)$, number k
Output: a locally optimal schedule σ_{act}

▷ Step 1: Construction of the initial schedule σ^0
Apply algorithm H_{54} to the sequence $(b_1, b_2, \ldots, b_n)$;
Call the obtained schedule σ^0;
$k \leftarrow 0$;
$\sigma_{act} \leftarrow \sigma_0$;
▷ Step 2: Iterative improvement of schedule σ_{act}
repeat
    $k \leftarrow k + 1$;
    $\sigma_{last} \leftarrow \sigma_{act}$;
    for $i \leftarrow n$ **downto** 2 **do**
        for $k \leftarrow i - 1$ **downto** 1 **do**
            if $(ind(\sigma_{last}, J_i) \neq ind(\sigma_{last}, J_k))$ **then** $\tau \leftarrow \sigma_{last}(J_i \leftrightarrow J_k)$;
            if $(\sum C_j(\sigma) - \sum C_j(\tau) > 0)$ **then** $\sigma_{act} \leftarrow \tau$;
    until $((\sum C_j(\sigma) - \sum C_j(\sigma_{last}) = 0) \vee (k > n))$;
▷ Step 3:
return σ_{act}.

The time complexity of algorithm H_{58} is $O(n^3)$, since in the worst case we have to check n times $O(n^2)$ possibilities of a mutual change of two jobs.

11.3.3 Experimental evaluation of algorithms H_{54} and H_{58}

In order to evaluate the quality of schedules generated by algorithms H_{54} and H_{58}, a computational experiment for $m = 3$ machines has been conducted. The obtained schedules were compared to schedules generated by algorithm H_{29}, presented in Sect. 9.2.

The coefficients $\beta_i = b_i + 1$ were randomly generated values from intervals $(2, 99)$ and $(1, 2)$. The results for $\beta_i \in (2, 99)$ and for $\beta_i \in (1, 2)$ are presented, respectively, in Tables 11.1 and 11.2 (cf. Gawiejnowicz et al. [102]). Each value in the tables is an average of results for 10 instances. In total, 160 instances have been tested.

Columns $R_{H_{29}}^{avg}$, $R_{H_{54}}^{avg}$ and $R_{H_{58}}^{avg}$ include an average of ratios $R_H^r(I)$ of the total completion time ($\sum C_j$) for algorithms H_{29}, H_{54} and H_{58}, respectively, calculated with respect to the optimal value of $\sum C_j$. Columns $R_{H_{29}}^{lb}$, $R_{H_{54}}^{lb}$ and $R_{H_{58}}^{lb}$ include an average of ratios $R_H^r(I)$ of $\sum C_j$ for algorithms H_{29}, H_{54} and H_{58}, respectively, calculated with respect to the lower bound of $\sum C_j$.

Tables 11.1 and 11.2 show that algorithm H_{58} is better than algorithm H_{29} for $\beta_i \in (2, 99)$, while for $\beta_i \in (1, 2)$ the algorithms are comparable. The quality of schedules generated by algorithm H_{54}, in comparison to H_{29}, is an open question that needs further research.

Table 11.1: Results of Computational Experiment for H_{54} and $H_{58}, \beta_i \in (2, 99)$

n	$R_{H_{29}}^{\text{avg}}$	$R_{H_{29}}^{\text{lb}}$	$R_{H_{54}}^{\text{avg}}$	$R_{H_{54}}^{\text{lb}}$	$R_{H_{58}}^{\text{avg}}$	$R_{H_{58}}^{\text{lb}}$
6	0.0	0.186729	0.167711	0.374054	0.0	0.186729
8	0.267105	0.639685	0.167173	0.493273	0.0	0.293706
10	0.366406	0.384822	0.121466	0.127353	0.016173	0.031644
12	0.116080	0.115118	0.459128	0.444614	0.003993	0.014459
14	-	0.476927	-	0.206961	-	0.090330
16	-	0.354809	-	0.237446	-	0.012126
18	-	0.052520	-	0.344081	-	0.054585
20	-	0.475177	-	0.161075	-	0.031898

Table 11.2: Results of Computational Experiment for H_{54} and $H_{58}, \beta_i \in (1, 2)$

n	$R_{H_{29}}^{\text{avg}}$	$R_{H_{29}}^{\text{lb}}$	$R_{H_{54}}^{\text{avg}}$	$R_{H_{54}}^{\text{lb}}$	$R_{H_{58}}^{\text{avg}}$	$R_{H_{58}}^{\text{lb}}$
5	0.0	0.003121	0.0	0.003121	0.0	0.003121
6	0.0	0.002508	0.000693	0.003204	0.0	0.002508
8	0.001603	0.005034	0.004348	0.007798	0.0	0.003425
10	0.001520	0.002659	0.014319	0.015473	0.000026	0.001163
12	0.001170	0.001809	0.020410	0.021059	0.003459	0.004098
14	-	0.002801	-	0.025815	-	0.005598
16	-	0.002348	-	0.031094	-	0.001261
18	-	0.001272	-	0.044117	-	0.013159
20	-	0.003101	-	0.049956	-	0.004320

11.3.4 Simulated annealing algorithms

The $C_{\max}$ criterion. For the problem $Pm|p_j = a_j + b_j t|C_{\max}$, Hindi and Mhlanga [134] proposed a simulated annealing algorithm. The algorithm will be called H_{59}. The schedule generated by algorithm H_{53} was selected in H_{59} as the initial schedule σ^0. The initial temperature T_0 was given by the formula

$$T_0 := \Delta^+ \left[\ln \left(\frac{m^+}{x m^+ - (1 - x)(m - m^+)} \right) \right]^{-1},$$

where m^+ is the number of cost increase moves found during the execution of algorithm H_{45}, Δ^+ is the average cost increase over these moves and $0 < x < 1$ is the acceptance ratio (the authors assumed $x := 0.95$).

The temperature decreased according to the formula

$$T_{i+1} := \frac{T_i}{1 + \kappa T_i},$$

where $\kappa \ll \frac{1}{U}$ with U being the largest absolute move value found during the execution of algorithm H_{53}.

The algorithm H_{59} was tested on a set of 320 instances in which deterioration rates b_j were randomly generated values from the interval $(0, 1)$ and the basic processing times a_j were chosen from a normal distribution, with a mean of 50 and standard deviation of 10. The experiment has shown that if we use the H_{59} algorithm in order to improve the schedule generated by algorithm H_{53}, the results are satisfactory (see [134, Sect. 5] for details).

Shiau et al. [257] proposed a simulated annealing algorithm for the problem $F2|p_{i,j} = b_{i,j}t| \sum C_j$. We will call the algorithm H_{60}.

In the algorithm, the solution space $\mathfrak{F}$ consists of all schedules that correspond to permutations of sequence $(1, 2, \ldots, n)$. New schedules from a neighbourhood $\mathcal{N}(\sigma_0)$ of a given schedule σ_0 are generated by a pairwise interchange of two randomly selected jobs. The probability $p(\sigma_0)$ of acceptance of a schedule σ_0 is generated from an exponential distribution. The number of iterations of algorithm H_{60} is limited to $50n$, where n is the number of jobs.

The algorithm has been tested on a number of instances with $10 \leq n \leq 100$ jobs, giving satisfactory results; see [257, Sect. 6] for details.

11.3.5 Evolutionary algorithms

We complete this section with some remarks concerning the application of evolutionary algorithms (EAs) to time-dependent scheduling problems.

The EAs can be constructed in many of ways. However, though all EAs share the same template, the main difficulty in developing a new EA is the effort which is needed to implement it in a programming language. There are various approaches which help overcome this difficulty. One of the approaches consists in the use of *libraries of classes*. The libraries allow programmers to construct new classes in a comfortable way, using mechanisms such as *encapsulation* and *inheritance* which are built in object programming languages. Libraries of classes, in particular, can be used for the construction of EAs.

Gawiejnowicz et al. [107] proposed a new library of this kind, developed in $C\#$ and designated for work on the .NET platform. The library, called $TEAC$ (*Toolbox for Evolutionary Algorithms in C#*), includes a number of classes, which allow to implement basic genetic operators applied in EAs. It also includes a few classes which implement genetic and evolutionary algorithms, in particular the $SimpleEvolutionaryAlgorithm$. We refer the reader to [107, Sect. 2] for more detailed description of the $TEAC$ library.

In order to evaluate the solutions generated by EAs implemented using the $TEAC$ library, a computational experiment has been conducted. Based on the classes defined in the $TEAC$ library, an EA was constructed for the problem $Jm||C_{\max}$ (cf. Sect. 4.1).

The algorithm was identical to the $SimpleEvolutionaryAlgorithm$ and it stopped after 2000 generations. The benchmark data files from Beasley's OR-Library [19] have been used in the experiment.

The solutions obtained by the EA for job shop problems are presented in Table 11.3. Column *Size* gives the size of a particular instance, $n \times m$, where n and m are the numbers of jobs and machines, respectively. Columns *OPT*, *AVR* and *Best* show the optimal, the average and the best found result, respectively. Column *Time* presents the running time of the EA (in seconds). All these results are average values of 10 independent runs of the algorithm.

Table 11.3: Results of Computational Experiment for Job Shop Problems

File	Size	OPT	AVR	Best	Time
FT06	6x6	55	55.0	55	40.50
FT10	10x10	930	979.1	955	130.44
FT20	20x5	1,165	1,228.8	1,202	199.44
LA01	10x5	666	666.0	666	66.02
LA16	10x10	945	972.9	956	141.70
LA20	10x10	902	914.7	907	141.08
LA21	15x10	1,046	1,111.6	1,097	288.40
LA25	15x10	977	1,032.5	1,019	253.66
LA28	20x10	1,216	1,302.2	1,286	412.35
LA29	20x10	1,152	1,266.9	1,248	409.20
LA39	15x15	1,233	1,320.3	1,291	384.13
LA40	15x15	1,222	1,312.4	1,288	385.36

The results were satisfactory, since suboptimal schedules were generated in reasonable time. The average error of the obtained solutions was from 0% (files FT06, LA01) to 9.97% (file LA29); see [107, Sect. 3] for details.

Since no special effort has been made in order to find optimal values of control parameters of the applied EA, in future experiments more time should be devoted to this aspect of the algorithm.

Gawiejnowicz and Suwalski [110] used the *TEAC* library for construction of an evolutionary algorithm for two- and three-machine time-dependent flow shop scheduling problems (cf. Sect. 4.1 and Sect. 5.2). We will call the algorithm H_{61}. As previously, the algorithm was identical to the *Simple Evolutionary Algorithm*, and exploited basic genetic operators (see [110, Sect. 4]).

The behaviour of the new EA has been tested in a number of experiments concerning the problems $Fm|p_{ij} = b_{ij}t|C_{\max}$ and $Fm|p_{ij} = b_{ij}(a + bt)|C_{\max}$, where $m \in \{2, 3\}$. For each problem, job deterioration rates were randomly generated integer values: $b_{ij} \in \langle 1, n - 1 \rangle$ and $a, b \in (1, \frac{n}{2})$.

In total, in the experiments, 360 random instances have been generated.

Some of results of the experiments are presented in Tables 11.4 and 11.5. Each value is an average of data of five distinct instances. Symbols $R_{H_{61}}^{\min}$, $R_{H_{61}}^{\mathrm{avg}}$ and $R_{H_{61}}^{\max}$ denote the minimum, average and maximum absolute ratio $R_{H_{61}}^{a}(I)$, respectively. Symbol T_{avg} denotes the computation time (in seconds). Symbol

$FmPn$ ($FmLn$) denotes an instance of the m-machine flow shop problem with n jobs with proportional (proportional-linear) job processing times.

The results of these experiments suggest main directions of further research. First, there is a real need to construct a set of benchmarks for intractable time-dependent scheduling problems. Second, it is worth to conduct similar experiments for the job shop problems. Finally, the experiments have shown that it is necessary to use floating-point arithmetic in order to avoid problems with the range of job completion times.

Table 11.4: EA Solutions vs. Exact Solutions for F2P/F2L Datasets

Dataset	$R_{H_{61}}^{\min}$	$R_{H_{61}}^{\text{avg}}$	$R_{H_{61}}^{\max}$	T_{avg}	Dataset	$R_{H_{61}}^{\min}$	$R_{H_{61}}^{\text{avg}}$	$R_{H_{61}}^{\max}$	T_{avg}
F2P05	1,000	1,000	1,000	0,880	F2L05	1,000	1,000	1,000	1,195
F2P06	1,000	1,000	1,000	1,056	F2L06	1,000	1,000	1,000	1,435
F2P07	1,000	1,000	1,000	1,259	F2L07	1,000	1,000	1,000	1,781
F2P08	1,000	1,000	1,000	1,477	F2L08	1,037	1,184	1,233	2,091
F2P09	1,000	1,109	1,175	1,699	F2L09	1,000	1,083	1,217	2,453
F2P10	1,000	1,000	1,000	1,978	F2L10	1,000	1,077	1,159	2,833
F2P11	1,027	1,000	1,000	2,252	F2L11	1,248	1,535	1,984	3,284
F2P12	1,094	1,018	1,024	2,526	F2L12	1,250	1,567	2,224	3,697

Table 11.5: EA Solutions vs. Exact Solutions for F3P/F3L Datasets

Dataset	$R_{H_{61}}^{\min}$	$R_{H_{61}}^{\text{avg}}$	$R_{H_{61}}^{\max}$	T_{avg}	Dataset	$R_{H_{61}}^{\min}$	$R_{H_{61}}^{\text{avg}}$	$R_{H_{61}}^{\max}$	T_{avg}
F3P05	1,000	1,000	1,000	1,186	F3L05	1,000	1,000	1,000	1,195
F3P06	1,000	1,000	1,000	1,444	F3L06	1,000	1,000	1,000	1,435
F3P07	1,000	1,000	1,000	1,781	F3L07	1,000	1,000	1,000	1,781
F3P08	1,000	1,000	1,000	2,105	F3L08	1,037	1,184	1,233	2,091
F3P09	1,000	1,000	1,000	2,440	F3L09	1,000	1,083	1,217	2,453
F3P10	1,000	1,003	1,013	2,822	F3L10	1,000	1,077	1,159	2,833
F3P11	1,027	1,122	1,249	3,260	F3L11	1,248	1,535	1,984	3,284
F3P12	1,094	1,227	1,319	3,722	F3L12	1,250	1,567	2,224	3,697

With these tables, we end the review of local search algorithms for time-dependent scheduling problems. This chapter also ends the third part of the book, which is devoted to the main classes of algorithms for computationally intractable time-dependent scheduling problems.

In the fourth part of the book, we will consider selected advanced topics in time-dependent scheduling.

11.4 Concluding remarks

Many time-dependent scheduling problems are computationally intractable. Therefore, we are interested in finding suboptimal schedules for them. In this chapter, we presented local search algorithms (see Table 11.6 for details) that have been proposed for intractable time-dependent scheduling problems.

Table 11.6: Local Search Algorithms for Time-Dependent Scheduling Problems

Algorithm	Complexity	Problem	Reference	This book
H_{53}	$O(n^2)$	$Pm\|p_j = b_j t\|C_{\max}$	[134]	Sect. 11.3, p. 273
H_{54}	$O(mn^2)$	$Pm\|p_j = b_j t\|\sum C_j$	[101]	Sect. 11.3, p. 274
H_{55}	$O(n^3)$	$1\|p_j = a_j + b_j t\|\sum w_j C_j$	[301]	Sect. 11.3, p. 275
H_{56}	$O(n^3)$	$1\|p_j = a_j + b_j t\|\sum w_j C_j$	[301]	Sect. 11.3, p. 275
H_{57}	$O(n^3)$	$1\|p_j = a_j + b_j t\|\sum w_j C_j$	[301]	Sect. 11.3, p. 276
H_{58}	$O(n^3)$	$Pm\|a_j + b_j t\|C_{\max}$	[134]	Sect. 11.3, p. 277
H_{59}	(a)	$Fm\|p_{ij} = b_{ij}(a + bt)\|C_{\max}$	[134]	Sect. 11.3, p. 278
H_{60}	(a)	$F2\|p_{i,j} = b_{i,j} t\|\sum C_j$	[257]	Sect. 11.3, p. 279
H_{61}	$O(n^3)$	$Pm\|p_j = 1 + b_j t\|\sum C_j$	[110]	Sect. 11.3, p. 280

(a) depends on the number of iterations

The results presented in the chapter suggest main directions of further research. First, there is a real need to construct a set of benchmarks for intractable time-dependent scheduling problems. Second, the experiments have shown that it is necessary to use floating-point arithmetic in order to avoid problems with the range of job completion times.

Local search time-dependent scheduling algorithms seem to be an attractive alternative for heuristic algorithms applied to such problems so far. However, though the local search algorithms generate suboptimal schedules in reasonable time, further experiments with larger instances are needed to make conclusions about the average performance of these algorithms.

Part IV

ADVANCED TOPICS

12

Matrix methods in time-dependent scheduling

Ⅰn the first three parts of the book, we have covered the basic topics related to time-dependent scheduling. The, present, fourth part of the book is devoted to selected advanced topics in the subject.

This part is composed of three chapters. In Chap. 12, we consider time-dependent scheduling problems that are formulated in terms of matrices and vectors. In Chap. 13, we discuss the problems of time-dependent scheduling with job precedence constraints. Finally, in Chap. 14, we study bicriteria time-dependent scheduling.

Chapter 12 is composed of five sections. In Sect. 12.1, we introduce the notation and an auxiliary result. In Sect. 12.2, we show how to formulate time-dependent scheduling problems in terms of vectors and matrices. In Sect. 12.3, we consider the problem of minimizing the l_p norm. In Sect. 12.4, we introduce the notion of equivalent time-dependent scheduling problems and show their properties. Concluding remarks and open problems are given in Sect. 12.5.

12.1 Preliminaries

In this section, we formulate the problem under consideration, introduce the notation and an auxiliary result.

12.1.1 Problem formulation

Throughout the chapter, we consider different versions of the following parallel-machine time-dependent scheduling problem.

We are given jobs $J_1, J_2, \ldots, J_n$ to be processed on $m \geq 1$ parallel identical machines $M_1, M_2, \ldots, M_m$, which are available at times $t_0^k \geq 0$, $1 \leq k \leq m$. Jobs are independent and no ready times nor deadlines are given. The processing time p_j of job J_j, $1 \leq j \leq n$, is in the form of (6.10), i.e., $p_j = a_j + b_j t$, where $a_j \geq 0$, $b_j > 0$ and $t \geq b_0^k := t_0^k$ for $1 \leq k \leq m$. The objective will be defined separately for each particular case.

We start with the case when $a_j = 1$ for $1 \leq j \leq n$ and $m = 1$. Notice that in this case, in view of the form of job processing times, the following recurrence equation holds:

$$
C_j = \begin{cases} 1, & j = 0, \\ C_{j-1} + p_j(C_{j-1}) = 1 + \beta_j C_{j-1}, & j = 1, 2, \ldots, n, \end{cases} \tag{12.1}
$$

where $\beta_j = 1 + b_j$ for $j = 0, 1, \ldots, n$.

12.1.2 Notation

Throughout the chapter, we will use the following notation. The vectors $(\beta_0, \beta_1, \ldots, \beta_n)$ and $(\beta_1, \beta_2, \ldots, \beta_n)$ will be denoted by $\hat{\beta}$ and β, respectively. By $\overline{x}$ we will denote the vector with the reverse order of components with respect to a vector x. The set of all $(p \times q)$-matrices over $\mathbb{R}$ will be denoted by $\mathcal{M}_{p \times q}(\mathbb{R})$. For a given matrix $H \in \mathcal{M}_{p \times q}(\mathbb{R})$, the transposed matrix and the transposed matrix in which rows and columns are in the reverse order will be denoted by H^T and $\overline{\overline{H^\mathsf{T}}}$.

12.1.3 Auxiliary result

We complete the section by the following result.

Lemma 12.1. (Gawiejnowicz et al. [97]) *Let $\phi(u, H, v) := u^\mathsf{T} H v$ be a function, where $u \in \mathbb{R}^p$, $v \in \mathbb{R}^q$ and $H \in \mathcal{M}_{p \times q}(\mathbb{R})$. Then, there holds the identity*

$$
\phi(u, H, v) = \phi(\overline{v}, \overline{\overline{H^\mathsf{T}}}, \overline{u}). \tag{12.2}
$$

Proof. First, note that we have $u^\mathsf{T} H v = v^\mathsf{T} H^\mathsf{T} u = (Pv)^\mathsf{T} P(H^\mathsf{T}) Q^\mathsf{T}(Qu)$, where $P \in \mathcal{M}_{p \times p}(\mathbb{R})$ and $Q \in \mathcal{M}_{q \times q}(\mathbb{R})$ are arbitrary permutation matrices. To complete the proof it is sufficient to take P and Q such that $Pv = \overline{v}$ and $Qu = \overline{u}$. Then $u^\mathsf{T} H v = \overline{v}^\mathsf{T} \overline{\overline{H^\mathsf{T}}} \overline{u}$. ∎

12.2 A matrix approach

In this section, we show how to represent in a matrix form a schedule for the time-dependent scheduling problem formulated in Sect. 12.1.

Remark 12.2. So far matrix methods have been applied to different areas of combinatorial optimization but are not popular in the scheduling theory. An exception is the so-called *Max-plus algebra*, applied in dynamic optimization (Bernhard [23]), theory of discrete event systems (Gunawardena [126]), railway timetabling (Goverde [121]) and scheduling (Bouquard et al. [30]).

12.2.1 The matrix form of single-machine schedules

Expanding formula (12.1) for $j = 0, 1, \ldots, n$ we have $C_0 = 1$, $C_1 = \beta_1 C_0 + 1$, $C_2 = \beta_2 C_1 + 1$, $\ldots$, $C_n = \beta_n C_{n-1} + 1$. These equations, in turn, can be rewritten in the following form:

$$
\begin{aligned}
C_0 &= 1, \\
-\beta_1 C_0 + C_1 &= 1, \\
-\beta_2 C_1 + C_2 &= 1, \\
&\ \ \vdots \qquad \vdots \ \ \vdots \\
-\beta_n C_{n-1} + C_n &= 1.
\end{aligned}
\tag{12.3}
$$

Rewriting equations (12.3) in the matrix form, we have

$$
\begin{bmatrix}
1 & 0 \ldots & 0 & 0 \\
-\beta_1 & 1 \ldots & 0 & 0 \\
0 & -\beta_2 \ldots & 0 & 0 \\
\vdots & \ldots & & \vdots \\
0 & 0 \ldots & -\beta_n & 1
\end{bmatrix}
\begin{bmatrix}
C_0 \\ C_1 \\ C_2 \\ \vdots \\ C_n
\end{bmatrix}
=
\begin{bmatrix}
1 \\ 1 \\ 1 \\ \vdots \\ 1
\end{bmatrix},
\tag{12.4}
$$

i.e., $A(\beta)C(\beta) = d(1)$, where $A(\beta)$ is the above matrix and both these vectors, $C(\beta) = [C_0(\beta), C_1(\beta), \ldots, C_n(\beta)]^\top$ and $d(1) = [1, 1, \ldots, 1]^\top$, belong to $\mathbb{R}^{n+1}$. Since $\det A(\beta) = 1$ for matrix $A(\beta)$ in (12.4), there exists an inverse matrix

$$
A^{-1}(\beta) =
\begin{bmatrix}
1 & 0 & \ldots 0 & 0 \\
\beta_1 & 1 & \ldots 0 & 0 \\
\beta_1\beta_2 & \beta_2 & \ldots 0 & 0 \\
\beta_1\beta_2\beta_3 & \beta_2\beta_3 & \ldots 0 & 0 \\
\vdots & \vdots & \ldots \vdots & \vdots \\
\beta_1\beta_2 \ldots \beta_n & \beta_2\beta_3 \ldots \beta_n & \ldots \beta_n & 1
\end{bmatrix}.
\tag{12.5}
$$

Knowing matrix $A^{-1}(\beta)$, given by (12.5), we can calculate the components of vector $C(\beta) = A^{-1}(\beta)d(1)$:

$$
C_k(\beta) = \sum_{i=0}^{k} \prod_{j=i+1}^{k} \beta_j,
\tag{12.6}
$$

where $k = 0, 1, \ldots, n$.

We illustrate the single-machine formulation on a numerical example.

Example 12.3. Consider three jobs with the following job processing times: $p_0 = 1 + 10t$, $p_1 = 1 + 2t$, $p_2 = 1 + 3t$.

For this set of jobs, we have $\hat{\beta} = (11, 3, 4)$ and $\beta = (3, 4)$. There are only two schedules possible, $\sigma^1 = (1, 2)$ and $\sigma^2 = (2, 1)$. In this case, we have

$$A(\beta_{\sigma^1}) = \begin{bmatrix} 1 & 0 & 0 \\ -3 & 1 & 0 \\ 0 & -4 & 1 \end{bmatrix} \quad \text{and} \quad A(\beta_{\sigma^2}) = \begin{bmatrix} 1 & 0 & 0 \\ -4 & 1 & 0 \\ 0 & -3 & 1 \end{bmatrix},$$

respectively.

The inverse matrices to the matrices $A(\beta_{\sigma^1})$ and $A(\beta_{\sigma^2})$ are as follows:

$$A^{-1}(\beta_{\sigma^1}) = \begin{bmatrix} 1 & 0 & 0 \\ 3 & 1 & 0 \\ 12 & 4 & 1 \end{bmatrix} \quad \text{and} \quad A^{-1}(\beta_{\sigma^2}) = \begin{bmatrix} 1 & 0 & 0 \\ 4 & 1 & 0 \\ 12 & 3 & 1 \end{bmatrix},$$

respectively. For the above data, we have

$$C(\beta_{\sigma^1}) = A^{-1}(\beta_{\sigma^1})d(1) = \begin{bmatrix} 1 & 0 & 0 \\ 3 & 1 & 0 \\ 12 & 4 & 1 \end{bmatrix} \begin{bmatrix} 1 \\ 1 \\ 1 \end{bmatrix} = [1, 4, 17]^{\top}$$

and

$$C(\beta_{\sigma^2}) = A^{-1}(\beta_{\sigma^2})d(1) = \begin{bmatrix} 1 & 0 & 0 \\ 4 & 1 & 0 \\ 12 & 3 & 1 \end{bmatrix} \begin{bmatrix} 1 \\ 1 \\ 1 \end{bmatrix} = [1, 5, 16]^{\top},$$

respectively. Hence, we have $C_0(\sigma^1) = C_0(\sigma^2) = 1$, $C_1(\sigma^1) = 4$, $C_1(\sigma^2) = 5$, $C_2(\sigma^1) = 17$, $C_2(\sigma^2) = 16$. ◆

12.2.2 The matrix form of parallel-machine schedules

Consider a system of linear equations $A(\beta)C(\beta) = D$ in the block form of

$$\begin{bmatrix} A_1 & O & & O \\ O & A_2 & & O \\ & & \ddots & \\ O & O & & A_m \end{bmatrix} \begin{bmatrix} C^1 \\ C^2 \\ \dots \\ C^m \end{bmatrix} = \begin{bmatrix} d^1 \\ d^2 \\ \dots \\ d^m \end{bmatrix}, \tag{12.7}$$

where $C(\beta) = [C^1, C^2, \dots, C^m]^T$ and $C^i = [C_0^i, C_1^i, \dots, C_{n_i}^i]$ is a vector of the completion times of the jobs assigned to machine M_i, $1 \le i \le m$. Moreover, $D = [d^1, d^2, \dots, d^m]^T$, where $d^i = (d, d, \dots, d) \in R^{n_i}$ and

$$A_i \equiv A(a^i) = \begin{bmatrix} 1 & 0 \dots & & 0 & 0 \\ -\beta_1^i & 1 \dots & & 0 & 0 \\ 0 & -\beta_2^i \dots & & 0 & 0 \\ \vdots & & \ddots & & \vdots \\ 0 & 0 \dots & -\beta_{n_i}^i & 1 \end{bmatrix}. \tag{12.8}$$

Since $\det(A(\beta)) = 1$, matrix $A(\beta)$ in (12.7) is non-singular. Its inverse, in block form, is as follows

$$A^{-1}(\beta) = \begin{bmatrix} A_1^{-1} & O & \dots & O \\ O & A_2^{-1} & \dots & O \\ \vdots & \vdots & & \vdots \\ O & O & \dots & A_m^{-1} \end{bmatrix}, \quad A^{-1}(a^i) = \begin{bmatrix} 1 & 0 & \dots 0 & 0 \\ \beta_1^i & 1 & \dots 0 & 0 \\ \beta_1^i\beta_2^i & \beta_2^i & \dots 0 & 0 \\ \vdots & \vdots & & \vdots & \vdots \\ \beta_1^i\dots\beta_{n_i}^i & \beta_2^i\dots\beta_{n_i}^i & \dots \beta_{n_i}^i & 1 \end{bmatrix},$$

where $A_i^{-1} \in \mathcal{M}_{p \times q}(\mathbb{R})$ and O denotes a zero matrix of a suitable size.

We illustrate the multi-machine formulation on a numerical example.

Example 12.4. Consider six jobs with the following job processing times: $p_0 = 1 + 10t$, $p_1 = 1 + 2t$, $p_2 = 1 + 3t$, $p_3 = 1 + 12t$, $p_4 = 1 + 5t$ and $p_5 = 1 + 4t$.

For this set of jobs and schedule $\sigma^1 = ((2,1)|(4,5))$ in which jobs J_0, J_2, J_1 and J_3, J_4, J_5 are assigned to machine M_1 and M_2, respectively, we have

$$A(\beta_{\sigma^1}) = \begin{bmatrix} 1 & 0 & 0 & 0 & 0 & 0 \\ -4 & 1 & 0 & 0 & 0 & 0 \\ 0 & -3 & 1 & 0 & 0 & 0 \\ 0 & 0 & 0 & 1 & 0 & 0 \\ 0 & 0 & 0 & -5 & 1 & 0 \\ 0 & 0 & 0 & 0 & -6 & 1 \end{bmatrix}$$

◆

So far, we only formulated a schedule in a matrix form. Since for a given β we can calculate by the matrix approach the vector $C(\beta)$, we can apply to the vector any criterion function that is a function of components of $C(\beta)$. In particular, we can apply the l_p norm (cf. Definition 1.18). This is the topic of the next section.

12.3 The l_p norm criterion

In this section, we consider the problem of single machine scheduling with deteriorating jobs and the criterion of minimization of the l_p norm.

12.3.1 Preliminaries

Our aim is to find an approximate solution of the single machine scheduling problem $1|p_j = 1 + b_j t| \|C\|_p$, where $C = [C_0, C_1, \dots, C_n]^\top \in \mathbb{R}^{n+1}$, $C_0 = 1$ and $\|C\|_p$ denotes the l_p norm of vector C.

Let $\gamma = \beta(\beta_i \leftrightarrow \beta_j)$ denote sequence β with components β_i and β_j mutually exchanged, $1 \le i \ne j \le n$. There holds the following result.

Lemma 12.5. (Gawiejnowicz et al. [105]) *Let $j = i + 1$, where $i = 1, 2, \ldots,$ $n - 1$. Then for $k = 1, 2, \ldots, n$ and $n \geq 2$ there holds the equality*

$$C_k(\gamma) - C_k(\beta) = \begin{cases} (\beta_i - \beta_{i+1})\beta_{i+2} \ldots \beta_k & , 1 \leq i < k, \\ (\beta_{i+1} - \beta_i) \sum_{l=0}^{i-1} \beta_{l+1} \ldots \beta_{i-1} & , i = k, \\ 0 & , i > k. \end{cases}$$

Moreover, $C_0(\gamma) - C_0(\beta) = 0$.

Proof. By applying formula (12.6) for $C_k = C_k(\beta)$, the result follows. ∎

12.3.2 Results

In this subsection, we present a few results concerning the l_p norm criterion.

Lemma 12.6. (Gawiejnowicz et al. [105]) *Let $1 \leq p < +\infty$ and $A(\beta)C(\beta) = A(\gamma)C(\gamma) = d(1)$, where $\gamma = \beta(\beta_i \leftrightarrow \beta_j)$ and $1 \leq i < j \leq n$. Then for $\delta(\theta) := \|C(\gamma)\|_p - \|C(\beta)\|_p$ there holds the equality*

$$\begin{aligned} \delta(\theta) &= \nabla\|C^\theta(\beta)\|_p (C(\gamma) - C(\beta)) \\ &= \sum_{k=0}^n \left(\frac{C_k^\theta(\beta)}{\|C^\theta(\beta)\|_p} \right)^{(p-1)} (C_k(\gamma) - C_k(\beta)), \end{aligned} \quad (12.9)$$

for some $\theta \in (0, 1)$.

Proof. Let $1 \leq p < +\infty$. The l_p−type norm $\|\cdot\|_p$ is a differentiable function in the interior of the positive cone of $\mathbb{R}^{n+1}$ and therefore the mean value theorem (cf. Remark 1.16) can be applied, i.e.,

$$\|y\|_p = \|x\|_p + \nabla\|x^\theta\|_p (y - x)$$

where $x, y \in \mathbb{R}^{n+1}$, $x, y > 0$ (coordinate-wise), $x^\theta = \theta x + (1 - \theta)y$ for some $\theta \in (0, 1)$. Let $\nabla\|x^\theta\|_p$ denote the gradient of function $\|\cdot\|_p$ at point x^θ. Since in the case under consideration $x = C(\beta)$ and $y = C(\gamma)$, there hold $x_i > 0, y_i > 0$ and hence, $x_i^\theta(a) > 0$. Finally, we conclude that for $1 \leq p < +\infty$ there holds the equality $\nabla\|C^\theta(\beta)\|_p = \left[\ldots, \left(\frac{C_i^\theta(\beta)}{\|C^\theta(\beta)\|_p} \right)^{(p-1)}, \ldots \right]_{(i=0,1,\ldots,n)}$. ∎

Formula (12.9) from Lemma 12.6 can be simplified if $j = i + 1$.

Lemma 12.7. (Gawiejnowicz et al. [105]) *Let $1 \leq p < +\infty, i = 1, 2, \ldots, n-1$, $\gamma = \beta(\beta_i \leftrightarrow \beta_{i+1})$, i.e. β_i and β_{i+1} are mutually exchanged in β. Then*

$$\begin{aligned} \delta(\theta) &= \left(\frac{C_i^\theta(\beta)}{\|C^\theta(\beta)\|_p} \right)^{(p-1)} (\beta_{i+1} - \beta_i) \cdot \\ &\left[\sum_{j=0}^{i-1} \beta_{j+1} \ldots \beta_{i-1} - \sum_{k=i+1}^n \left(\frac{C_k^\theta(\beta)}{C_i^\theta(\beta)} \right)^{(p-1)} \beta_{i+2} \ldots \beta_k \right], \end{aligned}$$

where $\beta_1, \beta_2, \ldots, \beta_{i-1}$ are included in the first sum, $\beta_{i+2}, \beta_{i+3}, \ldots, \beta_n$ are included in the second sum, and β_i and β_{i+1} are included only in the term $(\beta_{i+1} - \beta_i)$.

Proof. Similar to the proof of Lemma 12.6. □

Remark 12.8. Note that if for a given sequence $\beta = \beta_{\sigma^\star}$, where $\sigma^\star \in \mathfrak{S}_n$, the norm $\|C(\beta_{\sigma^\star})\|_p$ has a minimal value, then for each transposition $\gamma = \beta_{\sigma^\star} := \beta(\beta_i \leftrightarrow \beta_j)$ of sequence β, $1 \leq i < j \leq n$, there exists $\theta \in (0,1)$ such that $\delta(\theta) \geq 0$, since $\|C(\beta)\|_p \leq \|C(\gamma)\|_p$.

Theorem 12.9. (Gawiejnowicz et al. [105]) *Let $1 \leq p < +\infty$ and let $\beta = \beta_{\sigma^\star}$ for $\sigma^\star \in \mathfrak{S}_n$ be the sequence for which the minimum of the norm $\|C(\beta_{\sigma^\star})\|_p$ is obtained. Then for each $i = 1, 2, \ldots, n-1$, there exists $\theta \in (0,1)$ such that either*

(a) $\beta_{i+1} - \beta_i \leq 0$ and $C_{i-1}(\beta) \leq \sum_{k=i+1}^{n} \left(\frac{C_k^\theta(\beta)}{C_i^\theta(\beta)} \right)^{(p-1)} \beta_{i+2} \ldots \beta_k$

or

(b) $\beta_{i+1} - \beta_i \geq 0$ and $C_{i-1}(\beta) \geq \sum_{k=i+1}^{n} \left(\frac{C_k^\theta(\beta)}{C_i^\theta(\beta)} \right)^{(p-1)} \beta_{i+2} \ldots \beta_k,$

where $C_i^\theta(\beta) = \theta C_i(\beta) + (1-\theta) C_i(\gamma)$, $\gamma = \beta(\beta_i \leftrightarrow \beta_{i+1})$, $i = 0, 1, \ldots, n$.

Proof. The result is a consequence of Lemma 12.7. □

The next result states that expressions $\frac{C_i^\theta(\beta)}{C_k^\theta(\beta)}$, for $1 \leq k \leq n-1$, are uniformly (strictly) greater than 1. (We omit a technical proof.)

Lemma 12.10. (Gawiejnowicz et al. [105]) *There exists $\sigma_n(\beta) > 1$, not dependent on θ, i and k, such that*

$$1 + \sigma_n^{-1}(\beta) \; < \; \frac{C_i^\theta(\beta)}{C_k^\theta(\beta)} \; < \; \sigma_n(\beta)$$

for each $k = 1, 2, \ldots, n-1$, $i = k+1, k+2, \ldots, n$ and $\theta \in \langle 0, 1 \rangle$. In particular, $\sigma_n(\beta)$ can be determined by the formula $\sigma_n(\beta) = \max\{ C_n(\beta_\sigma) : \sigma \in \mathfrak{S}_n \}$.

The next result is a generalization of Theorem 6.120.

Theorem 12.11. (Gawiejnowicz et al. [105]) *If $1 \leq p \leq +\infty$, then there exists $p_1 > 1$ such that for all $p \geq p_1$ the problem $1|p_j = 1 + b_j t| \|C\|_p$ can be solved in $O(n \log n)$ time.*

Proof. From Theorem 12.9, it follows that for a sufficiently large p the optimal sequence $\beta = (\beta_1, \beta_2, \ldots, \beta_n)$ must be non-increasing and therefore it can be determined uniquely in $O(n \log n)$ time. ∎

The last result in this subsection is a generalization of Theorem 6.133.

Theorem 12.12. (Gawiejnowicz et al. [105]) *If $1 \leq p \leq +\infty$, then there exists p_0, $1 < p_0 \leq p_1$, such that for all $1 \leq p \leq p_0$ the optimal solution to the problem $1|p_j = 1 + b_j t| \|C\|_p$ must have a V-shape.*

Proof. Indeed, in view of Lemma 12.10, taking $1 < p_0$ small enough we can make $\sigma_n^{(p-1)}(\beta)$ sufficiently close to 1 for all $1 \leq p \leq r_0$. On the other hand, it is easy to see that for $p = 1$ either (a) or (b) from Theorem 12.9 must occur and, due to the continuity of the l_p norm, this is true for $1 \leq p \leq p_0$. ∎

In the next section, we consider the properties of pairs of time-dependent scheduling problems.

12.4 Equivalent problems

Some authors (see, e.g., Cheng and Ding [49, 52], Cheng et al. [55], Gawiejnowicz et al. [95]) noticed that there exist pairs of time-dependent scheduling problems that have similar properties. For example, the single-machine problem of scheduling jobs with processing times in the form of $p_j = b_j t$ and the $\sum C_j$ criterion is optimally solved by scheduling jobs in the non-decreasing order of b_j values, while the single-machine problem of scheduling jobs with processing times in the form of $p_j = 1 + b_j t$ and the $C_{\max}$ criterion is optimally solved by scheduling jobs in the non-increasing order of the b_j's, $1 \leq j \leq n$. The aim of the section is to explain the above phenomenon using the notion of *equivalent* problems.

12.4.1 The initial problem

Throughout this section, we consider a few different cases of the parallel-machine problem of minimizing the total weighted starting time of all jobs, $Pm|p_j = a_j + b_j t| \sum w_j S_j$, which will be called an *initial* problem.

Remark 12.13. Since $S_j = C_{j-1}$ for $1 \leq j \leq n$, the applied criterion $\sum w_j S_j$ can be replaced by a special version of the total weighted completion time criterion in which weight w_{j+1} is assigned to the completion time C_j, i.e., $\sum w_j S_j := \sum_{j=0}^{n} w_{j+1} C_j$. However, since such a form of the criterion $\sum w_j C_j$ may lead to misunderstanding, we only use the criterion $\sum w_j S_j$.

Now we describe the form of an arbitrary schedule for the initial problem, separately for single- and parallel-machine problems.

Single-machine problems. Let $\beta_j := 1 + b_j$ for $1 \leq j \leq n$ and let sequences $(\beta_1, \ldots, \beta_n)$, $(a_1, \ldots, a_n)$ and $(w_1, \ldots, w_n)$ be given. Any schedule for the problem will be identified with a sequence $\sigma = ((a_1, \beta_1, w_1), \ldots, (a_n, \beta_n, w_n))$ of the triples (a_i, β_i, w_i), $1 \leq i \leq n$. The minimization of the criterion $\sum w_j S_j$ will be carried over all $\sigma \in \mathfrak{S}_n(\sigma^\circ)$, where $\mathfrak{S}_n(\sigma^\circ)$ denotes the set of all permutations of the initial sequence σ° of triples.

Any schedule σ can be represented in another way by the following table:

$$T(\sigma) := \begin{bmatrix} a_0 & a_1 & a_2 & \dots & a_n \\ & \beta_1 & \beta_2 & \dots & \beta_n \\ & w_1 & w_2 & \dots & w_n & w_{n+1} \end{bmatrix}, \qquad (12.10)$$

in which $a_0 := t_0^1$ is the time at which the machine starts the processing of jobs and the weight w_{n+1} has a special meaning, which is defined in (12.15). Any other schedule $\sigma \in \mathfrak{S}_n(\sigma^\circ)$ can be obtained by a permutation of these columns of the table $T(\sigma)$ which correspond to triples (a_i, β_i, w_i), $1 \le i \le n$.

Given a schedule $\sigma \in \mathfrak{S}_n(\sigma^\circ)$, the completion times of jobs in the schedule are given by the recurrence equation $C_j(\sigma) = \beta_j C_{j-1}(\sigma) + a_j$, where $1 \le j \le n$ and $C_0(\sigma) := a_0$. Applying the matrix approach introduced in Sect. 12.2, the initial problem can be written as follows:

$$(P^1) \quad \begin{cases} \text{minimize } W_{P^1}(\sigma) := w^\mathsf{T} C(\sigma) \\ \text{subject to } A(\sigma) C(\sigma) = a, \ \sigma \in \mathfrak{S}_n(\sigma^\circ), \end{cases} \qquad (12.11)$$

where $w = (w_1, \dots, w_{n+1})^\mathsf{T}$, $a = (a_0, \dots, a_n)^\mathsf{T}$ and $C(\sigma) = (C_0, \dots, C_n)^\mathsf{T}$. The non-singular matrix $A(\sigma) \in \mathcal{M}_{(n+1) \times (n+1)}(\mathbb{R})$ is defined as in Sect. 12.1.

Our aim is to construct from any instance of the initial problem an instance of another problem, called the *transformed* problem, in such a way that both these problems are *equivalent* in the sense described below.

To obtain the transformed problem, we replace the formula $W_{P^1}(\sigma) := w^\mathsf{T} C(\sigma)$, where $A(\sigma) C(\sigma) = a$, by a dual formula, separately for every schedule $\sigma \in \mathfrak{S}_n(\sigma^\circ)$.

Parallel-machine problems. Now we pass to the parallel-machine problem $Pm|p_j = a_j + b_j t| \sum w_j S_j$. In this case, the schedule $\sigma = (\sigma^1, \dots, \sigma^m)$ is composed of subschedules σ^k, $1 \le k \le m$. The subschedule σ^k corresponds to machine M_k and it is in the form of $\sigma^k = ((a_1^k, \beta_1^k, w_1^k), \dots, (a_{n_k}^k, \beta_{n_k}^k, w_{n_k}^k))$, where $1 \le k \le m$ and $\sum_{k=1}^m n_k = n$.

The subschedule σ^k can be presented in another way by the table

$$T(\sigma^k) = \begin{bmatrix} a_0^k & a_1^k & a_2^k & \dots & a_{n_k}^k \\ & \beta_1^k & \beta_2^k & \dots & \beta_{n_k}^k \\ & w_1^k & w_2^k & \dots & w_{n_k}^k & w_{n_k+1}^k \end{bmatrix}, \qquad (12.12)$$

where $a_0^k := t_0^k \ge 0$, $1 \le k \le m$, is the time at which machine M_k starts the processing of jobs and the weight $w_{n_k+1}^k$ has a special meaning, defined in (12.20). Any other schedule for the problem can be obtained by permuting or mutually exchanging the triples $(a_i^k, \beta_i^k, w_i^k)$, $(a_j^l, \beta_j^l, w_j^l)$ for any possible $1 \le k, l \le m$, $1 \le i \le n_k$ and $1 \le j \le n_l$, including $k = l$.

Given a subschedule σ^k, $1 \le k \le m$, the completion times of jobs in the subschedule are given by the recurrence equation $C_j^k(\sigma^k) = \beta_j^k C_{j-1}^k(\sigma^k) + a_j^k$, where $1 \le j \le n_k$ and $C_0^k(\sigma^k) := a_0^k$.

In the matrix form, the problem can be written as follows:

$$(P^m) \quad \begin{cases} \text{minimize } W_{Pm}(\sigma) := w^\mathsf{T} C(\sigma) \\ \text{subject to } \mathcal{A}(\sigma) C(\sigma) = a, \ \sigma \in \mathfrak{S}_n(\sigma^\circ), \end{cases} \qquad (12.13)$$

where $\mathcal{A}(\sigma)C(\sigma) = a$ denotes the block system of equations defined as in Sect. 12.1.

12.4.2 The transformed problem

In this subsection, we describe how to transform instances of the initial problem into instances of the transformed problem.

Single-machine problems. Consider the problem $1|p_j = a_j + b_j t| \sum w_j S_j$ in the form of (P^1). Define problem (D^1), corresponding to (P^1), as follows:

$$(D^1) \quad \begin{cases} \text{minimize } W_{D^1}(\overline{\sigma}) := \overline{a}^\mathsf{T} C(\overline{\sigma}) \\ \text{subject to } A(\overline{\sigma})C(\overline{\sigma}) = \overline{w}, \ \overline{\sigma} \in \mathfrak{S}_n(\overline{\sigma}^\circ), \end{cases} \quad (12.14)$$

where $\overline{a} = (a_n, a_{n-1}, \ldots, a_0)^\mathsf{T}$ and $\overline{w} = (w_{n+1}, w_n, \ldots, w_1)^\mathsf{T}$.

For simplicity of further presentation, introduce the following definition.

Definition 12.14. (Equivalent problems)
Given a schedule $\sigma = ((a_1, \beta_1, w_1), \ldots, (a_n, \beta_n, w_n)) \in \mathfrak{S}_n(\sigma^\circ)$, let $\overline{\sigma} \in \mathfrak{S}_n(\overline{\sigma}^\circ)$ be a schedule such that $\overline{\sigma} = ((w_n, \beta_n, a_n), \ldots, (w_1, \beta_1, a_1))$. Then
(a) the correspondence $\sigma \longleftrightarrow \overline{\sigma}$ will be called a transformation of the schedule σ for the problem (P^1) into the schedule $\overline{\sigma}$ for the corresponding problem (D^1) and vice versa,
(b) both corresponding problems, (P^1) and (D^1), will be called equivalent problems.

Given a schedule $\sigma \in \mathfrak{S}_n(\sigma^\circ)$ described by the table $T(\sigma)$, the transformed schedule $\overline{\sigma}$ is fully described by the table

$$T(\overline{\sigma}) := \begin{bmatrix} w_{n+1} & w_n & w_{n-1} & \cdots & w_1 \\ & \beta_n & \beta_{n-1} & \cdots & \beta_1 \\ & a_n & a_{n-1} & \cdots & a_1 & a_0 \end{bmatrix}, \quad (12.15)$$

where $C_0(\overline{\sigma}) := w_{n+1}$ is the time at which the machine starts the processing of jobs in the problem (D^1), while $C_0(\sigma) := a_0$ is the time at which the machine starts the processing of jobs in the problem (P^1).

In Definition 12.14, we have defined an equivalence between the problems (P^1) and (D^1), based on the transformation $\sigma \longleftrightarrow \overline{\sigma}$. The equivalence is justified by the following result.

Theorem 12.15. (Gawiejnowicz et al. [97]) *Let $\sigma \in \mathfrak{S}_n(\sigma^\circ)$ and $\overline{\sigma} \in \mathfrak{S}_n(\overline{\sigma}^\circ)$, where $\sigma = ((a_1, \beta_1, w_1), \ldots, (a_n, \beta_n, w_n))$, $\overline{\sigma} = ((w_n, \beta_n, a_n), \ldots, (w_1, \beta_1, a_1))$. Then*
(a) if $\overline{\sigma}$ has been obtained by the transformation $\sigma \longleftrightarrow \overline{\sigma}$, then there holds the equality

$$W_{P^1}(\sigma) = w^\mathsf{T} C(\sigma) = \overline{a}^\mathsf{T} C(\overline{\sigma}) = W_{D^1}(\overline{\sigma}); \quad (12.16)$$

(b) $\sigma^\star$ is an optimal schedule for the problem (P^1) if and only if $\overline{\sigma}^\star$ is an optimal schedule for the problem (D^1); moreover, there holds the equality $W_{P^1}(\sigma^\star) = W_{D^1}(\overline{\sigma}^\star)$.

Proof. (a) The implication follows from Lemma 12.1 for $H \equiv A(\sigma)^{-1}$, $u \equiv w$ and $v \equiv a$ with $p = q = n + 1$.

(b) Let σ^* be an optimal schedule for the problem (P^1) and let there exist a schedule $\bar{p} \in \mathfrak{S}_n(\overline{\sigma^\circ})$ for the problem (D^1), $\bar{p} \neq \overline{\sigma^*}$, such that $W_{D^1}(\overline{\sigma^*}) > W_{D^1}(\bar{p})$. Consider a schedule $\rho \in \mathfrak{S}_n(\sigma^\circ)$ for (P^1), equivalent to $\bar{p}$. Then $W_{P^1}(\rho) = W_{D^1}(\bar{p}) < W_{D^1}(\overline{\sigma^*}) = W_{P^1}(\sigma^*)$. A contradiction. The converse implication can be proved in an analogous way. The equality $W_{P^1}(\sigma^*) = W_{D^1}(\overline{\sigma^*})$ follows from (a). ∎

We illustrate the results of this section with a numerical example.

Example 12.16. (Gawiejnowicz et al. [97]) Consider the following instance of the single-machine problem $1|p_j = b_j + a_j t| \sum w_j S_j$. We are given two jobs with processing times $p_1 = 1 + 2t$, $p_2 = 2 + 3t$ and weights $w_1 = 5$, $w_2 = 6$. Assume that the machine is available from $t_0 = 0$, and $w_3 = 1$.

Then there are only two possible schedules: $\sigma^1 = ((1,3,5)|(2,4,6))$ and $\sigma^2 = ((2,4,6)|(1,3,5))$. The tables $T(\sigma^i)$, $i = 1, 2$, are as follows:

$$T(\sigma^1) = \begin{bmatrix} 0 & 1 & 2 \\ 3 & 4 & \\ 5 & 6 & 1 \end{bmatrix} \quad \text{and} \quad T(\sigma^2) = \begin{bmatrix} 0 & 2 & 1 \\ 4 & 3 & \\ 6 & 5 & 1 \end{bmatrix}. \quad (12.17)$$

Given the table $T(\sigma^1)$, we can calculate that $C_0(\sigma^1) = 0$, $C_1(\sigma^1) = 3 \times 0 + 1 = 1$, $C_2(\sigma^1) = 4 \times 1 + 2 = 6$. Hence $\sum w_j S_j(\sigma^1) = 5 \times 0 + 6 \times 1 + 1 \times 6 = 12$.

Similarly, for $T(\sigma^2)$ we have $C_0(\sigma^2) = 0$, $C_1(\sigma^2) = 4 \times 0 + 2 = 2$, $C_2(\sigma^2) = 3 \times 2 + 1 = 7$. Hence $\sum w_j S_j(\sigma^2) = 6 \times 0 + 5 \times 2 + 1 \times 7 = 17$.

In the transformed problem, $1|p_j = w_j + a_j t| \sum b_j S_j$, we have two jobs with processing times $p_1 = 6 + 3t$, $p_2 = 5 + 2t$, with weights $w_1 = 2$, $w_2 = 1$. The machine is available from $t_0 = 1$, and $w_3 = 0$.

The tables $T(\overline{\sigma^i})$, $i = 1, 2$, are as follows:

$$T(\overline{\sigma^1}) = \begin{bmatrix} 1 & 6 & 5 \\ 4 & 3 & \\ 2 & 1 & 0 \end{bmatrix} \quad \text{and} \quad T(\overline{\sigma^2}) = \begin{bmatrix} 1 & 5 & 6 \\ 3 & 4 & \\ 1 & 2 & 0 \end{bmatrix}. \quad (12.18)$$

Since $C_0(\overline{\sigma^1}) = 1$, $C_1(\overline{\sigma^1}) = 4 \times 1 + 6 = 10$, $C_2(\overline{\sigma^1}) = 3 \times 10 + 5 = 35$, we have $\sum w_j S_j(\overline{\sigma^1}) = 2 \times 1 + 1 \times 10 + 0 \times 35 = 12 = \sum w_j S_j(\sigma^1)$.

Similarly, since $C_0(\overline{\sigma^2}) = 1$, $C_1(\overline{\sigma^2}) = 3 \times 1 + 5 = 8$, $C_2(\overline{\sigma^2}) = 4 \times 8 + 6 = 38$, we have $\sum w_j S_j(\overline{\sigma^2}) = 1 \times 1 + 2 \times 8 + 0 \times 38 = 17 = \sum w_j S_j(\sigma^2)$.

It is easy to see that to the optimal initial schedule, σ^1, there corresponds the transformed schedule, $\overline{\sigma^1}$, and vice versa. ◆

Parallel-machine problems. Consider the problem $Pm|p_j = a_j + b_j t| \sum w_j S_j$ in the form of (P^m). Define problem (D^m), corresponding to (P^m), as follows:

$$(D^m) \quad \begin{cases} \text{minimize } W_{D^m}(\overline{\sigma}) := \overline{a}^\mathsf{T} C(\overline{\sigma}) \\ \text{subject to } \mathcal{A}(\overline{\sigma}) C(\overline{\sigma}) = \overline{w}, \ \overline{\sigma} \in \mathfrak{S}_n(\overline{\sigma^\circ}), \end{cases} \quad (12.19)$$

where $\bar{a} = (\overline{a^1}, \ldots, \overline{a^m})^\mathsf{T}$ and $\bar{w} = (\overline{w^1}, \ldots, \overline{w^m})^\mathsf{T}$, $\overline{a^k} = (a_{n_k}^k, a_{n_k-1}^k, \ldots, b_0^k)^\mathsf{T}$ and $\overline{w^k} = (w_{n_k+1}^k, \ldots, w_1^k)$ for $1 \le k \le m$.

The following definition is an extension of Definition 12.14.

Definition 12.17. *Given a schedule $\sigma = (\sigma^1, \ldots, \sigma^m) \in \mathfrak{S}_n(\sigma^\circ)$, where $\sigma^k = ((a_1^k, \beta_1^k, w_1^k), \ldots, (a_{n_k}^k, \beta_{n_k}^k, w_{n_k}^k))$, let $\bar{\sigma} = (\bar{\sigma}^1, \ldots, \bar{\sigma}^m) \in \mathfrak{S}_n(\bar{\sigma}^\circ)$ be a schedule such that $\bar{\sigma}^k = ((w_{n_k}^k, \beta_{n_k}^k, a_{n_k}^k), \ldots, (w_1^k, \beta_1^k, a_1^k))$, $1 \le k \le m$. Then*

(a) the correspondence $\sigma \longleftrightarrow \bar{\sigma}$ will be called a transformation *of the schedule σ for the problem (P^m) into the schedule $\bar{\sigma}$ for the corresponding problem (D^m) and vice versa,*

(b) both corresponding problems, (P^m) and (D^m), will be called equivalent *problems.*

Given a schedule σ for the initial problem (P^m), described by the tables $T(\sigma^k)$, $1 \le k \le m$, the schedule $\bar{\sigma}$ for the transformed problem (D^m) is described by the tables

$$T(\bar{\sigma}^k) := \begin{bmatrix} w_{n_k+1}^k & w_{n_k}^k & w_{n_k-1}^k & \cdots & w_1^k \\ & \beta_{n_k}^k & \beta_{n_k-1}^k & \cdots & \beta_1^k \\ & a_{n_k}^k & a_{n_k-1}^k & \cdots & a_1^k & a_0^k \end{bmatrix}, \qquad (12.20)$$

where $1 \le k \le m$, $C_0^k(\bar{\sigma}^k) := w_{n_k+1}^k$ is the time at which the machine M_k starts the processing of jobs in the problem (D^m) and $C_0^k(\sigma^k) := a_0^k$ is the time at which the machine M_k starts the processing of jobs in the original problem (P^m), $1 \le k \le m$.

The next theorem, which concerns the equivalence of the problems (P^m) and (D^m), is a counterpart of Theorem 12.15 for the case of m machines.

Theorem 12.18. *(Gawiejnowicz et al. [97]) Let $\sigma = (\sigma^1, \ldots, \sigma^m)$, $\sigma^k = ((a_1^k, \beta_1^k, w_1^k), \ldots, (a_{n_k}^k, \beta_{n_k}^k, w_{n_k}^k))$, be an arbitrary schedule from $\mathfrak{S}_n(\sigma^\circ)$ and $\bar{\sigma} \in \mathfrak{S}_n(\bar{\sigma}^\circ)$ be the transformed schedule of σ in the form of $\bar{\sigma} = (\bar{\sigma}^1, \ldots, \bar{\sigma}^m)$, $\bar{\sigma}^k = ((w_{n_k}^k, \beta_{n_k}^k, a_{n_k}^k), \ldots, (w_1^k, \beta_1^k, a_1^k))$, $1 \le k \le m$. Then*

(a) if $\bar{\sigma}$ has been obtained by the transformation $\sigma \longleftrightarrow \bar{\sigma}$, then there holds the equality

$$W_{P^m}(\sigma) = w^\mathsf{T} C(\sigma) = \bar{a}^\mathsf{T} C(\bar{\sigma}) = W_{D^m}(\bar{\sigma}); \qquad (12.21)$$

(b) $\sigma^\star$ is an optimal schedule for the problem (P^m) if and only if $\bar{\sigma}^\star$ is an optimal schedule for the problem (D^m); moreover, there holds the equality $W_{P^m}(\sigma^\star) = W_{D^m}(\bar{\sigma}^\star)$.

Proof. (a) (b) Similar to the proof of Theorem 12.15. $\square$

12.4.3 Detailed results

In this subsection, we show several examples of results that can be obtained by the general transformations presented in this section.

We start with the result saying that in multi-machine problems the total weighted starting time criterion, $\sum w_j S_j$, is equivalent to the total machine load criterion, $\sum C_{\max}^{(k)}$. (Since in the considered case $\sum w_j S_j \equiv \sum w_j C_j$, we write $\sum w_j C_j$ instead of $\sum w_j S_j$.)

Theorem 12.19. (Gawiejnowicz et al. [97]) *Problems* $Pm|p_j = b_j t| \sum wC_j$ *and* $Pm|p_j = w + b_j t| \sum a_0^k C_{\max}^{(k)}$ *are equivalent.*

Proof. The result is a corollary from Theorem 12.18 for $w_j = w$, $a_j^k = 0$ and $a_0^k := t_0^k > 0$ for $1 \le j \le n_k$ and $1 \le k \le m$. $\square$

The next few results show common features of equivalent problems with $m \ge 2$ machines. First, such problems have the same lower bounds on the optimal value of a criterion.

Property 12.20. (Gawiejnowicz et al. [97]) *If* $w = 1$, $t_0^k = 1$ *for* $1 \le k \le m$, $h = \lfloor \frac{n}{m} \rfloor$ *and* $r = n - hm$, *the optimal total machine load for the problem* $Pm|p_j = 1 + b_j t| \sum C_{\max}^{(k)}$ *is not less than* $m \sum_{i=1}^{h} \sqrt[m]{\prod_{j=1}^{im+r} \beta_j} + \sum_{j=1}^{r} \beta_j$.

Proof. The result follows from Theorem 12.19 and the lower bound for the problem $Pm|p_j = b_j t| \sum C_j$ (Jeng and Lin [151]). $\square$

Second, equivalent problems have the same time complexity status.

Theorem 12.21. (Gawiejnowicz et al. [97]) *Let* $t_0^k = a_0 > 0$ *and* $w > 0$ *be common machine starting times for machines* M_k, $1 \le k \le m$, *in the problem* $Pm|p_j = b_j t| \sum wC_j$ *and in the problem* $Pm|p_j = w + b_j t| \sum a_0 C_{\max}^{(k)}$, *respectively. Then the first problem has the same time complexity as the second problem. In particular, if* $m \ge 2$, *both these problems are* $\mathcal{NP}$-*hard in the ordinary sense.*

Proof. The result follows from Theorem 12.19 and ordinary $\mathcal{NP}$-hardness of the problem $P2|p_j = b_j t| \sum C_j$ (Chen [43], Kononov [172]). $\square$

Third, equivalent problems have the same approximability status.

Theorem 12.22. (Gawiejnowicz et al. [97]) *There is no polynomial-time approximation algorithm with a constant worst-case bound for the problem* $P|p_j = w + b_j t| \sum a_0 C_{\max}^{(k)}$, *unless* $\mathcal{P} = \mathcal{NP}$.

Proof. The result follows from Theorem 12.19 and a theorem about non-approximability of the problem $P|p_j = b_j t| \sum C_j$ (Chen [43]). $\square$

Finally, if for one of equivalent problems there exists an FPTAS, then an FPTAS also exists for the second of these problems.

Theorem 12.23. (Gawiejnowicz et al. [97]) *For the problem* $Pm|p_j = \omega + b_j t| \sum a_0^k C_{\max}^{(k)}$ *there exists an FPTAS.*

Proof. The result follows from Theorem 12.19 and a theorem about existence of an FPTAS for the problem $Pm|p_j = b_j t| \sum w_j C_j$ (Woeginger [296]).   □

With this theorem, we end the presentation of applications of matrix methods in time-dependent scheduling. In Chaps. 13 and 14 we will consider time-dependent scheduling with job precedence constraints and time-dependent scheduling with two criteria, respectively.

12.5 Concluding remarks and open problems

In this chapter, we have shown that time-dependent scheduling problems can be formulated in terms of matrices and vectors. Using the matrix approach, we introduced the l_p norm as a schedule optimality criterion and we proved a number of results concerning minimization of this criterion. Finally, we have shown that there exist pairs of time-dependent scheduling problems, called equivalent problems, which have similar properties.

The results presented in the chapter have a few consequences. First, they explain the similarities between different time-dependent scheduling problems. Second, they allow to simplify the proofs of considered properties. Third, given a property of a time-dependent scheduling problem (e.g., its time complexity), we can establish a similar property of another time-dependent scheduling problem, if we know that the latter problem is equivalent to the first one. Fourth, if we know that some two $\mathcal{NP}$-hard time-dependent scheduling problems are equivalent, we can use the same heuristic algorithm for both these problems. Moreover, the performance of the algorithm will be the same in both cases. Gawiejnowicz et al. [102] proposed two such algorithms for two equivalent time-dependent scheduling problems. Finally, note that we assumed that jobs have neither ready times nor deadlines. Gawiejnowicz et al. [98] have shown how to extend the results concerning equivalent problems to the case when jobs have ready times and (or) deadlines.

Further research concerning the application of matrix methods to time-dependent scheduling problems may be focused on the status of the time complexity of the problem $1|p_j = 1 + b_j t| \, ||C||_p$ for $p \in (p_0, p_1)$, where p_0 and p_1 are constants defined as in Theorem 12.12 and Theorem 12.11, respectively.

13

Scheduling dependent deteriorating jobs

W e devoted the previous chapters to time-dependent scheduling problems with independent jobs. In this chapter, we consider single machine time-dependent scheduling problems with job precedence constraints.

Chapter 13 is composed of six sections. In Sect. 13.1, we introduce the notation and auxiliary results. In Sect. 13.2, we consider job precedence constraints in the form of a set of chains. In Sect. 13.3, we consider job precedence constraints in the form of a tree and a forest. In Sect. 13.4, we consider job precedence constraints in the form of a series-parallel digraph. In Sect. 13.5, we consider general precedence constraints. Concluding remarks, open problems and one table are given in Sect. 13.6.

13.1 Preliminaries

Throughout this chapter, we consider the problems $1|p_j = a_j + b_j t, \delta|C_{\max}$ and $1|p_j = b_j t, prec|f_{\max}$, where $\delta \in \{chain, chains, tree, ser\text{-}par\}$ and $a_j > 0$ and $b_j > 0$ for $1 \leq j \leq n$.

In this section, we introduce the notation and auxiliary results used in the chapter. Since basic definitions concerning graphs and digraphs were introduced in Sect. 1.2, we restrict our attention to only new notions.

For simplicity of further presentation, throughout the chapter, we identify job $J_j \in \mathcal{J}$ and its index j. We start with the following definition (cf. Gawiejnowicz et al. [93]).

Definition 13.1. (A module-chain)
Let G be graph of precedence constraints in a time-dependent scheduling problem. A chain $(n_1, \ldots, n_k)$ in the precedence graph G is called a module-chain *if for every job $j \in G \backslash \{n_1, \ldots, n_k\}$ there holds one of the following conditions:*
(a) there are no precedence constraints between an arbitrary job from the chain $(n_1, \ldots, n_k)$ and job j;
(b) job j precedes all jobs of the chain $(n_1, \ldots, n_k)$;
(c) job j follows all jobs of the chain $(n_1, \ldots, n_k)$.

We start with a lemma about the total processing time of a chain of jobs.

Lemma 13.2. (Gawiejnowicz et al. [93]) *Let there be given a chain* $(n_1, \ldots, n_k)$ *of k jobs with processing times form of* $p_j = a_j + b_j t, j = n_1, \ldots, n_k$, *and let these jobs be processed on a single machine sequentially and without idle times starting from time* $t_0 \geq 0$. *Then the total processing time* $P(n_1, \ldots, n_k)$ *of all jobs from this chain is equal to*

$$P(n_1, \ldots, n_k) := \sum_{j=n_1}^{n_k} p_j = \sum_{i=n_1}^{n_k} a_i \prod_{j=i+1}^{n_k} (1+b_j) + \left(\prod_{i=n_1}^{n_k} (1+b_i) - 1 \right) t_0.$$

$$(13.1)$$

Proof. We prove the lemma by mathematical induction with respect to k, the number of jobs in a chain. Formula (13.1) holds for a single job, since $P(n_1) = a_{n_1} + b_{n_1} t_0$. Assume that (13.1) holds for a chain of jobs $(n_1, \ldots, n_k)$. We should prove its validity for the chain $(n_1, \ldots, n_{k+1})$. We have

$$P(n_1, \ldots, n_{k+1}) = P(n_1, \ldots, n_k) + p_{n_{k+1}} =$$

$$P(n_1, \ldots, n_k) + a_{n_{k+1}} + b_{n_{k+1}}(t_0 + P(n_1, \ldots, n_k)) =$$

$$= \sum_{i=n_1}^{n_k} a_i \prod_{j=i+1}^{n_k} (1+b_j) + \left(\prod_{i=n_1}^{n_k} (1+b_i) - 1 \right) t_0 + a_{n_{k+1}} +$$

$$+ b_{n_{k+1}}(t_0 + \sum_{i=n_1}^{n_k} a_i \prod_{j=i+1}^{n_k} (1+b_j) + \left(\prod_{i=n_1}^{n_k} (1+b_i) - 1 \right) t_0) =$$

$$= \sum_{i=n_1}^{n_{k+1}} a_i \prod_{j=i+1}^{n_{k+1}} (1+b_j) + \left(\prod_{i=n_1}^{n_{k+1}} (1+b_i) - 1 \right) t_0.$$

■

Remark 13.3. Lemma 13.2 describes the first property of a chain of linearly deteriorating jobs: the processing time of the chain is a linear function of the starting time of the first job from this chain.

By Lemma 13.2, we can express the total processing time $P(n_1, \ldots, n_k)$ as a function of $A(n_1, n_k)$ and $B(n_1, n_k)$ coefficients, i.e.,

$$P(n_1, \ldots, n_k) := A(n_1, n_k) + B(n_1, n_k)t_0, \qquad (13.2)$$

where

$$A(n_1, n_k) := \sum_{i=n_1}^{n_k} a_i \prod_{j=i+1}^{n_k} (1+b_j) \qquad (13.3)$$

and

$$B(n_1, n_k) := \prod_{j=n_1}^{n_k} (1+b_j) - 1. \qquad (13.4)$$

Remark 13.4. If a chain is composed of only one job, we will write $P(n_1)$, $A(n_1)$ and $B(n_1)$ instead of $P(n_1, n_1)$, $A(n_1, n_1)$ and $B(n_1, n_1)$, respectively.

The next lemma indicates the importance of the ratio $\frac{B(n_1, n_k)}{A(n_1, n_k)}$ for a given chain of jobs $(n_1, \ldots, n_k)$.

Lemma 13.5. (Gawiejnowicz et al. [93]) *Let there be given two chains of jobs, $C_1 = (n_1, \ldots, n_k)$ and $C_2 = (n'_1, \ldots, n'_k)$, such that there are no precedence constraints between any job from C_1 and any job from C_2. Let σ' (σ'') denote the schedule in which all jobs from C_1 (C_2) are executed sequentially and without idle times, and are followed by all jobs from C_2 (C_1), and let the execution of the jobs start at the same time t_0 in both schedules. Then, $C_{\max}(\sigma') < C_{\max}(\sigma'')$ if and only if $\frac{B(n_1, n_k)}{A(n_1, n_k)} > \frac{B(n'_1, n'_k)}{A(n'_1, n'_k)}$.*

Proof. Let the execution of jobs start at time t_0. Calculate the length of both schedules, $C_{\max}(\sigma')$ and $C_{\max}(\sigma'')$. Using Lemma 13.2, we have

$$C_{\max}(\sigma') = t_0 + A(n_1, n_k) + B(n_1, n_k)t_0 + A(n'_1, n'_k) +$$

$$B(n'_1, n'_k)(t_0 + A(n_1, n_k) + B(n_1, n_k)t_0) = t_0 + A(n_1, n_k) + A(n'_1, n'_k) +$$

$$A(n_1, n_k)B(n'_1, n'_k) + (B(n_1, n_k) + B(n'_1, n'_k) + B(n_1, n_k)B(n'_1, n'_k))t_0$$

and

$$C_{\max}(\sigma'') = t_0 + A(n'_1, n'_k) + B(n'_1, n'_k)t_0 + A(n_1, n_k) +$$

$$B(n_1, n_k)(t_0 + A(n'_1, n'_k) + B(n'_1, n'_k)t_0) = t_0 + A(n_1, n_k) + A(n'_1, n'_k) +$$

$$B(n_1, n_k)A(n'_1, n'_k) + (B(n_1, n_k) + B(n'_1, n'_k) + B(n_1, n_k)B(n'_1, n'_k))t_0.$$

The difference between the length of schedules σ' and σ'' is then equal to

$$C_{\max}(\sigma') - C_{\max}(\sigma'') = A(n_1, n_k)B(n'_1, n'_k) - A(n'_1, n'_k)B(n_1, n_k)$$

and schedule σ' is shorter than σ'' if and only if $\frac{B(n_1, n_k)}{A(n_1, n_k)} > \frac{B(n'_1, n'_k)}{A(n'_1, n'_k)}$. ∎

Remark 13.6. Lemma 13.5 describes the second property of the considered problem: in the optimal schedule the chain $(n_1, \ldots, n_k)$ with the greatest ratio $\frac{B(n_1, n_k)}{A(n_1, n_k)}$ is scheduled as the first one.

Definition 13.7. (The ratio $R(n_1, n_k)$ for a chain $(n_1, n_2, \ldots, n_k)$ of jobs) *Given a chain $(n_1, n_2, \ldots, n_k)$ of jobs $J_{n_1}, J_{n_2}, \ldots, J_{n_k}$, the ratio*

$$R(n_1, n_k) := \frac{B(n_1, n_k)}{A(n_1, n_k)}$$

will be called the ratio $R(n_1, n_k)$ (ratio R, in short) for this chain.

Remark 13.8. Note that for chain $(n_1, \ldots, n_k)$ the following equations hold:

$$A(n_1, n_{i+1}) = A(n_1, n_i)(1 + b_{i+1}) + a_{i+1} \qquad (13.5)$$

and

$$B(n_1, n_{i+1}) = B(n_1, n_i)(1 + b_{i+1}) + b_{i+1}, \qquad (13.6)$$

where $i = 1, 2, \ldots, k - 1$. Hence, we can calculate the ratio R for a chain in time that is linear with respect to the length of this chain.

The next lemma describes a monotonicity property of the sequence of ratios $R(n_1, j)$ for $j = n_1, \ldots, n_k$.

Lemma 13.9. (Gawiejnowicz et al. [93]) *Let there be given the chain of jobs $(n_1, \ldots, n_k)$ and its two subchains, $(n_1, \ldots, n_l)$ and $(n_{l+1}, \ldots, n_k)$, where $1 \le l \le k - 1$. Then there holds the following implication:*

if $R(n_{l+1}, n_k) > R(n_1, n_l)$, then $R(n_{l+1}, n_k) > R(n_1, n_k) > R(n_1, n_l)$.

Moreover, similar implications hold, if we replace the sign '>' by either '<' or '=', respectively.

Proof. We prove the lemma for the case when $R(n_{l+1}, n_k) > R(n_1, n_l)$; the two remaining cases can be proved by similar reasoning.

Let the chain $(n_1, \ldots, n_k)$ be processed sequentially and without idle times and let l be any integer from the set $\{1, \ldots, k-1\}$. Then the chain $(n_1, \ldots, n_k)$ is divided by job l into two subchaines, $(n_1, \ldots, n_l)$ and $(n_{l+1}, \ldots, n_k)$. For simplicity, introduce the following notation: $A(n_1, n_l) = A_1$, $B(n_1, n_l) = B_1$, $A(n_{l+1}, n_k) = A_2$, $B(n_{l+1}, n_k) = B_2$, $A(n_1, n_k) = A$, $B(n_1, n_k) = B$.

From equations (13.3) and (13.4), we have

$$B = B(n_1, n_k) = \prod_{j=n_1}^{n_k} (1 + b_j) - 1 =$$

$$\prod_{j=n_1}^{n_l} (1 + b_j) \prod_{j=n_{l+1}}^{n_k} (1 + b_j) - 1 = (1 + B_1)(1 + B_2) - 1$$

and

$$A = \sum_{i=n_1}^{n_k} a_i \prod_{j=i+1}^{n_k} (1 + b_j) = \sum_{i=n_1}^{n_l} a_i \prod_{j=i+1}^{n_l} (1 + b_j) \prod_{j=n_{l+1}}^{n_k} (1 + b_j) +$$

$$\sum_{i=n_{l+1}}^{n_k} a_i \prod_{j=i+1}^{n_k} (1 + b_j) = A_1(1 + B_2) + A_2.$$

So, we have

$$A = A_1(1 + B_2) + A_2 \quad \text{and} \quad B = (1 + B_1)(1 + B_2) - 1. \tag{13.7}$$

First, we show that $R(n_{l+1}, n_k) > R(n_1, n_k)$. Since $\frac{B_2}{A_2} - \frac{B}{A} = \frac{AB_2 - BA_2}{A_2 A}$, by (13.7) we have

$$AB_2 - BA_2 = (1 + B_2)(A_1 B_2 - A_2 B_1). \tag{13.8}$$

But from assumption of this lemma we know that $\frac{B_2}{A_2} > \frac{B_1}{A_1}$. This implies that

$$A_1 B_2 - A_2 B_1 > 0. \tag{13.9}$$

Hence, by (13.8) and (13.9), we have

$$\frac{B_2}{A_2} = R(n_{l+1}, n_k) > \frac{B}{A} = R(n_1, n_k). \tag{13.10}$$

In a similar way, we prove the second inequality. Indeed, because

$$\frac{B}{A} - \frac{B_1}{A_1} = \frac{A_1 B - B_1 A}{A A_1} \tag{13.11}$$

and $A_1 B - AB_1 = A_1 B_2 - B_1 A_2$ we have, by (13.9) and (13.11), that

$$\frac{B}{A} = R(n_1, n_k) > \frac{B_1}{A_1} = R(n_1, n_l). \tag{13.12}$$

Therefore, by (13.10) and (13.12), $R(n_{l+1}, n_k) > R(n_1, n_k) > R(n_1, n_l)$. ∎

On the basis of Lemmata 13.2, 13.5 and 13.9, we get the following result, showing the role played by the element of a chain for which the maximum value of the ratio R is obtained.

Theorem 13.10. (Gawiejnowicz et al. [93]) *Let* $n_k = \arg \max\limits_{n_1 \leq j \leq n_k} \{R(n_1, j)\}$ *for a given module-chain* $(n_1, \ldots, n_k)$. *Then there exists an optimal schedule in which jobs of the chain* $(n_1, \ldots, n_k)$ *are executed sequentially, without idle times and such that no jobs from other chains are executed between the jobs of this chain.*

Proof. Let σ denote a schedule in which some jobs are executed between jobs of chain $(n_1, \ldots, n_k)$. We can represent schedule σ as the set of subchains $\mathcal{CL} = \{\mathcal{C}_1, \mathcal{L}_1, \mathcal{C}_2, \mathcal{L}_2, \ldots, \mathcal{C}_l, \mathcal{L}_l, \mathcal{C}_{l+1}\}$, $l = 1, \ldots, k-1$, where each subchain $\mathcal{C}_j$ and $\mathcal{L}_j$ contains, respectively, jobs only from and jobs only outside of chain $(n_1, \ldots, n_k)$. Denote by $R(\mathcal{C}_j)$, for $j = 1, \ldots, l+1$ ($R(\mathcal{L}_j)$, for $j = 1, \ldots, l$) the ratio R of subchain $\mathcal{C}_j$ ($\mathcal{L}_j$).

Definition 13.1 implies that there are no precedence constraints between any job from subchain $\mathcal{C}_j$ and any job from subchain $\mathcal{L}_i$. Hence, if we swap two successive subchains from the set $\mathcal{CL}$, then we get a feasible schedule again. We will show that we always can find two successive subchains such that one can swap these subchains without increasing the length of schedule. Contrary, suppose the opposite. Then, by Lemma 13.5, we have

$$R(\mathcal{C}_1) > R(\mathcal{L}_1) > R(\mathcal{C}_2) > R(\mathcal{L}_2) > \ldots > R(\mathcal{C}_l) > R(\mathcal{L}_l) > R(\mathcal{C}_{l+1}). \quad (13.13)$$

From (13.13), it follows that $R(\mathcal{C}_1) > R(\mathcal{C}_2) > \ldots > R(\mathcal{C}_l) > R(\mathcal{C}_{l+1})$. But then Lemma 13.9 implies that $R(\mathcal{C}_1) > R(n_1, n_k)$. We obtained a contradiction with the definition of n_k. Therefore, we can conclude that in schedule σ there exist two successive subchains of jobs, such that we can swap them without increasing the value of $C_{\max}(\sigma)$. This swap decreases the number l of subchains of jobs that do not belong to the chain $(n_1, \ldots, n_k)$.

Repeating this swapping procedure at most l times, we will obtain a schedule σ' in which jobs of the chain $(n_1, \ldots, n_k)$ are executed sequentially, without idle times and such that $C_{\max}(\sigma') \le C_{\max}(\sigma)$. ∎

Remark 13.11. Starting from now, we will assume that if there are two values for which the maximum is obtained, the function 'arg' chooses the job with the larger index.

Remark 13.12. Theorem 13.10 describes the third property of the considered problem: in the set of chains there exist some subchains which are processed in the optimal schedule like aggregated jobs, since inserting into these subchains either a separate job (jobs) or a chain (chains) makes the final schedule longer.

Remark 13.13. Some of the results presented in this section are given in another form by Wang et al. [285]. In particular, the authors give counterparts of the ratio $R(n_1, n_k)$ (see [285, p. 2687]), of Lemma 13.5 (see [285, Lemma 2]) and of Theorem 13.10 (see [285, Lemma 7]).

In subsequent sections, we study different forms of job precedence constraints in the problem under consideration. We start with the precedence constraints in the form of chains.

13.2 Chain precedence constraints

Assume that the precedence digraph G is a set of chains. Note that any chain in G is a module-chain. The results of Sect. 13.1 allow us to construct an exact polynomial-time algorithm for the problem $1|p_j = a_j + b_j t, chains|C_{\max}$.

We start the section with the formulation of an algorithm that constructs a partition $\mathcal{U}$ of the chain $\mathcal{C} = (n_1, \ldots, n_k)$ into subchains $\mathcal{C}_i = (n_{i,1}, \ldots, n_{i,k_i})$, $i = 1, 2, \ldots, l$. The partition $\mathcal{U}$ has the following properties:

Property 13.14. $\mathcal{C} = \cup_{i=1}^{l} \mathcal{C}_i$;

Property 13.15. $n_{i,k_i} = \arg \max_{n_{i,1} \le j \le n_{i,k_i}} \{R(n_{i_1}, j)\}$;

Property 13.16. $R(\mathcal{C}_1) > R(\mathcal{C}_2) > \ldots > R(\mathcal{C}_l)$.

The pseudo-code of the algorithm is as follows.

Algorithm A_{55} for the problem $1|p_j = a_j + b_j t, chain|C_{\max}$ ([93])

Input: sequences $(a_{n_1}, a_{n_2}, \ldots, a_{n_k})$, $(b_{n_1}, b_{n_2}, \ldots, b_{n_k})$
   chain $(n_1, n_2, \ldots, n_k)$
Output: an optimal schedule $\sigma^\star$

▷ Step 1:
   $i \leftarrow 1; j \leftarrow 1;$
   $\mathcal{C}_1 \leftarrow (n_1);$ Compute $R(\mathcal{C}_1);$
▷ Step 2:
   while $(j + 1 \leq k)$ **do**
      $i \leftarrow i + 1; j \leftarrow j + 1;$
      $\mathcal{C}_i \leftarrow (n_j);$ Compute $R(\mathcal{C}_i);$
      while $(R(\mathcal{C}_{i-1}) \leq R(\mathcal{C}_i))$ **do**
         $\mathcal{C}_{i-1} \leftarrow \mathcal{C}_{i-1} \cup \mathcal{C}_i;$
         Compute $R(\mathcal{C}_{i-1});$
         $i \leftarrow i - 1$
▷ Step 3:
   $\sigma^\star \leftarrow (\mathcal{C}_1, \mathcal{C}_2, \ldots, \mathcal{C}_i);$
   return $\sigma^\star.$

Remark 13.17. Algorithm A_{55} divides a chain (a module-chain) into subchains that are like 'big' independent jobs: none of such subchaines can be divided into smaller subchaines and jobs from different subchaines cannot be interleaved without increase of schedule length.

Remark 13.18. The subchains generated by algorithm A_{55} will be called *independent subchains*.

Lemma 13.19. (Gawiejnowicz et al. [93])
(a) *Algorithm A_{55} constructs a partition $\mathcal{U}$ of chain $\mathcal{C}$ in $O(k)$ time, $k = |\mathcal{C}|$.*
(b) *The partition $\mathcal{U}$ has Properties 13.14–13.16.*

Proof. (a) First, notice that algorithm A_{55} is composed of three steps and Step 1 is performed in a constant time. Second, the algorithm in Step 2 either generates a new subchain from an unconsidered vertex of chain $\mathcal{C}$ or joins two consecutive subchains. Third, each of these procedures can be executed in time at most k units and requires a fixed number of operations. From it follows that algorithm A_{55} runs in $O(k)$ time.

(b) Since in Step 2 we consider iteratively each vertex of chain $\mathcal{C}$, the union of obtained subchains covers $\mathcal{C}$ and Property 13.14 holds.

Now we will prove that there holds Property 13.15. Let $\mathcal{U} = (\mathcal{C}_1, \mathcal{C}_2, \ldots, \mathcal{C}_l)$ be the partition of chain $\mathcal{C}$ and let for a subchain $\mathcal{C}_r = (n_{r_1}, \ldots, n_{r_k})$ Property 13.15 does not hold, i.e., there exists index $p < r_k$ such that $R(n_{r_1}, n_p) > R(n_{r_1}, n_{r_k})$. Without loss of generality, we can assume that

$$n_p = \arg \max_{1 \leq j \leq k} \{R(n_{r_1}, n_{r_j})\}. \qquad (13.14)$$

Consider the subset of $\mathcal{U}$, obtained by algorithm A_{55} before considering element n_{p+1}, and denote it by $\mathcal{U}'$. Notice that the algorithm in Step 2 joins two last elements. Because in the final partition we have subchains $(\mathcal{C}_1, \mathcal{C}_2, \ldots, \mathcal{C}_{r-1})$, they are obtained before considering element n_{r_1} and, in conclusion, also belong to the set $\mathcal{U}'$. Let $\mathcal{C}_h = (n_{r_1}, n_p)$. Then, by (13.14), we have $\mathcal{U}' = ((\mathcal{C}_1, \mathcal{C}_2, \ldots, \mathcal{C}_{r-1}), \mathcal{C}_h)$. Contrary, assume the opposite. Let $\mathcal{U}' = ((\mathcal{C}_1, \mathcal{C}_2, \ldots, \mathcal{C}_{r-1}), \mathcal{C}_{h_1}, \ldots, \mathcal{C}_{h_p})$. Then, applying algorithm A_{55}, we have $R(\mathcal{C}_{h_1}) > R(\mathcal{C}_{h_2}) > \ldots > R(\mathcal{C}_{h_p})$. Hence, by Lemma 13.9, we have $R(\mathcal{C}_{h_1}) > R(n_{r_1}, n_p)$. A contradiction to equality (13.14).

Consider now the subset of $\mathcal{U}$, obtained by algorithm A_{55} before considering the element n_{r_k+1}, and denote it by $\mathcal{U}''$. From Lemma 13.9 and equality (13.14) it follows that $R(n_{r_1}, n_p) > R(n_{p+1}, n_j)$ for all $j = p+1, \ldots, r_k$. Hence checking the condition in the inner 'while' loop for iterations $p+1, \ldots, r_k$ gives a negative answer. Hence and since $p < r_k$, it follows that $\mathcal{U}''$ includes subchain $(n_{r_1}, \ldots, n_p)$ and, in conclusion, does not contain subchain $(n_{r_1}, \ldots, n_{r_k})$. From the pseudo-code of algorithm A_{55}, it is clear that such a chain cannot be obtained at subsequent iterations, either. A contradiction.

There remains to prove that Property 13.16 also holds. If for some $2 \leq i \leq l$ we have $R(\mathcal{C}_{i-1}) \leq R(\mathcal{C}_i)$, then algorithm A_{55} joins these subchains in Step 2 and from it follows that Property 13.16 holds. ∎

Let G, $\mathcal{C}_i$, where $1 \leq i \leq k$ and $\sum_i^k |\mathcal{C}_i| = n$, and $\sigma^\star$ denote the precedence graph, the i-th module-chain and the optimal job sequence, respectively. On the basis of algorithm A_{55}, we can formulate the following algorithm for the problem with precedence constraints in the form of a set of chains.

Algorithm A_{56} for the problem $1|p_j = a_j + b_j t, chains|C_{\max}$ ([93])

Input: sequences $(a_1, a_2, \ldots, a_n)$, $(b_1, b_2, \ldots, b_n)$, chains $\mathcal{C}_1, \mathcal{C}_2, \ldots, \mathcal{C}_k$
Output: an optimal schedule $\sigma^\star$

▷ Step 1:
   for $i \leftarrow 1$ to k do
        Apply algorithm A_{55} to chain $\mathcal{C}_i$;
▷ Step 2:
   Arrange G in the non-decreasing order of the R ratios
   ↪ of independent chains;
   Call the obtained sequence $\sigma^\star$;
▷ Step 3:
   return $\sigma^\star$.

Theorem 13.20. (Gawiejnowicz et al. [93]) *The problem $1|p_j = a_j + b_j t$, chains$|C_{\max}$ is solvable in $O(n \log n)$ time by algorithm A_{56}.*

Proof. The partition $\mathcal{U}$ constructed for each chain has Property 13.16. This implies the feasibility of the schedule generated by algorithm A_{56}. The optimality of the schedule follows from Theorem 13.10 and Lemma 13.5. Algorithm A_{55} runs in linear time with respect to the number of vertices. Hence the running time of algorithm A_{56} is determined by its Step 2. Since this step needs ordering of at most n elements, algorithm A_{56} runs in $O(n \log n)$ time. ∎

We illustrate the application of algorithm A_{56} by an example.

Example 13.21. We are given $n = 4$ jobs with the following processing times: $p_1 = 2 + t$, $p_2 = 1 + 3t$, $p_3 = 1 + t$, $p_4 = 3 + 2t$. Precedence constraints are as follows: job J_1 precedes job J_2, and job J_3 precedes job J_4 (see Fig. 13.1a).

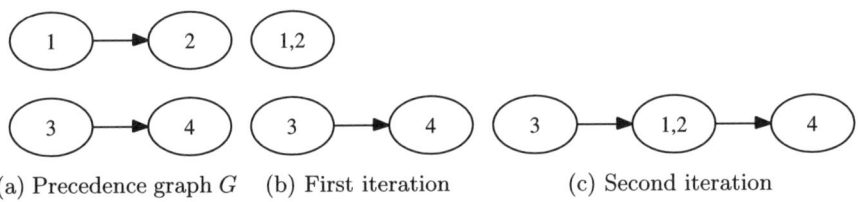

(a) Precedence graph G (b) First iteration   (c) Second iteration

Fig. 13.1: Precedence constraints in Example 13.21

Algorithm A_{55} for chain $(1, 2)$ starts with $i = 1$, $j = 1$, $C_1 = (1)$ and $R(C_1) = \frac{1}{2}$. Next, we have $i = 2$, $j = 2$, $C_2 = (2)$ and $R(C_2) = 3$. Since $R(C_1) < R(C_2)$, we join jobs 1 and 2, obtaining subchain $C_1 = (1, 2)$ with $R(C_1) = \frac{7}{9}$ (see Fig. 13.1b).

The execution of algorithm A_{55} for chain $(3, 4)$ runs similarly. First, we set $i = 1$, $j = 1$, $C'_1 = (3)$ and $R(C'_1) = 1$. Then $i = 2$, $j = 2$, $C'_2 = (4)$ and $R(C'_2) = \frac{2}{3}$. Since $R(C'_1) > R(C'_2)$, we cannot join vertices 3 and 4. Therefore, we get two subchains, $C'_1 = (3)$ and $C'_2 = (4)$.

After the completion of algorithm A_{55}, we obtain three independent subchains: $C_1 = (1, 2)$, $C_2 = (3)$ and $C_3 = (4)$.

Now we can arrange independent subchains in the non-decreasing order of the ratios R of subchains. Because $R(C_3) < R(C_1) < R(C_2)$, the optimal job sequence is $\sigma^* = (3, 1, 2, 4)$, see Fig. 13.1c. ♦

13.3 Tree and forest precedence constraints

In this section, we discuss the case when in considered problem the precedence constraints among jobs are in the form of a tree or a forest.

Let the precedence digraph G be an in-tree. Recall that v is called a *node* of in-tree G, if v has at least two immediate predecessors (cf. Definition 1.10).

Algorithm A_{57} for the problem $1|p_j = a_j + b_j t, in\text{-}tree|C_{\max}$ ([93])

Input: sequences $(a_1, a_2, \dots, a_n)$, $(b_1, b_2, \dots, b_n)$, in-tree G
Output: an optimal schedule $\sigma^\star$
▷ Step 1:
  while (G is not a single chain) **do**
    Choose $v \in G$ such that $Pred(v)$ is a union of module-chains;
    for each module-chain $\mathcal{C} \in Pred(v)$ **do**
      Apply algorithm A_{55} to $\mathcal{C}$;
    Arrange independent subchains of $Pred(v)$ in the non-increasing
    ↪ order of R ratios;
    Call the obtained sequence σ;
    Replace in G the set $Pred(v)$ by vertices corresponding to its
    ↪ independent subchains in sequence σ;
▷ Step 2:
  $\sigma^\star \leftarrow$ the order given by G;
  return $\sigma^\star$.

Theorem 13.22. (Gawiejnowicz et al. [93]) *The problem $1|p_j = a_j + b_j t$, in-tree$|C_{\max}$ is solvable in $O(n \log n)$ time by algorithm A_{57}.*

Proof. First, note that algorithm A_{57} generates a feasible schedule, since it always looks for subsequences of jobs that are feasible with respect to the precedence digraph G. Now we will show that the schedule generated by the algorithm is optimal.

Consider a vertex $v \in G$ such that set $Pred(v)$ is a union of module-chains, $Pred(v) = \mathcal{C}_1 \cup \mathcal{C}_2 \cup \dots \cup \mathcal{C}_k$, where $\mathcal{C}_i$ is a module-chain, $1 \le i \le k$. Note that if some job $j \notin Pred(v)$, then either any job from $Pred(v)$ precedes j or there is no precedence between j and any job from $Pred(v)$. Apply algorithm A_{55} to the set $Pred(v)$ and consider the final chain $\mathcal{C} = (\mathcal{C}_{n_1}, \mathcal{C}_{n_2}, \dots, \mathcal{C}_{n_s})$, where $R(\mathcal{C}_{n_1}) \ge R(\mathcal{C}_{n_2}) \ge \dots \ge R(\mathcal{C}_{n_s})$. We will show that there exists an optimal schedule in which all subchains are executed in the same order as in $\mathcal{C}$.

Let there exist an optimal schedule σ such that for some $i < j$ the jobs from $\mathcal{C}_{n_j}$ precede the jobs from $\mathcal{C}_{n_i}$. Denote by $\mathcal{L}$ the chain of jobs that are executed in σ, after jobs from $\mathcal{C}_{n_j}$ and before jobs from $\mathcal{C}_{n_i}$. Without loss of generality, we can assume that an intersection of $\mathcal{C}$ and $\mathcal{L}$ is empty, $\mathcal{C} \cap \mathcal{L} = \emptyset$. Indeed, if some $\mathcal{C}_{n_k} \in \mathcal{L}$, then either a pair $(\mathcal{C}_{n_k}, \mathcal{C}_{n_j})$ or a pair $(\mathcal{C}_{n_k}, \mathcal{C}_{n_j})$ violates the order of $\mathcal{C}$. Since $\mathcal{L}$ does not contain the jobs from $\mathcal{C}$, then there are no precedence constraints between any job of $\mathcal{L}$ and any job of $\mathcal{C}$.

Remind that we choosen $\mathcal{C}$ in such a way that if some job $j \notin \mathcal{C}$, then either any job from $\mathcal{C}$ precedes j or there are no precedence constraints between j and any job from $\mathcal{C}$. The first case is impossible, because σ is a feasible schedule. The same reasoning implies that C_{n_i} and C_{n_j} do not belong to the same module-chain in C. From that it follows that there are no precedence constraints between any job of C_{n_i} and any job C_{n_j}.

Remind that $R(C_{n_i}) \geq R(C_{n_j})$. Let us calculate $R(\mathcal{L})$. If $R(\mathcal{L}) \geq R(C_{n_j})$, then the schedule $(\mathcal{L}, C_{n_i}, C_{n_j})$ has the length that is at most equal to the length of σ. If $R(\mathcal{L}) < R(C_{n_j})$, then the schedule $(C_{n_i}, C_{n_j}, \mathcal{L})$ has the length that is at most equal to the length of σ. Repeating this reasoning for all $i < j$ such that the jobs of C_{n_j} precede the jobs of C_{n_i}, we get an optimal schedule $\sigma^\star$ in which all jobs from $\mathcal{L}$ are in the same order as the jobs from C.

So, we have shown that there exists an optimal schedule $\sigma^\star$ in which all jobs from $\mathcal{L}$ are in the order of jobs from C. Applying this procedure a finite number of times, we obtain from graph G a new graph $G^\star$ that is a single chain.

The reasoning for the case of an out-tree is similar.

If we apply 2–3 trees (cf. Remark 1.11), algorithm A_{57} can be implemented in $O(n \log n)$ time. ∎

We illustrate the application of algorithm A_{57} by an example.

Example 13.23. Let $n = 7$, $p_1 = 1 + t$, $p_2 = 2 + 3t$, $p_3 = 1 + 2t$, $p_4 = 2 + t$, $p_5 = 2 + t$, $p_6 = 1 + 3t$, $p_7 = 1 + t$. Precedence constraints are as in Fig. 13.2.

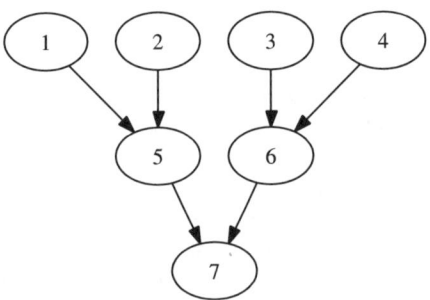

Fig. 13.2: Precedence constraints in Example 13.23

In Step 1 of algorithm A_{57}, we choose a set of module-chains that belong to the same vertex. In the case, we can choose chains (1), (2) that belong to vertex 5, or chains (3), (4) that belong to vertex 6.

Assume we choose chains (1) and (2). Since $R(2) = \frac{3}{2} > R(1) = 1$, job 2 has to precede job 1. Moreover, (2) and (1) are independent subchains. We transform precedence constraints into the ones given in Fig. 13.3a.

Now consider chains (3) and (4). After execution of Step 1, we obtain two independent subchains, (3) and (4). Since $R(3) = 2 > R(4) = \frac{1}{2}$, job 3 has to precede job 4. Moreover, (3) and (4) are independent subchains. The new form of precedence constraints is given in Fig. 13.3b.

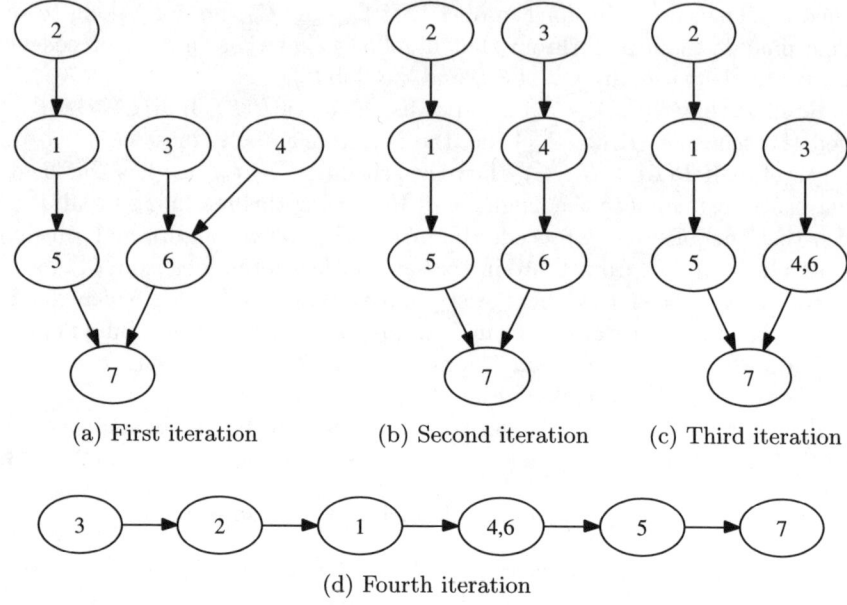

(a) First iteration   (b) Second iteration   (c) Third iteration

(d) Fourth iteration

Fig. 13.3: Subsequent iterations of algorithm A_{57} in Example 13.23

Now we have two module-chains $((2), (1), 5)$ and $((3), (4), 6)$. (We distin-guish by internal brackets the vertices corresponding to independent sub-chains.) Applying Step 1 twice, we find two new independent subchains, (5) and $(4, 6)$. Precedence constraints are as in Fig. 13.3c.

Finally, we arrange all independent subchains in the non-increasing order of R ratios, obtaining the chain given in Fig. 13.3d.

Now graph G is a single chain, the condition in loop 'while' is not satisfied and algorithm A_{57} stops. The optimal job sequence is $\sigma^\star = (3, 2, 1, 4, 6, 5, 7)$.

♦

Note that if for a vertex v the set $Succ(v)$ is a union of module-chains, then by replacing $Pred(v)$ by $Succ(v)$, we can apply algorithm A_{57} to an out-tree. Let us call the modified algorithm A_{57} by A_{58}.

Theorem 13.24. (Gawiejnowicz et al. [93]) *The problem* $1|p_j = a_j + b_j t, out-tree|C_{\max}$ *is solvable in* $O(n \log n)$ *time by algorithm* A_{58}.

Proof. Similar to the proof of Theorem 13.22.   □

We illustrate the application of algorithm A_{58} by an example.

Example 13.25. Consider Example 13.23 with reversed orientation of arcs. Precedence constraints are given in Fig. 13.4.

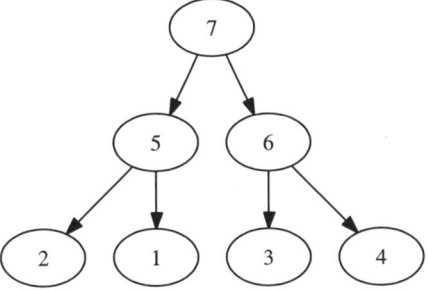

Fig. 13.4: Precedence constraints in Example 13.25

First, we choose a set of module-chains that belong to the same vertex. Again, we have two possibilities: we can choose chains (1) and (2) that belong to vertex 5, or chains (3) and (4) that belong to vertex 6. Assume we choose chains (1) and (2). Applying algorithm A_{58}, we see that job 2 has to precede job 1, since $R(2) = \frac{3}{2} > R(1) = 1$. Moreover, (2) and (1) are independent subchains. The new form of job precedence constraints is given in Fig. 13.5a.

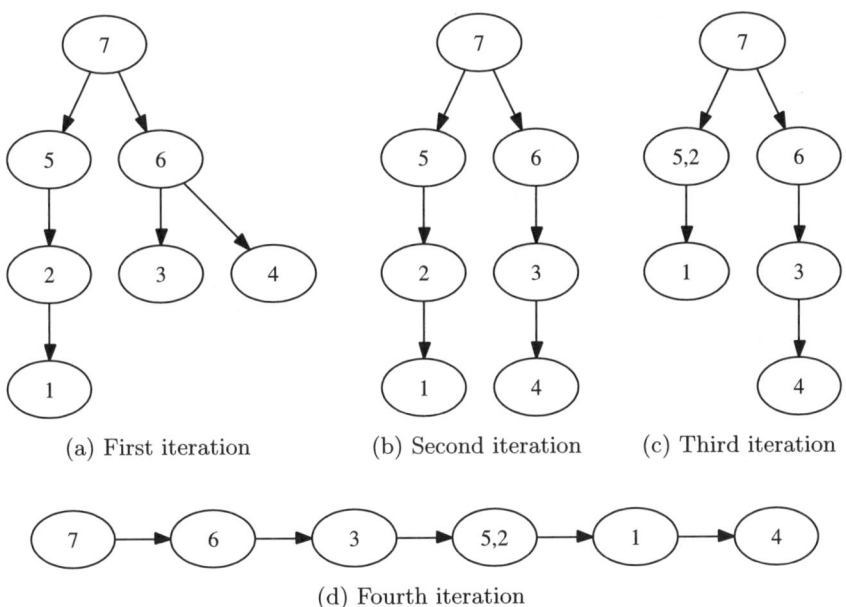

(a) First iteration   (b) Second iteration   (c) Third iteration

(d) Fourth iteration

Fig. 13.5: Subsequent iterations of algorithm A_{58} in Example 13.25

Now, consider chains (3) and (4). Since $R(3) = 2 > R(4) = \frac{1}{2}$, job 3 has to precede job 4. Moreover, (3) and (4) are independent subchains. The new digraph of job precedence constraints is given in Fig. 13.5b.

Now, we have two module-chains: $(5, (2), (1))$ and $(6, (3), (4))$. Applying algorithm A_{58}, we obtain two new independent subchaines, $(5, 2)$ and (6), with $R(5, 2) = \frac{7}{10}$ and $R(6) = 3$. The new digraph of job precedence constraints is given in Fig. 13.5c.

By arranging all independent subchains in the non-increasing order of the R ratios, we obtain the chain given in Fig. 13.5d.

Now, digraph G of job precedence constraints is a single chain and the optimal job sequence is $\sigma^\star = (7, 6, 3, 5, 2, 1, 4)$. ◆

If job precedence constraints are in the form of a forest, i.e., a set of trees, we can proceed in the following way. Assume that precedence constraints are in the form of an in-forest. By adding a dummy vertex 0, with processing time $p_0 = \epsilon = const > 0$, and by connecting roots of all in-trees with this vertex, we obtain a new in-tree. For this in-tree, we can apply algorithm A_{57} and ignore the dummy job in the final schedule. In a similar way, we can solve the case of an out-forest. Hence, we have the following result.

Theorem 13.26. (Gawiejnowicz et al. [93])
(a) *Problem* $1|p_j = a_j + b_j t, in\text{-}forest|C_{\max}$ *is solvable in* $O(n \log n)$ *time by algorithm* A_{57}.
(b) *The problem* $1|p_j = a_j + b_j t, out\text{-}forest|C_{\max}$ *is solvable in* $O(n \log n)$ *time by algorithm* A_{58}.

Proof. (a) (b) Similar to the proof of Theorem 13.22. □

13.4 Series-parallel constraints

In this section, we study the case when in the problem under consideration the job precedence constraints are in the form of a series-parallel digraph G. For simplicity, we assume that digraph G does not contain the arc (v, w) if there is a directed path from v to w not including (v, w).

We start the section with three properties of series-parallel digraphs. We omit simple proofs.

Property 13.27. (Gawiejnowicz et al. [93]) A chain is a series-parallel digraph.

Property 13.28. (Gawiejnowicz et al. [93]) The series composition of two chains is a chain.

Property 13.29. (Gawiejnowicz et al. [93]) If a node of the decomposition tree $T(G)$ of a series-parallel digraph G is parallel composition of two chains, then each of the chains is a module-chain in the digraph G.

Since digraph G from Property 13.29 is a union of module-chains, we can apply algorithm A_{32} to find the optimal sequence of vertices of this digraph. Recall that given a union of module-chains in digraph G, algorithm A_{33} finds an optimal sequence of vertices from the union. Hence, working from the bottom of the decomposition tree $T(G)$ upward and merging subsequences of vertices in an appropriate way, we find the optimal sequence.

The pseudo-code of the algorithm for the considered problem with series-parallel precedence constraints is as follows.

Algorithm A_{59} for the problem $1|p_j = a_j + b_j t, ser\text{-}par|C_{\max}$ ([93])

Input: sequences $(a_1, a_2, \ldots, a_n)$, $(b_1, b_2, \ldots, b_n)$,
    decomposition tree $T(G)$ of series-parallel digraph G
Output: an optimal schedule $\sigma^\star$

▷ Step 1:
  while (there exists $v \in T(G)$ such that $|Succ(v) = 2|$) **do**
      if (v has label P) **then**
          Apply the algorithm A_{56} to chains $C_1, C_2 \in Succ(v)$;
          Replace v, C_1 and C_2 in $T(G)$ by the obtained chain
      else Replace v, C_1 and C_2 in $T(G)$ by chain (C_1, C_2);
▷ Step 2:
  $\sigma^\star \leftarrow$ the order given by G;
  return $\sigma^\star$.

Theorem 13.30. (Gawiejnowicz et al. [93]) *The problem* $1|p_j = a_j + b_j t,$ *$ser\text{-}par|C_{\max}$ is solvable in $O(n \log n)$ time by algorithm A_{59}, provided the decomposition tree $T(G)$ of the precedence constraints digraph G is given.*

Proof. First, notice that algorithm A_{59} generates always a feasible job sequence, since it merges vertices of the decomposition tree $T(G)$. In order to show that the final sequence is optimal, we shall show how to obtain an optimal job sequence in the case of a parallel or series composition, given the already computed sequence.

Remember that if we find an optimal sequence for some job precedence digraph (subdigraph), we transform this digraph into a chain. Since each terminal node (leaf) of the tree $T(G)$ represents a single vertex (job), it is sufficient to show how to obtain an optimal sequnce of the jobs in a parallel or series composition if both arguments of the composition are chains.

First, consider the case when some node of $T(G)$ is a parallel composition of subgraphs G_1 and G_2. Let C_1 and C_2 be two chains that present an optimal sequence of the vertices in the subdigraphs G_1 and G_2. Applying algorithm A_{56} to the chains C_1 and C_2 we get an optimal sequence C for digraph G.

Let now some node of $T(G)$ be a series composition of subdigraphs G_1 and G_2. Without loss of generality, we can assume that G_1 precedes G_2. Let C_1 and

C_2 be two chains that present an optimal order of vertices in the subdigraphs G_1 and G_2. Setting the first vertex of C_2 after the last vertex of C_1, we get an optimal job sequence for digraph G.

By using 2–3 trees, algorithm A_{59} can be implemented in $O(n \log n)$ time. ∎

Remark 13.31. If the decomposition tree $T(G)$ of the series-parallel digraph G is not given, algorithm A_{59} must start with the step in which the tree is constructed (cf. Remark 1.13). Since this step needs $O(|V| + |E|) \equiv O(n^2)$ time, in this case the running time of algorithm A_{59} increases to $O(n^2)$ time.

The following example shows an application of algorithm A_{59}.

Example 13.32. Let $n = 6, p_1 = 2 + 3t, p_2 = 1 + t, p_3 = 2 + t, p_4 = 3 + 2t,$ $p_5 = 3 + 4t, p_6 = 2 + 5t$. The digraph G of precedence constraints is given in Fig. 13.6a. The decomposition tree $T(G)$ is given in Fig. 13.6b.

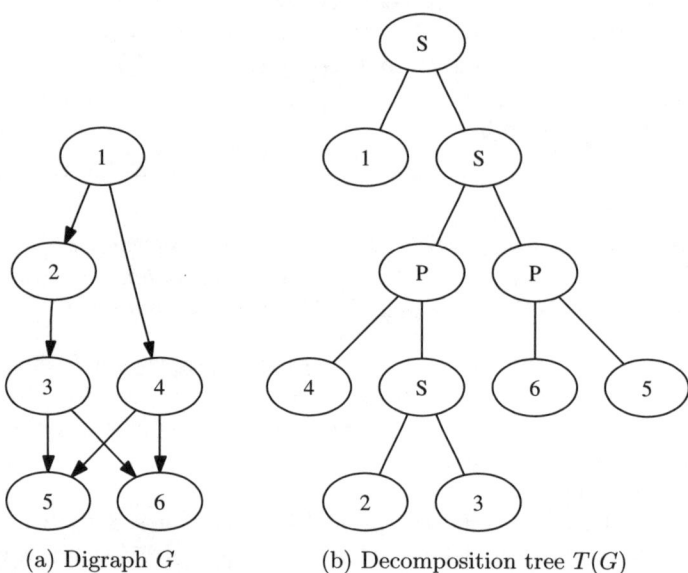

(a) Digraph G (b) Decomposition tree $T(G)$

Fig. 13.6: Precedence constraints in Example 13.32

We start with the vertex labelled S, whose immediate successors are vertices 2 and 3. We can replace these three vertices by chain $(2, 3)$. The new decomposition tree is given in Fig. 13.7a.

We proceed with the vertex labelled P, whose immediate successors are chain $(2, 3)$ and chain (4). Let us calculate R ratios for these vertices. We have

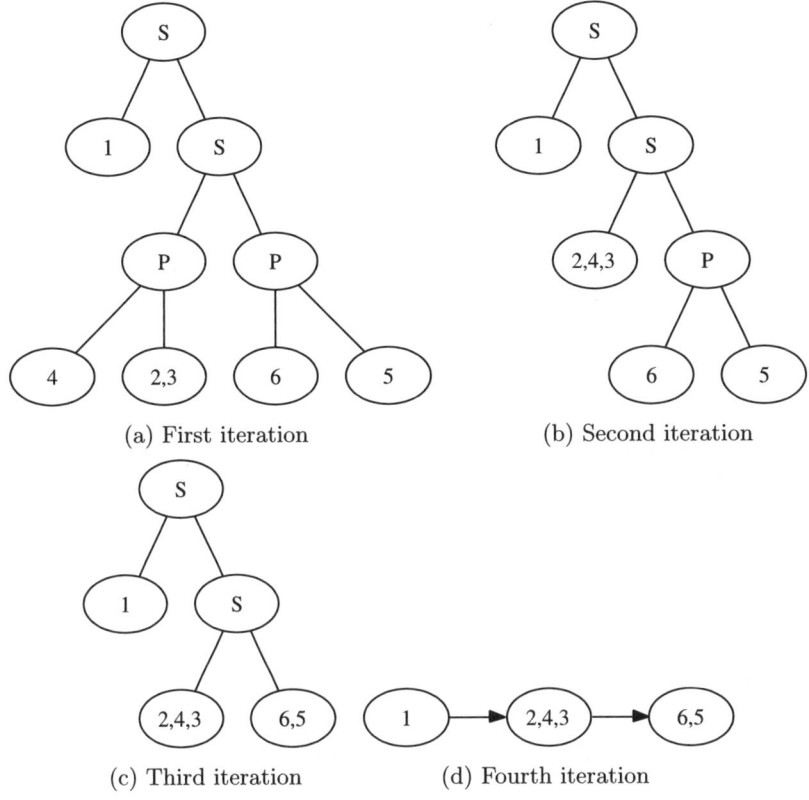

(a) First iteration

(b) Second iteration

(c) Third iteration

(d) Fourth iteration

Fig. 13.7: Subsequent iterations of algorithm A_{59} in Example 13.32

$R(2) = 1, R(3) = \frac{1}{2}$ and $R(4) = \frac{2}{3}$. Hence, (2), (3) and (4) are independent chains and their order is $((2), (4), (3))$. The new form of the decomposition tree is given in Fig. 13.7b.

Next, we choose vertices 5 and 6 that are immediate successors of the vertex labelled P. After calculations, we have $R(5) = \frac{4}{3} < R(5) = \frac{5}{2}$. Hence, the new form of the decomposition tree is as in Fig. 13.7c.

Because now all inner vertices are labelled S, the optimal job sequence is $\sigma^\star = (1, 2, 4, 3, 6, 5)$, see Fig. 13.7d.   ◆

Remark 13.33. A counterpart of algorithm A_{59} is given by Wang et al. [285, Algorithm 1].

13.5 General precedence constraints

In this section, we consider the problem $1|p_j = b_j t, prec|f_{\max}$. We assume that for job J_j, there is defined a non-decreasing cost function f_j that specifies a cost $f_j(C_j)$ that has to be paid at the completion time of the job,

$1 \leq j \leq n$. (Without loss of generality, we can assume that for a given function f_j, $1 \leq j \leq n$, the cost $f_j(C_j)$ can be computed in a constant time.) For simplicity, we assume also that $S_1 \equiv t_0 = 1$.

The main idea of the algorithm that solves the problem is the same as in Lawler's algorithm for the problem $1|prec|f_{\max}$ (Lawler [182]), i.e., from all jobs that do not have successors we choose these ones that will cause the smallest cost in the given position.

Let $NoSucc(G)$ denote the set of indices of jobs without immediate successors for a given digraph G of precedence constraints. The pseudo-code of the algorithm for the considered problem is as follows.

Algorithm A_{60} for the problem $1|p_j = b_j t, prec|f_{\max}$ ([93])

Input: sequences $(b_1, b_2, \ldots, b_n)$, $(f_1, f_2, \ldots, f_n)$,
      digraph G of precedence constraints
Output: an optimal schedule $\sigma^\star$

▷ **Step 1:**
   $\sigma^\star \leftarrow (\phi)$; $N_{\mathcal{J}} \leftarrow \{1, 2, \ldots, n\}$; $T \leftarrow \prod_{j=1}^n (b_j + 1)$;
▷ **Step 2:**
   while $(N_{\mathcal{J}} \neq \emptyset)$ **do**
      Construct the set $NoSucc(G)$;
      Find $k \in NoSucc(G)$ such that $f_k(T) = \min\{f_j(T) : j \in NoSucc(G)\}$;
      $\sigma^\star \leftarrow (\sigma^\star | k)$; $T \leftarrow \frac{T}{b_k + 1}$;
      $N_{\mathcal{J}} \leftarrow N_{\mathcal{J}} \setminus \{k\}$; $NoSucc(G) \leftarrow NoSucc(G) \setminus \{k\}$;
▷ **Step 3:**
   return $\sigma^\star$.

Remark 13.34. Algorithm A_{60} is a generalization of algorithm A_{29} presented in Sect. 6.4 (see Remark 6.203 for details).

Theorem 13.35. (Gawiejnowicz et al. [93]) *The problem $1|p_j = b_j t, prec|f_{\max}$ is solvable in $O(n^2)$ time by algorithm A_{60}.*

Proof. First, notice that without loss of generality, we can consider only schedules without idle times. Second, by Theorem 6.1 the value of $C_{\max}$ criterion for jobs with proportional processing times does not depend on the sequence of the jobs and is given by formula (6.2). Note also that the value of $C_{\max}$ can be calculated in $O(n)$ time.

Let $f_{\max}^\star(\mathcal{J})$ denote the value of the criterion $f_{\max}$ for an optimal schedule. Then the following two inequalities are satisfied:

$$f_{\max}^\star(\mathcal{J}) \geq \min_{j \in NoSucc(G)} \{f_j(C_n)\} \tag{13.15}$$

and

$$f_{\max}^\star(\mathcal{J}) \geq f_{\max}^\star(\mathcal{J} \setminus \{J_j\}) \tag{13.16}$$

for $j = 1, 2, \ldots, n$. Let job $J_k \in \mathcal{J}$ be such that

$$f_k(C_n) = \min\{f_j(C_n) : j \in NoSucc(G)\}$$

and let $f_k(\mathcal{J})$ denote the value of the criterion f_k, provided that job J_k is executed as the last one. Then,

$$f_k(\mathcal{J}) = \max\{f_k(C_n), f^\star_{\max}(\mathcal{J} \setminus \{J_k\})\}. \tag{13.17}$$

From (13.15), (13.16) and (13.17), it follows that $f^\star_{\max}(\mathcal{J}) \geq f_k(\mathcal{J})$ and there exists an optimal schedule in which job J_k is executed as the last one.

Repeating this procedure for the remaining jobs, we obtain an optimal schedule. Because in each run of the procedure we have to choose a job from $O(n)$ jobs, and there are n jobs, algorithm A_{60} runs in $O(n^2)$ time. ∎

Remark 13.36. Note that algorithm A_{60} can be easily modified for proportional-linear job processing times (6.5). Therefore, by Theorem 13.35, the problem $1|p_j = b_j(A+Bt), prec|f_{\max}$ is solvable in $O(n^2)$ time by the modified algorithm A_{60}.

We illustrate the application of algorithm A_{60} by an example.

Example 13.37. Let $n = 4$, $p_1 = t$, $p_2 = 3t$, $p_3 = 2t$, $p_4 = t$, $t_0 = 1$. Cost functions are in the form of $f_1 = 2C_1^2, f_2 = C_2^2 + 1, f_3 = C_3, f_4 = C_4 + 5$. Precedence constraints are given in Fig. 13.8.

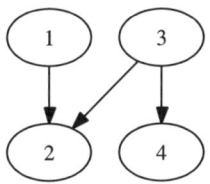

Fig. 13.8: Precedence constraints in Example 13.37

In Step 1, we have $C = 48$; in Step 2, we have $NoSucc(G) = \{2, 4\}$. Because $f_2(48) = 48^2 + 1 > f_4(48) = 48 + 5$, job J_4 is scheduled as the last one. Then $C = \frac{48}{2} = 24$ and $NoSucc(G) = \{2\}$. Therefore, job J_2 is scheduled as the second job from the end. Next, $C = \frac{24}{4} = 6$ and $NoSucc(G) = \{1, 2\}$. Because $f_1(6) = 2 \cdot 6^2 > f_3(6) = 6$, job J_3 is scheduled as the third one from the end. After that $C = \frac{6}{2} = 3$ and $NoSucc(G) = \{1\}$. Therefore, job J_1 is scheduled as the first one. The optimal schedule is $\sigma^\star = (1, 3, 2, 4)$. ◆

The linear case, $p_j = a_j + b_j t$, seems to be computationally intractable. Therefore, we state the following.

Conjecture 13.38. The decision version of the problem $1|p_j = a_j + b_j t, prec|C_{\max}$ is $\mathcal{NP}$-complete in the strong sense.

With this conjecture, we end the presentation of the results concerning time-dependent scheduling with job precedence constraints. In Chap. 14, we will consider time-dependent scheduling with two criteria.

13.6 Concluding remarks and open problems

In this chapter, we considered a single machine time-dependent scheduling problem with job precedence constraints. We proved that the problem with linear job processing times and precedence constraints in the form of a set of chains, a tree, a forest or a series-parallel graph can be solved in $O(n \log n)$ time. We also proved that the problem with proportional job processing times and arbitrary precedence constraints is solvable in $O(n^2)$ time. The algorithms presented in the chapter are summarized in Table 13.1.

Table 13.1: Polynomial-Time Algorithms for Time-Dependent Scheduling Problems with Dependent Jobs

Algorithm	Complexity	Problem	Reference	This book
A_{55}	$O(n \log n)$	$1\|p_j = a_j + b_j t,$ $chain\|C_{\max}$	[101]	Sect. 13.2, p. 305
A_{56}	$O(n \log n)$	$1\|p_j = a_j + b_j t,$ $chains\|C_{\max}$	[101]	Sect. 13.2, p. 306
A_{57}	$O(n \log n)$	$1\|p_j = a_j + b_j t,$ $in\text{-}tree\|C_{\max}$	[101]	Sect. 13.3, p. 308
A_{57}	$O(n \log n)$	$1\|p_j = a_j + b_j t,$ $in\text{-}forest\|C_{\max}$	[101]	Sect. 13.3, p. 308
A_{58}	$O(n \log n)$	$1\|p_j = a_j + b_j t,$ $out\text{-}tree\|C_{\max}$	[101]	Sect. 13.3, p. 310
A_{58}	$O(n \log n)$	$1\|p_j = a_j + b_j t,$ $out\text{-}forest\|C_{\max}$	[101]	Sect. 13.3, p. 310
A_{59}	$O(n \log n)$	$1\|p_j = a_j + b_j t,$ $ser\text{-}par\|C_{\max}$	[101]	Sect. 13.4, p. 313
A_{60}	$O(n^2)$	$1\|p_j = b_j t, prec\|C_{\max}$	[101]	Sect. 13.5, p. 316

Scheduling linearly deteriorating jobs was previously considered by other authors. The results mentioned in Table 13.1 are counterparts of the results presented by Tanaev et al. [264, Chap. 3], where two operations on an arbitrary acyclic digraph (the operation of identifying vertices and the operation of including an arc) have been introduced. Applying the notion of the priority-generating function (cf. Definition 1.19), Tanaev et al. [264] proved a number of results concerning different scheduling problems with precedence constraints, including the linear problems considered in the chapter.

Some of the results presented in the chapter (cf. Remarks 13.13 and 13.33) are also given by Wang et al. [285].

Further research in scheduling deteriorating jobs with precedence constraints may focus on the following problems. First, we can seek polynomial-time approximation algorithms for the linear case. It seems that here a certain role may play decomposition methods, like those discussed by Buer and Möhring [36] and Muller and Spinrad [222].

Second, we can seek other forms of job deterioration with simple precedence constraints, like chains or trees, which are solvable in polynomial time. However, it is disputable if there exist other than linear forms of job deterioration for which these precedence constrains are polynomially solvable.

Finally, one can consider other optimality criteria, e.g., $\sum C_j$ or $L_{\max}$.

14

Time-dependent scheduling with two criteria

The evaluation of schedule quality by a single criterion prevails in time-dependent scheduling. However, advanced scheduling problems require a multi-criteria approach. Therefore, in the last chapter of the book, we consider bicriteria time-dependent scheduling problems, in which a schedule is evaluated by two criteria that are minimized either in the Pareto or in the scalar sense.

Chapter 14 is composed of five sections. In Sect. 14.1, we formulate the considered problems and give some preliminary results. In Sects. 14.2 and 14.3, we present the results concerning bicriterion Pareto and bicriterion scalar optimality, respectively. In Sect. 14.4, we summarize the results of computational experiments related to the bicriterion Pareto optimality. Concluding remarks open problems are given in Sect. 14.5.

14.1 Preliminaries

In this section, we formulate two problems that we will consider. First, we state the assumptions, which are common for both problems.

14.1.1 Problems formulation

We are given a single machine and a set of $n + 1$ linearly deteriorating jobs to be processed on the machine. The processing times of jobs are in the form of $p_j = 1 + b_j t$, where $b_j > 0$ for $j = 0, 1, \ldots, n$. All the jobs are available for processing at time $t_0 = 0$.

Input data for the problems are described by the sequence $(b_0, b_1, \ldots, b_n)$ of job deterioration rates. For simplicity of further presentation, however, instead of the sequence $(b_0, b_1, \ldots, b_n)$, we use the sequence $\hat{\beta} = (\beta_0, \beta_1, \ldots, \beta_n)$, where $\beta_j = b_j + 1$ for $j = 0, 1, \ldots, n$. (The elements β_j will be called *deterioration coefficients* in order to distinguish them from deterioration rates b_j.)

The first problem is to find such a schedule $\beta^\star$ that the pair

$$\left(\sum_j C_j(\beta^\star), \max_j\{C_j(\beta^\star)\} \right)$$

of values of the total completion time and the maximum completion time criteria for this schedule is Pareto optimal, i.e.,

$$\beta^\star \equiv \hat{\beta}_{\pi^\star} = \arg\min_{\hat{\beta}_\pi} \left\{ \left(\sum_j C_j(\hat{\beta}_\pi), \max_j\{C_j(\hat{\beta}_\pi)\} \right) : \pi \in \mathfrak{S}_n \right\},$$

where the minimum, with respect to the order relation $\prec$, is taken in the sense of Pareto optimum.

Remark 14.1. The partial order relation $\prec$ and the Pareto optimality are introduced in Definition 1.4 and Definition 1.25, respectively.

Let $\|\cdot\|_{(\lambda)}$ denote a convex combination of the $\sum C_j$ and $C_{\max}$ criteria, i.e.,

$$\|C(\hat{\beta}_\pi)\|_{(\lambda)} := \lambda \sum_{j=0}^{n} C_j(\hat{\beta}_\pi) + (1-\lambda) \max_{0 \le j \le n}\{C_j(\hat{\beta}_\pi)\},$$

where $C(\hat{\beta}_\pi) = [C_0(\hat{\beta}_\pi), C_1(\hat{\beta}_\pi), \ldots, C_n(\hat{\beta}_\pi)]$ is the vector of job completion times for a given sequence $\hat{\beta}_\pi$ and $\lambda \in \langle 0, 1 \rangle$ is an arbitrary but fixed number.

The second problem is to find a schedule $\beta^\star$ for which the value of the criterion $\|\cdot\|_{(\lambda)}$ is minimal, i.e.,

$$\beta^\star \equiv \hat{\beta}_{\pi^\star} = \arg\min_{\hat{\beta}_\pi} \left\{ \|C(\hat{\beta}_\pi)\|_{(\lambda)} : \pi \in \mathfrak{S}_n \right\},$$

where the minimum is taken with respect to the ordinary relation $\le$.

We will refer to the first and the second problem as the *TDPS* (Time-Dependent Pareto-optimal Scheduling) and the *TDBS* (Time-Dependent Bi-criterion Scheduling) problem, respectively. Optimal schedules for these problems will be called *TDPS-optimal* and *TDBS-optimal* schedules, respectively.

14.1.2 Preliminary results

In this section, we prove some preliminary results that are used in the next two sections. The results refer to both problems under consideration. Introductory results concerning only one of the problems, either TDPS or TDBS, are presented in Sects. 14.2 and 14.3, respectively.

The following result is a generalization of Property 6.128.

Property 14.2. (Gawiejnowicz et al. [103]) Let $\beta_{\max} := \max\{\beta_j : 0 \le j \le n\}$ for a given sequence $\hat{\beta}$. Then in any TDPS-optimal (TDBS-optimal) schedule for $\hat{\beta}$, the job corresponding to deterioration coefficient $\beta_{\max}$ is scheduled as the first one.

Proof. Consider the TDPS problem. Assume that there exists a Pareto optimal schedule $\beta^\star$ for the TDPS problem in which the job corresponding to the greatest deterioration coefficient, $\beta_{\max}$, is not scheduled as the first one, $\beta_{[0]} \neq \beta_{\max}$. Let the first job in $\beta^\star$ have the coefficient β_m, $\beta_{[0]} = \beta_m$, where $\beta_m < \beta_{\max} = \beta_{[k]}$. Consider schedule β' obtained by switching in $\beta^\star$ the first job and the k-th job. Since the deterioration coefficient of the first scheduled job does not influence the values of the $\sum C_j$ and $C_{\max}$ criteria and, furthermore, by Remark 6.26 both these criteria are monotonically non-decreasing with respect to $\beta_{[j]}$ for $j = 1, 2, \ldots, n$, the above switching will decrease the values of $\sum C_j$ and $C_{\max}$. Thus, in view of Definition 1.25, schedule $\beta^\star$ cannot be optimal. A contradiction.

Consider now the TDBS problem. Since the value of the criterion $\| \cdot \|_{(\lambda)}$ decreases with the decreasing values of $\sum C_j$ and $C_{\max}$ criteria, then by applying similar reasoning as above, we complete the proof. ∎

Remark 14.3. From now on, we assume that $\beta_{[0]}$ has been established according to Property 14.2 and we denote sequence $\hat{\beta}$ without the maximal element $\beta_{[0]}$ by $\beta = (\beta_1, \beta_2, \ldots, \beta_n)$.

Remark 14.4. We assume that $n > 2$, i.e., sequence $\beta = (\beta_1, \beta_2, \ldots, \beta_n)$ contains at least three elements such that $\beta_j > 1$ for $j = 1, \ldots, n$.

Let $\beta(\beta_q \leftrightarrow \beta_r)$ denote sequence β with elements β_q and β_r mutually interchanged. The next preliminary result is the following lemma.

Lemma 14.5. (Gawiejnowicz et al. [103]) *Let $\beta' = \beta(\beta_q \leftrightarrow \beta_r)$ and $1 \le q < r \le n$. Then for $0 \le j \le n$, there holds the following equality:*

$$C_j(\beta') - C_j(\beta) = \begin{cases} \dfrac{\beta_q - \beta_r}{\beta_r} \displaystyle\sum_{i=q}^{r-1} \prod_{k=i+1}^{j} \beta_k \,, & 1 \le q < r \le j \le n, \\[2ex] \dfrac{\beta_r - \beta_q}{\beta_q} \displaystyle\sum_{i=0}^{q-1} \prod_{k=i+1}^{j} \beta_k \,, & 1 \le q \le j < r \le n, \\[2ex] 0 \,, & 0 \le j < q < r \le n. \end{cases}$$

Proof. The case $j = 0$ is clear, since $C_0(\beta) = C_0(\beta') = 1$. In the case when $0 < j < q < r \le n$, we have $C_j(\beta) = C_j(\beta') = \sum_{i=0}^{j} \prod_{k=i+1}^{j} \beta_k$, since $j < q$.

Let $r \le j \le n$ and $1 \le q < r \le n$. From formula (6.11), for $a_{\sigma_i} = 1$ for $1 \le i \le n$, it follows that

$$C_j(\beta) = \sum_{i=0}^{q-1} \beta_{i+1} \ldots \beta_q \ldots \beta_r \ldots \beta_j + \sum_{i=q}^{r-1} \beta_{i+1} \ldots \beta_r \ldots \beta_j + \sum_{i=r}^{j} \beta_{i+1} \ldots \beta_j.$$

Next, in the corresponding formula for $C_j(\beta')$, we have β_q and β_r mutually interchanged in the first sum. Clearly, in this case the first sum remains unchanged. In the second sum in $C_j(\beta')$, factor β_r must be replaced by factor β_q. Finally, in the third sum, we have no changes related to the transition from β to β'. Therefore, for $1 \leq q < r \leq j \leq n$, there holds the equality

$$C_j(\beta') - C_j(\beta) = \frac{\beta_q - \beta_r}{\beta_r} \sum_{i=q}^{r-1} \prod_{k=i+1}^{j} \beta_k.$$

The case when $1 \leq q \leq j < r \leq n$ can be proved in a similar way. ∎

Lemma 14.6. (Gawiejnowicz et al. [96, 103]) *Let* $\beta' = \beta(\beta_q \leftrightarrow \beta_{q+1})$, *where* $q = 1, 2, \ldots, n-1$. *Then for* $j = 0, 1, \ldots, n$, *there holds the following equality:*

$$C_j(\beta') - C_j(\beta) = \begin{cases} (\beta_q - \beta_{q+1}) \prod_{k=q+2}^{j} \beta_k & , 1 \leq q < j \leq n, \\ (\beta_{q+1} - \beta_q) \sum_{i=0}^{q-1} \prod_{k=i+1}^{q-1} \beta_k & , j = q, \\ 0 & , 0 \leq j < q. \end{cases}$$

Proof. The result follows from Lemma 14.5 by letting $r = q + 1$. ∎

Lemma 14.7. (Gawiejnowicz et al. [96, 103]) *Let* $\beta' = \beta(\beta_q \leftrightarrow \beta_{q+1})$, *where* $q = 1, 2, \ldots, n-1$. *Then there holds the following equality:*

$$\sum_{j=0}^{n} C_j(\beta') - \sum_{j=0}^{n} C_j(\beta) = (\beta_{q+1} - \beta_q) \left(\sum_{j=0}^{q-1} \prod_{k=j+1}^{q-1} \beta_k - \sum_{i=q+1}^{n} \prod_{k=q+2}^{i} \beta_k \right).$$

Proof. By Lemma 14.6, summing the differences $C_j(\beta') - C_j(\beta)$ for $0 \leq j \leq n$, we obtain the result. ∎

Lemma 14.8. (Gawiejnowicz et al. [96, 103]) *Let* $\beta' = \beta(\beta_q \leftrightarrow \beta_{q+1})$, *where* $q = 1, 2, \ldots, n-1$. *Then there holds the following equality:*

$$\max_{0 \leq j \leq n} \{C_j(\beta')\} - \max_{0 \leq j \leq n} \{C_j(\beta)\} = (\beta_q - \beta_{q+1}) \prod_{k=q+2}^{n} \beta_k.$$

Proof. Since $\max_{0 \leq j \leq n} \{C_j(\beta)\} = C_n(\beta)$, by letting $j = n$ in Lemma 14.6, the result follows. ∎

14.2 Pareto optimality

In this section, we consider the TDPS problem, i.e., the problem of finding a schedule that is Pareto optimal with respect to the $\sum C_j$ and $C_{\max}$ criteria.

Let X denote the set of all solutions of a bicriterion scheduling problem. In our case, $X = \{\hat{\beta}_\pi : \pi \in \mathfrak{S}_n\}$ is discrete and consists of all permutations of the original sequence $\hat{\beta}$. Recall also (cf. Definition 1.25) that for a given bicriterion optimization problem, X_{Par} ($X_{\text{w−Par}}$) denotes the set of all Pareto (weak Pareto) optimal solutions.

Notice that in view of Lemma 14.7 and Lemma 14.8, if $\beta' = \beta(\beta_q \leftrightarrow \beta_{q+1})$ is a pairwise transposition of sequence β, where $q = 1, 2, \ldots, n-1$, we have:

$$\|C(\beta')\|_1 - \|C(\beta)\|_1 = (\beta_{q+1} - \beta_q)\left(\sum_{j=0}^{q-1}\prod_{k=j+1}^{q-1}\beta_k - \sum_{i=q+1}^{n}\prod_{k=q+2}^{i}\beta_k\right) \quad (14.1)$$

and

$$\|C(\beta')\|_\infty - \|C(\beta)\|_\infty = -(\beta_{q+1} - \beta_q)\prod_{k=q+2}^{n}\beta_k. \quad (14.2)$$

First, we prove a sufficient condition for $\beta^\star \in X$ to be a (weakly) Pareto optimal solution to the TDPS problem. (A necessary condition for $\beta \in X$ to be a (weakly) Pareto optimal solution to the TDPS problem will be given later.)

Theorem 14.9. (Gawiejnowicz et al. [103])
(a) *A sufficient condition for sequence $\beta^\star \in X$ to be weakly Pareto optimal is that $\beta^\star$ is optimal with respect to the scalar criterion $\|\cdot\|_{(\lambda)}$, where $0 \leq \lambda \leq 1$.*
(b) *A sufficient condition for sequence $\beta^\star \in X$ to be Pareto optimal is that $\beta^\star$ is optimal with respect to the scalar criterion $\|\cdot\|_{(\lambda)}$, where $0 \leq \lambda < 1$. In particular, the sequence obtained by the non-increasing ordering of $\beta^\star$ is Pareto optimal for the TDPS problem.*

Proof. (a) The statement immediately follows from inclusion $X_{(\lambda)} \subset X_{\text{w−Par}}$, where $0 \leq \lambda \leq 1$.
(b) The statement follows from inclusion $X_{(\lambda)} \subset X_{\text{Par}}$, whenever $0 < \lambda < 1$. To end the proof it is sufficient to consider the case $\lambda = 0$, i.e., the case of criterion $\|\cdot\|_\infty$. Let $\beta^\star \in X$ be non-increasing. The sequence is Pareto optimal when relation

$$(\|C(\beta')\|_1, \|C(\beta')\|_\infty) \prec (\|C(\beta^\star)\|_1, \|C(\beta^\star)\|_\infty)$$

does not hold for any $\beta' \in X, \beta' \neq \beta^\star$. By Lemma 1.5, this means that there holds either the disjunction

$$\|C(\beta')\|_1 - \|C(\beta^\star)\|_1 > 0 \text{ or } \|C(\beta')\|_\infty - \|C(\beta^\star)\|_\infty > 0 \quad (14.3)$$

or the conjuction

$$\|C(\beta')\|_1 = \|C(\beta^\star)\|_1 \text{ and } \|C(\beta')\|_\infty = \|C(\beta^\star)\|_\infty. \quad (14.4)$$

If $\beta^\star$ contains only distinct elements, then $\|C(\beta')\|_\infty - \|C(\beta^\star)\|_\infty > 0$ and there holds the disjunction (14.3). In the opposite case, either $\|C(\beta')\|_\infty -$

$\|C(\beta^\star)\|_\infty > 0$, and then there holds the disjunction (14.3), or $\|C(\beta')\|_\infty - \|C(\beta^\star)\|_\infty = 0$, and then there holds the conjuction (14.4), since in this case we also have $\|C(\beta')\|_1 - \|C(\beta^\star)\|_1 = 0$ by the uniqueness of $\beta^\star$ up to the order of equal elements.

Concluding, a non-increasing $\beta^\star \in X$ must be Pareto optimal for the TDPS problem. ∎

Example 14.10. (Gawiejnowicz et al. [103]) Consider sequence $\hat\beta = (6, 3, 4, 5, 2)$. Then $\sum C_j(\hat\beta) = 281$ and $C_{\max}(\hat\beta) = 173$. By Theorem 14.9, we know that each non-increasing sequence is Pareto optimal for the TDPS problem. Thus $\beta' = (6, 5, 4, 3, 2)$ is Pareto optimal for this problem, with $\sum C_j(\beta') = 261$ and $C_{\max}(\beta') = 153$. ◆

Now we prove the necessary condition for $\beta^\star \in X$ to be a (weakly) Pareto optimal solution to the TDPS problem. We start with the following result.

Lemma 14.11. (Gawiejnowicz et al. [103]) *Let* $\Delta_q(\beta) := \sum\limits_{j=0}^{q-1} \prod\limits_{k=j+1}^{q-1} \beta_k - \sum\limits_{i=q+1}^{n} \prod\limits_{k=q+2}^{i} \beta_k$ *for a given sequence* β*. Then for any permutation of* β *there exists a unique number* q_0*,* $1 \le q_0 \le n-1$*, such that* q_0 *is the greatest number for which* $\Delta_q(\beta)$ $q = q_0$ *is negative, i.e.,* $\Delta_q(\beta) < 0$ *for* $1 \le q \le q_0$ *and* $\Delta_q(\beta) \ge 0$ *for* $q_0 < q \le n-1$*.*

Proof. First, note that

$$\Delta_1(\beta) = -\sum_{i=3}^{n} \prod_{k=3}^{i} \beta_k < 0$$

and

$$\Delta_{n-1}(\beta) = \sum_{j=0}^{n-3} \prod_{k=j+1}^{n-2} \beta_k > 0.$$

Moreover, sequence $\Delta_q(\beta)$ is strictly increasing for $q = 1, 2, \ldots, n-1$, since

$$\Delta_{q+1}(\beta) - \Delta_q(\beta) = \sum_{j=0}^{q} \prod_{k=j+1}^{q} \beta_k - \sum_{i=q+2}^{n} \prod_{k=q+3}^{i} \beta_k - \sum_{j=0}^{q-1} \prod_{k=j+1}^{q-1} \beta_k + \sum_{i=q+1}^{n} \prod_{k=q+2}^{i} \beta_k$$

$$= (\beta_q - 1) \sum_{j=0}^{q-1} \prod_{k=j+1}^{q-1} \beta_k + (\beta_{q+2} - 1) \sum_{i=q+2}^{n} \prod_{k=q+3}^{i} \beta_k + 2 > 0.$$

Thus, there must exist a maximal integer q_0 such that $\Delta_q(\beta) < 0$ for $1 \le q \le q_0$ and $\Delta_q(\beta) \ge 0$ for $q_0 < q \le n-1$. ∎

Now, assume that $\beta^\star$ is a Pareto optimal solution of the TDPS problem, $\beta^\star \in X_{\text{Par}}$. Then, by Lemma 14.11, there exists a $q_0^\star$ such that

$$\Delta_q(\beta^\star) < 0 \quad \text{for} \quad 1 \le q \le q_0^\star \tag{14.5}$$

and

$$\Delta_q(\beta^\star) \geq 0 \quad \text{for} \quad q_0^\star < q \leq n - 1. \tag{14.6}$$

Knowing that there exists a $q_0^\star$ which is the point of change of sign of $\Delta_q(\beta^\star)$, we can prove the following result.

Theorem 14.12. (Gawiejnowicz et al. [103]) *Let* $\beta^\star = (\beta_1^\star, \beta_2^\star, \ldots, \beta_n^\star) \in X_{\mathrm{Par}}$ *and* $q_0^\star$ *be specified by conditions (14.5) and (14.6). Then for* $q = 1, 2, \ldots, n-1$ *there holds the inequality* $\beta_q^\star \geq \beta_{q+1}^\star$ *or, if* $q_0^\star < q \leq n - 1$, *the inequality* $\beta_q^\star \leq \beta_{q+1}^\star$.

Proof. Let $\beta^\star = (\beta_1^\star, \beta_2^\star, \ldots, \beta_n^\star)$ be Pareto optimal solution of the TDPS problem. Then there does not exist $\beta' \in X$ such that

$$(\|C(\beta')\|_1, \|C(\beta')\|_\infty) \prec (\|C(\beta^\star)\|_1, \|C(\beta^\star)\|_\infty)$$

or, equivalently, for each $\beta' \in X$ there does not hold the relation

$$(\|C(\beta')\|_1 - \|C(\beta^\star)\|_1, \|C(\beta')\|_\infty - \|C(\beta^\star)\|_\infty) \prec 0.$$

In particular, this relation does not hold for transpositions $\beta' = \beta^\star(\beta_q^\star \leftrightarrow \beta_{q+1}^\star)$ of the optimal sequence for $q = 1, 2, \ldots, n - 1$. Now, applying Lemma 1.5, (14.1) and (14.2), we see that for $q = 1, 2, \ldots, n - 1$ there holds either the alternative

$$\|C(\beta')\|_1 - \|C(\beta^\star)\|_1 = (\beta_{q+1}^\star - \beta_q^\star)\Delta_q(\beta^\star) > 0 \tag{14.7}$$

or

$$\|C(\beta')\|_\infty - \|C(\beta^\star)\|_\infty = -(\beta_{q+1}^\star - \beta_q^\star) \prod_{k=q+2}^{n} \beta_k^\star > 0 \tag{14.8}$$

or the conjuction

$$\|C(\beta')\|_1 - \|C(\beta^\star)\|_1 = 0 \text{ and } \|C(\beta')\|_\infty - \|C(\beta^\star)\|_\infty = 0. \tag{14.9}$$

If $\beta^\star$ contains distinct elements only, then the conjuction (14.9) cannot be satisfied. Thus, there must hold the disjunction (14.7) or (14.8). Hence, for $q = 1, 2, \ldots, n - 1$, there holds the inequality $\beta_q^\star > \beta_{q+1}^\star$ or, if $q_0^\star < q \leq n - 1$, the inequality $\beta_q^\star < \beta_{q+1}^\star$.

If not all elements of $\beta^\star$ are distinct, then apart from the disjunction (14.7) or (14.8) also the conjuction (14.9) can be satisfied. Hence, for $q = 1, 2, \ldots, n - 1$ there holds the inequality $\beta_q^\star \geq \beta_{q+1}^\star$ or, if $q_0^\star < q \leq n - 1$, the inequality $\beta_q^\star \leq \beta_{q+1}^\star$. ∎

We now introduce a definition (cf. Gawiejnowicz et al. [103]) that allows us to formulate the previous result in a more concise way.

Definition 14.13. (A weakly V-shaped sequence)
A sequence $\beta = (\beta_1, \beta_2, \ldots, \beta_n)$ *is said* to have a weak V-shape (to be weakly V-shaped) with respect to $\Delta_q(\beta)$, *if β is non-increasing for indices q for which inequality $\Delta_q(\beta) < 0$ holds.*

Note that from Definition 14.13 and from properties of function $\Delta_q(\beta)$ it follows that, in general, weakly V-shaped sequences are non-increasing for $1 \leq q \leq q_0$ and can vary in an arbitrary way for $q_0 < q \leq n - 1$, with appropriate $1 \leq q_0 \leq n - 1$.

Applying Definition 14.13, we can now reformulate Theorem 14.12 in the following way.

Theorem 14.12′. (Gawiejnowicz et al. [103]) *A necessary condition for sequence $\beta^\star \in X$ to be a Pareto optimal solution to the TDPS problem is that $\beta^\star$ must be weakly V-shaped with respect to $\Delta_q(\beta^\star)$.*

We illustrate applications of Theorem 14.12′ by two examples (cf. [103]).

Example 14.14. Let sequence $\hat{\beta} = (\beta_0, \beta_1, \ldots, \beta_n)$ be such that $\beta_0 = \max_i\{\beta_i\}$ and $\beta_1 = \min_i\{\beta_i\}$. Since $\Delta_1(\hat{\beta}) < 0$ (by Lemma 14.11), we have $q_0 \geq 1$ and by Theorem 14.12′ no sequence which is in the form of $(\beta_0, \beta_1, \beta_{\pi_2}, \beta_{\pi_3}, \ldots, \beta_{\pi_n})$, where $\beta_{\pi_i} \in \{\beta_2, \beta_3, \ldots, \beta_n\}$ for $2 \leq i \leq n$ and $\pi_i \neq \pi_j$ for $i \neq j$, can be Pareto optimal. ◆

Example 14.15. Let $\hat{\beta} = (7, 3, 2, 4, 5, 6)$. To check if $\hat{\beta}$ can be a solution to the TDPS problem, we must determine the value of q_0. After calculations we have: $\Delta_1(\hat{\beta}) = -144$, $\Delta_2(\hat{\beta}) = -32$, $\Delta_3(\hat{\beta}) = 2$. Hence, $q_0 = 2$ and sequence $\hat{\beta}$, according to Theorem 14.12′, cannot be a Pareto optimal solution to the TDPS problem. Moreover, all schedules which are in the form of $(7, 3, 2, \beta_{\pi_3}, \beta_{\pi_4}, \beta_{\pi_5})$, where $\beta_{\pi_i} \in \{4, 5, 6\}$ for $3 \leq i \leq 5$ and $\pi_3 \neq \pi_4 \neq \pi_5$, cannot be Pareto optimal, either. There are 6 such schedules: $(7, 3, 2, 4, 5, 6)$, $(7, 3, 2, 4, 6, 5)$, $(7, 3, 2, 5, 4, 6)$, $(7, 3, 2, 5, 6, 4)$, $(7, 3, 2, 6, 4, 5)$ and $(7, 3, 2, 6, 5, 4)$. ◆

14.3 Scalar optimality

In this section, we consider the TDBS problem of finding a schedule that is optimal for the criterion $\| \cdot \|_{(\lambda)}$.

Recall that the criterion is a convex combination of the total completion time $\sum C_j$ and the maximum completion time $C_{\max}$, i.e.,

$$\|C(\beta)\|_{(\lambda)} := \lambda \sum_{j=0}^{n} C_j(\beta) + (1 - \lambda) \max_{0 \leq j \leq n} \{C_j(\beta)\}, \tag{14.10}$$

where $\lambda \in \langle 0, 1 \rangle$ is arbitrary but fixed. Our aim is to find such a sequence $\beta^\star$ for which the value of $\|C(\beta^\star)\|_{(\lambda)}$ is minimal.

Remark 14.16. Notice that if we are interested in minimizing the combination of the $\sum C_j$ and $C_{\max}$ criteria with arbitrary weights, the criterion $\|\cdot\|_{(\lambda)}$ is general enough, since for any real numbers $\alpha > 0, \beta > 0$, we have $\alpha \sum C_j + \beta C_{\max} = (\alpha + \beta) \left(\frac{\alpha}{\alpha+\beta} \sum C_j + \frac{\beta}{\alpha+\beta} C_{\max} \right) = (\alpha + \beta) \|\cdot\|_{(\lambda)}$ with $\lambda = \frac{\alpha}{\alpha+\beta}$.

Remark 14.17. Note also that since the criterion $\|\cdot\|_{(\lambda)}$ is a convex combination of the $\sum C_j$ and $C_{\max}$ criteria, which are particular cases of the l_p norm (cf. Definition 1.18), $\|\cdot\|_{(\lambda)}$ is also a norm.

In other words, the sequence $\beta^\star = (\beta_1^\star, \beta_2^\star, \ldots, \beta_n^\star)$ is optimal for the problem TDBS, if $\|C(\beta^\star)\|_{(\lambda)} = \min\{\|C(\beta_\pi\|_{(\lambda)} : \pi \in \mathfrak{S}_n\}$, i.e., for all $\pi \in \mathfrak{S}_n$ there holds the inequality

$$0 \leq \|C(\beta_\pi)\|_{(\lambda)} - \|C(\beta^\star)\|_{(\lambda)}. \tag{14.11}$$

Remark 14.18. Note that we can modify (14.10); since $C(\beta)$ is non-decreasing, we have $\max_{1 \leq j \leq n} \{C_j(\beta)\} = C_n(\beta)$. Thus, to define the norm $\|\cdot\|_{(\lambda)}$ we can also use the formula $\|C(\beta)\|_{(\lambda)} = \lambda \sum_{j=0}^{n-1} C_j(\beta) + C_n(\beta)$.

In view of the form of $\|\cdot\|_{(\lambda)}$, there hold some relations between this criterion and the $\sum C_j$ and $C_{\max}$ criteria.

Lemma 14.19. (Gawiejnowicz et al. [104]) *There hold the inequalities:*
(a) $\|C(\beta)\|_{(\lambda)} - \|C(\beta)\|_\infty \leq \lambda(n-1)\|C(\beta)\|_\infty$,
(b) $0 \leq \frac{\|C(\beta)\|_{(\lambda)} - \|C(\beta)\|_\infty}{\|C(\beta)\|_\infty} \leq \lambda(n-1)$,
(c) $0 \leq \frac{\|C(\beta)\|_1 - \|C(\beta)\|_{(\lambda)}}{\|C(\beta)\|_1} \leq (1-\lambda)\frac{n-1}{n}$.

Proof. (a) By (14.10) and since $\|C\|_\infty \leq \|C\|_1 \leq n\|C\|_\infty$, the inequality follows.
   (b) The inequalities follow from the definition of the criterion $\|\cdot\|_{(\lambda)}$ and from (a).
   (c) Similar to the proof of (a).   □

Theorem 14.20. (Gawiejnowicz et al. [104]) *If $\lambda \in \langle 0, 1 \rangle$ and n is any natural number, then there hold inequalities:*

$$\|C(\beta)\|_{(\lambda)} \leq \|C(\beta)\|_1 \leq \frac{n}{1 + \lambda(n-1)} \|C(\beta)\|_{(\lambda)}.$$

Proof. The result follows from Lemma 14.19.   □

From (14.10), Lemma 14.7 and Lemma 14.8 we get a formula describing the behaviour of $\|C(\beta)\|_{(\lambda)}$ under transpositions $\beta' = \beta(\beta_q \leftrightarrow \beta_{q+1})$.

Theorem 14.21. (Gawiejnowicz et al. [103]) *Let $\beta' = \beta(\beta_q \leftrightarrow \beta_{q+1})$. Then for $q = 1, 2, \ldots, n-1$ there holds the following equality:*

$$\|C(\beta')\|_{(\lambda)} - \|C(\beta)\|_{(\lambda)} =$$

$$(\beta_{q+1} - \beta_q) \left(\lambda \left(\sum_{j=0}^{q-1} \prod_{k=j+1}^{q-1} \beta_k - \sum_{j=q+1}^{n-1} \prod_{k=q+2}^{j} \beta_k \right) - \prod_{k=q+2}^{n} \beta_k \right).$$

Proof. From (14.10), it follows that

$$\|C(\beta')\|_{(\lambda)} - \|C(\beta)\|_{(\lambda)} = \lambda \left(\sum_{j=0}^{n} C_j(\beta') - \sum_{j=0}^{n} C_j(\beta) \right) +$$

$$(1 - \lambda) \left(\max_{0 \le j \le n} \{C_j(\beta')\} - \max_{0 \le j \le n} \{C_j(\beta)\} \right).$$

Consequently, in view of Lemma 14.7 and Lemma 14.8, we obtain

$$\|C(\beta')\|_{(\lambda)} - \|C(\beta)\|_{(\lambda)} = \lambda(\beta_{q+1} - \beta_q) \left(\sum_{j=0}^{q-1} \prod_{k=j+1}^{q-1} \beta_k - \sum_{j=q+1}^{n-1} \prod_{k=q+2}^{j} \beta_k \right) +$$

$$(1 - \lambda)(\beta_q - \beta_{q+1}) \prod_{k=q+2}^{n} \beta_k =$$

$$= (\beta_{q+1} - \beta_q) \left(\lambda \left(\sum_{j=0}^{q-1} \prod_{k=j+1}^{q-1} \beta_k - \sum_{j=q+1}^{n-1} \prod_{k=q+2}^{j} \beta_k \right) - \prod_{k=q+2}^{n} \beta_k \right). \quad \blacksquare$$

Now, we will prove that for infinitely many values of the parameter $\lambda \in \langle 0, \lambda_0 \rangle$, for some $0 < \lambda_0 < 1$, the TDBS problem can be solved in $O(n \log n)$ time. We will also show that there exist infinitely many values of $\lambda \in \langle \lambda_1, 1 \rangle$, for some λ_1, where $\lambda_0 < \lambda_1 < 1$, such that the optimal schedule for this problem has a V-shape.

Let $q = 1, 2, \ldots, n-1$ and let $\lambda \in \langle 0, 1 \rangle$ be arbitrary but fixed. For a given $\beta = (\beta_1, \beta_2, \ldots, \beta_n)$, define function $\Lambda_q(\lambda)$ as follows:

$$\Lambda_q(\lambda) := \lambda \left(\sum_{j=0}^{q-1} \prod_{k=j+1}^{q-1} \beta_k - \sum_{j=q+1}^{n-1} \prod_{k=q+2}^{j} \beta_k \right) - \prod_{k=q+2}^{n} \beta_k. \quad (14.12)$$

The behaviour of function $\Lambda_q(\lambda)$ is crucial for further considerations. We begin with a necessary condition for the sequence $\beta = (\beta_1, \beta_2, \ldots, \beta_n)$ to be optimal with respect to the criterion $\| \cdot \|_{(\lambda)}$.

Lemma 14.22. (Gawiejnowicz et al. [103]) *Let sequence* $\beta^\star = (\beta_1^\star, \beta_2^\star, \ldots, \beta_n^\star)$ *be optimal with respect to the criterion* $\| \cdot \|_{(\lambda)}$ *and let* $\beta' = \beta^\star(\beta_q^\star \leftrightarrow \beta_{q+1}^\star)$. *Then for* $q = 1, 2, \ldots, n-1$ *the following inequality holds:*

$$0 \le \|C(\beta')\|_{(\lambda)} - \|C(\beta^\star)\|_{(\lambda)} = (\beta_{q+1}^\star - \beta_q^\star)\Lambda_q(\lambda). \quad (14.13)$$

Proof. In view of (14.11), (14.12) and Theorem 14.21, the result follows. □

In view of Lemma 14.22, it is important to know the behaviour of the sign of function $\Lambda_q(\lambda)$ for $q = 1, 2, \ldots, n-1$ and $\lambda \in \langle 0, 1 \rangle$, since then, by (14.13), we can control the sign of difference $\|C(\beta')\|_{(\lambda)} - \|C(\beta)\|_{(\lambda)}$.

Note that $\Lambda_1(\lambda)$ is always strictly less than 0, while the sign of $\Lambda_q(\lambda)$, for $q = 2, \ldots, n-1$, depends on λ. In fact, from definition of $\Lambda_q(\lambda)$, we obtain the following lemma.

Lemma 14.23. (Gawiejnowicz et al. [103]) *Let $\lambda \in \langle 0, 1 \rangle$ be arbitrary but fixed. Then $\Lambda_1(\lambda) < 0$ and the following inequalities hold:*

$$\Lambda_1(\lambda) \le \Lambda_2(\lambda) \le \ldots \le \Lambda_{n-1}(\lambda).$$

Proof. Indeed, these inequalities hold since for $q = 1, 2, \ldots, n-1$ we have

$$\Lambda_{q+1}(\lambda) - \Lambda_q(\lambda) = \lambda \left((\beta_q - 1) \sum_{j=0}^{q-1} \prod_{k=j+1}^{q-1} \beta_k + 1 \right) + $$
$$\lambda \left((\beta_{q+2} - 1) \sum_{j=q+2}^{n-1} \prod_{k=q+3}^{j} \beta_k + 1 \right) + $$
$$\prod_{k=q+3}^{n} \beta_k (\beta_{q+2} - 1) \ge 0.$$

To end the proof, it is sufficient to note that for $q = 1$ there holds

$$\Lambda_1(\lambda) = -\lambda \sum_{j=3}^{n-1} \prod_{k=3}^{j} \beta_k - \prod_{k=3}^{n} \beta_k < 0.$$

■

In view of these results, given the sequence $\beta = (\beta_1, \beta_2, \ldots, \beta_n)$, the fundamental problem is to determine λ_0 and λ_1, $0 < \lambda_0 < \lambda_1 < 1$, such that $\Lambda_q(\lambda) \le 0$ for all $\lambda \in \langle 0, \lambda_0 \rangle$ and $q = 1, 2, \ldots, n-1$, and $\Lambda_{n-1}(\lambda) \ge 0$ for all $\lambda \in \langle \lambda_1, 1 \rangle$. In the first case, sequence $\Lambda_q(\lambda)$ has only non-positive elements and the non-increasing ordering of sequence β is, by (14.13), a necessary condition for optimality of β. In the second case, there is the change of sign in sequence $\Lambda_q(\lambda)$, and the sequence β must have a V-shape.

We now prove the following result in which strongly restrictive formulae for λ_0 and λ_1 are used. We start with a definition (cf. [103]).

Definition 14.24. (Numbers λ_0 and λ_1)
Let $\bar\beta := \max\{\beta_1, \beta_2, \ldots, \beta_n\}$ and $\underline\beta := \min\{\beta_1, \beta_2, \ldots, \beta_n\}$. Define λ_0 and λ_1 as follows:

$$\lambda_0 := \frac{\bar\beta - 1}{\bar\beta^{n-1} - 1}. \tag{14.14}$$

and

$$\lambda_1 := \frac{\underline\beta - 1}{\underline\beta^{n-1} - 1}. \tag{14.15}$$

Lemma 14.25. (Gawiejnowicz et al. [103]) *Let there be given a sequence* $\beta = (\beta_1, \beta_2, \ldots, \beta_n)$. *Then for* $q = 1, 2, \ldots, n-1$ *there holds the inequality* $\Lambda_q(\lambda) \leq 0$, *where* $\lambda \in \langle 0, \lambda_0 \rangle$ *and* $\lambda_0 > 0$ *is defined by* (14.14).

Proof. Note that function $\Lambda_q(\lambda)$ is non-decreasing for each $\lambda \in \langle 0, 1 \rangle$ and $q = 1, 2, \ldots, n-1$. Therefore, $\Lambda_q(\lambda) \leq 0$ for $q = 1, 2, \ldots, n-1$ if and only if $\Lambda_{n-1}(\lambda) \leq 0$. This, in turn, is equivalent to

$$0 \leq \lambda \leq \frac{1}{\sum\limits_{j=0}^{n-2} \prod\limits_{k=j+1}^{n-2} \beta_k}.$$

Since

$$\sum_{j=0}^{n-2} \prod_{k=j+1}^{n-2} \beta_k \leq \sum_{i=0}^{n-2} \bar{\beta}^i = \frac{\bar{\beta}^{n-1} - 1}{\bar{\beta} - 1} \equiv \frac{1}{\lambda_0},$$

it is sufficient for $\Lambda_q(\lambda) \leq 0$ that $\lambda \in \langle 0, \lambda_0 \rangle$, where $q = 1, 2, \ldots, n-1$. ∎

Lemma 14.26. (Gawiejnowicz et al. [103]) *Let there be given a sequence* $\beta = (\beta_1, \beta_2, \ldots, \beta_n)$. *Then for each* $\lambda \in \langle \lambda_1, 1 \rangle$ *there holds* $\Lambda_{n-1}(\lambda) \geq 0$, *where* $\lambda \in \langle \lambda_1, 1 \rangle$ *and* $\lambda_1 < 1$ *is defined by formula* (14.15).

Proof. $\Lambda_{n-1}(\lambda) \geq 0$ if and only if

$$\lambda \geq \frac{1}{\sum\limits_{j=0}^{n-2} \prod\limits_{k=j+1}^{n-2} \beta_k}.$$

Since there holds

$$\sum_{j=0}^{n-2} \prod_{k=j+1}^{n-2} \beta_k \geq \sum_{i=0}^{n-2} \underline{\beta}^i = \frac{\underline{\beta}^{n-1} - 1}{\underline{\beta} - 1} \equiv \frac{1}{\lambda_1},$$

it is sufficient for $\Lambda_{n-1}(\lambda) \geq 0$ that $\lambda \in \langle \lambda_1, 1 \rangle$. ∎

The following result is a corollary from Lemmata 14.22, 14.25 and 14.26.

Theorem 14.27. (Gawiejnowicz et al. [103]) *Let sequence* $\beta^\star = (\beta_1^\star, \beta_2^\star, \ldots, \beta_n^\star)$ *be optimal with respect to the criterion* $\| \cdot \|_{(\lambda)}$ *and let* λ_0 *and* λ_1 *be defined by formulae* (14.14) *and* (14.15), *respectively. Then* $0 < \lambda_0 \leq \lambda_1 < 1$ *and there hold the following implications:*
(a) *if* $\lambda \in \langle 0, \lambda_0 \rangle$, *then* $\beta^\star$ *is non-increasing;*
(b) *if* $\lambda \in \langle \lambda_1, 1 \rangle$, *then* $\beta^\star$ *has a V-shape.*
Moreover, if sequence $\beta^\star$ *contains distinct elements, then* $\lambda_0 < \lambda_1$.

Proof. (a) Let sequence $\beta^\star = (\beta_1^\star, \beta_2^\star, \ldots, \beta_n^\star)$ be optimal with respect to the criterion $\| \cdot \|_{(\lambda)}$, let λ_0 be defined by formula (14.14) and $\lambda \in \langle 0, \lambda_0 \rangle$ be arbitrary but fixed. Then, by Lemma 14.25, for $q = 1, 2, \ldots, n - 1$ there holds inequality $\Lambda_q(\lambda) \leq 0$. But this means, by Lemma 14.22, that for $q = 1, 2, \ldots, n - 1$ we have $\beta_{q+1}^\star - \beta_q^\star \leq 0$. Hence, the sequence $\beta^\star$ is non-increasing.

(b) Let sequence $\beta^\star = (\beta_1^\star, \beta_2^\star, \ldots, \beta_n^\star)$ again be optimal with respect to the criterion $\| \cdot \|_{(\lambda)}$, let λ_1 be defined by (14.15) and $\lambda \in \langle \lambda_1, 1 \rangle$ be arbitrary but fixed. Then, by Lemma 14.26, the inequality $\Lambda_{n-1}(\lambda) \geq 0$ holds. But we know, by Lemma 14.23, that $\Lambda_1(\lambda) < 0$ and the sequence $\Lambda_q(\lambda)$, for $q = 1, 2, \ldots, n - 1$, is non-decreasing. Hence, there must exist $1 < r < n - 1$ such that $\Lambda_{r-1}(\lambda) \leq 0$ but $\Lambda_r(\lambda) \geq 0$. But this implies, by Lemma 14.22, that for $q = 1, 2, \ldots, r - 1$ there holds inequality $\beta_{q+1}^\star - \beta_q^\star \leq 0$ and for $q = r, r+1, \ldots, n-1$ there holds inequality $\beta_{q+1}^\star - \beta_q^\star \geq 0$. Thus the sequence $\beta^\star$ must have a V-shape.

To end the proof, it is sufficient to notice that if $\beta^\star$ contains distinct elements, then $\underline{\beta}^\star \neq \bar{\beta}^\star$ and $\lambda_0 < \lambda_1$. ∎

We can formulate a stronger version of Theorem 14.27, which gives more precise conditions for the monotonicity and the V-shapeness of the optimal sequence $\beta^\star$. The version requires, however, $O(n \log n)$ additional operations to determine the respective values of λ_0 and λ_1.

Before we prove the main result, we introduce a definition (cf. [103]).

Definition 14.28. (Numbers $\lambda_\bullet$ and $\lambda^\bullet$)
Given a sequence $\beta = (\beta_1, \beta_2, \ldots, \beta_n)$, define $\lambda(\beta)$ as follows:

$$\lambda(\beta) := \frac{1}{\sum\limits_{j=0}^{n-2} \prod\limits_{k=j+1}^{n-2} \beta_k}. \tag{14.16}$$

Moreover, let

$$\lambda_\bullet := \min\{\lambda(\beta_\pi) : \pi \in \mathfrak{S}_n\}$$

and

$$\lambda^\bullet := \max\{\lambda(\beta_\pi) : \pi \in \mathfrak{S}_n\}.$$

Theorem 14.29. (Gawiejnowicz et al. [103]) *Let sequence $\beta^\star = (\beta_1^\star, \beta_2^\star, \ldots, \beta_n^\star)$ be optimal with respect to the criterion $\| \cdot \|_{(\lambda)}$, and let λ_0 and λ_1 be defined by formulae (14.14) and (14.15), respectively. Then there hold the following implications:*
(a) *if $\lambda \in \langle 0, \lambda_\bullet \rangle$, then $\beta^\star$ is non-increasing;*
(b) *if $\lambda \in \langle \lambda^\bullet, 1 \rangle$, then $\beta^\star$ has a V-shape.*
Moreover, $0 < \lambda_0 \leq \lambda_\bullet$ and $\lambda^\bullet \leq \lambda_1 < 1$, and these inequalities are strict, whenever sequence $\beta^\star$ contains only distinct elements.

Proof. Similar to the proof of Theorem 14.27. □

Remark 14.30. Note that if $n = 2$, then we have $\lambda_0 = \lambda_1 = 1$. Moreover, for $n = 2, 3$ there hold equalities $\lambda_0 = \lambda_\bullet$ and $\lambda_1 = \lambda^\bullet$. Note also that calculating the minimum (the maximum) in the definition of $\lambda_\bullet$ ($\lambda^\bullet$) needs only $O(n \log n)$ time, since by Lemma 1.2 (a), the denominator in formula (14.16) is maximized (minimized) by ordering β non-decreasingly (non-increasingly). Finally, there exists only one sequence (up to the order of equal elements) that maximizes (minimizes) this denominator.

Now we present a few examples (cf. Gawiejnowicz et al. [103]) that illustrate some consequences and applications of Theorem 14.27.

Example 14.31. Consider the sequence $\hat{\beta} = (5, 3, 2, 4)$. Then $\underline{\beta} = 2$, $\bar{\beta} = 4$ and $\lambda_0 = \lambda_\bullet = \frac{1}{5}$, $\lambda_1 = \lambda^\bullet = \frac{1}{3}$. By Theorem 14.27, for any $\lambda \in \langle 0, \frac{1}{5} \rangle$ the optimal schedule for the TDBS problem is non-increasing, $\beta^\star = (5, 4, 3, 2)$, while for any $\lambda \in \langle \frac{1}{3}, 1 \rangle$ the optimal schedule for the problem has a V-shape.   ◆

Theorem 14.27 is also useful in the case when the form of criterion $\| \cdot \|_{(\lambda)}$ is known in advance and we want to check if a given sequence is optimal with respect to this particular criterion.

Example 14.32. Let $\|C(\hat{\beta})\|_{(\lambda)} = \frac{1}{7} \sum C_j(\hat{\beta}) + \frac{6}{7} C_{\max}(\hat{\beta})$ and $\hat{\beta} = (2, 3, 4, 5)$. Then the sequence $(5, 4, 3, 2)$, by Theorem 14.27, is optimal for the criterion $\| \cdot \|_{(\lambda)}$, since $\lambda = \frac{1}{7} < \lambda_0 = \frac{1}{5}$.   ◆

Example 14.33. Let $\|C(\hat{\beta})\|_{(\lambda)} = \frac{6}{7} \sum C_j(\hat{\beta}) + \frac{1}{7} C_{\max}(\hat{\beta})$ and $\hat{\beta} = (2, 3, 4, 5)$. Then, since $\lambda = \frac{6}{7} > \lambda_1 = \frac{1}{3}$, any optimal solution is a V-shaped sequence. There are three such V-shaped sequences: $(5, 4, 2, 3)$, $(5, 2, 3, 4)$, and $(5, 3, 2, 4)$. The first sequence is the optimal solution.   ◆

The next example shows the main difference between the values of λ_0 and λ_1 and the values of $\lambda_\bullet$ and $\lambda^\bullet$: in order to calculate $\lambda_\bullet$ and $\lambda^\bullet$, we must know all elements of sequence $\hat{\beta}$ whereas, to calculate λ_0 and λ_1, we need only the values of $\underline{\beta}$ and $\bar{\beta}$.

Example 14.34. Let $\hat{\beta} = (1.5, 1.3, 1.1, 1.2, 1.4)$. Then we have $\underline{\beta} = 1.1$, $\bar{\beta} = 1.4$ and $\lambda_0 = 0.23 < \lambda_\bullet = 0.24$, $\lambda^\bullet = 0.28 < \lambda_1 = 0.30$. (Note that we have the same values of λ_0 and λ_1 for all sequences with $n = 4$ elements, in which $\bar{\beta} = 1.4$ and $\underline{\beta} = 1.1$.) If we knew only the values of $\underline{\beta}$ and $\bar{\beta}$, we still could calculate λ_0 and λ_1 but we would not be able to calculate $\lambda_\bullet$ and $\lambda^\bullet$.   ◆

The results presented in Theorem 14.27 and Theorem 14.29 were *necessary* conditions, i.e., we assumed that β is an optimal sequence and we showed its properties. Now we give a *sufficient* condition for sequence $\beta = (\beta_1, \beta_2, \ldots, \beta_n)$ to be optimal solution to the TDBS problem.

Theorem 14.35. (Gawiejnowicz et al. [103]) *A sufficient condition for a sequence* $\beta = (\beta_1, \beta_2, \ldots, \beta_n)$ *to be optimal with respect to the criterion* $\| \cdot \|_{(\lambda)}$, $\lambda \in \langle 0, 1 \rangle$, *is that* β *is non-increasing and* $0 \leq \lambda \leq \lambda_\bullet$.

Proof. Let $0 \leq \lambda \leq \lambda_{\bullet}$. Then, by Theorem 14.29, any sequence β which is optimal with respect to criterion $\| \cdot \|_{(\lambda)}$ must be non-increasing. Moreover, there exists only one (up to the order of equal elements) such optimal sequence. Thus, since this monotonic sequence is unique (again up to the order of equal elements), it must coincide with the optimal sequence.   ∎

14.4 Computational experiments

In this section, we present selected results of computational experiments related to the TDPS problem.

In the first computational experiment, the average behaviour of q_0 (cf. Lemma 14.11) was investigated. In the experiment, 100 random sequences β have been generated, each with $n = 20$ elements taken from the interval $\langle 4, 30 \rangle$.

The results of this experiment (see Fig. 14.1, cf. [103]) suggest that for random sequences composed of elements generated from uniform distribution, the values of q_0 concentrate in the middle part of the domain of $\Delta_q(\beta)$.

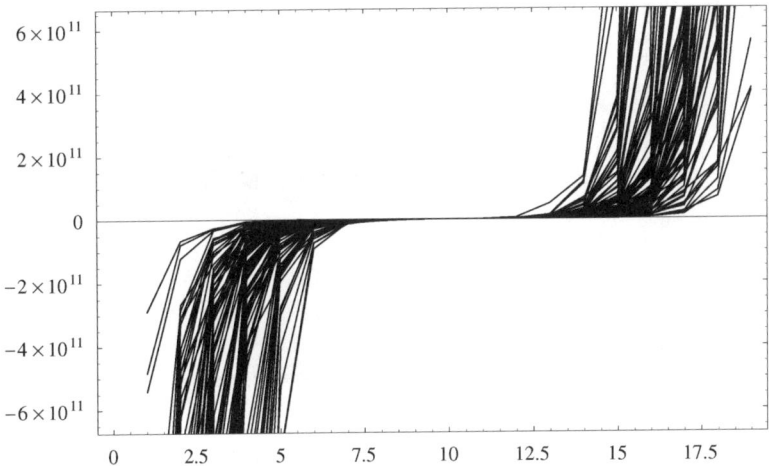

Fig. 14.1: Behaviour of q_0 for 100 random sequences

In order to obtain some insight into the structure of the set $Y = f(X)$ of all solutions of the TDPS problem, where $f = (\Sigma C_j, C_{\max})$ and $X = \{\hat{\beta}_{\pi} : \pi \in \mathfrak{S}_n\}$ for a given $\hat{\beta}$, several other computational experiments have been conducted.

Examples of such a structure are given in Figs. 14.2 and 14.3 (cf. [103]). The box '□' denotes a V-shaped solution, the diamond '◆' denotes a weakly V-shaped solution, the circle '○' denotes a Pareto optimal solution and the symbol '×' denotes a solution that is neither V-shaped nor Pareto optimal.

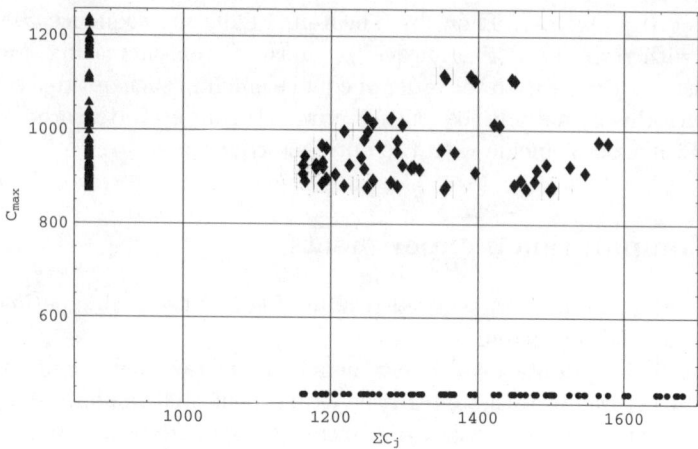

Fig. 14.2: The structure of set Y for $\hat{a} = (2, 3, 4, 5, 6)$

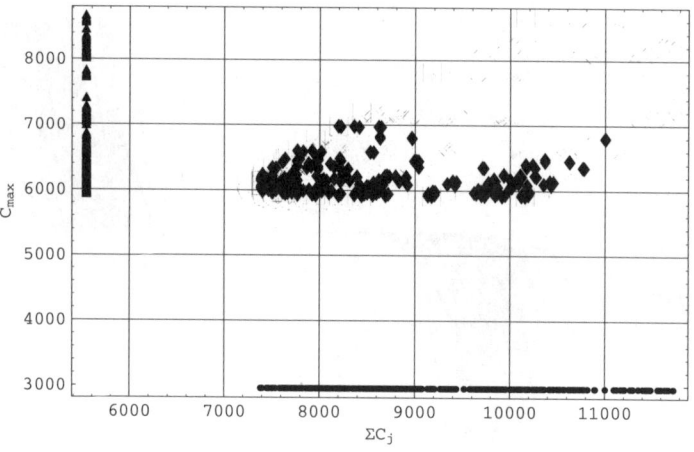

Fig. 14.3: The structure of set Y for $\hat{a} = (2, 3, 4, 5, 6, 7)$

The results of these experiments suggest that Pareto optimal schedules can be found only in the triangle with vertices in points $(\sum C_j(\beta^\star), C_{\max}(\beta^\star))$, $(\sum C_j(\beta^\bullet), C_{\max}(\beta^\bullet))$ and $(\sum C_j(\beta^\star), C_{\max}(\beta^\bullet))$, where $\beta^\star$ and $\beta^\bullet$ denote an optimal schedule for the $\sum C_j$ and $C_{\max}$ criterion, respectively.

With these figures, we end the presentation of the results concerning time-dependent scheduling with two criteria. Also, this chapter ends the book.

14.5 Concluding remarks and open problems

In this chapter, we considered two bicriterion time-dependent scheduling problems, TDPS and TDBS. For the TDPS problem, we have given necessary and sufficient conditions for a schedule to be Pareto optimal. For the TDBS problem, we have shown that there exists $0 \le \lambda_0 \le 1$ such that for all $\lambda \in \langle \lambda_0, 1 \rangle$ the problem is solvable in $O(n \log n)$ time. We proved that the optimal schedule for the TDBS problem has a V-shape for all $\lambda \in \langle \lambda_1, 1 \rangle$, where $\lambda_1 > \lambda_0$. We also proved a few properties of the criterion applied in the TDBS problem.

There remain some open problems to solve. The first open problem is the *hierarchical* minimization of the $\sum C_j$ and $C_{\max}$ criteria. On the one hand, a single machine scheduling problem with job processing times in the form of $p_j = 1 + b_j t$ and the $C_{\max}$ criterion is solvable in $O(n \log n)$ time by arranging jobs in the non-increasing order of deterioration rates b_j, while the optimal schedule for the same problem but with the $\sum C_j$ criterion must have a V-shape. On the other hand, these criteria are not agreeable with each other. In conclusion, the problem of the hierarchical minimization of the $\sum C_j$ and $C_{\max}$ criteria seems to be non-trivial. We suppose that it is at least $\mathcal{NP}$-hard in the ordinary sense.

The second open problem is establishing the time complexity of the TDPS problem. We conjecture that the TDPS problem is also intractable.

Finally, an interesting open problem is establishing the status of the time complexity of the TDBS problem in the interval $\langle \lambda_0, \lambda_1 \rangle$.

Afterword

Although almost 30 years have elapsed since time-dependent scheduling was originated, this theory is still developing. Without exaggeration we can say, however, that the youth of time-dependent scheduling has passed away, and the theory slowly comes into a mature age. The aim of this afterword is to summarize the present state of this research area and to delineate its possible development in the future.

Time-dependent scheduling as a research area

First, one can observe a certain asymmetry in the development of time-dependent scheduling. The asymmetry manifests itself in the number of results concerning different forms of job deterioration vs. job shortening: most results concern problems with unbounded job deteriorating processing times.

Second, most results for problems with deteriorating and shortening job processing times concern the linear case. This, it seems, is caused by the fact that linear functions are easier to study than non-linear ones.

Third, at present the status of time complexity of the most important time-dependent scheduling problems is known. The exceptions, like the problem $1|p_j = 1 + b_j t| \sum C_j$, are very rare.

Concluding, we can state that although there remain some unclear places, the general landscape of time-dependent scheduling is known today.

Time-dependent scheduling algorithms

First, since most non-trivial time-dependent scheduling problems are at least $\mathcal{NP}$-hard in the ordinary sense, only a few polynomial-time algorithms for such problems exist. Therefore, a considerable research effort in the area is directed at the construction of polynomial-time suboptimal algorithms.

Second, most of the time-dependent scheduling algorithms are adaptations of the algorithms known in the classic scheduling. Only a few algorithms have been constructed specifically for scheduling problems with deteriorating or shortening job processing times.

Third, the main tool used in order to evaluate the algorithms is the experimental analysis, since the worst-case analysis for time-dependent scheduling problems is a difficult task. Hence, only a few worst-case ratios are known for time-dependent scheduling algorithms.

Finally, most heuristic algorithms for time-dependent scheduling problems construct the final schedule in a step-by-step manner. Time-dependent scheduling heuristics based on metaheuristics have been proposed by only a few authors. Since metaheuristics have been successfully applied to many $\mathcal{NP}$-hard combinatorial problems, it is possible that their average behaviour will be acceptable also in the case of time-dependent scheduling problems.

Concluding, we can state that in most cases it is known which algorithm to apply for a given time-dependent scheduling problem, though the average behaviour of the algorithm may remain an open question.

Future of time-dependent scheduling

The future development of time-dependent scheduling may go in two directions. The first direction is connected with possible modifications of the basic assumption of time-dependent scheduling, concerning job processing time. For example, some authors assume that job processing time is a function of time and the position of the job in the schedule (cf. Remark 5.4) or a function of the waiting time of the job (cf. Sect. 5.5).

The second direction of research is to consider, in the framework of time-dependent scheduling, the counterparts of the problems known in the classic scheduling. For example, we can consider time-dependent scheduling problems on machines with non-availability periods (cf. Sect. 6.1.1), time-dependent scheduling with batching (cf. Sect. 6.1.1), time-dependent versions of the common due-date assignment problem (cf. Sect. 6.4.3) or time-dependent scheduling with two criteria (cf. Chap. 14).

Time-dependent scheduling has a clear future. It is a very attractive research field, since there are still many problems that are awaiting a solution. The author hopes that this book will contribute to increased interest in this branch of modern scheduling theory. If this hope becomes true, the main aim of this book will have been achieved.

References

1. E. Aarts and J. Korst, *Simulated Annealing and Boltzmann Machines: A Stochastic Approach to Combinatorial Optimization and Neural Computing.* New York: Wiley 1989.
2. A.V. Aho, J.E. Hopcroft and J.D. Ullman, *The Design and Analysis of Computer Algorithms.* Reading: Addison-Wesley 1974.
3. S. Albers, Online algorithms. In: D. Goldin, S.A. Smolka and P. Wegner (eds.), *Interactive Computation: The New Paradigm*, Berlin-Heidelberg: Springer 2006, pp. 143–164.
4. B. Alidaee, A heuristic solution procedure to minimize makespan on a single machine with non-linear cost functions. *Journal of the Operational Research Society* **41** (1990), no. 11, 1065–1068.
5. B. Alidaee and F. Landram, Scheduling deteriorating jobs on a single machine to minimize the maximum processing times. *International Journal of Systems Science* **27** (1996), no. 5, 507–510.
6. B. Alidaee and N.K. Womer, Scheduling with time dependent processing times: review and extensions. *Journal of the Operatational Research Society* **50** (1999), no. 6, 711–720.
7. A. Allahverdi, J.N.D. Gupta and T. Aldowaisan, A review of scheduling research involving setup considerations. *Omega* **27** (1999), no. 2, 219–239.
8. M.J. Atallah (ed.), *Algorithms and Theory of Computation Handbook.* Boca Raton-Washington: CRC Press 1999.
9. G. Ausiello, P. Crescenzi, G. Gambosi, V. Kann, A. Marchetti-Spaccamela and M. Protasi, *Complexity and Approximation: Combinatorial Optimization Problems and Their Approximability Properties.* Berlin-Heidelberg: Springer 1999.
10. A. Bachman, T-C.E. Cheng, A. Janiak and C-T. Ng, Scheduling start time dependent jobs to minimize the total weighted completion time. *Journal of the Operational Research Society* **53** (2002), no. 6, 688–693.
11. A. Bachman and A. Janiak, Minimizing maximum lateness under linear deterioration. *European Journal of Operational Research* **126** (2000), no. 3, 557–566.
12. A. Bachman and A. Janiak, Scheduling jobs with position-dependent processing times. *Journal of the Operational Research Society* **55** (2004), no. 3, 257–264.

13. A. Bachman, A. Janiak and M.Y. Kovalyov, Minimizing the total weighted completion time of deteriorating jobs. *Information Processing Letters* **81** (2002), no. 2, 81–84.
14. K.R. Baker, *Introduction to Sequencing and Scheduling*. New York: Wiley 1974.
15. K.R. Baker and G.D. Scudder, Sequencing with earliness and tardiness penalties: a review. *Operations Research* **38** (1990), no. 1, 22-36.
16. E. Bampis and A. Kononov, On the approximability of scheduling multiprocessor tasks with time-dependent processor and time requirements. In: *Proceedings of the 15th International Symposium on Parallel and Distributed Processing*, Los Angeles, IEEE Press 2001, pp. 2144–2151.
17. M.S. Barketau, T-C.E. Cheng, C-T. Ng, V. Kotov and M.Y. Kovalyov, Batch scheduling of step deteriorating jobs. *Journal of Scheduling* **11** (2008), no. 1, 17–28.
18. T. Bäck, *Evolutionary Algorithms in Theory and Practice: Evolution Strategies, Evolutionary Programming, Genetic Algorithms*. Oxford: Oxford University Press 1996.
19. J.E. Beasley, OR-Library. See http://msmcga.ms.ic.ac.uk/info.html.
20. R. Bellman, *Dynamic Programming*. Princeton: Princeton University Press 1957.
21. R. Bellman and S.E. Dreyfus, *Applied Dynamic Programming*. Princeton: Princeton University Press 1962.
22. C. Berge, *Graphs and Hypergraphs*. Amsterdam: North-Holland 1973.
23. P. Bernhard, Max-Plus algebra and mathematical fear in dynamic optimization. *Set-Valued Analysis* **8** (2000), no. 1–2, 71-84.
24. D. Biskup, A state-of-the-art review on scheduling with learning effects. *European Journal of Operational Research* **188** (2008), no. 2, 315-329.
25. C. Blum and A. Roli, Metaheuristics in combinatorial optimization: overview and conceptual comparison. *ACM Computing Surveys* **35** (2003), no. 3, 268–308.
26. J. Błażewicz, W. Cellary, R. Słowiński and J. Węglarz, *Scheduling under Resource Constraints: Deterministic Models*. Basel: Baltzer 1986.
27. J. Błażewicz, K.H. Ecker, E. Pesch, G. Schmidt and J. Węglarz, *Handbook on Scheduling*. Berlin-Heidelberg: Springer 2007.
28. A. Borodin and R. El-Yaniv, *Online Computation and Competitive Analysis*, Cambridge: Cambridge University Press 1998.
29. A. Bosio and G. Righini, A dynamic programming algorithm for the single-machine scheduling problem with deteriorating processing times. *Electronic Notes in Discrete Mathematics* **25** (2006), 139–142.
30. J-L. Bouquard, C. Lente and J-C. Billaut, Application of an optimization problem in Max-Plus algebra to scheduling problems. *Discrete Applied Mathematics* **154** (2006), no. 15, 2064–2079.
31. D.P. Bovet and P. Crescenzi, *Introduction to Theory of Complexity*. Harlow: Prentice-Hall 1994.
32. A. Brandstädt, L. Van Bang and J.P. Spinrad, *Graph Clases: A Survey*. Philadelphia: SIAM 1999.
33. S. Browne and U. Yechiali. Scheduling deteriorating jobs on a single processor. *Operations Research* **38** (1990), no. 3, 495–498.
34. P. Brucker, *Scheduling Algorithms*, 5th edition, Berlin-Heidelberg: Springer 2007.

35. P. Brucker and S. Knust, Complexity results for scheduling problems. See http://www.mathematik.uni-osnabrueck.de/research/OR/index.shtml.

36. H. Buer and R.H. Möhring, A fast algorithm for the decomposition of graphs and posets. *Mathematics of Operations Research* **8** (1983), no. 2, 170–184.

37. P.S. Bullen, D.S. Mitrinović and P.M. Vasić, *Mean and Their Inequalities.* Dordrecht: Reidel 1988.

38. P. Cai, J-Y. Cai and A.V. Naik, Efficient algorithms for a scheduling problem and its applications to illicit drug market crackdowns. *Journal of Combinatorial Optimization* **1** (1998), no. 4, 367–376.

39. J-Y. Cai, P. Cai and Y. Zhu, On a scheduling problem of time deteriorating jobs. *Journal of Complexity* **14** (1998), no. 2, 190–209.

40. P. Calégari, G. Coray, A. Hertz, D. Kobler and P. Kuonen, A taxonomy of evolutionary algorithms in combinatorial optimization. *Journal of Heuristics* **5** (1999), no. 2, 145–158.

41. V. Chakaravarthy, V. Pandit, G. Parija and S. Roy, Scheduling critical tasks. Unpublished manuscript, IBM India Research Laboratory, August 2007.

42. Z-L. Chen, A note on single-processor scheduling with time-dependent execution times. *Operations Research Letters* **17** (1995), no. 3, 127–129.

43. Z-L. Chen. Parallel machine scheduling with time dependent processing times. *Discrete Applied Mathematics* **70** (1996), no. 1, 81–93. (Erratum: *Discrete Applied Mathematics* **75** (1996), no. 1, 103.)

44. M-B. Cheng and S-J. Sun, Two scheduling problems in group technology with deteriorating jobs. *Applied Mathematics Journal of Chinese Universities*, Series B, **20** (2005), no. 2, 225–234.

45. M-B. Cheng and S-J. Sun, The single-machine scheduling problems with deteriorating jobs and learning effect. *Journal of Zhejiang University*, Science A, **7** (2006), no. 4, 597–601.

46. M-B. Cheng and S-J. Sun, A heuristic MBLS algorithm for the two semi-online parallel machine scheduling problems with deterioration jobs. *Journal of Shanghai University* **11** (2007), no. 5, 451–456.

47. M-B. Cheng, S-J. Sun and L-M. He, Flow shop scheduling problems with deteriorating jobs on no-idle dominant machines. *European Journal of Operational Research*, **183** (2007), no. 1, 115-124.

48. T-C.E. Cheng and Q. Ding, The complexity of single machine scheduling with two distinct deadlines and identical decreasing rates of processing times. *Computers and Mathematics with Applications* **35** (1998), no. 12, 95–100.

49. T-C.E. Cheng and Q. Ding, The complexity of scheduling starting time dependent tasks with release times. *Information Processing Letters* **65** (1998), no. 2, 75–79.

50. T-C.E. Cheng and Q. Ding, The time dependent makespan problem is strongly NP-complete. *Computers and Operations Research* **26** (1999), no. 8, 749–754.

51. T-C.E. Cheng and Q. Ding, Single machine scheduling with step-deteriorating processing times. *European Journal of Operational Research* **134** (2001), no. 3, 623–630.

52. T-C.E. Cheng and Q. Ding, Single machine scheduling with deadlines and increasing rates of processing times. *Acta Informatica* **36** (2000), no. 9–10, 673–692.

53. T-C.E. Cheng and Q. Ding, Scheduling start time dependent tasks with deadlines and identical initial processing times on a single machine. *Computers and Operations Research* **30** (2003), no. 1, 51–62.

54. T-C.E. Cheng, Q. Ding, M.Y. Kovalyov, A. Bachman and A. Janiak, Scheduling jobs with piecewise linear decreasing processing times. *Naval Research Logistics* **50** (2003), no. 6, 531–554.

55. T-C.E. Cheng, Q. Ding and B.M.T. Lin, A concise survey of scheduling with time-dependent processing times. *European Journal of Operational Research* **152** (2004), no. 1, 1–13.

56. T-C.E. Cheng, Y. He, H. Hoogeveen, M. Ji and G. Woeginger, Scheduling with step-improving processing times. *Operations Research Letters* **34** (2006), no. 1, 37–40.

57. T-C.E. Cheng, L-Y. Kang and C-T. Ng, Due-date assignment and single machine scheduling with deteriorating jobs. *Journal of the Operational Research Society* **55** (2004), no. 2, 198–203.

58. T-C.E. Cheng, L-Y. Kang and C-T. Ng, Due-date assignment and parallel-machine scheduling with deteriorating jobs. *Journal of the Operational Research Society* **58** (2007), no. 8, 1103–1108.

59. T-C.E. Cheng, L-Y. Kang and C-T. Ng, Single machine due-date scheduling of jobs with decreasing start-time dependent processing times. *International Transactions in Operational Research* **12** (2005), no. 4, 355–366.

60. P. Chrétienne, E.G. Coffman, jr, J.K. Lenstra and Z. Liu (eds.), *Scheduling Theory and its Applications*. New York: Wiley 1995.

61. E.G. Coffman, jr, (ed.), *Computer and Job-shop Scheduling Theory*. New York: Wiley 1976.

62. R.W. Conway, W.L. Maxwell and L.W. Miller, *Theory of Scheduling*. Reading: Addison-Wesley 1967.

63. S.A. Cook, The complexity of theorem-proving procedures. In: *Proceedings of the 3rd ACM Symposium on the Theory of Computing*, New York, SIGACT ACM 1971, pp. 151–158.

64. W.J. Cook, W.H. Cunningham, W.R. Pulleyblank and A. Schrijver, *Combinatorial Optimization*. New York-Toronto: Wiley 1998.

65. I.M. Copi, *Introduction to Logic*, 7th ed. New York-London: MacMillan 1986.

66. T.H. Cormen, C.E. Leiserson and R.L. Rivest, *Introduction to Algorithms*, 13th printing. Cambridge: MIT Press 1994.

67. M. Davies, *Computability and Undecidability*. New York: McGraw-Hill 1958.

68. M. Davies, *The Undecidable*. New York: Raven Press 1965.

69. M.A.H. Dempster, J.K. Lenstra and A.H.G. Rinnooy Kan, *Deterministic and Stochastic Scheduling*. Dordrecht: Reidel 1982.

70. N. Deo, *Graph Theory with Applications to Engineering and Computer Science*. Englewood Cliffs: Prentice-Hall 1974.

71. P. Dileepan and T. Sen, Bicriterion static scheduling research for a single machine. *Omega* **16** (1988), no. 1, 53–59.

72. M. Dror, H.I. Stern and J.K. Lenstra, Parallel machine scheduling: Processing rates dependent on number of jobs in operation. *Management Science* **33** (1987), no. 8, 1001–1009.

73. M. Drozdowski, Scheduling multiprocessor tasks – an overview. *European Journal of Operational Research* **94** (1996), no. 2, 215–230.

74. M. Drozdowski, Scheduling parallel tasks – algorithms and complexity. In: J.Y-T. Leung (ed.), *Handbook of Scheduling*. Boca Raton: Chapman and Hall/CRC 2004.

75. J. Du and J.Y-T. Leung, Minimizing total tardiness on one machine is NP-hard. *Mathematics of Operations Research* **15** (1990), no. 3, 483–495.

76. M. Ehrgott, *Multicriteria Optimization*. Lecture Notes in Economics and Mathematical Systems **491**. Berlin-Heidelberg: Springer 2000.

77. A.E. Eiben, J.E. Smith and J.D. Smith, *Introduction to Evolutionary Computing*. Berlin-Heidelberg: Springer 2003.

78. S.E. Elmaghraby (ed.), *Symposium on the Theory of Scheduling and its Applications*. Lecture Notes in Economics and Mathematical Systems **86**. Berlin-Heidelberg: Springer 1973.

79. A. Fiat and G. Woeginger (eds.), *Online Algorithms: The State of the Art*. Lecture Notes in Computer Science **1442**. Berlin-Heidelberg: Springer 1998.

80. G. Finke and H. Jiang, A variant of the permutation flow shop model with variable processing times. *Discrete Applied Mathematics* **76** (1997), no. 1–3, 123–140.

81. G. Finke, M.L. Espinouse, H. Jiang. General flowshop models: Job dependent capacities, job overlapping and deterioration. *International Transactions in Operational Research* **9** (2002), no. 4, 399–414.

82. G. Finke and A. Oulamara, Total completion time in a two-machine flowshop with deteriorating tasks. *Journal of Mathematical Modelling and Algorithms* **6** (2007), no. 4, 563–576.

83. L.J. Fogel, *Intelligence Through Simulated Evolution: Forty Years of Evolutionary Programming*. New York: Wiley 1999.

84. S. French, *Sequencing and Scheduling: An Introduction to the Mathematics of the Job-Shop*. Chichester: Horwood 1982.

85. M.R. Garey and D.S. Johnson, *Computers and Intractability: A Guide to the Theory of NP-Completeness*. San Francisco: Freeman 1979.

86. M.R. Garey, D.S. Johnson and R. Sethi, The complexity of flowshop and job-shop scheduling. *Mathematics of Operations Research* **1** (1976), no. 2, 117–129.

87. S. Gawiejnowicz, Brief survey of continuous models of scheduling. *Foundations of Computing and Decision Sciences* **21** (1996), no. 2, 81–100.

88. S. Gawiejnowicz, A note on scheduling on a single processor with speed dependent on a number of executed jobs. *Information Processing Letters* **57** (1996), no. 6, 297–300.

89. S. Gawiejnowicz, *Scheduling jobs with varying processing times*. Ph.D. dissertation, Poznań University of Technology, Poznań 1997, 102 pp. (in Polish).

90. S. Gawiejnowicz, Minimizing the flow time and the lateness on a processor with a varying speed. *Ricerca Operativa* **21** (1997), no. 83, 53–58.

91. S. Gawiejnowicz, Scheduling deteriorating jobs subject to job or machine availability constraints. *European Journal of Operational Research* **180** (2007), no. 1, 472–478.

92. S. Gawiejnowicz and A. Kononov, Scheduling resumable deteriorating jobs. Report 135/2007, Faculty of Mathematics and Computer Science, Adam Mickiewicz University, Poznań, Poland, February 2007.

93. S. Gawiejnowicz, A. Kononov and T-C. Lai, Scheduling deteriorating jobs subject to precedence constraints. Report 112/2001, Faculty of Mathematics and Computer Science, Adam Mickiewicz University, Poznań, Poland, June 2001.

94. S. Gawiejnowicz, W. Kurc and L. Pankowska, A greedy approach for a time-dependent scheduling problem, in: R. Wyrzykowski et al. (eds.), *Parallel Processing and Applied Mathematics*. Lecture Notes in Computer Science **2328**. Berlin-Heidelberg: Springer 2002, pp. 79–86.

95. S. Gawiejnowicz, W. Kurc and L. Pankowska, Analysis of a time-dependent scheduling problem by signatures of deterioration rate sequences. *Discrete Applied Mathematics* **154** (2006), no. 15, 2150–2166.

96. S. Gawiejnowicz, W. Kurc and L. Pankowska, Bicriterion approach to a single machine time-dependent scheduling problem. In: P. Chamoni et al. (eds.), *Operations Research Proceedings 2001*. Berlin-Heidelberg: Springer 2002, pp. 199–206.

97. S. Gawiejnowicz, W. Kurc and L. Pankowska, The equivalence of linear time-dependent scheduling problems, Report 127/2005, Adam Mickiewicz University, Faculty of Mathematics and Computer Science, Poznań, December 2005.

98. S. Gawiejnowicz, W. Kurc and L. Pankowska, Equivalent time-dependent scheduling problems. *European Journal of Operational Research* (2008), doi: 10.1016/j.ejor.2008.04.040.

99. S. Gawiejnowicz, W. Kurc and L. Pankowska, Greedy scheduling of time-dependent jobs with arithmetic or geometric deterioration rates. In: S. Domek and R. Kaszyński (eds.), *Proceedings of the 12th IEEE International Conference on Methods and Models in Automation and Robotics*, pp. 1119–1124.

100. S. Gawiejnowicz, W. Kurc and L. Pankowska, A greedy algorithm and an FP-TAS for a linear time-dependent scheduling problem. Report 132/2006, Adam Mickiewicz University, Faculty of Mathematics and Computer Science, Poznań, December 2006.

101. S. Gawiejnowicz, W. Kurc and L. Pankowska, Minimizing time-dependent total completion time on parallel identical machines. In: R. Wyrzykowski et al. (eds.), *Parallel Processing and Applied Mathematics*. Lecture Notes in Computer Science **3019**. Berlin-Heidelberg: Springer 2004, pp. 89–96.

102. S. Gawiejnowicz, W. Kurc and L. Pankowska, Parallel machine scheduling of deteriorating jobs by modified steepest descent search. In: R. Wyrzykowski et al. (eds.), *Parallel Processing and Applied Mathematics*. Lecture Notes in Computer Science **3911**. Berlin-Heidelberg: Springer 2006, pp. 116–123.

103. S. Gawiejnowicz, W. Kurc and L. Pankowska, Pareto and scalar bicriterion scheduling of deteriorating jobs. *Computers and Operations Research* **33** (2006), no. 3, 746–767.

104. S. Gawiejnowicz, W. Kurc and L. Pankowska, Polynomial-time solutions in a bicriterion time-dependent scheduling problem. Report 115/2002, Adam Mickiewicz University, Faculty of Mathematics and Computer Science, Poznań, October 2002.

105. S. Gawiejnowicz, W. Kurc, L. Pankowska and C. Suwalski, Approximate solution of a time-dependent scheduling problem for l_p-norm-based criteria. In: B. Fleischmann et al. (eds.), *Operations Research OR2000*. Berlin-Heidelberg: Springer 2001, pp. 372–377.

106. S. Gawiejnowicz, T-C. Lai and M-H. Chiang, Polynomially solvable cases of scheduling deteriorating jobs to minimize total completion time. In: P. Brucker et al. (eds.), *Extended Abstracts of the 7th Workshop on Project Management and Scheduling*. Osnabrück: University of Osnabrück 2000, pp. 131–134.

107. S. Gawiejnowicz, T. Onak and C. Suwalski, A new library for evolutionary algorithms. In: R. Wyrzykowski et al. (eds.), *Parallel Processing and Applied Mathematics*. Lecture Notes in Computer Science **3911**. Berlin-Heidelberg: Springer 2006, pp. 414–421.

108. (a) S. Gawiejnowicz and L. Pankowska, Scheduling jobs with variable processing times. Report 020/1994, Faculty of Mathematics and Computer Science, Adam Mickiewicz University, Poznań, October 1994 (in Polish).

(b) S. Gawiejnowicz and L. Pankowska, Scheduling jobs with variable process-ing times. Report 027/1995, Faculty of Mathematics and Computer Science, Adam Mickiewicz University, Poznań, January 1995 (in English).

109. S. Gawiejnowicz and L. Pankowska, Scheduling jobs with varying processing times. *Information Processing Letters* **54** (1995), no. 3, 175–178.

110. S. Gawiejnowicz and C. Suwalski, Solving time-dependent scheduling prob-lems by local search algorithms. Report 133/2006, Faculty of Mathemat-ics and Computer Science, Adam Mickiewicz University, Poznań, November 2006.

111. M. Gendreau and J-Y. Potvin, Metaheuristics in combinatorial optimization. *Annals of Operations Research* **140** (2005), no. 1, 189–213.

112. K.D. Glazebrook, Single-machine scheduling of stochastic jobs subject to deterioration or delay. *Naval Research Logistics* **39** (1992), no. 5, 613–633.

113. F.W. Glover and G.A. Kochenberger (eds.), *Handbook of Metaheuristics*. In-ternational Series in Operations Research & Management Science **57**. Berlin-Heidelberg: Springer 2003.

114. F.W. Glover and M. Laguna, *Tabu Search*. Dordrecht: Kluwer 1996.

115. T. Gonzalez and S. Sahni, Open shop scheduling to minimize finish time. *Journal of the Association for Computing Machinery* **23** (1976), no. 4, 665–679.

116. V.S. Gordon, Some properties of series-parallel graphs. *Izvestiya Akademii Nauk BSSR* **1** (1981), no. 1, 18-23 (in Russian).

117. V.S. Gordon, C.N. Potts, V.A. Strusevich and J.D. Whitehead, Single ma-chine scheduling models with deterioration and learning: handling prece-dence constraints via priority generation. *Journal of Scheduling* (2008), doi: 10.1007/s10951-008-0064-x.

118. V.S. Gordon, J-M. Proth and C. Chu, A survey of the state-of-the-art of com-mon due date assignment and scheduling research. *European Journal of Oper-ational Research* **139** (2002), no. 1, 1-25.

119. V.S. Gordon, J-M. Proth and C. Chu, Due date assignment and scheduling: SLK, TWK and other due date assignment models. *Production Planning and Control* **13** (2002), no. 2, 117-132.

120. V.S. Gordon, J-M. Proth and V.A. Strusevich, Single machine scheduling and due-date assignment under series-parallel precedence constraints. *Central Eu-ropean Journal of Operations Research* **13** (2005), no. 1, 15–35.

121. R.M.P. Goverde, Railway timetable stability analysis using max-plus system theory. *Transportation Research*, Part B, **41** (2007), no. 2, 179–201.

122. R.L. Graham, Bounds for certain multiprocessing anomalies. *Bell System Tech-nology Journal* **45** (1966), no. 2, 1563–1581.

123. R.L. Graham, Bounds on multiprocessing timing anomalies. *SIAM Journal on Applied Mathematics* **17** (1969), no. 2, 416–429.

124. R.L. Graham, Bounds on the performance of scheduling algorithms. In: E.G. Coffman, jr, (ed.), *Computer and Job-shop Scheduling Theory*. New York: Wiley 1976.

125. R.L. Graham, E.L. Lawler, J.K. Lenstra and A.H.G. Rinnooy Kan, Optimiza-tion and approximation in deterministic sequencing and scheduling: a survey. *Annals of Discrete Mathematics* **5** (1979), 287–326.

126. J. Gunawardena, From max-plus algebra to nonexpansive mappings: A nonlinear theory for discrete event systems. *Theoretical Computer Science* **293** (2003), no. 1, 141–167.

127. A-X. Guo and J-B. Wang, Single machine scheduling with deteriorating jobs under the group technology assumption. *International Journal of Pure and Applied Mathematics* **18** (2005), no. 2, 225–231.

128. J.N.D. Gupta and S.K. Gupta, Single facility scheduling with nonlinear processing times. *Computers and Industrial Engineering* **14** (1988), no. 4, 387–393.

129. S.K. Gupta, A.S. Kunnathur and K. Dandapani, Optimal repayment policies for multiple loans. *Omega* **15** (1987), no. 4, 323–330.

130. F. Harary, *Graph Theory*. Reading: Addison-Wesley 1969.

131. G.H. Hardy, J. Littlewood and G. Polya, *Inequalities*. Cambridge: Oxford Press 1934.

132. E. Hart, P. Ross and D. Corne, Evolutionary scheduling: a review. *Genetic Programming and Evolvable Machines* **6** (2005), no. 2, 191–220.

133. S. Hartmann, *Project Scheduling under Limited Resources*. Lecture Notes in Economical and Mathematical Systems **478**. Berlin-Heidelberg: Springer 1999.

134. K.S. Hindi and S. Mhlanga, Scheduling linearly deteriorating jobs on parallel machines: a simulated annealing approach. *Production Planning and Control* **12** (2001), no. 12, 76–80.

135. K.I-J. Ho, J.Y-T. Leung and W-D. Wei, Complexity of scheduling tasks with time-dependent execution times. *Information Processing Letters* **48** (1993), no. 6, 315–320.

136. D. Hochbaum (ed.), *Approximation Algorithms for NP-hard Problems*. Boston: PWS Publishing 1998.

137. H. Hoogeveen, Multicriteria scheduling. *European Journal of Operational Research* **167** (2005), no. 3, 592-623.

138. J.A. Hoogeveen, *Single Machine Bicriteria Scheduling*. Amsterdam: Centrum voor Wiskunde en Informatica 1992.

139. J.E. Hopcroft and J.D. Ullman, *Introduction to Automata Theory, Languages and Computation*. Reading: Addison-Wesley 1979.

140. Y-C. Hsieh, A new algorithm to schedule exponentially deteriorating jobs on a single machine. *Proceedings of the 1998 CIIE National Conference*, Hsin-Hua, Taiwan, 1998, pp. 48–51.

141. Y-C. Hsieh and D.L. Bricker, Scheduling linearly deteriorating jobs on multiple machines. *Computers and Industrial Engineering* **32** (1997), no. 4, 727–734.

142. Y.S. Hsu and B.M.T. Lin, Minimization of maximum lateness under linear deterioration. *Omega* **31** (2003), no. 6, 459–469.

143. S. Irani and A. Karlin, Online computation. In: D. Hochbaum (ed.), *Approximation Algorithms for NP-hard Problems*. Boston: PWS Publishing 1998.

144. A. Janiak, Time-optimal control in a single machine problem with resource constraints. *Automatica* **22** (1986), no. 6, 745–747.

145. A. Janiak, General flow-shop scheduling with resource constraints. *International Journal of Production Research* **26** (1988), no. 6, 1089–1103.

146. A. Janiak and M.Y. Kovalyov, Job sequencing with exponential functions of processing times. *Informatica* **17** (2006), no. 1, 13–14.

147. A. Janiak and M.Y. Kovalyov, Scheduling in a contaminated area: a model and polynomial algorithms. *European Journal of Operational Research* **173** (2006), no. 1, 125-132.

148. A. Janiak and M.Y. Kovalyov, Scheduling jobs in a contaminated area: a model and heuristic algorithms. *Journal of the Operational Research Society* **59** (2008), no. 7, 977–987.

149. A. Janiak, Y.M. Shafransky and A.V. Tuzikov, Sequencing with ordered criteria, precedence and group technology constraints. *Informatica* **12** (2001), no. 1, 61–88.

150. A.A-K. Jeng and B.M-T. Lin, A note on parallel-machine scheduling with deteriorating jobs. *Journal of the Operational Research Society* **58** (2007), no. 6, 824–826.

151. A.A-K. Jeng and B.M-T. Lin, Makespan minimization in single-machine scheduling with step-deterioration of processing times. *Journal of the Operational Research Society* **55** (2004), no. 3, 247–256.

152. A.A-K. Jeng and B.M-T. Lin, Minimizing the total completion time in single-machine scheduling with step-deteriorating jobs. *Computers and Operations Research* **32** (2005), no. 3, 521–536.

153. M. Ji and T-C.E. Cheng, An FPTAS for scheduling jobs with piecewise linear decreasing processing times to minimize makespan. *Information Processing Letters* **102** (2007), no. 2–3, 41-47.

154. M. Ji and T-C.E. Cheng, Parallel-machine scheduling with simple linear deterioration to minimize total completion time. *European Journal of Operational Research,* **188** (2008), no. 2, 342–347.

155. M. Ji, Y. He and T-C.E. Cheng, Scheduling linear deteriorating jobs with an availability constraint on a single machine. *Theoretical Computer Science* **362** (2006), no. 1–3, 115–126.

156. M. Ji, Y. He and T-C.E. Cheng, A simple linear time algorithm for scheduling with step-improving processing times. *Computers and Operations Research* **34** (2007), no. 8, 2396–2402.

157. D.S. Johnson, A catalog of complexity classes. In: J. Van Leeuwen (ed.), *Handbook of Theoretical Computer Science: Algorithms and Complexity.* Amsterdam-Cambridge: Elsevier/MIT Press 1990.

158. D.S. Johnson, The NP-completeness column: an ongoing guide. *Journal of Algorithms* **2** (1982), no. 4, 393–405.

159. S.M. Johnson, Optimal two and three stage production schedules with setup times included. *Naval Research Logistics Quarterly* **1** (1954), no. 1, 61–68.

160. L-Y. Kang, T-C.E. Cheng, C-T. Ng and M. Zhao, Scheduling to minimize makespan with time-dependent processing times. *Lecture Notes in Computer Science* **3827**. Berlin-Heidelberg: Springer 2005, pp. 925–933.

161. L-Y. Kang and C-T. Ng, A note on a fully polynomial-time approximation scheme for parallel-machine scheduling with deteriorating jobs. *International Journal of Production Economics* **109** (2007), no. 1–2, 180–184.

162. R.M. Karp, Reducibility among combinatorial problems. In: R.E. Miller and J.W. Thatcher (eds.), *Complexity of Computer Computations.* New York: Plenum Press 1972, pp. 85–103.

163. H. Kellerer, U. Pferschy and D. Pisinger, *Knapsack Problems.* Berlin-Heidelberg: Springer 2004.

164. F.P. Kelly, A remark on search and sequencing problems. *Mathematics of Operations Research* **7** (1982), no. 1, 154–157.

165. K. Klamroth and M. Wiecek, A time-dependent multiple criteria single-machine scheduling problem. *European Journal of Operational Research* **135** (2001), no. 1, 17–26.

166. P.N. Klein and N.E. Young, Approximation algorithms. In: M.J. Atallah (ed.), *Algorithms and Theory of Computation Handbook.* Boca Raton-Washington: CRC Press 1999.

167. D.E. Knuth, *The Art of Computer Programming*, vols. 1–3. Reading: Addison-Wesley 1967–1969.

168. W.H. Kohler and K. Steiglitz, Enumerative and iterative computational approaches. In: E.G. Coffman, jr, (ed.), *Computer and Job-shop Scheduling Theory*. New York: Wiley 1976.

169. A. Kononov, Combinatorial complexity of scheduling jobs with simple linear deterioration. *Discrete Analysis and Operations Research* **3** (1996), no. 2, 15–32 (in Russian).

170. A. Kononov, *On the complexity of the problems of scheduling with time-dependent job processing times*. Ph.D. dissertation, Sobolev Institute of Mathematics, Novosibirsk 1999, 106 pp. (in Russian).

171. A. Kononov, On schedules of a single machine jobs with processing times nonlinear in time. In: A.D. Korshunov (ed.), *Operations Research and Discrete Analysis*. Dordrecht: Kluwer 1997, pp. 109–122.

172. A. Kononov, Scheduling problems with linear increasing processing times. In: U. Zimmermann et al. (eds.), *Operations Research 1996*. Berlin-Heidelberg: Springer 1997, pp. 208–212.

173. A. Kononov, A single machine scheduling problems with processing times proportional to an arbitrary function. *Discrete Analysis and Operations Research* **5** (1998), no. 3, 17–37 (in Russian).

174. A. Kononov and S. Gawiejnowicz, NP-hard cases in scheduling deteriorating jobs on dedicated machines. *Journal of the Operational Research Society* **52** (2001), no. 6, 708–718.

175. B. Korte, L. Lovász and R. Schrader, *Greedoids*. Berlin-Heidelberg: Springer 1991.

176. M.Y. Kovalyov and W. Kubiak, A fully polynomial approximation scheme for minimizing makespan of deteriorating jobs. *Journal of Heuristics* **3** (1998), no. 4, 287–297.

177. W. Kubiak and S.L. van de Velde, Scheduling deteriorating jobs to minimize makespan. *Naval Research Logistics* **45** (1998), no. 5, 511–523.

178. A.S. Kunnathur and S.K. Gupta, Minimizing the makespan with late start penalties added to processing times in a single facility scheduling problem. *European Journal of Operational Research* **47** (1990), no. 1, 56–64.

179. W-H. Kuo and D-L. Yang, Single-machine scheduling problems with start-time dependent processing time. *Computers and Mathematics with Applications* **53** (2007), no. 11, 1658-1664.

180. W-H. Kuo and D-L. Yang, A note on due-date assignment and single-machine scheduling with deteriorating jobs. *Journal of the Operational Research Society* **59** (2008), no. 6, 857–859.

181. C. Lahlou and S. Dauzère-Pérès, Single-machine scheduling with time window-dependent processing times. *Journal of the Operational Research Society* **57** (2006), no. 2, 133-139.

182. E.L. Lawler, Optimal sequencing of a single machine subject to precedence constraints. *Management Science* **19** (1973), no. 5, 544–546.

183. E.L. Lawler, *Combinatorial Optimization: Networks and Matroids*. New York: Holt, Rinehart and Winston 1976.

184. E.L. Lawler, A "pseudopolynomial" algorithm for sequencing jobs to minimize total tardiness. *Annals of Discrete Mathematics* **1** (1977), 331–342.

185. E.L. Lawler, Sequencing jobs to minimize total weighted completion time subject to precedence constraints. *Annals of Discrete Mathematics* **2** (1978), 75–90.

186. E.L. Lawler, J.K. Lenstra, A.H.G. Rinnooy Kan, D.B. Shmoys. Sequencing and scheduling: algorithms and complexity. In: S.C. Graves, A.H.G. Rinnooy Kan, and P.H. Zipkin (eds.), *Logistics of Production and Inventory*. Handbooks in Operations Research and Management Science **4**. Amsterdam: North–Holland 1993, pp. 445–522.

187. E.L. Lawler and D.E. Woods, Branch-and-bound methods: a survey. *Operations Research* **14** (1966), no. 4, 699–719.

188. C-Y. Lee, Machine scheduling with an availability constraint. *Journal of Global Optimization* **9** (1996), no. 3–4, 363–382.

189. C-Y. Lee, Machine scheduling with availability constraints. In: J.Y-T. Leung (ed.), *Handbook of Scheduling*. Boca Raton: Chapman and Hall/CRC 2004.

190. C-Y. Lee, L. Lei and M. Pinedo, Current trends in deterministic scheduling. *Annals of Operations Research* **70** (1997), no. 1, 1–41.

191. C-Y. Lee and G. Vairaktarakis, Complexity of single machine hierarchical scheduling: a survey. In: P.M. Pardalos (ed.), *Complexity in Numerical Optimization*. Singapore: World Scientific 1993, pp. 269–298.

192. W-C. Lee, A note on deteriorating jobs and learning in single-machine scheduling problems. *International Journal of Business and Economics* **3** (2004), no. 1, 83–89.

193. W-C. Lee and C-C. Wu, Multi-machine scheduling with deteriorating jobs and scheduled maintenance. *Applied Mathematical Modelling* **32** (2008), no. 3, 362–373.

194. W-C. Lee, C-C. Wu and Y-H. Chung, Scheduling deteriorating jobs on a single machine with release times. *Computers and Industrial Engineering* **54** (2008), no. 3, 441–452.

195. W-C. Lee, C-C. Wu and H-C. Liu, A note on single-machine makespan problem with general deteriorating function. *International Journal of Advanced Manufacturing Technology* (2008), doi: 10.1007/s00170-008-1421-9.

196. W-C. Lee, C-C. Wu, C-C. Wen and Y-H. Chung, A two-machine flowshop makespan scheduling problem with deteriorating jobs. *Computers and Industrial Engineering* **54** (2008), no. 4, 737–749.

197. J.K. Lenstra, A.H.G. Rinnooy Kan and P. Brucker, Complexity of machine scheduling problems. *Annals of Discrete Mathematics* **1** (1977), 343–362.

198. J.Y-T. Leung (ed.), *Handbook of Scheduling*. Boca Raton: Chapman and Hall/CRC 2004.

199. J. Y-T. Leung, C-T. Ng and T.C-E. Cheng, Minimizing sum of completion times for batch scheduling of jobs with deteriorating processing times. *European Journal of Operational Research* **187** (2008), no. 3, 1090–1099.

200. L.A. Levin, Universal sorting problems. *Problemy Peredachi Informatsii* **9** (1973), no. 3, 115–116 (in Russian). English translation: *Problems of Information Transmission* **9** (1973), no. 3, 265–266.

201. A. Lew and H. Mauch, *Dynamic Programming: A Computational Tool*. Berlin-Heidelberg: Springer 2007.

202. H.R. Lewis and C.H. Papadimitriou, *Elements of the Theory of Computation*, 2nd ed. Upper Saddle River: Prentice-Hall 1998.

203. B.M-T. Lin and T-C.E. Cheng, Two-machine flowshop scheduling with conditional deteriorating second operations. *International Transactions in Operations Research* **13** (2006), no. 2, 91-98.

204. Yu.V. Matijasevich, Solution to the tenth problem of Hilbert, *Matematikai Lapok* **21** (1970), 83–87.

205. K. Maurin, *Analysis*. Dordrecht: Reidel 1976.
206. I. Meilijson and A. Tamir, Minimizing flow time on parallel identical processors with variable unit processing time. *Operations Research* **32** (1984), no. 2, 440–448.
207. O.I. Melnikov and Y.M. Shafransky, Parametric problem of scheduling theory. *Kibernetika* **6** (1979), 53–57 (in Russian). English translation: *Cybernetics* **15** (1980), 352–357.
208. Z. Michalewicz, *Genetic Algorithms + Data Structures = Evolution Programs*. Berlin-Heidelberg: Springer 1994.
209. Z. Michalewicz and D.B. Fogel, *How to Solve It: Modern Heuristics*. Berlin-Heidelberg: Springer 2004.
210. W. Michiels, E. Aarts and J. Korst, *Theoretical Aspects of Local Search*. Monographs in Theoretical Computer Science. An EATCS Series. Berlin-Heidelberg: Springer 2007.
211. L. Mitten, Branch-and-bound methods: general formulation and properties. *Operations Research* **18** (1970), no. 1, 24–34. (Errata: *Operations Research* **19** (1971), no. 2, 550.)
212. D.S. Mitrinović, J.E. Pečarić and A.M. Fink, *Classical and New Inequalities in Analysis*. Dordrecht-Boston: Kluwer 1993.
213. J. Moore, An n job, one machine sequencing algorithm for minimizing the number of late jobs. *Management Science* **15** (1968), no. 1, 102–109.
214. T.E. Morton and D.W. Pentico, *Heuristic Scheduling Systems with Applications to Production Systems and Project Management*. New York: Wiley 1993.
215. G. Mosheiov, V-shaped policies for scheduling deteriorating jobs. *Operations Research* **39** (1991), no. 6, 979–991.
216. G. Mosheiov, Scheduling jobs under simple linear deterioration. *Computers and Operations Research* **21** (1994), no. 6, 653–659.
217. G. Mosheiov, Scheduling jobs with step-deterioration; Minimizing makespan on a single- and multi-machine. *Computers and Industrial Engineering* **28** (1995), no. 4, 869–879.
218. G. Mosheiov, Λ-shaped policies to schedule deteriorating jobs. *Journal of the Operational Research Society* **47** (1996), no. 9, 1184–1191.
219. G. Mosheiov, Multi-machine scheduling with linear deterioration. *Infor* **36** (1998), no. 4, 205–214.
220. G. Mosheiov, Complexity analysis of job-shop scheduling with deteriorating jobs. *Discrete Applied Mathematics* **117** (2002), no. 1–3, 195–209.
221. R.H. Möhring, Computationally tractable classes of ordered sets. In: I. Rival (ed.), *Algorithms and Order*. Dordrecht: Kluwer 1989, pp. 105–193.
222. J.H. Muller and J. Spinrad, Incremental modular decomposition. *Journal of Association for Computing Machinery* **36** (1989), no. 1, 1–19.
223. J.F. Muth and G.L. Thompson, *Industrial Scheduling*. Englewood Cliffs: Prentice-Hall 1963.
224. A. Nagar, J. Haddock and S. Heragu, Multiple and bicriteria scheduling: a literature survey. *European Journal of the Operational Research* **81** (1995), no. 1, 88–104.
225. G.L. Nemhauser and L.A. Wolsey, *Integer and Combinatorial Optimization*. New York: Wiley 1988.
226. C-T. Ng, T-C.E. Cheng, A. Bachman and A. Janiak, Three scheduling problems with deteriorating jobs to minimize the total completion time. *Information Processing Letters* **81** (2002), no. 6, 327–333.

227. E. Nowicki and S. Zdrzałka, Optimal control of a complex of independent operations. *International Journal of Systems Sciences* **12** (1981), no. 1, 77–93.

228. K. Ocetkiewicz, Polynomial case of V-shaped policies in scheduling deteriorating jobs. Report ETI-13/06, Gdańsk University of Technology, Gdańsk, Poland, November 2006.

229. D. Oron, Single machine scheduling with simple linear deterioration to minimize total absolute deviation of completion times. *Computers and Operations Research* **35** (2008), no. 6, 2071–2078.

230. I.H. Osman and J.P. Kelly, *Meta-Heuristics: Theory and Applications*. Dordrecht: Kluwer 1996.

231. J.G. Oxley, *Matroid Theory*. Oxford: Oxford University Press 1992.

232. S.S. Panwalkar, M.L. Smith and A. Seidmann, Common due-date assignment to minimize total penalty for the one machine scheduling problem. *Operations Research* **30** (1982), no. 2, 391–399.

233. C.H. Papadimitriou, *Computational Complexity*. Reading: Addison-Wesley 1994.

234. C.H. Papadimitriou and K. Steiglitz, *Combinatorial Optimization: Algorithms and Complexity*. Englewood Cliffs: Prentice-Hall 1981.

235. R.G. Parker, *Deterministic Scheduling Theory*. London: Chapman and Hall 1995.

236. S. Phillips and J. Westbrook, On-line algorithms: competitive analysis and beyond. In: M.J. Atallah (ed.), *Algorithms and Theory of Computation Handbook*. Boca Raton-Washington: CRC Press 1999.

237. M. Pinedo, *Scheduling: Theory, Algorithms, and Systems*. Englewood Cliffs: Prentice-Hall 1995.

238. C.N. Potts and M.Y. Kovalyov, Scheduling with batching: a review. *European Journal of Operational Research* **120** (2000), no. 2, 228–249.

239. K. Pruhs, E. Torng and J. Sgall, Online scheduling. In: J.Y-T. Leung (ed.), *Handbook of Scheduling*. Boca Raton: Chapman and Hall/CRC 2004.

240. N.P. Rachaniotis and C.P. Pappis, Scheduling fire-fighting tasks using the concept of "deteriorating jobs". *Canadian Journal of Forest Research* **36** (2006), no. 3, 652–658.

241. H. Rasiowa, *Introduction to Contemporary Mathematics*. Amsterdam: North-Holland 1973.

242. J.G. Rau, Minimizing a function of permutations of n integers. *Operations Research* **19** (1971), no. 1, 237–240.

243. C-R. Ren and L-Y. Kang, An approximation algorithm for parallel machine scheduling with simple linear deterioration. *Journal of Shanghai University* **11** (2007), no. 4, 351–354.

244. A.M. Revyakin, Matroids. *Journal of Mathematical Sciences* **108** (2002), no. 1, 71–130.

245. A.H.G. Rinnooy Kan, *Machine Scheduling Problems: Classification, Complexity and Computations*, The Hague: Nijhoff 1976.

246. A.H.G. Rinnooy Kan, On Mitten's axioms for branch-and-bound, *Operations Research* **24** (1976), no. 6, 1176–1178.

247. H. Rogers, jr, *Theory of Recursive Functions and Effective Computability*. New York: McGraw-Hill 1967.

248. W. Rudin, *Principles of Mathematical Analysis*, 3rd ed. New York: McGraw-Hill 1976.

249. P. Salamon, P. Sibani and R. Frost, *Facts, Conjectures, and Improvements for Simulated Annealing*. SIAM Monographs on Mathematical Modeling and Computation **7**. Philadelphia: SIAM 2002.
250. J.E. Savage, *The Complexity of Computing*. New York: Wiley, 1976.
251. G. Schmidt, Scheduling with limited machine availability. *European Journal of the Operational Research* **121** (2000), no. 1, 1–15.
252. P. Schuurman and G. Woeginger, Approximation schemes – A tutorial. See http://wwwhome.cs.utwente.nl/~woegingergj/papers/ptas.pdf.
253. D.M. Seegmuller, S.E. Visagie, H.C. de Kock and W.J. Pienaar, Selection and scheduling of jobs with time-dependent duration. *Orion* **23** (2007), no. 1, 17–28.
254. S. Shakeri and R. Logendran, A mathematical programming-based scheduling framework for multitasking environments. *European Journal of the Operational Research* **176** (2007), no. 1, 193–209.
255. N.V. Shakhlevich and V.A. Strusevich, Preemptive scheduling problems with controllable processing times. *Journal of Scheduling* **8** (2005), no. 3, 233–253.
256. P. Sharma, Permutation polyhedra and minimisation of the variance of completion times on a single machine. *Journal of Heuristics* **8** (2002), no. 4, 467-485.
257. Y-R. Shiau, W-C. Lee, C-C. Wu and C-M. Chang, Two-machine flowshop scheduling to minimize mean flow time under simple linear deterioration. *International Journal on Advanced Manufacturing Technology* **34** (2007), no. 7–8, 774–782.
258. M. Sipser, *Introduction to the Theory of Computation*. Boston: PWS Publishing 1997.
259. R. Słowiński and M. Hapke (eds.), *Scheduling under Fuzziness*. Heidelberg: Physica-Verlag 2000.
260. W.E. Smith, Various optimizers for single-stage production. *Naval Research Logistics Quarterly* **3** (1956), no. 1–2, 59–66.
261. C. Sriskandarajah and S.K. Goyal, Scheduling of a two-machine flowshop with processing time linearly dependent on job waiting-time. *Journal of the Operational Research Society* **40** (1989), no. 10, 907–921.
262. D.R. Sule, *Industrial Scheduling*. Boston: PWS Publishing 1997.
263. P.S. Sundararaghavan and A.S. Kunnathur, Single machine scheduling with start time dependent processing times: some solvable cases. *European Journal of Operational Research* **78** (1994), no. 3, 394–403.
264. V.S. Tanaev, V.S. Gordon and Y.M. Shafransky, *Scheduling Theory. Single-Stage Systems*. Dordrecht: Kluwer 1994.
265. V.S. Tanaev, M.Y. Kovalyov and Y.M. Shafransky, *Scheduling Theory. Group Technologies*. Minsk: Institute of Technical Cybernetics, National Academy of Sciences of Byelarus 1998 (in Russian).
266. V.S. Tanaev, Y.N. Sotskov and V.A. Strusevich, *Scheduling Theory. Multi-Stage Systems*. Dordrecht: Kluwer 1994.
267. V. T'kindt and J-C. Billaut, *Multicriteria Scheduling*. Berlin-Heidelberg: Springer 2002.
268. M.D. Toksarı and E. Güner, Minimizing the earliness/tardiness costs on parallel machine with learning effects and deteriorating jobs: a mixed nonlinear integer programming approach. *International Journal of Advanced Manufacturing Technology*, doi: 10.1007/s00170-007-1128-3.
269. M.A. Trick, Scheduling multiple variable-speed machines. *Operations Research* **42** (1994), no. 2, 234–248.

270. A. Turing, On computable numbers, with an application to the Entscheidungsproblem. *Proceedings of the London Mathematical Society* **42** (1936), no. 1, 230–265. (Erratum: *Proceedings of the London Mathematical Society* **43** (1937), no. 6, 544–546.)

271. J. Valdes, R.E. Tarjan and E.L. Lawler, The recognition of series-parallel digraphs. *SIAM Journal on Computing* **11** (1982), no. 3, 298–311.

272. V.V. Vazirani, *Approximation Algorithms*, 2nd ed. Berlin-Heidelberg: Springer 2003.

273. R. Vicson, Two single-machine sequencing problems involving controllable job processing times. *IEE Transactions* **12** (1980), no. 3, 258–262.

274. S. Voss, S. Martello, I.H. Osman and C. Roucairol (eds.), *Meta-Heuristics: Advances and Trends in Local Search Paradigms for Optimization*. Dordrecht: Kluwer 1999.

275. T.G. Voutsinas and C.P. Pappis, Scheduling jobs with values exponentially deteriorating over time. *International Journal of Production Economics* **79** (2002), no. 1–3, 163–169.

276. K. Wagner and G. Wechsung, *Computational Complexity*. Berlin: Deutscher Verlag der Wissenschaften 1986.

277. W. Wajs, Polynomial algorithm for dynamic sequencing problem. *Archiwum Automatyki i Telemechaniki* **31** (1986), no. 3, 209–213 (in Polish).

278. M. Walk, *Theory of Duality in Mathematical Programming*. Berlin: Akademie-Verlag 1989.

279. S. Walukiewicz, *Integer Programming*. Warszawa: Polish Scientific Publishers 1991.

280. J-B. Wang, Flow shop scheduling problems with decreasing linear deterioration under dominant machines. *Computers and Operations Research* **34** (2007), no. 7, 2043-2058.

281. J-B. Wang, Single-machine scheduling problems with the effects of learning and deterioration. *Omega* **35** (2007), no. 4, 397–402.

282. J-B. Wang and T-C.E. Cheng, Scheduling problems with the effects of deterioration and learning. *Asia-Pacific Journal of Operational Research* **24** (2007), no. 2, 245–263.

283. J-B. Wang, A-X. Guo, F. Shan, B. Jiang and L-Y. Wang, Single machine group scheduling under decreasing linear deterioration. *Journal of Applied Mathematics and Computing* **24** (2007), no. 1–2, 283–293.

284. J-B. Wang, L. Lin and F. Shan, Single-machine group scheduling problems with deteriorating jobs. *International Journal of Advanced Manufacturing Technology* (2008), doi: 10.1007/s00170-007-1255-x.

285. J-B. Wang, C-T. Ng and T-C.E. Cheng, Single-machine scheduling with deteriorating jobs under a series-parallel graph constraint. *Computers and Operations Research* **35** (2008), no. 8, 2684–2693.

286. J-B. Wang, C-T. Ng, T-C.E. Cheng and L-L. Liu, Minimizing total completion time in a two-machine flow shop with deteriorating jobs. *Applied Mathematics and Computation* **180** (2006), no. 1, 185–193.

287. J-B. Wang and Z-Q. Xia, Flow shop scheduling with deteriorating jobs under dominating machines. *Omega* **34** (2006), no. 4, 327–336.

288. J-B. Wang and Z-Q. Xia, Flow shop scheduling problems with deteriorating jobs under dominating machines. *Journal of the Operational Research Society* **57** (2006), no. 2, 220–226.

289. J-B. Wang and Z-Q. Xia, Scheduling jobs under decreasing linear deterioration. *Information Processing Letters* **94** (2005), no. 2, 63–69.

290. X. Wang and T-C.E. Cheng, Single-machine scheduling with deteriorating jobs and learning effect to minimize the makespan. *European Journal of Operational Research* **178** (2007), no. 1, 57–70.

291. S.T. Webster and K.R. Baker, Scheduling groups of jobs on a single machine. *Operations Research* **43** (1995), no. 4, 692–703.

292. D.J.A. Welsh, *Matroid Theory*. London: Academic Press 1976.

293. N. White (ed.), *Matroid Applications*. Cambridge: Cambridge University Press 1992.

294. R.J. Wilson, *Introduction to Graph Theory*, 2nd ed. London: Longman Group Ltd. 1979.

295. G.J. Woeginger, Scheduling with time-dependent execution times. *Information Processing Letters* **54** (1995), no. 3, 155–156.

296. G.J. Woeginger, When does a dynamic programming formulation guarantee the existence of an FPTAS? *INFORMS Journal on Computing* **12** (2000), no. 1, 57–73.

297. G.J. Woeginger, Exact algorithms for NP-hard problems: a survey. *Lecture Notes in Computer Science* **2570**. Berlin-Heidelberg, Springer 2003, pp. 187–205.

298. C-C. Wu and W-C. Lee, Scheduling linear deteriorating jobs to minimize makespan with an availability constraint on a single machine. *Information Processing Letters* **87** (2003), no. 2, 89–93.

299. C-C. Wu and W-C. Lee, Two-machine flowshop scheduling to minimize mean flow time under linear deterioration. *International Journal of Production Economics* **103** (2006), no. 2, 572–584.

300. C-C. Wu, Y-R Shiau and W-C. Lee, Single-machine group scheduling problems with deterioration consideration. *Computers and Operations Research* **35** (2008), no. 5, 1652–1659.

301. C-C. Wu, W-C. Lee and Y-R. Shiau, Minimizing the total weighted completion time on a single machine under linear deterioration. *International Journal of Advanced Manufacturing Technology* **33** (2007), no. 11–12, 1237-1243.

302. F. Xu, A-X. Guo, J-B. Wang and F. Shan, Single machine scheduling problem with linear deterioration under group technology. *International Journal of Pure and Applied Mathematics* **28** (2006), no. 3, 401–406.

303. C-L. Zhao and H-Y. Tang, Single machine scheduling problems with deteriorating jobs. *Applied Mathematics and Computation* **161** (2005), no. 3, 865–874.

304. C-L. Zhao, Q-L. Zhang and H-Y. Tang, Scheduling problems under linear deterioration. *Acta Automatica Sinica* **29** (2003), no. 4, 531–535.

305. C-L. Zhao, Q-L. Zhang and H-Y. Tang, Single machine scheduling with linear processing times. *Acta Automatica Sinica* **29** (2003), no. 5, 703–708.

List of Figures

List of Tables

Author Index

Subject Index

Symbol Index

K. Jensen
Coloured Petri Nets
Basic Concepts, Analysis Methods
and Practical Use, Vol. 1
2nd ed.

K. Jensen
Coloured Petri Nets
Basic Concepts, *Analysis Methods*
and Practical Use, Vol. 2

K. Jensen
Coloured Petri Nets
Basic Concepts, Analysis Methods
and *Practical Use*, Vol. 3

A. Nait Abdallah
The Logic of Partial Information

Z. Fülöp, H.Vogler
Syntax-Directed Semantics
Formal Models Based
on Tree Transducers

A. de Luca, S. Varricchio
**Finiteness and Regularity
in Semigroups
and Formal Languages**

E. Best, R. Devillers, M. Koutny
Petri Net Algebra

S.P. Demri, E.S. Orlowska
**Incomplete Information:
Structure, Inference,
Complexity**

J.C.M. Baeten, C.A. Middelburg
Process Algebra with Timing

L.A. Hemaspaandra, L. Torenvliet
**Theory of Semi-Feasible
Algorithms**

E. Fink, D. Wood
Restricted-Orientation Convexity

Zhou Chaochen, M.R. Hansen
Duration Calculus
A Formal Approach to Real-Time
Systems

M. Große-Rhode
**Semantic Integration
of Heterogeneous Software
Specifications**

H. Ehrig, K. Ehrig, U. Prange,
G. Taentzer
**Fundamentals of Algebraic
Graph Transformation**

W. Michiels, E. Aarts, J. Korst
**Theoretical Aspects
of Local Search**

D. Bjørner, M.C. Henson (Eds.)
**Logics of Specification
Languages**

J. Esparza, K. Heljanko
Unfoldings
A Partial-Order Approach to Model
Checking

Stanisław Gawiejnowicz
Time-Dependent Scheduling